# Advanced Structured Materials

Volume 152

Common engineering materials reach in many applications their limits and new developments are required to fulfil increasing demands on engineering materials. The performance of materials can be increased by combining different materials to achieve better properties than a single constituent or by shaping the material or constituents in a specific structure. The interaction between material and structure may arise on different length scales, such as micro-, meso- or macroscale, and offers possible applications in quite diverse fields.

This book series addresses the fundamental relationship between materials and their structure on the overall properties (e.g. mechanical, thermal, chemical or magnetic etc.) and applications.

The topics of *Advanced Structured Materials* include but are not limited to

- classical fibre-reinforced composites (e.g. glass, carbon or Aramid reinforced plastics)
- metal matrix composites (MMCs)
- micro porous composites
- micro channel materials
- multilayered materials
- cellular materials (e.g., metallic or polymer foams, sponges, hollow sphere structures)
- porous materials
- truss structures
- nanocomposite materials
- biomaterials
- nanoporous metals
- concrete
- coated materials
- smart materials

Advanced Structured Materials is indexed in Google Scholar and Scopus.

More information about this series at http://www.springer.com/series/8611

Francesco dell'Isola · Simon R. Eugster ·
Mario Spagnuolo · Emilio Barchiesi
Editors

# Evaluation of Scientific Sources in Mechanics

## Heiberg's Prolegomena to the Works of Archimedes and Hellinger's Encyclopedia Article on Continuum Mechanics

Springer

*Editors*
Francesco dell'Isola
DICEAA and International Research Center on Mathematics and Mechanics of Complex Systems
University of L'Aquila
L'Aquila, Italy

Mario Spagnuolo
DICAAR
University of Cagliari
Cagliari, Italy

Simon R. Eugster
Institute for Nonlinear Mechanics
University of Stuttgart
Stuttgart, Germany

Emilio Barchiesi
International Research Center on Mathematics and Mechanics of Complex Systems
University of L'Aquila
L'Aquila, Italy

ISSN 1869-8433 ISSN 1869-8441 (electronic)
Advanced Structured Materials
ISBN 978-3-030-80549-4 ISBN 978-3-030-80550-0 (eBook)
https://doi.org/10.1007/978-3-030-80550-0



This Springer imprint is published by the registered company Springer Nature Switzerland AG
The registered company address is: Gewerbestrasse 11, 6330 Cham, Switzerland

# Preface

This volume is the result of a collaboration among scholars having different competences, albeit the majority of them can be classified to be applied mathematicians or mechanicians.

It is really pitiful how knowledge is being lost so quickly and so systematically in the span of few scholars' generations. A text written in Latin, or even in a modern language different from the modern Lingua Franca *English*, cannot be read by the greatest majority of scholars that are active in present times. This means that a wide part of human knowledge risks to be forgotten and lost, in the worse case, or rediscovered several times, in different places, times and languages, in the best case. In our opinion, this circumstance risks to be very detrimental for the advancement of human knowledge in general, and could cause some regressions in both the human technological capacities and quality of life.

The eldest editor did experience the very sad sensation of realizing how big are the risks of regression in science when he heard an otherwise very clever scholar arguing that «something not written in English is virtually non-existing». Being Italian, he is accustomed to see his mother language treated as a kind of lost dialect whose knowledge is useful only for understanding some ancient songs. For this reason one can understand why the revolutionary works by Gabrio Piola are nearly completely ignored in the modern literature of mechanical sciences. Piola wrote his works in Italian: his ideology considered the concept of Nation so important that he was ready to sacrifice the diffusion of his ideas for it. He could have written them in French, so that the audience of his works would have been somehow larger. However, a reader may comment that French is not English and, therefore, the previously mentioned scholar, whose mother language is American English, would have similarly argued that also if Piola had written in French, unavoidably, his work would have been, very similarly, ignored. And this reader could be considered to be right, because to the same editor it happened to see how a historical paper about mechanics was rejected by a journal with the following argument: «there are too many French excerpts inside it». Needless to say: the editor tried to explain to the Editor-in-Chief that the French sentences were already duly translated into English in the submitted manuscript to prevent any problem that could arise with readers who cannot read French. The answer was: «it is not interesting to discuss about what Lagrange wanted to say, arriving to examine his own words, by checking if his thought was faithfully translated into English». The problem, most likely, was that Lagrange wrote in French and because of that his ideas cannot be so interesting. Even though such a statement was not explicitly expressed, it was implicitly assumed. We believe, instead, that every interesting contribution must be studied, independent of the language it has been written in. This work wants to send the following message: *In Mechanics and also in other sciences, there are very interesting ideas written in languages different from English. These ideas deserve to be translated into English and should not be forgotten.*

There are also two more interesting points that have attracted our attention. The first point concerns the study of the origins of scientific theories as a tool for understanding how novel scientific theories must be formulated. Since we cannot teach to younger generations an infallible method for formulating well-posed and efficient theories capable to predict observed and not yet observed phenomena, we must behave as the ancient Renaissance Maestri teaching an art to their pupil. We must show them how available theories were invented, hoping that this lesson will guide them.

The second point concerns the role of scholars that, while not fully understanding a specific theory, still actively participate to the process of transmitting it. The role of Tartaglia in the transmission of Archimedes' works to posterity is examined as a prototype of many similar behaviors, as observed in many scholars in every age, époque, place and generation. We could list many modern epigones of Tartaglia; but this will be considered to be gossip, or an act of academic political battles. We will refrain from this kind of disputes, as we want to describe the following phenomenon. A scholar, aiming for the sinecure represented by an academic position, strives to prove the world that his intellectual work deserves to be paid by a public institution. Therefore, he tries to make the other scholars believe to have done a great job with his contributions. If he is not as clever as he believes, then he needs to *reformulate*, *translate* or *make precise* what had been written by his predecessors. Tartaglia himself declared, in the title of one of his works, that «here I make clear what was not possible to understand in the original Greek works». Now, Heiberg, in his monumental work gathering all the available opus of Archimedes, proved that Tartaglia was not able to write in correct Latin. Therefore, we believe that it is almost impossible that Tartaglia could have translated from the Doric Greek of Archimedes to Latin a text that, in addition, is very difficult, as it contains complex mathematical concepts. Unfortunately, this argument was buried in the Prolegomena of Heiberg's Archimedes Edition.

In this volume, we present an annotated translation of Heiberg's Prolegomena together with a description of the sociological phenomena involved in the transmission of knowledge. The importance of these phenomena is enormous if we want to understand how novel theories were formulated at first, in order to train the younger generations of scientists. The phenomenon of science transmission has many interesting aspects and we cannot hope to deal with all of them. A particularly important one concerns the role of encyclopedias. They allow for a synthetic account for large bodies of knowledge and are very precious for younger generations of scholars, when a global understanding of the state of the art in a scientific field is required. Hellinger's encyclopedia article describing the state of the art of continuum mechanics in 1913 is astonishing. It proves that continuum mechanics had been blocked in its development by the fact that the article was written in German and that no translation into English was available. In fact, the lucid analysis by Hellinger had been ignored by too many scholars in Mechanics simply because it was written in German. Probably the fact that the author was Jewish increased the speed of erasure of his contribution from the consciousness of scholars in Mechanics. The great scholarly work by Hellinger could have given a stronger momentum to continuum mechanics if it had been properly evaluated by the scientific community. Unfortunately, it had been ignored and mentioned only in a critical way by the few authors who believed that this reference was necessary. The motivations of this sociological phenomenon deserves to be understood, if one wants to organize Academia in a more efficient way.

Rome, Stuttgart
May 2021

*Francesco dell'Isola*
*Simon R. Eugster*
*Mario Spagnuolo*
*Emilio Barchiesi*

# Contents

# Chapter 1
# The study of the genesis of novel mathematical and mechanical theories provides an inspiration for future original research

Mario Spagnuolo, Francesco dell'Isola and Antonio Cazzani

## 1.1 Introduction

This introductory Chapter is conceived in order to make explicit the motivations that led the authors and the editors to work on this volume. The reader will find additional arguments and considerations on some of the epistemological and methodological questions discussed here in the Chapter referred to in [41]: we will however try to present self-consistent reasonings, so that one is not expected to complement this chapter with other readings, if she/he does not wish. The question we want to face is simply stated as follows: Can we find a meta-theory teaching us to formulate a set of specific theories each of them being suitable to describe a well-precise set of phenomena? Unfortunately, it seems that to this question there are not fully satisfactory answers yet. There is not, in fact, any kind of *algorithm* following which one can construct a reasonably efficient theory: whatever it may be said by the supporters of *Data Science* it is not possible, in this moment of the scientific development, to replace the creativity act of a scholar in formulating a model with any kind of *Big Data* algorithm. We do not want to say that such a possibility is precluded to humankind: after having invented robots that relieved us from the greatest part of manual work, it is possible, if not likely, that in future we will be relieved by Artificial Intelligence from the greatest part, or maybe from all, the intellectual work. What has to be clear is that, notwithstanding the trends and pretensions of many, the ambitious program of replacing human mind in its formulation of mathematical models is by

Mario Spagnuolo, e-mail: mario.spagnuolo@unica.it
DICAAR, Università degli Studi di Cagliari, Cagliari, Italy.
International Research Center on Mathematics and Mechanics of Complex Systems, University of L'Aquila, L'Aquila, Italy.
Francesco dell'Isola, e-mail: francesco.dellisola.aquila@gmail.com
DICEAA and International Research Center on Mathematics and Mechanics of Complex Systems, University of L'Aquila, L'Aquila, Italy.
Antonio M. Cazzani, e-mail: antonio.cazzani@unica.it
DICAAR, Università degli Studi di Cagliari, Cagliari, Italy.



F. dell'Isola et al. (eds.), *Evaluation of Scientific Sources in Mechanics*,
Advanced Structured Materials 152, https://doi.org/10.1007/978-3-030-80550-0_1

far out of the reach of present times science. We will not dare to give some timeline indication concerning the occurrence of such a gigantic innovation, as the second author did remember very well when a famous scientist, who was his professor in electronics, announced that it was not conceivable the construction of a computer which could beat a human champion in a chess game. It was not earlier than 1983, and *Deep Blue* in 1996 did manage in the endeavor. We want to underline that we are, however, confident that such a progress will occur, and, that it will revolutionize our life, possibly our species biology and, surely, it will open a new era in *Natural History*. However, exactly as Eugenics did not represent any true advancement of science (nothing even barely comparable with modern molecular genetics, whose successes seem to be limitless), present time Data Science seems to be a fashion that is simply exploited by some scholars who are trying to get more academic (and maybe economical) power. The situation, as realistically presents itself in the current historical moment, is really clear: one can teach to young generations how to formulate a scientific model in only one way, that is by showing them how successful scientific models were at first formulated. This aim motivates the entire content of this work.

## 1.2 The process of knowledge transmission: a sociological problem that needs to be studied by using the scientific method

The present work has been produced by the collaboration of scholars whose competences are relatively varied. However, all of them never accepted a deleterious concept that is at the basis of modern organization of scientific research: that which lead to fragmentation of knowledge into hyper-specialized sub disciplines rigidly divided by sharp boundaries.

### *Fragmentation of culture brings to inability to deal with complexity*

Albeit nearly all authors and editors can be defined to be (Applied) Mathematicians, Physicists (or Mechanicians) or Engineering Scientists, all of them were exposed to humanistic culture and greatly value multidisciplinar studies. They all agreed that it is really dangerous, and surely pitiful, that, in the modern more fashionable attitudes of academic milieux, the fundamental unity of human knowledge is being unrecognized as a founding and strong feature of scholarly activity. As a consequence, quickly and systematically, Western Culture has experienced a decay of the capacity of addressing complex problems with a unitary vision of all their various facets.

In the short span of few scholars' generations there was a dramatic change in the perception of the role of a scholar: in present times the fields of expertise are being more and more carefully delimited by boundaries that are more and more difficult to trespass. Therefore one may wonder if, in present days, a scholar like Johan Ludvig Heiberg (1854-1928) would have been produced by the current academic system. Heiberg not only mastered fully the Ancient Greek language, in its main versions, including Doric Greek, together with Latin. What makes his personality really unique is that he has systematically shown to master some not easy parts of mathematics, as he could perfectly understand very deep texts by Archimedes, those contained in the famous Palimpsest. Postponing to the subsequent chapters of this work some relevant and more detailed discussion about this text, we start recalling here that it is a parchment codex palimpsest, containing, after having scratched the first written text, some prayers. How it could be possible that a scholar did manage to abrade a text by the greatest recognized scientist who ever wrote in Greek language has to be studied carefully, and is one of the problems that we signal for future scientific investigations. In fact, the original text is a Byzantine Greek copy of a compilation of works by many authors among whom Archimedes seems to be the most prominent. This original text contained two previously unknown very important texts by Archimedes i.e. the *Stomachion* and the *Method of Mechanical Theorems* (which is shortly referred to as *The Method*). Moreover, it contains also the only surviving original Greek manuscript of the celebrated Archimedean work on *On Floating Bodies*. Heiberg, following the tradition of Western Culture, did translate the Greek Archimedean text into Latin, whose role of *lingua franca* had been recognized nearly universally. Unfortunately, after the 1906 Heiberg's discovery and the subsequent publication of his *Archimedis Opera Omnia* (i.e. The complete works of Archimedes), and after a short period in which French seemed to have replaced Latin, English has established itself as the modern, possibly even more universal, *lingua franca*.

As a consequence of this sudden change, a text written in Latin (but unfortunately also if it was written in any living language different from English) became not readable by nearly every modern scholar (sometimes even by some professor of Latin Language!). As it always happened when a change of the used language in science occurred, there is a very likely phenomenon that systematically occurs: a large part of the knowledge accumulated in the old language is forgotten and lost. Another part of this knowledge resurfaces in the new dominant language (the part of Archimedean results that resurfaced thanks to Tartaglia gives an example of such a phenomenon, see the following chapters) and is *rediscovered* several times. These rediscoveries occur in several different space locations, in different times (many anachronisms may be explained in this way) and also in different languages. Useless to say, this process of systematic rediscovery slows down a lot the advancement of science and is really detrimental, as it systematically causes regressions in technology.

An example of the rediscovery of a body of knowledge lost because of linguistic barriers that we will examine concerns the works by Gabrio Piola (see [46, 47, 40, 42, 43]). Piola's work were nearly completely ignored for a long period and had been recovered because of a series of fortuitous events. Piola's works were written in Italian and because of the wrong choice of the used language their diffusion was strongly limited. In this work, we will prove that there are, also in mechanical science, very interesting ideas that were originally written in different languages than English.

The aim of the present work is to prove that, differently from what has been too often conjectured, scientific knowledge transmission is not a simple process: the vision of science as a continuous and endless progress from less advanced to more advanced stages has been falsified even too many times in the history of science. Surely there are the problems related to linguistic barriers, when the *lingua franca* changes because of one of the many possible social reasons. Many scientific ideas were lost in translation! However, there are also some psychological and barely *survival* mechanisms that causes erasures, loss and deformation of the scientific knowledge in its transmission process. These mechanisms play a crucial role in the advancement of science, whatever may be believed by some *right-thinking* scholars. These scholars believe that one has to avoid the consideration, when studying knowledge transmission processes, of the social phenomena related to jealousy, revenge, inflated self-esteem, bare ignorance, arrogance, need of earning from academic positions, every form of nepotism and use of scientific knowledge for getting any form of power. Many are embarrassed when the existence of these social and psychological mechanism are evoked and when one expresses the opinion that they may play a crucial role in the rise of any form of Dark Ages. In fact, their consideration is considered not politically correct and trying to take into account their influence in history of science a form of mental disorder of the kind of paranoia. Instead, exactly as Alfred Kinsey has scientifically shown how important is sexuality in human life and in the psychopathology of humankind, we believe that the social forces that are shaping human psychology are of great relevance in the mechanisms that produce scientific research. Such an obvious statement, as a similarly obvious consequence, implies that it is possible that a deep scientific theory, a useful body of knowledge or an effective mathematical model may be erased, lost, or, in the best case, forgotten for a while in a scientific group, simply because of a series of socio-psychological reasons which are completely unrelated to their absolute scientific merit. Aforementioned *right-thinking* scholars will claim that science is *objective* and that even considering the possibility of any influence of the dark side of human mind on its development is harmful for humankind. This reasoning may be considered equivalent to believing that one can defeat an epidemics simply ignoring its existence: an action whose consequences are well-known. The story of the struggles of Tartaglia (1499-1557) to persuade all his contemporaries that he could understand and translate the Archimedean works, as reconstructed objectively by Heiberg, gives us a paradigmatic and incontrovertible evidence that our thesis is very well-grounded.

## *Why to try to establish how and when a scientific theory was first formulated? Difficulties in this endeavour*

In the discussion that we will develop in our work, we will focus on at least two important aspects of the considered question. The first aspect concerns the importance of study of the true origins of the scientific theories. One may argue that the value of a scientific theory resides in its predictive capacity, and that it is enough to supply a whatsoever rigorous and precise formulation for it. If one accepts this point of view then, when a theory is formulated in a way or in another equivalent way then she/he can choose the preferred way based on any reasonable and useful criterion. Our point of view is, instead, that if one wants to learn how to formulate a completely new theory, a theory that was never formulated before, she/he has to learn, in absence of the meta-theory invoked and dreamed before, how the successful theories have been formulated first, and how they were subsequently developed. To see how old and established theories were born may be of use in the process of inventing a completely new one. Indeed, we do not have, presently, any way to supply to younger generations any other well-working method for teaching them how to build theories that are efficiently capable to give the correct predictions for both observed and not-yet-observed phenomena.

As a consequence we are obliged to follow the educational methods of those ancient *Renaissance Maestri*, who trained their pupils to sculpture or painting simply by showing them as the *Maestro* was painting or sculpting. Unfortunately, there are very few great *Maestri* alive in a certain historical moments and, moreover, their workshops are already full of pupils. Therefore one has to show to those young scholars, who aspire to invent something original, how the available theories, relevant in the chosen disciplines, were first conceived and developed: in this way we hope that the lesson given by great scholars example will guide new generations. For this reason a presentation of available theories must follow, as carefully as possible, the original invention process that led their inventors to get them[1].

The second aspect, on which our analysis will particularly focus, concerns the process of transmission of knowledge from competent scholars to competent scholars via intermediate scholars who are not so competent. Albeit the transmission of science is based on written texts, the role of the scholars participating to the editing of the texts and using them as textbooks for their young pupils cannot be neglected. When the books were handwritten, their relatively enormous economical value introduced a further selection filter in knowledge transmission: the economical costs imposed a selection of what could be copied and what deserved oblivion. In this choice the Archimedean Palimpsest was sacrificed for a book of prayers against diseases, a subject that seemed more "practical" than abstract mathematics. A scholar choosing what kind of textbook deserves to be transmitted plays a relevant role also in the era of printed books: many books are not reprinted and remain in fewer and fewer exemplars in the storages of libraries, virtually disappearing from the attention of younger generations. In our (unfortunate) époque of citations metrics another method has been conceived to condemn to oblivion certain textbooks, authors or theories: it is enough to forget to cite them, and soon nobody will find these works in the *mare magnum* of modern literature, which is literally overflown with too many repetitive and not original papers and textbooks.

Finally, an influential compiler of a textbook, having many students may influence many of them with his biased choices. In the milieux of mechanical sciences there are many textbooks that were very successful in transmitting the correct ideas to

[1] The second author is greatly indebted to Prof. Roberto Stroffolini (Università di Napoli Federico II) for having shown him how such a teaching method has to be pursued.

clever students, albeit it is clear that their compilers did not understand very much the scientific results that they had carefully copied from reliable sources. There are, also, examples of textbooks that deformed the true intent of their sources, imposing to too many younger scholars wrong points of view or making for them every original research extremely difficult. We will fully describe, under the guidance of the authoritative Heiberg's analysis, how Tartaglia did manage to have a relevant role in the *translation* in the *language* used by Western Science of some of the most relevant works by Archimedes. Albeit this may seem rather simple (and most likely also very useful), we will not try to establish any relationship between the publishing (and survival) strategy chosen by Tartaglia and that chosen by (too) many more modern scientists. The need of getting a salary seems to allow for any kind of deplorable choice, while Tartaglia features a "representative" scholar, belonging to a specific kind. This kind of scholar is observed nearly ubiquitously in history of science: one can find examples of it in any social group, language, scientific discipline, historical period, geographical location and economical and political organization.

Instead of looking for specific examples of such kind of scholar, we will try to phenomenologically describe their behavior, the effects of their existence on science transmission and on its accumulation and loss. We will try to apply the scientific method in our phenomenological description and in our first efforts of looking for a model of it. The phenomenology can be shortly resumed as follows: in the competition that they need to accept in order to have recognized their own scientific capacities, many scholars systematically want to ignore any signal indicating that they are not original enough to deserve an academic position. They badly need the sinecure that they believe to be associated to it, and therefore try to prove, in any possible way, that they do deserve highly ranked positions. If they meet somebody indicating how weak their scientific skills are, then they may react in two different ways: i) they start believing that there is a conspiracy against them or ii) albeit they may understand that the criticism against them is well-founded, they manage to persuade themselves that since there are so many incompetent scholars, then their own exclusion from academia is not moral. These scholars, either if they are conscious of their weaknesses or if they sincerely believe to be clever enough for their ambitions, try to make their best to persuade all other scholars that they can be considered original thinkers. Sometimes, exactly as it was done by Tartaglia, these self-proclaimed scientists *reformulate*, *make rigorous*, *translate*, *clarify* or *make more precise* works that they have found in the literature. Exactly as Tartaglia included in the title of one of his presumed translations the following statement: "here I make clear what was not possible to understand in Archimedes works", his epigones manage to declare that they "clarified" the previously "obscure" theories, while in fact they are completely misunderstanding them.

## *To unveil the real contribution of Tartaglia (and his encyclopaedic or polymath epigones) to science is not easy*

The capacity of some scholars in avoiding any discussion about the merit of their scientific contributions is legendary. They manage to bend even mathematical argument to their aims, making any discussion about what they claim to have discovered completely useless. One has to avoid any effort in trying to prove that a single specific scholar is not producing any original contribution or any original view in presenting already known results. Instead it is very useful to describe from a general point of view the kind of effect that the existence of the aforementioned type of scholars has on science transmission and development. If this phenomenology is understood then, most likely, some countermeasures can be acted to limit the unavoidable impact of such scholars on the destinies of science.

Albeit this information seems to have been somehow forgotten, Heiberg happened to discover, while reordering and preparing for his edition the whole available texts by Archimedes, that, in reality, the only merit one can attribute to Tartaglia, for what concerns the appreciation of Archimedes work, is purely propagandistic. Tartaglia contributed to revive the interest in Archimedes. Heiberg, while prefacing his Complete Works by Archimedes, gathered all necessary evidence to prove that Tartaglia's capacity in writing in a correct Latin was rather scarce. One can deduce therefore that he could never have the possibility to translate, from the Doric Greek used by Archimedes into Latin, a complex text of advanced mathematics.

Heiberg's argument seems to us very detailed, serious and careful: unfortunately, this argument was buried in the Prolegomena of the famous Archimedes Edition. We could say it was buried since this Prolegomena (as well as the whole translation of Greek text) was written in Latin. While there are many valuable translations into English of Heiberg's Latin text, the Prolegomena, to our knowledge, were never translated into any modern language. Therefore, we were motivated to translate in this work aforementioned Prolegomena and to add our own comments to it, in order to highlight those aspects of the phenomenology of knowledge to which we are particularly interested. The sociological and cultural phenomena that are surfacing from this reading deserve, in our opinion, a great attention.

Their importance cannot, indeed, be underestimated: if one wants to describe carefully the process of birth of a novel theory she/he must establish exactly when, how and in which formulation, it was first conceived. This description is essential for pedagogical aims: younger generations of scientists must learn how to formulate novel theories by looking at the invention process of the most successful ones. The phenomenon of science transmission is rather complex and manifold: one can find many of its aspects that are of great relevance. One that plays an important role concerns the systematizing and paradigmatic role of Encyclopaedias and encyclopaedic compilations. Because of their true nature, they gather many important aspects of knowledge into a well-organized and unitary way, by using a common formalism and vision. Moreover, they give a synthetic account of all human knowledge, in the most ambitious projects, or for a specific group of disciplines, in other cases. Encyclopaedias supply a precious support for subsequent generations of scholars, as they supply a global understanding of the state of the art, in a given group of scholars, place and époque. By sacrificing some technical details, they resume large bodies of knowledge in an agile presentation and indicate where the interested scholar can find the details that she/he may need. However, the existence of encyclopaedic summaries makes more difficult to understand if a certain scholar did really master her/his discipline, or if she/he did simply adsorb superficially one of the available Compendia.

Our attention has been attracted, in this context, by the 1913 Hellinger's Entry of the German Encyclopaedia of Mathematics whose aim was to give an overview of the then current state of the art in Continuum Mechanics and list some research perspectives that seemed promising to the author. This text has not been translated into English until recently (see [69, 70, 71]) and proves that, in fact, Continuum Mechanics has been "frozen" because of the establishment of English as the novel *lingua franca*, and by the incapacity of the community of experts in Mechanics to read French, Italian or German.

The summary and the analysis presented by Hellinger is really clear and far reaching. The research perspectives, read by somebody in 2021, seem to be even visionary: only recently some of them are being developed. It is remarkable that Hellinger could forecast the main directions of future development of Continuum Mechanics with such a large anticipation. The question therefore is: why Hellinger's work has been removed by the list of the most used sources of the 20th century by the great majority of scholars in Mechanics? A partial answer is that it was written in German. Moreover, the author was Jewish and, unfortunately, this did not help the diffusion of his work in the milieu of German speaking Mechanicians, at least until the end of Second World War.

Such an erasure resulted in a great damage to the advancement of Mechanical Sciences. The loss of the consciousness of Hellinger's analysis in the German speaking Mechanics community had rather singular effects. Indeed, while many authors showed to be aware of the results presented in his work, the information about the fundamental fact that these results were, for the first time, obtained by using variational principles was lost. Therefore, exactly as it was done by Tartaglia, the secondary sources from Hellinger (some of them emigrated in the USA, together with their authors) presented some reworks of Hellinger's Compendium in such a way that it was impossible to get from them any hint about the heuristic method used for finding the presented results. These Compendia were presented as if they were a completely original contribution of their authors, who seemed to have had an *out of the blue* inspiration. This feeling impresses on the readers the false belief that science is an *epic* endeavor where few, particularly gifted scientists, *wake up one morning* and without any apparent cause, simply because they are geniuses, manage to invent a novel theory. In fact, any theory is the result of a choral work of generations of scholars: what is found in some modern textbooks in Continuum Mechanics is the elaboration, hiding the variational procedure first used for finding them, of the contribution to the discipline given by many scholars, starting from Lagrange [108, 107, 20], Piola [46, 47, 40, 128, 151], the Cosserat brothers [38, 4, 8, 58, 120], and continuing with Sedov [137, 135, 136], Toupin [144] and Mindlin [116], among many others [68, 76, 88, 87, 105, 72].

The Entry by Hellinger represents a deep scientific contribution to Mechanics, as it originally reorganizes, with the rigor of a gifted mathematician, all results available up to 1913. It could have given an impressive impulse to the development of 20th century Continuum Mechanics if only it had been understood by the scientific community.

It has to be said that there is a possible misuse of the Encyclopaedic Entries, and this misuse concerned also that by Hellinger: indeed, the results presented in this kind of Compendia may be *adsorbed* and *reworked* by Tartaglia's epigones, who will present them from different, and sometimes twisted, points of view. The existence of Encyclopaedic Entry make possible the existence of so-called polymath scholars: these scholars, who probably have the access to Encyclopaedia Entry, are claimed to have a universal knowledge. Instead, most probably, they simply had access to a, very often lost, Encyclopaedic Entry. In particular Hellinger's work is surely the starting point of the reworking of Continuum Mechanics as presented by those scholars who do refuse Variational Postulation. Knowing in advance the correct results it is easy to deduce them by a series of ad hoc postulates, claiming that they are *induced* by experimental evidence. We will more diffusely present this point in the following sections of this Chapter. Of course this misuse was not intended by Helinger when he conceived his Entry. Unfortunately, until very recently, as a direct source this Entry was completely ignored. We could find a few fugitive mentions of it, where it has been rather harshly criticized. What we have just described is another of the sociological phenomena that must be studied and understood. Understanding it will have an important consequence: thanks to the obtained insight one can find operative methods for organizing the recruitment of academic bodies in a more efficient way.

## *Archimedes: "The Method of Mechanical Theorems" is an authoritative source confirming our thesis*

To our knowledge, Archimedes is the first known scientist who described explicitly a heuristic way for finding novel theories, theorems and mathematical models. Archimedes' mastering of the concept of "model" of physical reality by using mathematical deductive theories has pushed us to conjecture that his epistemological vision may be considered, in essence, to be that of a falsificationist.

This statement may need further deepening: here we consider sufficient to quote, once more, and to give some further few comments, what Archimedes wrote at the beginning of his "The Method of Mechanical Theorems, for Eratosthenes". The Archimedean text was written in Doric Greek, and it is a difficult issue to decide which English translation transmits more faithfully the original ideas and spirits. It seems that the scholarly work of those who are capable to understand mathematics, physics, model theory and Doric Greek is very useful also nowadays. The English text which we are going to reproduce here is the final result of many transformations: the Greek text found by Heiberg was translated by Heiberg himself into (Modern) Latin (in his celebrated Edition of Archimedes' Works). Heiberg's Latin text was then translated into Dutch by E. J. Dijksterhuis in 1938 and then into English by C. Dikshoorn in 1956 (see [56]). Notwithstanding this subsequent translations, we manage to see clearly the original ideas of Archimedes. Instead, Hellinger's ideas [69, 70, 71] have been reformulated into English [146] without citing them directly, in a way that blurs the original spirit. There are also some hints about the way in which Hellenistic culture organized science in this text. Archimedes starts his "cover letter" by recognizing to Eratosthenes a scholarly preeminence but only as a "manager of scientific research" and as "editor-in-chief" of the publications and manuscripts produced by the library of Alexandria: "Since, as I said, I know that you are diligent, an excellent teacher of philosophy, and greatly interested in any mathematical investigations that may come your way, I thought it might be appropriate to write down and set forth for you in this same book a certain special method, by means of which you will be enabled to recognize certain mathematical questions with the aid of mechanics."

This preamble may be interpreted as a kind of *captatio benevolentiae*. Now, from all sources we know how great was the fame that Archimedes enjoyed also during his life. Why did he need to be so careful in sending his paper to Eratosthenes? One can conjecture that also in Hellenistic scientific milieux it was possible to observe a phenomenon that to a much larger extent has been developed later: the diffusion of culture happens to be controlled by few powerful scholars, whose decision can greatly influence the destiny of any scientific work, including those written by outstanding persons, as Archimedes was already recognized to be. The existence of "well-established" scientific personalities who had the power to control what can be published or what must be bound to oblivion seems to be therefore attested already at the époque of the library of Alexandria, and seems to be an unavoidable side effect of any form of organization of Big Science.

Eratosthenes of Cyrene (c. 276 – c. 195/194 BC) was probably one of the most influential personality of Hellensitic science. Obviously, having the control and full access to the biggest source of scientific knowledge of antiquity, he is often described as a polymath. It is interesting to remark here that the etymology of the word "polymath" goes back to ancient Greek. The Greek word πολυμαθής can be translated as follows: "[somebody] having learned much". The translation that has been more often used in Latin is: *homo universalis*. i.e., "universal man". We believe that too often polymaths are simply scholars who managed to better reorganize the results found by other, more original, scientists. Very often the compiler of Compendia or Encyclopaedia Entries are this kind of erudite polymath. The most skilled among polymaths, however, are very precious: they allow for the diffusion of specialist theories among a wider set of scholars: we believe that Hellinger, being an original mathematician himself, when accepting to write an Entry about Continuum Mechanics did cleverly master the subject and then could give the best indications about its future paths of development.

Eratosthenes' interests apparently spanned mathematics, poetry, geography, astronomy and music theory. In fact, most likely he was an erudite who managed to persuade the Pharaoh Ptolemy III Euergetes to nominate him as a "chief-librarian" at the Library of Alexandria in the year 245 BC. One has to consider that the choice was really appropriate: as head of such an institution one needs indeed a true and gifted polymath. He was the leader of the group of scientists and technicians that founded scientific geography and he is best known for having directed the group of scholars that obtained a careful calculation of the circumference of the Earth and the tilt of the Earth's axis. He introduced the first global planar projection of the world, by using parallels and meridians. Most likely he has also calculated the distance from the Earth to the Sun and understood the need of the leap day for a precise Calendar. In number theory, the sieve of Eratosthenes, an efficient algorithm for calculating prime numbers is attributed to him. In the entry of the Suda[2] concerning Eratosthenes it is reported that his critics called him Beta (that is: the "second", as beta is the second letter of the Greek alphabet). This scornful attribute had been chosen to underline that he was the second biggest expert in all his domains of competence. On the other hand, without denying this circumstance and even confirming it, his supporters called him *Pentathlos* after the Olympian Athletes competing in the pentathlon, i.e. athletes being "well-rounded" in five different sports. Eratosthenes' approach to science can be positively interpreted by stating that he tried to dominate the complexities of reality (in fact his appointment at the Library required this kind of skills!) and, for this reason, he had to prove to have talents in a large variety of disciplines. He was capable to understand many things and wanted to use every kind of information which he could achieve. As a consequence he could not be the best expert in anything, but he could play a role in transmitting knowledge from one discipline to another. In fact, as reported by Strabo: Eratosthenes was regarded to be a mathematician among geographers and a geographer among mathematicians[3].

His skills placed Eratosthenes in a very privileged position: he could decide what had to be published becoming a book stored in the library of Alexandria, and therefore, considering the importance of this library, which book could be transmitted to future generations. Without any doubt, Eratosthenes belonged to the *timocratic scientific élite*, i.e. the dominant group of intellectuals of his epoch. Archimedes, who usually could not hide his great self-esteem (see [56, 97]), was obliged to treat with great reverence such an important person. He, therefore, called him a "diligent", "an excellent teacher of philosophy", and "greatly interested in any mathematical investigations that may come your way". Archimedes, as modern scholars are often doing when submitting a paper, writes clearly to the editor-in-chief about its motivations:

> *I am convinced that this* [heuristic method] *is no less useful for finding the proofs of these same theorems. For some things, which first became clear to me by the mechanical method,*

---

[2] The Suda is a Byzantine encyclopedia, written during the 10th-century after Christ. It is a Greek lexicon, having 30,000 entries and including many drawings copied from ancient sources, sources which have been, unfortunately, subsequently lost. The name derives probably from the Byzantine Greek word "souda", which means "stronghold [of knowledge]". Eustathius, misunderstanding the etymology of the title, declared that Suda was a deformation of the name Suidas, that was his author's name.

[3] This destiny is bounded to modern mathematical physicists: they are neither mathematicians nor physicists. However they can be useful in allowing for the communication among the two groups.

*were afterwards proved geometrically, because their investigation by the said method does not furnish an actual demonstration.*

The reader must remember here that the expression "proved geometrically" is a precise calque of the Greek original expression. It has to be understood, in modern language, as follows: "proved with mathematical rigor". Archimedes has a great standard of mathematical rigor. He states that something is "proven" only when he finds a logically precise sequence of statements which can be deduced, one after the other, from his axioms. A heuristic reasoning is NOT a theorem, for every mathematician since the Greek invention of rigorous mathematics. The use of the word "geometry" in Archimedes' text is simply related to the fact that, in Hellenistic science, the theory of real numbers was formulated in terms of geometrical entities like segments, areas and volumes (see e.g. [37]). The argument of Archimedes continues as follows:

*For it is easier to supply the proof when we have previously acquired, by the method, some knowledge of the questions than it is to find it without any previous knowledge. That is the reason why, in the case of the theorems, the proofs of which Eudoxus was the first to discover, viz. on the cone and the pyramid, that the cone is one-third* [of the volume] *of the cylinder and the pyramid one-third of the prism having the same base and equal height, no small share of the credit should be given to Democritus, who was the first to state the fact about the said figure, though without proof.*

Archimedes is aware of the importance of both the heuristic, creative invention act which leads to the conjecture of a mathematical result and the technical rigorous demonstration which is needed to state that such a theorem is true. He distinguishes between the inventor of a mathematical proof and the discoverer, who is aware of a well-conceived conjecture, whose result is left to be proven. Then, he discusses the specific heuristic procedure, based on his understanding of a problem of mechanics, which led him to calculate the area of a parabolic section:

*My own experience is also that I discovered the theorem now published, in the same way as the earlier ones* [the theorems conjectured by Democritus and proven by Eudoxus]. *I now wish to describe the method in writing, partly, because I have already spoken about it before, that I may not impress some people as having uttered idle talk*

Archimedes wants to underline that his creative work has to be split into two parts: i) the conjecture of the statement of the theorem, based on a heuristic argument, and ii) the rigorous proof of the theorem, based on a logical procedure, starting from the axioms he has accepted. It has to be remarked here explicitly that Archimedes calculates the aerea of a parabolic section by what will be called later an integration method. For doing so, he needs the rigorous definition of the set of real numbers, which Archimedes attributes to Eudoxus of Cnidus. On the other hand he conjectures that the area of the parabolic section has a certain value by means of an experimental measure. Archimedes, following a habit that is unfortunately too often spread among pure mathematicians, communicated his rigorous proof without any reference to his heuristic mental process. However, he had spoken about it while discussing with his colleagues: he feels the need to describe it in a written form. He is doing this in order to keep his reputation of serious scientist, who is not talking in vain. To keep his own high reputation is not the only reason for which he discloses his way of reasoning:

*partly because I am convinced that it will prove very useful for mathematics; in fact, I presume there will be some among the present as well as future generations who by means of the method here explained will be enabled to find other theorems which have not yet fallen to our share.*

Archimedes wants to show to future generations how a theorem is conjectured: he is not happy to give the rigorous proof of it, only. As he has not a technique of discovery which can be formally presented to the reader, he explains his own mental process, based on a clear understanding of mechanical phenomena. Finally, he gives us the specific technical details concerning his theorem

*We will now first write down what first became clear to us by the mechanical method, viz. that any segment of an orthotome[4] is larger by one-third than the triangle which has the same base and equal height, and thereafter all the things that have become clear in this way. At the end of the book we will give the geometrical proofs of the theorems whose propositions we sent you on an earlier occasion.*

The few sentences cited above were considered by Heiberg, their modern discoverer, as possibly the most important ones uttered by Archimedes. Archimedes transmits to us the mental process which occurred in his mind during his mathematical creation. Rather seldom such a clear perspective is given in a mathematical text. Hellenistic Mathematics, and also all subsequent mathematical tradition, is characterized and founded on the logical rigor of the presentation. The economy of thought and its precise formulation are considered the prevalent criterion when presenting mathematical results. A mathematical text, since Hellenistic mathematicians, is a sequence of logical conclusions, obtained with correct deduction rules, starting from the accepted hypotheses, conceived in such a way that the theses are related to the hypotheses by a irrefutable reasoning. While this demand of rigor is essential for the development of hard sciences, it is also undoubtedly true that this style of presentation, giving the synthetic final result of the process of demonstration, is ignoring the equally important demand of understanding the reasons which led the mathematician to the presented demonstration and the heuristic method using which this demonstration was found for the first time. Risking to spoil the myth of his own genius, Archimedes reveals spontaneously how himself, before even starting to try to prove his theorems, conjectured their theses and managed to be persuaded that they were true.

## 1.3 An epistemological intermezzo: inductivism versus falsificationism

Without any hope to succeed in presenting an exhaustive report of the epistemological knowledge that led us to understand how scientific theories are built, for seek of

[4] An old name first used by Menaechmus to designate the particular conic section resulting from cutting a right-angled cone by a plane which is perpendicular to its surface, thus producing a parabola

self-consistency, we sketch here those most fundamental ideas that should guide a mathematical physicist in his scientific practice.

We have been sometimes very surprised in discovering that otherwise very gifted scholars may have a too naive vision about the epistemological concepts which are needed for correctly guiding their scientific research. In general, for what concerns the postulation scheme used in Continuum Mechanics, we have seen too many presentations in which a series of ad hoc postulates are accepted based on *experimental evidence* or even claiming that they are *induced by experience*. These approaches led to an occlusion of Continuum Mechanics in a stage that was already recognized to be too particular in the works of Gabrio Piola [46, 47, 40, 43, 42, 122, 128].

In order to get rid of the limiting scheme of Continuum Mechanics as elaborated by Cauchy and imposed in Engineering Sciences by its undoubted successes in predicting deformative behavior of bodies, it is necessary to resort to a truly falsificationist approach in the comparison of different mathematical models used for *describing* reality.

## *Relation between Science and Technology: a view back through History of Science*

For the kind of analysis we want to conduct, it is of primary importance to ask what is the effective relationship between Science and Technology. Is there a theory that describes the birth, growth and decay of Scientific Theories and Scientific Technology? To get answers in this direction, it is necessary to refer to concepts that are the specific object of History and Philosophy of Science. If thinking about History of Science does not confuse us, because we can easily recognize in it the ordered set of observed facts, discussing Philosophy of Science may induce misunderstandings. We refer to Philosophy of Science as that meta-theory which, by organizing the set of available information about the way in which well-established theories were constructed, tries to supply efficient methodologies apt to formulate new theories. In the perspective of a mathematical physicist, therefore, a Philosophy of Science is indispensable.

But let's go back to the original question that we believe has a basic importance: what is the relationship between the development of an organized Science and the technological progress of a society? The answer to this question is extremely complex, but we can already get a clear idea by considering on an imaginary time line the focal points of human technological development and then, on the same time line, place the cornerstones of scientific development. What would immediately appear is that for about two million years man has used chops and more or less polished stones for hunting, working skins, cutting wood and other subsistence activities. A few thousand years before Christ, man began to build the first instruments. Gradually technological advances have increased, but there has been an incredible acceleration in correspondence with the birth of Hellenistic Science: the ballista, the Syracusia ship, the astronomical calculator of Antikythera, just to name a few. It has to be remarked that the existence of the disk of Nebra (we will give details about it in the following) seems to indicate that, albeit we do not have any written evidence about it, the great development of the technology related to the Bronze Age may be related to a first elaboration of a proto-Science.

One can follow this imaginary timeline up to the present day by observing how the relationship between scientific development and technological development is inextricably linked. A society that abandons Science, after a suitable time-delay, goes through three successive phases:

i. it no longer produces any kind of new technological development,
ii. it loses the knowledge related to the use of technological tools developed in a previous era of scientific flowering and
iii. transforms (in the best case) such tools into religious objects.

We will see how this decline of Science, and consequently of Technology, is inexorable when certain conditions are created in a given society. One can possibly explain the fall of Western Roman Empire relating it to the loss of awareness about the importance of Hellenistic Science and the related slower, but equally inexorable, loss of technological capacity. We believe that there is an exemplary case that deserves to be shortly discussed here: we mean the use of gravity aqueduct. Hellenistic hydraulics did know a form of the law that has been named after Bernoulli. This theoretical knowledge allows for the conception and construction of the cheaper pressurized aqueduct. In fact, in Pompei we can see a network of pipes distributing the water in the city with a small local pressurized aqueduct. However, building a large aqueduct is not a very frequent need. In the Pergamon Museum in Berlin important parts of a large pressurized aqueduct serving the Pergamon Acropolis are shown. We do not know when the needed theoretical knowledges of hydraulics were lost: for serving Rome, unfortunately, engineers who ignored hydraulics built gravity aqueducts, causing a large economical loss. A sum of such losses most likely made the difference of the destiny between Western and Eastern Roman Empires. One may consider that for some unknown reason the advanced topographic knowledge needed for building a gravity aqueduct were not lost in the passage between Hellenistic and Roman cultures: the reasons for which Romans did manage to preserve a part of Engineering Sciences (Topography) while loosing another part (Hydraulics) may be related to an arbitrary choice of a librarian who could not understand the mathematically difficult arguments in Hydraulics while could catch the simpler reasonings used in Topography, probably because this last can be synthesized using drawings and simple Euclidean Geometry.

An interesting philosophical question that arises spontaneously when we try to organize the phenomenology of scientific progress of human societies is to wonder if the path of human history is a progression of stages that has been repeated many times, independently by different groups, in the same order or if each progress has occurred only once and then it has consequently widespread. This distinction between social determinism and diffusionism finds its basis in the thought of Giambattista Vico, who wrote

*Similar ideas that originate from entire peoples unknown to each other must have a common basis of truth.*

We tend, differently by what appears in Vico's thought, towards a diffusionist approach. This approach explains better the phenomena related to the scientific flowering which occurred in the Renaissance. Is there really anyone who can believe that the Renaissance evolved in a completely autonomous way? Can anyone really continue to deny the very strong influences that Hellenistic thought had on Renaissance thought? And if there are still few who deny such influences, why, instead, are there still so many who deny the importance of Hellenistic Science and even deny it a classification as a truly "modern" Science?

The library of Cardinal Bessarion is the first fundamental part of Marcian Library in Venice and was constituted mainly by Greek codices. Based on the *transport* of Hellenistic Science via Greek manuscripts arriving in Europe, the main characters of Italian Renaissance started the re-discovery of ancient Science not always recognizing their debt towards their sources.

## Approaches to Science: Falsificationism or Inductivism?

In the formulation of a scientific theory at least two alternative approaches can be used. Following the standard nomenclature in the literature, they are called inductivism and falsificationism. We are aware of the fact that more sophisticated conceptual frames have been adopted in Philosophy of Science. However, discussing only these two approaches is enough for our aims. Both of these visions can be traced back throughout the History of Science. As far as we will discuss in the following of this chapter, we are interested in how they were declined in Hellenistic thought, as we will analyze the development and decline of the models introduced for the description of the motion of the planets, and how they were used within the group of scientists who in the 19th century and later developed modern Continuum Mechanics.

As for Hellenistic Science, as we shall see, unfortunately surviving sources are so rare that it is difficult to tell in which form the debate on inductivism and falsificationism took place among Hellenistic scientists. The echoes of this debate, however, are resonating in a significantly later period: Proclus (412-485) discusses the nature of epicycles (we will see below the details of the deferent-epicle model) and asks himself whether they exist or are pure mathematical hypotheses in his treatise Hypotyposis (i.e. Exposition of Astronomical Hypotheses). As we will see, for scientists of the Hellenistic age, as Apollonius of Perga who first introduced planet models using deferents and epicycles, it was obvious that these were simple mathematical objects and that they are not objects in the physical world. They loose every meaning if not contextualized in the model where they were introduced. As Proclus is a post-scientific philosopher, he seems to report about an ancient debate and, being completely unable to fully understand its content, he manages to deny the validity of both positions. However, Proclus claims to be a follower of the philosophical thought of the Platonic school: therefore, he should be able to see a difference between mathematical and physical objects, albeit believing that one can experience, in the world of mathematical ideas, some experiences leading mathematicians to mathematical theorems.

When Platonism is adopted in the development of mathematical thought, then extreme positions are generated. In fact, according to Hardy [96], mathematical platonism is based on the statement

> *Mathematical reality lies outside of us and our function is to discover and observe it and the theorems we prove* [...] *are simply the accounts of our observations.*

According to mathematical Platonism, then, physicists discover physical reality while mathematicians deal with mathematical reality. As we will see with examples taken from both the development of models for the motion of the planets and the development of modern Continuum Mechanics, it is very dangerous to confuse, or even identify, mathematical entities with the physical entities of which they are assumed to be models. Moreover, there are some mathematical entities for which one cannot find any physical correspondence: these mathematical entities are useful only in the logical development of the formulated mathematical model. When one confuses the mathematical model with the physical reality, it may happen that, instead of concluding that the specific model is not suitable to describe physical evidence, one could believe that reality is not self-consistent and may arrive at the conclusion that nature is intrinsically paradoxical. This ontological point of view should be avoided if one wants to have any hope to describe and predict physical phenomenology.

The confusion between models and physical reality is carefully avoided by Platonic mathematicians: therefore, such a philosophical position is not impeaching the needed distinction between mathematical objects and the physical objects they are modeling. Once one has distinguished between mathematical models and real objects, it is easy to confute the so-called inductivist vision of Philosophy of Science.

Inductivism has been considered for too long time as the true scientific method that has to be practiced by diligent scientists. Unfortunately, it is still a commonplace view in many scientific milieux to believe that one can induce from many observations some physical laws, that belong to physical reality and can be established once forever. Such a vision of the scientific method is not efficient and effective to develop scientific theories, as an efficient process like *induction of a physical law* cannot be established. In fact, inductivism is based on the belief that a systematic research approach exists, that involves an inductive reasoning (whatever it may mean) enabling scientists, when applied with due diligence, to *objectively discover* the unique true theory describing every phenomenon. The prescription of inductivism, when examined attentively, presents a very ambiguous clause: the scientist must apply *due diligence*. Therefore, when an induced physical law reveals some limits, naive inductivists are simply stating that the scientist formulating it was not diligent enough. Such a point of view is not at all scientific: how can a scientist know which is the *due diligence* necessary for being sure that his law is "true"? The position of naive inductivists has been ridiculed by Bertrand Russel with his famous anecdote about the inductivist chicken [131, Ch. 6, p. 47]:

> *Domestic animals expect food when they see the person who usually feeds them. We know that all these rather crude expectations of uniformity are liable to be misleading. The man who has fed the chicken every day throughout its life at last wrings its neck instead, showing that more refined views as to the uniformity of nature would have been useful to the chicken.*

In a more picturesque way, Chalmers in [34] reformulates it as follows:

> [We present] *a gruesome example attributed to Bertrand Russell. It concerns a turkey who noted on his first morning at the turkey farm that he was fed at 9 am. After this experience had been repeated daily for several weeks the turkey felt safe, in drawing the conclusion "I am always fed at 9 am". Alas, this conclusion was shown to be false in no uncertain manner when, on Christmas eve, instead of being fed, the turkey's throat was cut. The turkey's argument led it from a number of true observations to a false conclusion, clearly indicating the invalidity of the argument from a logical point of view.*

More seriously and shortly, but maybe in a more effective way, Einstein also criticizes inductivism:

> *Any amount of experiments may prove that I am right; a single experiment can prove that I am wrong* [A. Einstein, letter to Max Born on the December 4th of 1926].

In conclusion, *the idea that theories can be derived from, or established on the basis of, facts* is a statement with an empty meaning, and we believe that the same use of the world "theory" is not appropriate. In fact, a theory is, etymologically a sequence of statements deduced logically from a conjectured set of postulates. The commonplace statement which we have quoted before should be rephrased by introducing instead the world "physical laws" if one could give a meaning to such an expression.

Inductivism was formulated, in our opinion, while misunderstanding Hellenistic sources that stressed the importance of the experimental verification of formulated mathematical theories. Inductivism was developed during four centuries and Francis Bacon was one of its champions. Western Europe's prevailing epistemological approach, in the époque of Bacon, was the so-called scholasticism. Also scholasticism was based presumably on a misunderstanding of Hellenistic sources: the philosophers of this school believed that, based on preconceived beliefs, one could, without any interrogation of experimental evidence, forecast the behavior of physical phenomena. Clearly, scholasticism was accepting only partially what we presume was the true formulation of ancient falsificationism. The falsificationist approach, which consists in conjecturing a model having the aim of describing a set of observed facts, verifies only a posteriori how much can be predicted on the basis of the assumed conjecture.

In fact, falsificationism bases its analysis of natural phenomena, and the corresponding formulation of theories, on the conjecture of some basic postulates from which the scientist must deduce consequences, to be used, when possible, to predict physical phenomena. Therefore, while the stress of scholasticism was presumably focused on the first part of the process of scientific invention as described by ancient falsificationism and neglected the important required check obtained by experiments, inductivism stressed only on experimental evidence, by loosing the deductive part so highly considered in ancient falsificationism. It is clear that the scholars of Middle Ages, having a partial understanding of their sources, could catch only a part of the original complex epistemological vision. This vision has been completely reconstructed only at the beginning of 20th century, when it was necessary in order to formulate really novel physical theories like Quantum Mechanics or General Relativity.

A falsificationist does not try to induce his postulates, he only checks that all the logical consequences of his postulates, for which this is possible, are verified experimentally. Falsificationism has shown to be extremely advantageous in the advancement of scientific progress, compared to a naive inductivism. We claim that one of the first implicit expositions of falsificationism can be found in Archimedes' *Treatise on Method*, in which the Syracusan scientist provides guidance on how to proceed in conjecturing new theories correctly. If it were not for the fact that modern Science is Archimedes' progeny, we could say that Archimedes has all the characteristics of a modern scientist!

Contrary to what History of Science has shown so far, i.e. that only a scientific knowledge produces advances in Technology (therefore, we claim that the only possible way to produce new technological advances is to develop new theories that allow us to observe phenomena never observed before), unfortunately today scientific progress appears to be stuck in the pointless debate on a *data driven* or *theory driven* Science. This debate represents the modern rephrasing of the debate between inductivism and falsificationism, that seems to have been evoked by Proclus.

Proponents of the *data driven* strategy, strengthened by the fact that today there is a relative overabundance of data available and computing capacity, argue that the description of reality can be simply induced by means of the manipulation of experimentally collected data. We will see, in the following, a fundamental example of how even the modern critical interpretation of Hellenistic Science is sometimes given in a *data driven* key. In fact, while Hipparchus of Nicaea conjectured a priori the precession motion of the rotation axis of the Earth, today's modern inductivists, who are data driven, let us believe that Hipparchus *induced* the precession law from a comparison of the positions of certain stars as measured by him and those reported in a star catalog compiled 150 years before him. We believe, and we will describe extensively the reasoning that leads us to this belief, that, instead, Hipparchus first conjectured Earth's axis precession and only after then, based on his conjecture, explained the discrepancies between the two catalogs. Albeit we do not have the relevant sources available (imagine if we could find Hipparchus counterpart of Archimedes' *On the Method*!), we can suppose that, after having seen the motion of a spinning top (see below for more details), Hipparchus, knowing what he was looking for, checked the star catalog for obtaining a confirmation of his conjecture.

The debate between inductivists and falsificationists is being repeated nowadays, for instance, also in the research field devoted to the invention of new materials with properties which are not observed *spontaneously* (i.e. not too frequently) in nature. In this area, which is also discussed extensively in other chapters of this work, a "data driven" strategy is not only impractical, but also conceptually wrong and economically disadvantageous. Therefore, we claim that an awareness of epistemological basic concepts is needed also in nowadays researchers studying basic problems in Engineering Sciences.

## *Underdetermination of Scientific Theories: a problem for Inductivism?*

In the conceptual framework we have discussed up to now, when formulating a new theory, a fundamental role is played by the basic hypotheses, or physical postulates. In the falsificationist approach, starting from the basic hypotheses (postulates), and using rigorous logical procedures, one can deduce consequences that can be confronted with experimental data. It has no sense wondering a priori whether hypotheses are true or false: hypotheses can be only judged on the basis of the comparison between the whole set of their consequences and available experience. Moreover, hypotheses have to be contextualized in the model for which they are formulated. It is a very common misunderstanding the confusion between the hypotheses of a specific model and the hypotheses of another model treating a different aspect of the same physical system. Also if two models are describing the same physical entity, this does not imply that one has to assume the same hypotheses in both of them, if the phenomena to be described are sufficiently different. We present here some paradigmatic examples of this underdetermination of scientific theories.

We do not believe into the inductivist approach, because, obviously, a collection of phenomena concerning a physical system does not uniquely determine *the true and only scientific theory* to be used for describing it. In fact, and as we have stressed before, the used hypotheses may change when choosing a model or another model for the same physical object. A very famous example of the underdetermination of scientific theories is given by Archimedean study of the mechanical behavior of Oceans.

Let us start from a strong ontological statement, clearly accepted by Archimedes: oceans exists and are always the same physical object where tides occur and on which vessels float! Now Archimedes knows that the phenomena involving the floating of vessels can be described by the model of planar surface of oceans. Indeed, Archimedes uses the hypothesis that the surface of seawater is a horizontal plane (in the treatise *On floating bodies*) as a basic one when he wants to establish the stability conditions for ships hull in the vertical configuration. Archimedes had to develop his famous buoyancy law to found this specific theory. However, somewhere else (we conjecture this happened when he was preparing the model for describing tides, that we know has been developed by Seleucus) Archimedes also proves, starting from other postulates, that the surface of the Oceans has to be spherical!

He knew how to use different hypotheses, depending on the different type of phenomena he wanted to describe. Can we find a contradiction between the two models for the surface of Oceans? Is Archimedes, as it is claimed by some modernistic historians of Science, a primitive and confused scholar? In fact, the two visions of Nature, as described by the two Archimedean models, can be reconciled. The floating phenomena of vessels, actually, can be described either by assuming a planar ocean surface or by a spherical surface with a much bigger Earth radius than the dimensions of the vessel. We can, then, easily agree with the fact that a collection of phenomena does not uniquely determine a scientific theory and that the basic hypotheses may change when considering different models formulated for describing different phenomena involving the same physical object.

How can we decide if the Earth surface is not more complicated than a sphere? The ancient Greek observation that one sees at a distance the sails of a ship before seeing its hull can be explained in different ways, attributing to the surface of the Earth different shapes. An application of Occam razor suggests that it is wise to start with the simplest conjecture: that it is a sphere. However, every surface locally similar to a sphere can, in principle, be adopted. Once more naive inductivism seems to find an insurmountable obstacle.

Another useful example is given by the many different models introduced for describing the physical objects planets (and specifically the Earth). The possibility that one can model Earth as a moving material point (as it is done in Celestial Mechanics), or as a rigid sphere (in elementary Astronomical Geography), or as a rigid geoid (in advanced Astronomical Geography), or as a deformable geoid (in Seismology), or as a multi-phase deformable solid (in Geochronology), implies that there are not preferential *true* hypotheses to adopt, but that for a given set of phenomenological evidences a most *suitable* mathematical model is conceivable and that the discussed underdetermination can be solved with a kind of *minimization principle*, that is Occam razor.

In conclusion, we share the belief that (i) the basic postulates of a theory are statements whose truth value can be uniquely posed a priori and (ii) only their being false can be determined once for all. The previous statement is the essence of falsificationist approach, while naive inductivism believes that the basic postulates of a theory can be proven to be true by means of a series of experiments. To believe into inductivism is a (negative) change of perspective dating back to more recent times (i.e. Newton) with respect to the Hellenistic view. This perspective change, we believe, corresponds to a diminution of epistemological awareness.

In fact, Archimedes accepts that a certain theory is valid to describe the phenomena of buoyancy and understands that for this theory to be predictive it is necessary that a certain theorem be true, starting from some basic postulates. So he commits himself to prove this theorem with mathematical rigor.

The example, to which we refer, requires the application of the law of buoyancy and the demonstration of a theorem, which is given by Archimedes by an argument of exhaustion. Archimedes understands that formulating postulates is an important step in the procedure of developing any scientific theory and that experimental evidence cannot be used to prove theorems, that is the consequences of the accepted postulates.

The clarification of the role of mathematical deduction from postulates and of their comparison with experiment represents the main ideas contained in his treatise *On the Method*. The epistemological ideas at the basis of that treatise are manifestly more modern than many contained in works that claim to be milestones in modern Science. Paraphrasing Archimedes, we can say that the fact that the law of buoyancy produces some predictions that can be experimentally verified (using modern language) does not imply that the equality is mathematically true, on the contrary it must be, in fact, proven starting from the mathematical definition of the set of real numbers. Archimedes is confident of the descriptive capacity of his model in explaining buoyancy phenomena. Therefore, he is ready to assume that the entire mathematical architecture needed in the deductive part of his theory is correct, and that the predictivity of his model points the way to a demonstration of the mathematical theorem that *must be true*.

In doing so, we believe that Archimedes proves to be a falsificationist. Moreover, he is so aware of the importance of his Method that, as we have already previously remarked, he claims:

> *I am persuaded that it* [the Method of Mechanical Theorems] *will be of no little service to mathematics; for I apprehend that some, either of my contemporaries or of my successors, will, by means of the method when once established, be able to discover other theorems in addition, which have not yet occurred to me.*

History of Science teaches: developing mathematical models for describing new phenomena can lead to unexpectedly useful results not only in inventing new technological artifact, and predicting the existence of new phenomena, but also in conjecturing new mathematical theorems. This point will be made clearer in the next sections.

## 1.4 From the world reality to its mathematical model and from the model to the replacement of the world reality

In this section we present two paradigmatic cases of how several times human society has seen the birth and subsequent decline of Science. A very interesting aspect lies in the fact that the state of decline is generally not universally recognized except by a few voices that are however isolated and, if possible, silenced. The picture that comes out from the analysis of many cases of decline that have affected human society is disconcerting: it could seem that this decline is the result of an extremely organized operation rather than the result of a series of unhappy choices, of either political or social nature. The question arises spontaneously: who would benefit most from the decline of a scientific society? Who would have the courage to condemn the human society to a sort of Dark Ages in order to favor their own interest?

The answer is not uniquely determined. Certainly when in human society a few groups of unscrupulous individuals assume the leadership and replace in power people who are prepared and work for the common good, the decline is already at a very advanced stage. One aspect which is common to all moments of decline is the relative importance that bureaucrats acquire. Bureaucrats who should limit themselves to facilitating the choices of politicians replace them and ensure that society remains entangled in useless discussions. When, in the late Byzantine era, the highest scientific-philosophical discussion of the intelligentsia of the time concerned the sex of angels, society had already been in decline since long time and the conquest of Constantinople with the consequent collapse of the Eastern Roman Empire in 1453 represented only the formal end of an era that had already ended long before.

The process that determines the decline of a society does not merely ensure its end at a given historical moment, but often also ensures that not enough traces of its civilization survive to determine a new cultural and scientific flowering at a later time. This results in a veritable erasure of certain theories or of the name of their founders.

A notable example that reaches us from Greek antiquity is given by the case of Archytas of Tarentum, who was several times *strategos* (i.e. general) of Tarentum (and therefore his name could not be erased completely from history). He was the first to introduce the Principle of Virtual Work in the study of mechanical systems, but knowledge of this was lost until a few years ago when the treatise *Mechanica Problemata* historically attributed to Aristotle was recognized to be likely authored by Archytas, according to T. N. Winter (2007). On the other hand, it was not lost the information that he had invented a mechanical bird and a toy called ratchet. It is interesting to see that not only his name was erased from the *Mechanica Problemata* (which, by the way, could be an exercise book associated with a much deeper theoretical text), but it was transmitted to us only that his main contribution in Mechanics was the invention of toys. Instead, we believe that he was considering these toys as a way for explain the basic mechanical principles exactly as Heron of Alexandria did later in his *Mechanica* and *Automata*. The process of cancellation is systematic: not only it does eliminate all original sources that it can, but when it cannot manage to eliminate them altogether it makes them sound less authoritative. It is not easy to establish if the erasure process which cancels the name of great scientists and deforms or removes completely their theories is conscious or a consequence of the lack of intelligence and capacity of understanding. This dilemma appears also when discussing the motivations of those politicians mentioned before, whose choices produce the cultural and scientific collapse of the societies that they lead. Most likely the behavior of both scholars and politicians whose disastrous choices were mentioned before can be described by a famous Friedrich Schiller's quote[5]:

> *Against stupidity the very gods themselves contend in vain.*

One of the most frequent phenomena occurring in the phase of degeneration of the scientific culture in a social group consists in the systematic confusion of a mathematical model with the physical object that this mathematical model is aimed to describe. Of course, this confusion is deadly because it poses a series of apparent paradoxes which may lead to believe that the predictive limits of the model represent instead an intrinsic self-contradicting nature of reality. The destructive ontological consequences of these phenomena may lead to a violent reaction against the process

[5] *Die Jungfrau von Orleans* (The Maid of Orleans) (1801), Act III, sc. vi (as translated by Anna Swanwick)

of mathematical modeling, that can be exemplified by the skeptic philosophy that led Sextus Empiricus to abjure Hellenistic Science.

We can recognize a repetitive pattern of growth, decline and collapse of scientific theories and scientific cultures, so that we tend to generalize the analysis by Giambattista Vico, originally limited to the cyclic repetition of social structures, also to history of Science.

In order to show how the decline of a scientific society generally occurs, we propose in the following two examples of models that, in the progress of time, have been confused with reality itself and have, therefore, generated atrocious misunderstandings. The first case consists in the observation of how the mathematical model for the motion of the planets, formulated in an extremely accurate way and also by means of advanced mathematics in the Hellenistic age, has been abused and completely misunderstood until producing in the Middle Ages the idea that the planets actually moved on metallic guides placed in the heavens. The second case that we will deal with is that of Continuum Mechanics, where starting from a certain époque, the concept of force, which was introduced only to simplify the mathematical formulation of the Principle of Virtual Work, has assumed a completely unjustified fundamental role in the postulation of basics Mechanics principles. We observe here that it is possible to recognize a process of *materialization* or *transformation into a real object* for the completely abstract concept of force. In a kind of Platonistic delirium many scholars managed to persuade themselves, and to persuade their pupils, that forces are real objects that one can meet in everyday life: the resulting confusion between physical objects and mathematical objects used in a model for describing real world phenomena is extremely misleading. Those who believe in the *reality* of forces want to give at any cost to this object a wrong ontological essence.

## 1.5 Reconstruction, partly conjectural, of the birth and decline of the mathematical models for planetary motion

We now want to mention a reconstruction, clearly partly conjectural, of the evolution of the mathematical models for the motion of the planets. It is necessary to make two premises: (1) the purpose of what we will describe is not the in-depth historical study of given scientific theories (for this there are several texts available in the literature [37, 35]), but rather to show a sociological aspect of the transmission of scientific culture, which, of course, can be studied only by resorting to the development of the models in non-negligible periods of time; (2) the reconstruction that we will present of the evolution of the motion of the planets is obviously conjectural, in the sense that not all sources are available, but, from the few sources that have come down to us and from secondary sources, it is possible to conjecture the scientific panorama of the Hellenistic age to obtain a vision about Hellenistic Science that is, in many aspects, really surprising.

## *An example of precursory proto-model before Hellenistic Astronomy*

Before properly analyzing Greek astronomy, it is appropriate that we mention an object dating back to the Bronze Age recently found near the German town of Nebra. It is a bronze disc with gold applications representing the sky, which was most likely used in the period 2100–1700 BC. Specifically one can recognize the crescent Moon, the Pleiades and a disk that could be the Full Moon or the Sun. Two arches are affixed to the edges of the disk. In more recent times it has been added a small arch that could represent a solar boat, typical of a religious representation and also found in other cultures such as the Egyptian one.

It has been conjectured that the disk could be used to precisely determine the equinoxes and solstices, aligning it with the stars at certain times of the year and taking into account the orography of the place where it was found. So it would be a rudimentary scientific instrument used for determining the calendar of agricultural activities: it is therefore one of the first available examples of a technological tool developed on the basis of a predictive model about the universe but used for practical applications. The subsequent affixing of the solar boat suggests that the scientific instrument has been transformed into an object of cult and then was finally buried in a tomb. The story of the disk of Nebra is the story of a scientific society, obviously in its embryonic state, that arises and produces useful instruments and then declines making what is no longer understood to become a religious cult. The fact that the fate of a scientific instrument, which is no longer useful because it has been transformed into a cultic object, is the tomb is very explanatory.

The disk of Nebra gives a strong support to the vision of history of Science that considers cyclic cultural declines as frequent social phenomena. It supports Giambattista Vico's vision of cyclicality of social phenomena and completely falsifies the belief that human progress is only proceeding towards higher cultural consciousness.

## *Eudoxus and the model of homocentric spheres*

The first known scientific model describing the motion of the planets is due to Eudoxus of Cnidus (408–355 BC). Eudoxus was a mathematician and astronomer. He is one of the fathers of mathematics. Pupil of Archytas of Tarentum, among other things he studied the problem of finding the algorithm (with ruler and compass) for the duplication of the cube. The problem of the duplication of the cube is an absolutely non-trivial problem, since to be treated properly one must master the concept of irrational numbers, that before Eudoxus most likely had not been developed. It has to be remarked that, when Pythagoreans discovered that the hypothenuse of an isosceles right triangle is incommensurable with the catheti, the first reaction was to believe that nature was paradoxical. In fact, Pythagoreans did confuse the mathematical model *rational numbers* with the concept of *length of a segment*: when it was proven that the above-mentioned hypothenuse could not be represented by a fraction they were led to believe that such hypothenuse did not exist. The reader is invited to consider this as a paradigmatic example of the disastrous potential consequences of the epistemological and ontological mistake which is intrinsic in the confusion of a mathematical model with the physical object that one intends to describe.

It is not a coincidence, then, that Archimedes attributes to Eudoxus the invention of the concept of real number in its geometric definition. Unfortunately, all Eudoxus' works are lost: we have only secondary sources. But these are enough to give us an idea of the level of depth of Eudoxus' discoveries. Among the secondary sources, we recall the treatise of Theodosius of Bithynia *Sphaericae*, which is probably based on his work. Of his other works we have received only the titles: *Eclipses of the Sun*, *Octaeterides* (solar lunar cycle of eight years), *Phenomena* and *Entropon* (spherical astronomy based on observations made in Egypt and Cnidus), *In motion*. As mechanicians we were ready to pay a very high price for having a copy of this last text, as it could give us a clear vision of the first true scientific stage of our discipline and could guide us in the development of novel models. Eudoxus' passion for astronomy was not, of course, only theoretical, but had significant practical implications and, in fact, he built an astronomical observatory.

Eudoxus' fame is related to the model of homocentric spheres. This model describes a universe divided into spheres having a single center of rotation. At the center Eudoxus put the Earth surrounded by spheres in uniform circular motion. The outermost sphere contained the fixed stars. On the other spheres moved the planets.

To better understand the mechanics of the model of Eudoxus, we use the words of G. V. Schiaparelli [134]:

> "*Eudoxus thus imagined, almost as Plato had done before him, that every celestial body was set in motion by a sphere revolving over two poles, and endowed with uniform rotation; he further supposed that the body was attached to a point of the equator of this sphere, so as to describe, during the rotation, a maximum circle, placed in the plane perpendicular to the axis of rotation of the same. To account for the variations in the speed of the planets, their retrograde motion, and their deviation to the right and left in the direction of latitude, this hypothesis was not sufficient, and it was necessary to suppose that the planet was moved by several movements analogous to the first, which overlapped and produced that unique movement, apparently irregular, which is what is observed. Eudoxus therefore established that the poles of the sphere carrying the planet were not immobile, but were carried by a larger sphere, concentric to the first, rotating itself in turn with uniform motion and with its own speed around two poles different from the first ones. And since even with this supposition it was not possible to represent the observations of any of the seven celestial bodies, Eudoxus attached the poles of the second sphere inside a third one, concentric to the first two and larger than them, to which he also attributed other poles and another speed of its own. And where three spheres were not enough, he added a fourth sphere, including in itself the first three, carrying in itself the two poles of the third, and also rotating with its own speed around its own poles. And examining the effects of these movements combined, Eudoxus found that, choosing conveniently the positions of the poles and the speeds of rotation, the movements of the Sun and Moon could be represented well, assuming each of them carried by three spheres; the more variegated movements of the planets he found required four spheres each. The driving spheres of each celestial body he assumed to be independent of those that served to move the others.* [...]

*Thus the total number of moving spheres was 26, plus one for the fixed stars. What was the cause of these rotating movements, and how they communicated from one sphere to another, is not found that Eudoxus had looked for; nor what was the material and size of the spheres themselves; nor what were their diameters and their intervals.* [...] *Eudoxus therefore totally omitted to research what did not matter to his main problem, the geometrical representation of phenomena; and in this we see another proof of his sober and rigorous genius. He did not care at all to connect the driving spheres with those of the planet immediately above and the planet immediately below, and assumed that the spheres involved in the movement of each planet formed an isolated system independent of the rest. In short, everything leads to believe that the spheres were for him the elements of a mathematical hypothesis, not physical entities; from which he was wrongly reproached for having closed the universe in crystal vaults, and for having multiplied them without necessity.*"

Eudoxus was not a mere observer of the sky: certainly it was by observing the sky that he formulated his conjecture at the basis of the model of the homocentric spheres. In fact, he was a great mathematician: this last characterization leads us to conjecture that he was probably aware that his model was not reality, but only an attempt to describe it. This conjecture is strongly supported by the recognition found in Archimedes sources about Eudoxus invention of irrational numbers: only a sophisticated epistemological understanding could have led Eudoxus to his solution of Pythagoreans apparent paradox. It is significant that many scientists even today are unable to distinguish their model from the reality they claim to describe it.

The geocentric model of Eudoxus did not succeed in any case to explain completely the planets retrograde motions and also failed to give an explanation of the variation of brightness of the planets during their motion (which instead is obvious if we consider that the distance of the given planet from the Earth is variable in time). Remaining within a geocentric model, the system was refined by Apollonius of Perga (262–190 BC) who first introduced the concept of deferents and epicycles (which we will discuss in more detail when we present the algorithm of Claudius Ptolemy). Apollonius considered the motion of the planets as a composition of several uniform circular motions and in this way he was able to approximate the retrograde motions and to give a convincing explanation of the variation of apparent brightness. Also in this case, as for Eudoxus, we can say with some confidence that the model with deferents and epicycles was perceived by Apollonius as a mere mathematical model and that he was not confusing his model with physical reality. Unfortunately, the same cannot be said about Ptolemy.

## *Aristarchus: an ancient Copernicus? Or more likely Copernicus is the modern Aristarchus?*

The progress in the development of a model for the description of the motion of the planets obtained by Eudoxus is the basis of the huge advances made by Aristarchus of Samos (310–230 BC). As we mentioned in the previous section, probably Eudoxus knew that his model did not coincide with reality, but that it was only a description of it, a certainly imperfect and obviously perfectible description: the attempt to describe in some way the retrograde motion of the planets by adding extra spheres represents the most evident proof that Eudoxus had a clear idea of the concept of successive (mathematical) approximations of reality. This idea will be fully developed by the sources of Ptolemy as it is evident by inspecting his computation method based on the introduction of deferents and epicycles. If one wants to build a model to describe the motion of the planets, the first step is to obtain kinematical estimates that are consistent (if not overlapping) with observational data. This is what Eudoxus did. Aristarchus goes a step further and introduces the first heliocentric model. He wonders how well the representation of the cosmos given by Eudoxus closely describes reality. Certainly today everyone should be able to agree with the fact that to pass from the geocentric model to the heliocentric one is a simple change of reference and that, once fixed the correct transformation from a reference to the other, there is absolutely no difference in using one reference or the other one. Actually, another possible, if not preferable, choice would be to place the reference in the center of mass of the solar system and consider the motion of all celestial bodies, including the Sun, around this center of mass. In fact, for the scientist of the third century BC the change of reference is an absolutely not trivial conceptual step. We will see how in the Archimedes' planetary stolen by Marcellus and described by Cicero this change of observer was included in the mechanism.

Let us try to reconstruct the various stages that occur in the research of the Hellenistic scientist to arrive at the heliocentric model. The first observation, the most obvious one, concerns the motion of the Sun and Moon, which describe an arc in the sky during the day and the night. We consider already overcome any kind of religious conception that can come out from such observations and we consider already established the knowledge of the sphericity of the Earth (since Parmenides onwards this was well known to the Greeks!). Based simply on the observations of the positions of the Sun and Moon, it is then licit, for the Hellenistic scientist, to imagine that these two celestial bodies rotate around the Earth, which instead remains fixed. The big jump in quality of mathematical modeling is made in the attempt to explain the retrograde motions of the planets (to observe and record which quite advanced technologies are already required, because it is impossible to think that with the bare eye one can record with precision the positions of all planets and constellations). A second fundamental observation in the path towards heliocentrism concerns the motion of the fixed stars, which, without apparently changing their inter-distance on the sky, rotate all together. As we will see, the fixed stars constitute a problem for heliocentrism (but Aristarchus responds extremely lucidly to the objection made to him, see below).

So, this is the picture from which the Hellenistic scientist starts:

i. Sun and Moon follow arcs of circumference;
ii. planets show regular and retrograde motions;
iii. fixed stars have an immutable reciprocal inter-distance on the *celestial sphere* that rotates instead on a yearly basis cycle.

The phenomena (i) and (iii) are perfectly described by Eudoxus' model of homocentric spheres. Retrograde motions require a complexified explanation by means of various spheres, with contained relative motions, associated to the same celestial body. To be predictive, Eudoxus' model becomes very cumbersome. In addition, the tendency to the search of the most economical logical reasoning, typical of the Hellenistic scientist, who was trained on Euclid's Geometry and therefore is accustomed to reasoning as simple as possible, cannot explain why a few celestial bodies (planets) behave differently from the other celestial bodies and go back and forth in the sky. We can imagine Aristarchus' astonishment when he realizes that fixing the reference system on the Sun and not on the Earth, the motions of the planets become all nearly-circular (or possibly elliptical): from a complex and cumbersome description, modified *ad hoc* for each celestial body, this Hellenistic scientist is passing to a unified description that treats all the motions of the celestial bodies in the same way. Once heliocentrism is introduced, it will not be possible anymore to come back to other models!

It remains to be settled, in the proposed model, the question of the fixed stars: if the observation is made from the Earth, which according to the heliocentric model is itself in rotation around the Sun, why should the fixed stars appear to have a fixed relative distance? Aristarchus, who, like all his contemporary scientists, knew deeply Geometry, answered in an ingenious and at the same time obvious way: the distance between the Earth and the fixed stars is enormously greater than the diameter of the Earth's orbit, so that, for what concerns our measurements of relative distances of very distant stars, it makes absolutely no difference to fix the observer reference on the Earth or on the Sun. This will be understood again in modern age with Giordano Bruno and Galileo Galilei only.

Obviously, as we will underline in the following discussion about Hipparchus of Nicea, the relative positions between the so-called fixed stars are not at all immutable, but vary on a time scale much longer than the life of a man because of their natural motion. The reason for the very slow variation of the apparent-from-Earth relative distances is related to the enormous distance between the Solar System and these stars when compared with the diameter of the Solar System: exactly the same explanation given by Aristarchus to establish that heliocentrism and geocentrism are equivalent models for what concerns the description of the phenomenology concerning the motion of fixed stars. We believe that this explanation must have been obviously understandable for those who first came to formulate the heliocentric theory, but we have no sources available to know Aristarchus' thought regarding this issue.

Unfortunately, the work of Aristarchus on the heliocentric theory has been lost and we have available only some fragments reported by secondary sources. The only work which survived is *On the dimensions and distances of the Sun and Moon*. In this work Aristarchus gives another proof of the high level reached by Hellenistic science. With an extremely simple reasoning he succeeds in deducing dimensions and distances ratios from the Earth to the Moon and to the Sun based on the powerful results of Euclidean Geometry and his own results which are based on trigonometric functions.

The whole reasoning of Aristarchus is based on the fact that, when the Moon is in quadrature, i.e. it is illuminated by half, it forms a right triangle with the Earth and the Sun. By measuring in this condition the angle between the Earth-Sun direction and the Earth-Moon direction it is possible to calculate the ratio between their distances using trigonometric arguments.

To calculate dimensions and distance ratios, Aristarchus is forced to invent a way to approximate the calculation of the tangent of the angle. The tangent of an angle is a function that assumes values throughout the whole set of the real numbers, eventually diverging. In fact, as the angle approaches a right angle, the tangent function tends to diverge, i.e. small changes in the angle correspond to huge changes in its tangent. This implies that if the angle in question is almost a right angle then small errors in the measurement of the angle produce large errors in the calculation of its tangent and therefore in the estimation of distance ratios. The estimate of Aristarchus was, in fact, wrong by several orders of magnitude. Nevertheless, the algorithm invented by Aristarchus for the calculation of the tangent of an angle is correct and it will be very useful for the subsequent development of Hellenistic Science (and of Science *tout court*).

A final note on how many paradoxes may arise while describing the process of transmission of Science is needed. In many modern texts Aristarchus, who first introduced the heliocentric model, is referred to as the ancient Copernicus (who lived almost two thousand years later). As we have repeatedly seen in other chapters of this work, often in the history of Science those who come after claim authorship of an idea, even if this idea was developed by others long before. In the present case, obviously, it was not Copernicus, who probably knew very well the works of Hellenistic Science, to claim the paternity of heliocentrism. In fact, Copernicus clearly attributes to Aristarchus the formulation of such important mathematical hypothesis (see [97]). The causes of this absurd misunderstanding are to be found in the works of modern scholars. Why this reversal of ideas? Why not calling Copernicus *the modern Aristarchus* but, instead, doing the opposite? It may appear as if Aristarchus had in some way wanted to refer to the ideas of Copernicus. The reasons of this aberrant time-reversal are found, in our opinion, in that very modern attitude that sees with extreme disregard the ancient Science (and indeed, many contemporary scholars warn that one should never speak of *science* in antiquity!) and that wants to show how well we can manage ignoring our past. But it should be considered that if removed from the shoulders of giants the dwarfs will fall into the void.

## *A mature Science is sometimes too complex to be transmitted to posterity: Hipparchus's explanation of the precession of the equinoxes*

Once the heliocentric model has been acquired, Hellenistic Science continued to refine its models by focusing on the study of further available phenomenology. There is, as we have extensively emphasized in the section on Philosophy of Science and Epistemology, an important requirement that a theory must fulfill: not only it must be able to reproduce available phenomenology, but also it has to allow, giving directions to experimental research, for new discoveries by indicating where and how new measurements have to be made. Precisely framed in this panorama, Hipparchus of Nicaea (200–120 BC) is the first who was able to accurately predict the eclipses of the Sun and the Moon and, demonstrating profound and pronounced skills as a mathematical physicist, as we would say today, he could explain the discrepancies found between the star catalog compiled at the turn of the fourth and third centuries BC by Timocharis (of Alexandria) and Aristyllus (which were based on previous measurements of the Babylonian Chaldeans) and his own star catalog: indeed, between the two catalogs there is a time-lapse of about 150 years and, for this reason, the apparent positions of the stars on the sky show small variations.

Aristarchus, following the hypothesis first suggested by Heraclides Ponticus (c. 390 – c. 310 BC), whose works are however lost, had attributed to the Earth, in addition to the motion of revolution around the Sun, also a motion of rotation around its own axis. He had, moreover, established that, to take into account the alternation of the seasons, it was sufficient that the axis of the Earth's rotation was inclined with respect to the plane of the orbit around the Sun (also known as the *ecliptic* plane). While Aristarchus probably had no idea that the direction of this axis was not constant in time, Hipparchus of Nicaea conjectured the presence of one between the two motions nowadays attributed to the Earth's axis. Hipparchus, in fact, introduced the Earth's axis precession motion, which consists in the rotation of the Earth's axis around the normal to the ecliptic plane: today we also introduce the nutation motion, which consists in a further periodic oscillation of the Earth's axis during the precession motion.

The reasons why Aristarchus assumed the inclination of the Earth's axis should be nowadays part of general culture, albeit they are not at all trivial, as it is needed to explain the alternation of seasons. On the other hand understanding how Hipparchus was able to deduce the motion of precession is an extremely challenging question, which deserves some explanations here.

As we said before, Hipparchus compares two stellar catalogs, that of Timocharis and Aristyllus (based on data already collected by the Babylonian Chaldeans) and his own. The two catalogs have differences in the measured position of some stars (for example Spica). It is possible that Hipparchus formulated his hypothesis of Earth's axis precession from the discrepancies between these measurements. We believe that such a precise hypothesis cannot be the result of this comparison alone, without the aid of a complex modeling procedure and postulation. In fact, once it has been established by Aristarchus that the Earth rotates around its axis and that this axis is inclined with respect to the ecliptic plane, the Hellenistic scientist probably tried to formulate a model of the motion of the planet Earth around its axis, by conceptually separating this motion from that of the rest of the universe. If this simplifying hypothesis is well-grounded, then the model-seeking scientist will as a first step attempt to represent the Earth's axis motion as a superposition of simpler motions. In this aspect, Hipparchus works in continuity with the Hellenistic tradition that represents celestial motions using sub-sequent epicycles. By conceptually isolating the Earth in its motion, it is likely that Hipparchus could have established a parallelism with the motion of a spinning top. It is well attested the use of spinning tops in

Hellenistic époque and possibly earlier. Callimachus from Cyrene (c. 310/305–240 BC) reports the use of spinning tops as toys in his first Epigram [Call. Epigr. 1, 9-10]:

*Those, some children, played with rapid spinning tops twirling them in the wide crossroads.*

It is also attested, as proven by the role of Archytas (c. 435/410 – c. 360/350 BC) as inventor of pedagogical toys for children, that Hellenistic scientists aimed to exemplify physical phenomena by means of toys, a tradition which was also continued by Heron of Alexandria (c. 10 – c. 70 AD). It can be assumed, therefore, that Hipparchus, knowing that analogous mathematical descriptions can be used to describe different physical phenomena (today, following Feynman [75], we would say that the same equations may model different phenomena), was bound to attempt the description of the physical system "Earth rotating around an inclined axis" using the knowledge acquired in the already known areas of the Science of his time, i.e. "spinning top rotating around its axis". It is widely known that the Greeks knew the spinning top even before Callimachus: in the VII book of the *Iliad* Homer (late eighth or early seventh century BC) describes the motion of a stone thrown by Ajax Telamonius against Hector as the motion of a spinning top

*An even bigger stone* [...]
*Telamonius grasped and his strong*
*right hand twirled it like a stone thrown from a slingshot.*

If Homer could describe a stone thrown by Ajax as a spinning top, why Hipparchus could not think of modeling the Earth rotating around its axis as a spinning top? Which is exactly what Maxwell will do in his treatise on spinning tops [115]. If one observes a spinning top rotating then he will see: (i) the rotation of the spinning top around its axis, (ii) the direction of the axis changing in time (precession); (iii) the variation of the inclination of the axis due to a certain oscillation (nutation).

The question we have to ask ourselves now is: is it easier to deduce the precession motion by observing the discordance of the measures or to conjecture it by observing the motion of a spinning top thrown by children playing in the street?

It is attested that Hipparchus did conjecture the nutation motion of the Earth's axis. We claim that the genius of Hipparchus consists in imagining the similitude between the spinning top and the Earth and, consequently, in interpreting the discrepancies between the measures reported in the two catalogs not as an indication that the oldest measures could be wrong but as a proof that the Earth could actually be described as a spinning top. It is clear that an essential prerequisite for the advancement of knowledge is that one generation of scientist can rely on the results obtained and transmitted to them by the previous generation. It is therefore to be blamed the modernistic attitude of considering everything coming from the past as unavoidably primitive.

The hypothesis of Hipparchus, contained in his lost work *On the displacement of the solstitial and equinoctial signs*, is applied to the analysis of the longitude of the apparent position of the star Spica during a lunar eclipse. The method adopted by Hipparchus to measure the longitude is known because it was reported by Claudius Ptolemy (c. 100 – c. 170 AD) in his *Almagest*. After the measurement, Hipparchus compared it with the longitude of Spica reported in the catalog of Timocharis and

Aristyllus and he noted that this longitude had varied by 2° in about 150 years. From this observation, he made the hypothesis that the fixed stars have shifted with time and estimated a precession of 48" per year. It is remarkable that the precession measured by Hipparchus with the instruments of his epoch is so close to the value measured with today's instruments and expressed as 50.26" per year. It is singular that Hipparchus's estimate is also considerably better than that obtained by Claudius Ptolemy (36" per year) about three centuries later.

The measurements made by Hipparchus to validate his hypothesis of precession of the equinoctial points obviously require the use of instruments, both theoretical and practical, which are extremely accurate. This is how trigonometry and the astrolabe were born. As far as trigonometry is concerned, Aristarchus had already introduced some basic concepts, very much linked to the formulation in terms of Euclidean geometry. In Hipparchus, we find trigonometry in its modern formulation, except for the use of a different symbology. In fact, the symbology used in modern times, as it is well known, was introduced only by Euler (1707 – 1783).

## *Other achievements of Hellenistic Science in the study of the motion of the planets: Seleucus' explanation of ocean tides and the Antikythera calculator*

So far, we have described the genesis and development of the heliocentric model, but except for the indirect evidence we have mentioned, neither Aristarchus nor Hipparchus had given a *demonstration* of it. According to Plutarch (46 AD – after 119 AD), Seleucus of Seleucia (floruit 150 BC) gave a formal proof of the heliocentric theory. We believe that Plutarch with the word *demonstration* meant the deduction from more fundamental postulates. If our interpretation is correct this could imply that Seleucus had invented a form of dynamics. We will see that Middle Ages echoes of dynamical theories seem to support our conjecture. Another indirect support for it can be found in the explanation, also attributed to Seleucus, of the complex phenomenology involved in ocean tides.

In fact, from a reconstruction based on secondary sources (because even of Seleucus nothing has come down to us) it can be said that the greatest contribution of Seleucus to Hellenistic Science consists in the in-depth study of the tides. Now, while it can be simply understood that the tides are related to the combined interaction of Sun, Moon and Earth, the specific phenomenology, especially in its quantitative aspects, requires a very detailed analysis. Indeed, if one wants to reproduce with some accuracy the experimental observations, the extremely simplified vision where the tidal phenomenon is described by static interactions with celestial bodies is not sufficient. Today, we use an extremely complex model based on a dynamic approach, introduced by Laplace (1749-1827), which takes into account also the inertial effect of the ocean motion relative to the Earth. In that formulation, the well-known Coriolis force needs also to be introduced.

An interesting aspect resulting from the few secondary sources of Seleucus' thought is that he related tides not only to the position of the Moon and the Sun, but also to the motions of the Earth. The main sources from which we get information about Seleucus are Strabo (64/63 BC – 24 AD) and Aetius (1st or 2nd century AD), and the latter reports, in an extremely confused way, this idea, which in some ways recalls the dynamic model of Laplace. In this regard, Galileo Galilei (1564-1642) had already tried to give a dynamic interpretation of the phenomenon, but produced not very clear results. We believe that both Galilei and Laplace were at least inspired by the words of Seleucus (probably not by the confusing version of Aetius, but by another clearer source that has not reached us). Some authors, with philological evidence, have tried to interpret the text reported by Aetius and to relate it to the information referred by Plutarch about the presumed demonstration of the heliocentric theory presented by Seleucus, but we will not delve here into this subject.

The scientific progress of the Hellenistic age was not only theoretical but also had strong practical implications. One of the most paradigmatic proofs of the technological development induced by the theoretical advancement of Hellenistic astronomy is represented by the calculator of Antikythera (150–100 BC). This famous astronomical calculator, which even for the complexity of the gears that compose it gives fundamental information on the high level reached by Greek metallurgy, was able to predict an enormous number of celestial phenomena, as well as provide a series of calendars. By turning the crank on the side, it was not only possible to calculate the exact position of each planet and the phases of the moon, but also the eclipses of the Sun and Moon. Since, of course, information about the positions of the planets was given relative to the latitude of a chosen point on the surface of the Earth, some have speculated that, using a sort of inverse method, the calculator could be used on sea voyages to estimate latitude based on a comparison of the positions of the planets in the sky and those determined using the astronomical calculator.

The astronomical calculator of Antikythera shows the position of the planets in a reference centered on the Earth. In this case the choice of the geocentric model is justified by the fact that the scientific instrument has a specific purpose and, if it is true the hypothesis that the calculator provided the latitude by comparison with the sky, then it is logical that the represented system be geocentric. Obviously, in order to describe the complexity of the apparent motions of the celestial bodies in the geocentric system, it was necessary to have an extremely precise and reliable computational algorithm. The algorithm used and realized by means of numerous gears was that due to Apollonius of Perga, who decomposed the motion of the planets in circular motions on deferents and epicycles. For each planet the Antikythera calculator has a series of gears for the deferent and for the various epicycles.

The discovery of the astronomical calculator of Antikythera has shown us an aspect of the geocentric system that is rarely emphasized. To the question of why the ancients had begun to study the sky and the motion of celestial bodies, the right answer is not, as it is often trivially suggested, the wonder that the uncultured ancient man felt in observing the starry sky. This is a romantic view that we should learn to circumstantiate. The fundamental reason why it was necessary to study the sky lies in the fact that until before the invention of the compass this was the only reliable way to get orientation. So it is also understandable why, although it was already clear with Aristarchus that the heliocentric system was more effective in describing phenomena than the geocentric one, the geocentric description of planetary motions has never been abandoned and, indeed, has been gradually refined: it is absolutely necessary to obtain precise estimates of the positions of the planets in the reference centered on the Earth and thus be able to orient. In short, we could say that the Hellenistic scientists knew very well that at the center of the planetary system there was the Sun, but they needed to put the Earth at the center of the solar system for using the theory to obtain practical results.

Using modern language, the study of heliocentric theory represents pure research, while the study of apparent positions of celestial bodies in a geocentric reference frame represents applied research. Archimedes (287–212 BC) did succeed in making theory and practice dialogue fruitfully and it is not a coincidence if in the famous planetarium belonging to Archimedes, as Cicero reports, one could, depending on the needs, fix the Sun or the Earth and observe directly the motions of the planets from the heliocentric view-point or from the geocentric one.

## *The death of Archimedes as a metaphor of the beginning of the end: the slow decline leading to Dark Ages did begin with the end of Hellenistic Science*

Archimedes was one of the greatest scientists and mathematicians of human history. In several parts of this work we have spoken of his outstanding scientific discoveries and especially of his way of approaching the scientific research, which, while remaining strongly connected to physical reality, had the merit of being lucidly formulated in precise and rigorous mathematical terms. As for the topic we discuss in this section, we limit our attention to two aspects, one of technical and the other of historical nature, which are related to the death of the great scientist, after the end of the Roman siege of Syracuse in 212 BC led by consul Marcellus. The technical aspect is reported by Cicero (and we have mentioned it previously): Marcellus brought in the booty of war taken in Syracuse the famous planetarium of Archimedes. Cicero speaks of this planetarium several times, in the *De Re Publica* and in the *Tusculanae Disputationes*. In the latter work he reports:

> *"Nam cum Archimedes lunae solis quinque errantium motus in sphaeram inligavit, effecit idem quod ille, qui in Timaeo mundum aedificavit, Platonis deus, ut tarditate et celeritate dissimillimos motus una regeret conversio. Quod si in hoc mundo fieri sine deo non potest, ne in sphaera quidem eosdem motus Archimedes sine divino ingenio potuisset imitari."* [Cicero, Tusculanae Disputationes I, 63]

> "In fact, when Archimedes bound in a sphere the motions of the Moon, the Sun, and the five errant planets, he obtained the same result as [the Demiurge] who in Timaeus constructed the universe, i.e. the Plato's god, so that a single revolution governed motions very different from each other in slowness and speed. If it is not possible for this to happen in this world without the intervention of a god, certainly not even in his sphere Archimedes would have been able to imitate the same movements without a divine intelligence."

Archimedes' planetarium probably represents the highest point of Hellenistic Science, and today we can only have a vague idea of it by looking at the astronomical calculator of Antikythera, which was in all probability a portable version of the planetarium.

The siege of Syracuse is also sadly known because it was during this siege that Archimedes lost his life. Plutarch, in his *Life of Marcellus*, reports three different versions of the death of Archimedes: all versions agree in the fact that he died by the hand of a Roman soldier, although the Syracuse scientist is said to have been extremely appreciated by Marcellus, who seemed to be grieved by his death and gave him an honorable burial.

The death of Archimedes marks a symbolic point of no return for Hellenistic Science: after this event the phase of decline begins. As we will see, the decline is not immediately recognizable as such, but is usually preceded by a phase of mannerist fashion in which there are no more original ideas, but only repetitions and progressive refinements of pre-existing ideas. We believe that the siege of Syracuse and the consequent decline of Hellenistic Science started an inexorable process that, centuries later, will lead to the Dark Ages.

In the artistic domain, the Renaissance was followed by Mannerism, which in its most negative sense is depicted as the artistic current in which the artist no longer seeks inspiration in nature, but limits himself to attempting to imitate the works of the three great Renaissance artists, Leonardo, Michelangelo and Raphael (thus losing the instinct of originality that had characterized the Renaissance artist). Similarly, Roman art limited itself to copy and reproduce Hellenistic masterpieces. In the same way, and we could say cyclically, in every stage of history of Science one recognizes a phase of maximum development followed by a *mannerist* phase, which preludes to an imminent decline eventually followed by another growth stage: we believe to be followers of Giambattista Vico's doctrine. The great scientific advances of the Hellenistic period, which in the restricted field of Astronomy the available sources attribute mainly to Eudoxus, Aristarchus and Hipparchus, are followed by a phase of stagnation in which attention is focused on computational aspects and loses, therefore, that originality which had characterized the scientific revolution of the 4th-3rd century BC.

In this mannerist framework stands Claudius Ptolemy (100–170 AD), whose greatest contribution to ancient Science consists in the refinement of the algorithm, originally due to Apollonius of Perga, that allowed to calculate precisely the positions of the planets of the Solar System. Ptolemy worked in Alexandria when probably the Library still existed and therefore had at his disposal the largest database in the world to which one could have access in that time. It is peculiar that, while Ptolemy was concentrated purely on the problem of calculating apparent motions of stars in a geocentric reference, his successors attributed to him the choice of the geocentric model of Eudoxus. We do not believe that Ptolemy had consciously refused the heliocentric model of Aristarchus: like many modern engineers he was only interested in practical calculations, and spent all his time in describing calculation algorithms. In any case, it is necessary to point out that, as we mentioned when discussing the Antikythera calculator, the calculation by deferents and epicycle had been introduced by Apollonius of Perga centuries before Ptolemy. From this consideration Ptolemy appears to be a compiler of already known results rather than the inventor of something new.

Ptolemy's algorithm turns out however to be a computational tool more precise than the algorithm developed by Apollonius of Perga and capable of giving estimates of the positions of the planets with sufficient precision for the astronomy of his time. We stress that his time is quite different from the centuries in which Eudoxus, Aristarchus and Hipparchus operated and in fact, for example, the estimate given by Hipparchus of precession is significantly more accurate than that made by Ptolemy three centuries later. The algorithm is based on a system of successive approximations made of compositions of uniform circular motions on circles of different sizes and with the centers located at *ad hoc* chosen points. An epicycle is a circumference whose center is placed on the circumference of a larger circle called deferent. In the model of Apollonius of Perga, therefore, the planetary orbits are represented as a composite motion of the revolution of the planet along the epicycle and of the epicycle along the deferent. By increasing the number of epicycles, one can obtain more and more accurate estimates of the orbits of the planets: one can conjecture that Apollonius of Perga was aware of the fact that increasing the number of epicycles one could reduce the error in the estimates of the kinematics of planets. We doubt that Ptolemy had this awareness. This multiplication of epicycles has been widely criticized in the past by the followers of Copernicus (1473-1543), against the opinion of some Jesuit erudites: the fact that more precise estimates could be obtained by increasing the number of epicycles was seen as an unnecessary complication of the model. In fact, the controversy between Copernicans and some Jesuits was based on a fundamental misunderstanding: while Copernicans considered the number of circumferences involved in the mathematical description of Solar System as a part of a postulation scheme, and therefore wanted to reduce it using Occam razor, their Jesuit opponents stressed the mathematical aspect of the question, remarking that periodic motions can be approximated better and better by increasing the number of epicycles.

As reported by Gallavotti [80], since Schiaparelli's analysis [134] the approximation technique via epicycles for the periodic motion of planets can be recognized as an initial form of Fourier analysis. As it is well known, Fourier (1768-1830) joined Napoleon's Egyptian campaign in 1798. The development of his analysis, conversely, dates from 1822. We have, in the present work, repeatedly conjectured, sometimes even demonstrated, that in the history of the transmission of scientific thought the often unmentioned source of works that are perceived as revolutionary and forerunners for modern Science is to be found in works of the Hellenistic age, which are nowadays (perhaps not by accident) lost.

It is clear that Fourier could have simply been inspired by what was already known about this technique. It is purely speculative to believe that he could have found other sources while campaigning in the place where the largest Library in the ancient world had risen. It is also clear that Apollonius' model and Ptolemy's algorithm were known, at the expense of Aristarchus' heliocentrism, throughout the Middle Ages and were considered basic until Kepler (1571-1630).

## The materialization of Eudoxus' model

It is remarkable that during Dark Ages a choice among available models for the Universe was made. Soon one model was confused with reality. In fact, it was the simplest, and less predictive, model to be confused with reality for at least six centuries (the time interval between the fall of the Western Roman Empire, i.e. 476 AD, and the small Renaissance of Frederick II Hohenstaufen, who lived between 1194 and 1250). The complexity of all models formulated after Eudoxus was totally out of the understanding possibilities of nearly every intellectual of that Ages. Therefore, what could not be understood because of its sophistication and complexity was rejected as false, a useless and empty complex philosophy, soon associated with useless mathematics. Instead, naive and primitive models were promoted to crystal clear truths, that one could not dispute without risking to be considered heretical.

Moreover, the concept of mathematical description of phenomena and the role of mathematical entities used to predict them were completely lost and therefore, while describing Eudoxus' model, it was felt necessary to materialize the hinged rotating spheres assuming (as also it has been recalled explicitly by Schiaparelli) that the Universe was closed by crystal vaults mechanically interconnected one to the other.

One can get a clear idea of how much the thinking has regressed with respect to the Hellenistic period by considering that Bede the Venerable (672-735), one of the greatest scholars of the period immediately following the collapse of the Western Roman Empire, is remembered for having invented a method of counting up to a million with the fingers of the hands. So great is the devastation following the end of Hellenistic Science that mankind had to learn again how to count!

The lowest point in the scientific understanding during the Dark Ages occurs with the materialization of Eudoxus' model of homocentric spheres. Paradoxically, in opposition to the state of intellectual disruption produced by this materialization, from an artistic point of view the distorted view of Eudoxus' model generates a series of masterpieces in figurative art that perhaps had, at least, the merit of inspiring the efforts of Renaissance scholars to restart the systematic study of the problems addressed by Hellenistic scientists.

The materialization of the model of the homocentric spheres leads to two misunderstandings, the former of a purely scientific nature and the latter of a socio-cultural nature. The first misunderstanding concerns the vision of the universe that the man of the Middle Ages has: the Sun and the planets not only rotate around the Earth, but also they are stuck on metal rings (or crystal vaults) hinged to each other and rotating around the center of the Earth. This abnormal misunderstanding is generated by the literal interpretation that the majority of the medieval intellectuals were able to give of the drawings representing the homocentric spheres or of their practical realization in the ancient Greek armillary spheres (and in fact the armillary spheres begin to be spread again in Europe in the Late Middle Ages).

The second misunderstanding, as we said above, is of socio-cultural character and concerns the perception that the modern History of Science has of Eudoxus and his model. As we have repeatedly emphasized, relying also on the opinion of

Schiaparelli, Eudoxus was fully aware that his model was not the physical reality and that, for example, the homocentric spheres represented only the elements of a mathematical model of reality and absolutely not the reality itself. Instead, partly because of medieval misunderstandings about it, the common perception of much of the modern scientific world is that Eudoxus, and Hellenistic scholars in general, had a very naive idea of reality and of its mathematical modeling. With due differences, it is as if in a thousand years from now our descendants will report that we were convinced that electrons are yellow balls with an arrow stuck along one of their diameters just because in some physics textbooks similar images are proposed to give an approximate idea of the spin. Of course, there are in present times some physicists who have such a belief: however, nobody attributes it to Wolfgang Pauli (1900-1958)! Similarly, we should respectfully appreciate Eudoxus' vision of Science.

We can attribute to two factors of fundamental importance the fact that after a thousand years of darkness the flowering of the Renaissance revived scientific interest. The first and inescapable factor consists in the fact that during all the Middle Ages the only scientific discipline that continued to be taught and transmitted from *maestro* to *pupil* was Euclidean Geometry. The presence of Euclidean Geometry in the cultural background of the first humanists certainly allowed them to appreciate the importance of the content of the ancient Greek texts of the Hellenistic school and to be able to read them. Obviously, not all humanists had the same skills and the same preparation and, for example, as discussed in detail in other chapters of this work, Tartaglia is not able to fully understand the reasoning of Archimedes and therefore modifies Archimedes' figures considering them wrong.

The second very important factor is given by the Byzantine cultural school, which, differently from the Western one, had remained active until the fall of the Byzantine Empire, which occurred in 1453 with the fall of Constantinople. One of the most important intellectuals of the 9th century Byzantium is Leo the Mathematician or the Geometer (790-869). This erudite had all the skills which will be found in the future humanist and, indeed, we due to him and to his farsightedness, probably, the first spark of Humanism and Renaissance in Europe. Leo the Mathematician commissioned the copy of many Hellenistic scientific manuscripts and, among the others, of the works of Archimedes. At least three manuscript containing the works of Archimedes were produced under his responsibility, today known as codices A, B and C. When Byzantium was sieged and conquered by the Crusaders in 1204, Leo's library was dismembered and a part of the manuscripts stored in it was brought to Europe. The presence of all these Hellenistic works in Europe gave rise to a strong revival of interest in science, and for the first time in a thousand years scientific progress started again. The Renaissance had begun.

## 1.6 The postulations of Mechanics, forces and their materializations

From now on we will focus on another materialization of mathematical concepts which is still occurring in many scientific milieux. While armillary spheres are not believed to be real anymore, there are too many contemporary scholars who managed to persuade themselves about the *reality* of forces. These scholars talk about forces as if they were objects one can observe in real world: forces seem to have, in somebody's words, the same ontological reality as walls, boats, wind, pulleys.

### *The recovery of ancient Hellenistic Mechanics: Middle Ages mechanicians*

It has to be remarked that many historians talk about the "Renaissance of the 12th century" as a period in which in Western Europe the Latin translation of Greek and Arabic works, especially in Natural Science, Philosophy and Mathematics, greatly changed the cultural standing of Latin-speaking culture. In this period, and after it, we meet many erudites and scholars who tried to recover the lost Hellenistic knowledge.

While we have some ancient sources about the Mechanics of material points, we can only conjecture the existence of a Hellenistic Mechanics of deformable bodies. The most meaningful hint indicating its existence can be found in the works of Galileo Galilei [19]. Exactly as he tries to reconstruct Seleucus' theory of tides, and as he tries to reconstruct the theory of planetary system by Hipparchus, Galilei also tries to understand the theory of deformable beams: though, we must say, without great success [19]. In fact, Galilei did not manage to understand how bending stiffness of a beam depends on the geometry of the beam's cross-section: his deduction starts from a wrong conjecture about the deformation field inside the section. By the way, the fact that there is an evidence that Leonardo da Vinci (1452-1519) tried to understand the theory of beams is another hint about the existence of an Hellenistic source in the subject, as Leonardo is known to be a great estimator of Greek Science.

The difficult point that Galileo did not manage to fully understand concerns the deduction of a theory of a 1-dimensional continuum (the simplest being Euler-Bernoulli beam theory) from a more detailed 3-dimensional continuum theory. In fact, the process of micro-macro identification has been fully developed only when the variational postulations of Mechanics have been recovered [23, 26, 18, 16, 15]. Galileo did not conjecture the right linear dependence of the contact force intensity on the distance from the neutral axis in the beam theory: clearly, it is extremely useful, if one wants to develop generalized beam theories, to understand how the progenitor theory has been formulated [104, 143, 145, 102, 103, 33, 31, 148, 150, 152, 147, 151].

The existence of more ancient (and sometimes partially lost) sources may contribute to explain the reasons why in Mechanical sciences one observes very often, especially when considering Middle Age texts, some oddities in the diachrony of Mechanics development. One observes more advanced texts which are precedent to less advanced ones and to definitely primitive others. The existence of linguistic and social barriers does not seem enough to explain the mentioned observed evidence: we believe that some scholars could access to sources that were very faithful to the original Hellenistic thought while others had access to worse sources or, even, could not understand really the content of such sources.

The strangeness in the diachrony of the development of Mechanics leads us to conjecture the following hypothesis: the distribution and accessibility of the ancient Hellenistic Science texts, often available in a single copy, modulate the speed and readiness of the rebirth of scientific thought. With the fall of the Western Roman Empire, the few remaining elements of unity in scientific thought collapsed, and the enormous progresses made in the 4th and 3rd centuries BC were relegated to the monasteries that owned the only medieval libraries. And in the monastery libraries these texts were sometimes lost and the genealogy of scholars, with a Maestro explaining to the pupils the content of the ancient texts, was broken. The practically null scientific preparation of medieval westerner scholars decreed the loss of many Greek scientific works. One who does not understand what she/he reads prefers to believe that what she/he is reading is, at least, useless and, therefore, unworthy of being transmitted to posterity. Thus, unlike works of literature, philosophy, historiography and other non-scientific disciplines, not only scientific texts were not copied, but often the very expensive parchment on which they were written was reused. This is precisely what unfortunately happened to codex C of Archimedes' works: the parchment was scraped off to make space for prayers against the flu.

It is not surprising that one of the greatest intellectuals of this period was Bede the Venerable (672-735)! The only significant aspect in the scientific sphere of this period is that the habit of studying Geometry remained almost intact and that Euclid's Elements remained one of the essential texts even during the Middle Ages: in fact, geometry was taught to all scholars. It seems that one of the few copies of the Elements of Euclid was preserved to posterity by the family Hohenstaufen and, in particular, by Frederick the Second. We can make the following conjecture: it was only thanks to the education in abstract thought provided by Geometry that a return to Science was possible in the Renaissance. Or, at least, this return required less time than it would have been necessary if Euclidean Geometry had not been taught during the Middle Ages.

Perhaps the clearest sign of the scientific regression of the Middle Ages is the theory of the nine medieval heavens, which is directly induced by the complete misunderstanding of the theory of the motion of the planets. At the beginning of the previous section we showed the sad fate of a proto-scientific theory, the one that was supposed to be at the basis of Nebra's disc: with the demise of the society that produced the theory and the birth of a non-scientific society, the scientific content of an abstract theory is completely lost and subsequently distorted into a religious belief.

After the *small Renaissance* of Frederick II Hohenstaufen there was a slow rediscovery of the distinction between model and real object: the basis of this slow process was the return to the study of Logic, which provided a structure for the subsequent return to scientific thought. An important role in this rediscovery process was played by William of of Occam (1285-1347), whose *Summa logicae* (c. 1323) constitutes a kind of meta-theory necessary to formulate theories. In fact, an important first step towards the rediscovery of scientific thought is represented by the so-called Occam's razor[6]:

*Pluralitas non est ponenda sine necessitate*[7]

It was only after 1100 AD that Latin translations of textbooks on Logic developed in Arab cultural circles started to arrive in Christian Europe. In Arab Science, in fact, we find the eminent scholar Avicenna (980-1037) that tried to propose a proto-inductivist system of Logic which is alternative to the Aristotelian one. Avicenna's Logic also influenced Western thought and we can say that, in a certain sense, Avicenna can be considered one of the founders of scientific inductivism, which will be discussed shortly below.

If we want to get an idea of how much the collapse of the Western Roman Empire influenced the collapse of scientific thought, we can consider the apparent anachronism observed in the cultural milieux of the capital of the Eastern Roman Empire, Byzantium (i.e. Costantinople), where the cultural ferment that had characterized Hellenistic circles survived for a few centuries. Byzantine intellectuals, whose works would only be rediscovered later in Western Europe, provided the first westerner humanists with a key for decoding Greek scientific thought, and, in particular, Greek Mechanics. John Philoponus (490-570), for example, proposed the concept of *impetus*, which seems strongly related to the concept of inertia, about a thousand years before Galileo and Newton. Moreover, Byzantium also represents a sort of *reduced Alexandria*, collecting the knowledge of the time and organizing it for later dissemination. Among the various examples of this Byzantine ferment, we cannot forget the already mentioned Leo the Mathematician, promoter of the renaissance of mathematical studies and of the rediscovery of Archimedes' scientific personality. Here, we will limit ourselves to recall that it is to him that we owe the rescue and transmission of many of Archimedes' texts, which only reached our hands thanks to the copies he commissioned. Several wars and an unfortunate Crusade that diverted Christians from the liberation of the Holy Sepulchre to the sack of Constantinople, which occurred in 1204, later, disseminated throughout Europe the famous Archimedean codices A, B and C. Albeit he was living several centuries before, Leo the Mathematician can be considered a true Renaissance man.

In fact, the role of the scientific ferment of Byzantium is crucial in the development of the Italian Renaissance, albeit it would still need almost five hundred years for westerner intellectuals to stop discussing about the sex of angels and to devote themselves to problems of a less elevated nature, perhaps, but certainly more useful to the progress of humanity. On the path to the Renaissance, we can at least mention

[6] Some modern scholars misunderstood Occam razor spirit and believed that it was forbidding theories in which too many parameters appear: in this way they exclude, a priori, any possibility to model complex mechanical systems, as those studied, for instance, in [55, 54, 68, 124, 88].

[7] Plurality has not to be posed without necessity.

some scholars who characterized the slow rediscovery of mature scientific thought. Thomas Bradwardine (*Doctor Profundus*, floruit 1330) distinguished kinematics from dynamics, introduced the concept of instantaneous velocity and discussed the law of falling bodies. Nicolas d'Oresme (c. 1320/1325-1382), institutor of the Dauphin of France, studied the Universe with mathematical methods: in particular he formulated a Galilean invariance principle and established the foundations of Analytical Geometry. Many others would deserve to be cited: however, we simply want to stress here that we know for sure that at least in one époque the main work of scholars consisted in reading ancient books whose content was perceived as very profound albeit resulting very obscure.

In fact, we believe to recognize in modern times the slow rediscovery of ancient theories. Albeit this rediscovery occurs in restricted disciplinary subgroups of scholars, the features of the sociological process seem to be the same. Of course, the fact that other contemporary groups had not lost the knowledge which is being recovered for sure helps in the rediscovery endeavor. We believe that in theoretical Continuum Mechanics the rediscovery of Lagrange-Piola postulation for generalized continua in the group of scholars following Truesdell orthodoxy has been held by the existence of Landau textbook in Theoretical Physics[8].

### *Fundamental concepts and frequent misconceptions in the field of Mechanics of materials*

In the previous sections we dealt with the Mechanics of material points and its applications to the description of the planetary motion, from now on we will focus on the Mechanics of deformable bodies. However, we will base our analysis on available sources, which are much more modern. It will be clear that the same sociological phenomena involved in the transmission of knowledge observed in the transmission of the Mechanics of material points through the centuries occur also in the transmission of the Mechanics of deformable bodies. Our description of historical development of the Mechanics of deformable bodies starts with the works of Gabrio Piola (1794-1850) [46, 47] and continues until contemporary times: all sources are fully available.

Mechanics of deformable bodies studies how the equilibrium shapes of bodies change because of their interactions with the external world. A given body is assumed to be constituted, in every of its material points, by a specific material. The current shape of a body is kinematically modeled, since the fundamental work by Lagrange (1736-1813) [107], by means of a placement function. Each material is mathematically modeled, in its range of elastic deformation, by the corresponding deformation energy density, depending objectively on the gradient of placement.

[8] Richard Toupin admitted (personal communication) that, since his studies on Landau's lecture notes, he always believed that Mechanics had to be founded on variational principles, notwithstanding what advocated by Truesdellians.

Further constitutive functions and kinematical descriptors need to be introduced for modeling damage, plastic phenomena, etc.

In this context, it seems absolutely meaningless the expression *natural material.* One may argue, in fact, that human activity did modify everything in the world (think, as an example, about forests: almost all of them have their present shape as the result of a human design). From an Engineering point of view, we can only talk about materials that have a simple microstructure (i.e. the more often used, up to now, in Engineering) and materials that have a complex microstructure. We do not share the primitivistic belief that natural is equivalent to simple, also because the definition of simplicity depends on the particular historical period.

Every material which exists is natural. Of course, we may ask ourselves if it is possible to find an existing material whose behavior can be described by certain constitutive functions. It is, therefore, meaningful to establish some *physical admissibility* criteria for *logically conceivable* constitutive functions. For instance, by introducing constitutive functions for a material which allows some deformative cycles that produce energy, one can get a mechanical system which contradicts the Principle of conservation of energy. Clearly, such constitutive functions would not be physically admissible. In [27], it is clearly stated that, unlikely what believed by Truesdellian school, there are no elastic materials which are not also hyperelastic. Truesdell wants to try that there are relationships between stress and strain, in first gradient materials, which do not derive from a principle of minimum of energy in a stable equilibrium configuration. He wants to prove that a postulation based on the laws of balance of forces and moments of forces is more general than a postulation based on the Principle of Virtual Work. This effort, as we will discuss later, is vain as Gabrio Piola has proven [46, 47, 55] that in every generalized theory of continua balance of forces and moments of forces are necessary conditions for the validity of the Principle of Virtual Work, while there are generalized continua (for instance, second gradient continua [82, 83, 57, 84, 85, 86, 137, 135, 136, 144, 2, 3, 51, 52, 138, 139, 140]) for which the balance of forces and moments of forces *are not sufficient conditions* to ensure the validity of the Principle of Virtual Work. This principle seems to be the most fundamental one in Mechanics. Therefore, the important question "what is a natural material?" can have a simple answer: it is a material which may exist.

Therefore, the theory of metamaterials, if one wants to avoid ontological paradoxes, cannot be defined as the theory of those materials which are not natural, because otherwise we were dealing with non-existing materials. Another possibility is to define metamaterials as those materials whose mechanical behavior is "exotic". Now the obvious question arises: what is an exotic mechanical behavior? The answer could be: an exotic mechanical behavior is a behavior which has not been yet observed. Of course, what is exotic in a certain historical moment may become standard in another one. For instance, Lamé (1795-1870), Navier (1785-1836), Cauchy (1789-1857), Poisson (1781-1840), all considered a material with negative Poisson's ratio as very exotic, and some scholars of their époque did even believe that such a material was unphysical (see [19, 73, 74]) and could not exist. Instead, auxetic metamaterials do exist and play a relevant role in modern Engineering.

Also in the group of scholars in Mechanics, albeit this theory is the eldest one in mathematical physics, some epistemological misconceptions are rather common. The main among these misconceptions are:

i. confusing a mathematical model for a material with the physical material itself (the same ontological misconception occurred to Eudoxus' model);
ii. believing that particular assumptions accepted for describing particular phenomena are universally valid in every physical situation (the same extreme platonistic or inductivistic misconception occurred in the school of Truesdell, where it is believed that every existing material must be modeled by first gradient continua, whose properties have been induced with experiments);
iii. refusing the Principle of Occam Razor by constructing theories with a series of ad hoc assumptions guided by experience (naive inductivism);
iv. believing that, simply manipulating a lot of data without any postulated model, one can predict, maybe using large computers, the behavior of physical systems (a modernistic form of naive inductivism, by means of which many want to find new metamaterials by simply divining in a random way metamaterial microstructures).

Concerning the confusion of a mathematical model postulated for a material with the physical material itself (ontological misconception), we must say that this is an old misconception that is very often met in history of Science. The example about the models of planetary systems can be considered a prototypal social phenomenon of this kind, because entire groups of scholars fall in this mistake. Usually, we have heard in debates among experts of Continuum Mechanics the following *wrong* statement: *Second gradient materials do not exist because used materials in Engineering do not show their properties and standard theoretical framework does not forecast them.* In this statement, one can find many layers of misunderstanding based on the following misconceptions:

(a) confusing first gradient continuum model (a mathematical model) with existing materials in nature (a physical object);
(b) believing that the standard theoretical framework, which has been paradigmatic in a school of Mechanics, includes every conceivable phenomena (this misunderstanding is induced by naive inductivism);
(c) confusing the standard first gradient continuum model with all used materials in Engineering (that includes both presently used and all usable in future physical objects);
(d) believing that, without having a theory describing it, one cannot use a material even when such a material is in her/his hands, with the paradoxical consequence that the material would not exist.

## *Mathematics designs the world: metamaterials, a change of paradigm*

The mathematical modeling of physical phenomena has shaped the world, notwithstanding what *practical people* may believe. In fact, in Engineering Sciences the following phenomenon occurs: a theory is formulated, it applies to a specific set of physical objects and physical situations, therefore in Engineering practice only these objects and situations are considered for Engineering artifacts. The fact that one does not have a model describing the behavior of a physical system, or some physical situations where a physical system can occur, or can be observed, implies that physical systems and situations which are not described, have to be carefully avoided in Engineering applications. For instance, if one has the capacity of calculating the deformed shape of a body only when linearized equations apply, then she/he limits the functioning regime of the artifacts which are built according to the above-mentioned theory to small loads, small deformations and small displacements.

Many engineers declare as a consequence that non-linear phenomena are not of interest in Engineering, with a typical process of removal of the complexity. In Engineering, non-linear phenomena are important; however, when they could not be fully studied with available mathematical tools, then they are avoided.

Therefore, the limits of our mathematical capacities limit consequently our predictive capacity and then our design capacity. For instance, sky-scrapers could not be built until Structural Mechanics became sophisticated enough to be able to design them. What can be mathematically conceived by means of a model can be transformed into an Engineering artifact, while every data-driven series of subsequent trials never produced any functioning Engineering solution. Data-driven research has produced, maybe, some interesting technical software solutions: however, when not guided by a clear modeling vision, it could not predict novel phenomena and seems to be a modern version of naive inductivism. On the other hand, in general, Engineering Sciences choose among available and conceivable systems those which can be mathematically modeled and limit its designing efforts to those for which mathematical predictions are possible, given the available computing tools. In few situations, on the contrary, Engineering Sciences attack a very difficult problem: that happens, for instance, when, given a mathematical model, the goal is to find a physical system which can be carefully described by that model.

On the basis of what we have discussed so far, it should be clear that a good theory is useless without suitable computational tools. This concept, which may perhaps seem trivial, assumes considerable importance if contextualized in the historical perspective that we have presented in the previous sections. The epistemological appreciation of the quality of a theory cannot prescind from the availability of suitable computational tools that allow for its use in getting predictions. A very detailed theory that cannot produce quantitative predictions is useless. If one theoretically tries to take into account too many phenomena, without considering the technical and computing difficulties which are found when applying such a detailed theory, then he/she does not supply the Engineering practice with a useful tool: being potentially capable to predict *everything* leads to the incapacity to predict anything. The classical example is given by the efforts of Navier [19] to develop a theory for predicting the deformation of a beam by starting from a *molecular* model: such a detailed model could not produce any prediction, due to calculation difficulties. Therefore, Navier was obliged to homogenize his discrete equations, for obtaining a computable model: his averaging hypotheses led him to believe that Poisson's ratio for isotropic materials could only assume the value 0.3, which is clearly against evidence.

Not only simplicity in the involved computing process must be required to a modeling effort, but also conceptual simplicity in the model formulation, that implies dramatic simplifications in the prediction process. Consider, as an example for this last statement, the relationship between the predictive capacity of Eudoxus' model of homocentric spheres and that of Aristarchus' heliocentric model: as we have seen, Eudoxus was unable to explain correctly, even by greatly increasing the number of spheres associated with a given planet, its retrograde motion, while it could be explained, instead, extremely clearly by Aristarchus.

Quantitative predictive capacity is inescapable in mathematical modeling and in its applications to Engineering Sciences [28, 29, 30, 32, 148, 150, 12, 15, 81]: in fact, there is no scientific designing without accurate quantitative prediction. Furthermore, from a technological point of view, it is certainly easier to build a mechanism for getting predictions by using a model where all planets travel more regular orbits around the Sun, than a model in which the planets move seemingly randomly in the sky, traveling very irregular orbits, although the latter model, being geocentric, seems more faithful to observational reality.

In general, a model producing some *theoretically correct* or *physically intuitive* equations that cannot be efficiently solved is technologically (and scientifically!) of little significance.

A similar example is provided by what happens much later, when the Copernican system replaced the Ptolemaic system. Behind the process of substituting one model for another there is a technical consideration: predictions are obtained via computations and model development is constrained by available computational tools. The Copernican system did not give much more accurate predictions than Ptolemaic system. In fact, by adding a suitable number of epicycles, as rigorously proven by Gallavotti [80], one can approximate the apparent motion of planets as seen from the Earth as accurately as possible. Moreover, the kinematics of both systems are based similarly on the principle of the composition of circular motions. But the Copernican system is enormously simpler conceptually and allows for less laborious calculations, as Cicero did observe when describing Archimedes' planetarium mechanism. Technological capacities, in a sense, introduce a hierarchical ordering in the set of models: those models, for which simpler computing methods are available, become preferable.

We believe that both Eudoxus and Aristarchus did have falsificationist points of view when they formulated their models and that the debate inside Hellenistic Science about their competing models did involve only the models predictivity capacity. It was only after the decline of Hellenistic Science that models started to be confused with reality and that *true* models were opposed to *false* models: the loss of epistemological consciousness led scientists towards a vain search for ultimate truth. Therefore, after many centuries in which scholars were looking for ultimate truth and believed that such truth could be attained, the contraposition between geocentric and heliocentric models developed the characteristics of a religion war. Instead of debating about the predictive capacity of one model as compared with the other, the scholarly debate was involved in scientifically irrelevant questions concerning the role of religion, ethics and vision of life in Science. On the contrary, we believe that the true debate between scholars was not about the ultimate truth of geocentrism or heliocentrism but about the real aim of scientific research: is Science looking for ultimate truth (assuming that such a truth can be established once forever) or is Science formulating one after the other a series of conjectures to be tested with experimental evidence and possibly changed when such evidence requires it? Paradoxically, Cardinal Bellarmino (1542-1621), whose intention was to reaffirm that only theology had the capacity to reach ultimate truths, following the orthodoxy of St. Thomas Aquinas (1225-1274) and St. Augustine of Hippo (354-430), tried to get from Galileo a simple falsificationist statement about heliocentrism, while Galileo remained in an inductivist position, albeit formally changing his position in order to avoid to be condemned to be burned at the stake.

In a sense, a contemporary version of the debate involving geocentrism and heliocentrism is represented by the debate between the supporters of Cauchy postulation and d'Alembert postulation for the foundations of Mechanics. The Truesdellian supporters of Cauchy postulation believe that it is an ultimate and experimentally proven truth, which cannot but be improved by adding some epicycles, i.e. small corrections. Their attitude blocked the growth of generalized models in Continuum Mechanics. Another important circumstance to be taken into account, both when describing the paradigmatic change between geocentric and heliocentric models or between Cauchy continua and Generalized continua, is the development and improvement of computing tools that occurred during the change. While in Hellenistic times the only computing tools were based on a geometric understanding of the concept of real numbers, so producing mechanical computing devices like the Antikythera mechanism or the Archimedean planetarium, after the Renaissance of Science, first Copernicus rediscovered ancient heliocentrism. Subsequently, Kepler could exploit the method of calculations based on Napier (1550-1617) tables of logarithms and finally Newton, by using Cartesian geometry, could get a prediction of the planetary motions without computing mechanisms. Therefore, it seems that, while being initially blocked in an inductivistic epistemological view point, modern Science could improve its understanding of the planetary phenomenology, when compared with Hellenistic Science, only because the development of modern computation tools, based on algebra.

Coming back to Continuum Mechanics, we limit ourselves, here, to describe some fundamental points in the process that led to the introduction of Generalized Continuum Mechanics [114, 1, 10, 6, 7, 4, 8, 9, 11, 13, 63].

As we will see, Gabrio Piola introduced in 1848 the generalized continuum model based on the use of deformation energies depending on the $n$–th gradients of displacement, being aware of the conjectural nature of such mathematical models

[122]. However, Cauchy and his followers did try to formulate the ultimate continuum model, based on induced true properties of matter, at macroscopic level. It is paradigmatic, in this context, the unconditional acceptance by Cauchy and his followers of the so-called *Cauchy postulate*, stating that contact forces, inside continua, can be only forces per unit area which, moreover, depend only on the normal to Cauchy cuts. Even though Piola was well aware of the limits of this conjecture, whose applicability is limited only to a particular class of materials, and even though Piola himself wrote clearly that Cauchy postulate had to be regarded as a constitutive equation, in a large group of scholars Cauchy postulate has been accepted as a religious ultimate truth that cannot be doubted. It is remarkable that Gurtin, who had started from an orthodox Cauchy-Truesdellian viewpoint, in his subsequent papers [93, 94] changed his fundamental postulation approach and, albeit ignoring Lagrange, attributes, with a typical modernist attitude, to an explicit Lagrangian follower (i.e. Toupin) the choice of what seems to be the most appropriate postulation of Mechanics. Piola's works were reappreciated only at the end of the 20th century, while his models had been rediscovered already 50 years before and had became the object of in-depth study in view of their potential technological application [82, 83, 116, 137, 135, 136, 144]. What has changed in the century and half that separates Piola's pioneering work from his (slow) rediscovery? Why did the Continuum Mechanics of the Cauchy school ignore (and in part still tries to ignore) Piola's results for over a century?

In Cauchy's version of Continuum Mechanics a number of *ad hoc* limitations are inserted, including the fact that the deformation energy of a continuum medium can only depend objectively on the first gradient of the displacement field. A priori, nothing would limit a dependence on higher order gradients, but the simplest choice, consistent with the phenomenology disclosed by Cauchy continuum model, is to limit oneself to the first gradient of the displacement. Piola, as we have said, introduces, for a purely logical need, the higher displacement gradients in the calculation of the deformation energy, and argues to characterize those microstructures for which homogenized models must be of this more general kind. Unfortunately, the differential geometric tools available to Piola did not allow him to characterize internal contact forces in the case of second and higher gradient continua: instead, he did manage to do so in the case of first gradient continua. It is not a coincidence that exactly when Piola succeeded in finding a representation for contact forces in first gradient continua, Cauchy (who probably met Piola in Italy during his exile following French July Revolution of 1830) developed his postulation scheme based on balance of forces and balance of moment of forces. It is only after more than a century that Paul Germain showed, in his fundamental work [122], which is the structure of contact forces in second gradient continua, by remarking that so-called Cauchy postulate is not valid for these continua and that edge contact forces may arise (see also [52, 51, 53]). Moreover, in [119, 23, 60, 44] it is proven that models where the second gradient of displacement acquires a non-negligible role, at macroscopic level, are obtained by homogenization starting from a microstructure, or architecture, at a lower scale in a continuum medium where high contrasts of stiffnesses are present. We believe that Piola had guessed this result: see [46, 47]. Therefore, in order to become able to evaluate and reveal experimentally the effects of the presence of the second gradient of the displacement field, it is necessary to have a technology which is capable of producing a microstructured material [126, 5, 23, 26, 62, 118, 154] and, above all, a material whose microstructure shows the suitable highly contrasted stiffness fields, so that, at the macroscopic level, the terms used by Piola and Germain in the deformation energy do appear (examples in which the required technology has successfully produced such microstructured materials can be found in [17, 16, 110, 149]). This is a very clear example of how the limited technological capacity of an epoch can indeed block also its scientific development. As long as the lack of technological capacity does not reach a level where the results introduced in the new theories can be tested, the new theories will remain blocked, ignored and, definitely, unusable. The absurdity of contemporary situation lies in the fact that despite the technological ability to produce materials whose behavior is described by Piola's theory (and cannot be described in the framework of Cauchy models), there are still scholars who are obstinate in denying its usefulness.

The mathematical challenge that researchers in the field of Continuum Mechanics face today, therefore, is to design metamaterials that can be described within the framework of a generalized theory [10, 6, 11, 50, 109, 117]. These materials, as we shall see, are conceived in order to possess mechanical properties that are significantly more performing than those of the commonly called *natural materials*.

Therefore, the fundamental problem in modern theory of metamaterials consists in the problem of synthesis of microstructures producing a specific desired macrobehavior [2, 18, 119, 121, 125, 129]. In fact, the modern challenge in the theory of metamaterials consists in finding that microstructure, or that micro-architecture, which, at the macroscopic level, produces a specifically required mechanical behavior. In this context, the most difficult problem to face from a mathematical point of view is to connect micro-structures and macro-behaviors. So, given a macroscopic theory (appropriate action functionals and consequent stationary conditions) one wants to find an algorithm to calculate the microstructure that, once homogenized, at the macroscopic level is described by the chosen macroscopic model [87, 133, 153, 77, 79, 24, 78]. In this context it is important to remark that a major change in the research tools has been induced by the use of powerful computers to find suitable motions for minimizing postulated action functionals: in fact, especially in non-linear regimes, it is simply inconceivable to find closed form solutions and therefore only by means of suitably conceived algorithms it is possible to design and to predict the behavior of novel metamaterials [90, 39, 91, 30, 113, 92, 89].

The basic ideas in the field of the synthesis of metamaterials may be borrowed from the *ancient* theory of synthesis of analog circuits. In this theory, it was possible to prove that every passive linear $n$-port element is algorithmically synthesized using inductors, capacitors, resistors and transformers [106, 22, 21]. In fact, it has been proven that, given the desired passive impedance, one can build a graph and can find for each branch of the graph a suitable circuital element such that the resulting discrete circuit has the chosen impedance.

The classical theory by Kron and McNeal [106, 112] allows us, given quadratic Lagrangian and Rayleigh's potentials, to algorithmically determine the graph structure of the searched electric circuit and its elements *synthesizing* the given linear

$n$-port element. This latter is mathematically characterized by its Lagrangian and Rayleigh's potentials. Therefore, at least in the theory of circuits, by starting from a finite number of basic microstructures and reproducing them at different length-scales, it is possible to devise a most general microstructure. The big challenge, now, consists in conjecturing that this method can be also applied to the synthesis of non-linear mechanical (and multiphysics) systems [22]. Some papers [49, 61, 141, 142] can be referred to for well understanding the fundamental role of the synthesis process in metamaterial theory.

It has to be remarked that the theory of analog circuits has been partially lost and generally forgotten. The reason is that in the early '60s digital computing methods became *dominant* and analog computers were considered obsolete. As a consequence, it is becoming more and more difficult to find the sources in the theory of synthesis of analog circuits and a sudden change of paradigm occurred also in the textbooks of Mechanics and, in particular, of Structural Mechanics. In fact, many textbooks in Structural Mechanics in the '50s were full of schemes of analog circuits, considered very important for *practical applications*. One could have believed, by consulting such textbooks, that a structural engineer could not become a skilled professional without knowing the theory of circuits. It is ironic that after few decades the great majority of civil engineers do ignore even the existence of inductors and capacitors, not to mention transformers. This sociological phenomenon, that occurred in an époque when books are not easily lost and when a large number of scholars are active, proves three important theses:

i. loss of knowledge is a sociological process, which is always active in every group of scholars and in every society;
ii. in Science every knowledge may be useful in every other research field;
iii. there is no such thing as *obsolete* knowledge!

## *The Principle of Virtual Work and its correct application produces (generalized) Continuum Mechanics*

Let us now proceed to examine in greater detail the distortion of sources and modification of basic principles that occurred in Continuum Mechanics. Unlike the conjectural study we have made of the development of planetary models in Hellenistic Science, in the case of Continuum Mechanics all modern (since d'Alembert's *Traité de dynamique*, 1743 [111]) sources are available and therefore the reconstruction that we present here is not conjectural. However, in the evaluation of modern sources, we still have to consider a problem that may be considered logically absurd, and that yet, unfortunately, is having, also now, a considerable weight in the development of Continuum Mechanics: some fundamental sources in this field are not written in English (e.g. they are written instead in French and Italian) and, as a consequence, some scholars believe to be allowed to ignore them. This point can be fully developed when one details the study of Gabrio Piola's contribution to Continuum Mechanics.

Continuum Mechanics has been based by d'Alembert on the Principle of Virtual Work only. This principle allows for the calculation of the equations of equilibrium of a continuum and it is easily connected to the Principle of the Minimum Potential Energy for a stable equilibrium. As we will see, starting from Cauchy, Navier and Poisson, a very strong current of thought has been developed which has, in fact, replaced this fundamental principle with the independent postulation of the balance of forces and of the moments of forces, introducing some auxiliary concepts such as forces and attributing to them a fundamental role in Mechanics and in the phenomenology that it aims to describe. One must, however, agree on the fact that, especially when the physical system under examination is very complex, the Principle of Virtual Work is not only easier to apply, but it is also applicable when the balances of forces and moments of forces are not sufficient to characterize equilibrium.

We want to stress, in this context, that it is not by chance that mechanical systems, of interest in Engineering Sciences, were first studied through the application of this principle, introduced by Archytas of Tarentum in his *Mechanica Problemata*. While we do not know how Archytas had formulated the Principle of Virtual Work, it is evident that in his *opus* he uses it to study problems of applicative relevance such as the functioning of machines and levers (which are sometimes still studied in middle schools based on this principle, even before the concept of force itself is introduced).

An interesting problem related to the *Mechanica Problemata* is given by the following question: should it be considered as an exercise book whose reading had to be combined with a more theoretical work? Could the theoretical work have been lost? To give a definite answer to this question is not possible: however, we can make some conjectures, by considering the analogy with other pre-Hellenistic authors. It is, in fact, now widely accepted that, for instance, the production of Plato (c. 428/427 – c. 348/347 BC) was of a twofold nature: one part of the works, those dedicated to his pupils, was of an extremely technical nature and specific for *experts*, with a level of complexity equal to the surviving works of Aristotle; a second part of Plato's works was of a rather popular nature. The latter were written in the form of dialogues and were thus more accessible to the general public. For what concerns Plato, the most technically difficult works were lost and only the popularizing ones were transmitted to us, while the opposite happened to Aristotle (384–322 BC). We remark here that one finds an enormous body of critical works commenting philosophical, historical and literary ancient production, while such an analysis is not dedicated to ancient scientific texts, so that the reasons for this kind of selection of transmitted scientific works were not deeply investigated. In any case, it is a reasonable conjecture to assume that Archytas may have written a simplified and applied version (i.e. the *Mechanica Problemata*) of a more complex work, in which, for instance, the Principle of Virtual Work was formulated in a more explicit way. If this conjecture is true (remember that there are still many texts of Hellenistic Science that are preserved in libraries and remain forgotten because today it is rare to find a scholar who knows mathematics, Greek, Armenian, etc.), it would imply that the presumably lost text of Archytas was extremely more abstract than the *Mechanica Problemata* and, therefore, that constituted a masterpiece in the field of Mechanics.

Its importance could consist in the clarification of the mental process that led the first scholars in Mechanics to find its conceptual bases.

However, we cannot exclude the other possibility to be considered about Archytas' text: that it was an autonomous and self-contained work. This would support a different hypothesis of historical and epistemological relevance: perhaps, in the early phases of Mechanics, theory and exercises were mixed up in a single treatise. Perhaps the approach to complex problems was deliberately simplified by the proposal of a series of applied examples. This approach is the one preferred by some modern textbooks in Physics [95], based on the idea that a student will understand general concepts by inducing them on the basis of many examples. This approach is, instead, considered not efficient by those who study Mechanics from a deductive postulation point of view [57, 82, 83, 84, 86, 137, 135, 136, 116]. The reader will understand that knowing how the Principle of Virtual Work was first formulated could be very important to settle this controversy. In any case, the text of the *Mechanica Problemata* that has come down to us is already rather abstract. In some places it seems to refer to concepts already known to the reader, just as one often reads in a modern text of solved exercises! For this reason we believe that the possibility of a second work that dealt with a complete and rigorous treatment of the theory behind the practical examples cannot be excluded.

We now want to discuss, in an obviously simplified and concise way, some aspects of fundamental importance in the formulation of the Principle of Virtual Work. A first aspect to underline lies in the fact that, as we also mentioned before, it is well-known and universally accepted, at least since the works of Archimedes, that to a stable equilibrium configuration corresponds a minimum of the total energy (Total Potential Energy Minimum Principle). It can be also demonstrated, using some mathematical reasonings, that the Total Potential Energy Minimum Principle implies a stationary condition (the first variation of Total Potential Energy is zero in its minima) that, on its turn, can be regarded as a particular form of the Principle of Virtual Work. Therefore, the validity of this form of the Principle of Virtual Work can be deduced as a consequence of the Total Potential Energy Minimum Principle. The Total Energy Minimum Principle can be formulated as follows:

> **Total Potential Energy Minimum Principle:** *The stable equilibrium configurations are the only ones for which the total potential energy has a local minimum.*

A necessary condition for stable equilibrium can be formulated if the total potential energy is differentiable with respect to the variation of configuration.

> **Necessary condition for equilibrium:** *Starting from a stable equilibrium configuration, the first variation of the total energy corresponding to each virtual displacement is zero.*

A virtual displacement is simply a small variation (more precisely, an *infinitesimal* one) of the body's configuration (but respectful of internal constraints and of kinematic boundary conditions) to be added to the tentative minimum energy configuration in the verification process aiming that such tentative minimum energy configuration is effectively of equilibrium. This formulation requires the deduction of a number of non-trivial mathematical results, which make the treatment of

Mechanics by means of variational principles complex, whose esoteric content is reserved to scholars having a deep knowledge of complex mathematical theories. It is very presumable that this is the fundamental reason why some scholars decide to ignore completely this approach to Mechanics and turn, instead, to the simpler, but somehow incomplete (and surely unfit for the discovery of novel models) formulation based on the postulation of balance of forces and moments of forces. To give an idea of the mathematical difficulties implied by the postulation of Mechanics based on variational principles [48, 18, 39, 98, 123, 124], one must think that there is a whole branch of mathematics, the Calculus of Variations, which was developed to supply the needed conceptual tools to Mechanicians. It is suggestive to think that Calculus of Variations has deep roots in Hellenistic Science, as witnessed by the fact that isoperimetric problems are traditionally called also *Dido's problems*. As it is reported by [132], our conjecture is not too much daring. In fact, in the *Synagoge* by Pappus of Alexandria (c. 290 – c. 350 AD), as well as in the commentary by Theon of Alexandria (c. 335 – c. 405 AD) on Ptolemy, which both were transmitted to us, Zenodorus (c. 200 – c. 140 BC) treated isoperimetric plane problems in a treatise which was lost. It is remarkable that Dido's problem was formulated and solved by Zenodorus, albeit we do not know the methods that he had used.

We can make a list of the main mathematical difficulties to be faced when deciding to resort to a formulation of Mechanics based on the Total Potential Energy Minimum Principle; indeed to this aim it is necessary:

i. to introduce the concept of infinitesimal variation of a configuration (otherwise called *small displacement*);
ii. to introduce the concept of work done by an interaction on a virtual displacement and the concept of virtual displacement itself;
iii. to define the first variation of a functional in terms of Taylor series developments (which, despite the simplicity and elegance of this powerful mathematical tool, appears to be indigestible to many scholars).

As mentioned above, the Total Potential Energy Minimum Principle implies the more general Principle of the Virtual Work, which can be also formulated in a simpler way from a mathematical point of view. Probably, in order to be able to understand in detail the efficacy of the variational approach to Mechanics, and, at the same time, in order not to be discouraged by the mentioned difficulties, it can be useful to refer directly to the formulation of the Principle of the Virtual Work given by d'Alembert, who was the first, in the modern age, to found Mechanics on it. In fact, in his treatise of 1743, d'Alembert formulated the Principle of Virtual Work in a more modern language with respect to that found in Archytas' *Mechanica Problemata*. d'Alembert's formulation generalized the previous formulation of stationary condition implied by Total Potential Energy Minimum Principle, and, because of its greater generality, it allows for a better focus on the key points of the variational approach:

**Principle of Virtual Work (d'Alembert, 1743):** *A system is in equilibrium in a given configuration when the total work done by all interactions involving the system is zero for each virtual displacement from that configuration.*

From a correct application of the Principle of Virtual Work, one can obtain the equations of equilibrium of a mechanical system, also known as its Euler-Lagrange equations. It is to Lagrange that we owe the application of this principle to a wider class of mechanical systems. In the last version of his *Mécanique Analytique*, Lagrange formulated the Principle of Virtual Work for a continuum system and applied it to the study of the motion of fluids. In his nomenclature, Lagrange called *power* what will later be called force, and *momentum* what we know today as power. He claimed to prefer this nomenclature as it had been previously chosen by Galileo Galilei: we agree with his motivations, and we regret that unfortunately his suggestion has not been accepted in mechanical literature. Some, rather naively, from this different nomenclature used for the mathematical objects used by Lagrange, *deduce* that Lagrange did not understand the problem he was formulating. Once again, we observe this modernist attitude that wants to judge the past by current conventions. Quoting Shakespeare[9]:

> *"What's in a name? That which we call a rose*
> *By any other name would smell as sweet".*

To show how the ideas of d'Alembert were elaborated and improved by Lagrange, it is very useful, finally, to introduce the formulation given by Lagrange[10]:

> **Principle of Virtual Velocities (Lagrange):** *If a system constituted by bodies or points, each of which is pushed by any power, is in equilibrium and if a small movement is given to this system, by virtue of which each point will cover an infinitesimally small distance that will express its virtual velocity, then the sum of the powers multiplied by the distance covered by the points where it is applied along the line of application of this same power will be equal to zero, if we consider as positive the small distances covered in the same direction of the power and as negative the distances covered in the opposite direction.*

Although a modern formulation of this principle usually includes the use of concepts from functional analysis, tensor algebra and mathematical analysis, one has to agree on these points:

i. Lagrange's formulation seems so general that it includes all the versions that have been formulated so far;
ii. this formulation uses the minimum possible mathematical concepts (i.e. only concepts from Euclidean geometry) that are sufficient to rigorously express the principle in its full generality.

We can, therefore, conclude that it has been correct to call, in the past, mathematical physicists with the attribute of *Geometricians*, as it was the geometrical language that allowed for the first formulation of mechanical theories.

The life-long work by Piola consisted in completing the work that Lagrange had left to be completed after his death. Piola, also formulating a micro-macro identification procedure, in 1848 published a fundamental work [122] where:

---

[9] W. Shakespeare, Romeo and Juliet, Act II, Scene I.

[10] The translation has been performed by the authors of this Chapter from the Lagrange's original text.

i. he deduced a macroscopic model describing the overall behavior of a system of a large number of interacting particles, obtaining also macroscopic constitutive equations in terms of microscopic geometric and mechanical properties;
ii. he introduced $n$th-gradient continua, determining the conditions for which they must be used in order to describe correctly the behavior of microscopically complex mechanical systems;
iii. he determined the structure of contact forces for first gradient continua, as studied by Cauchy, and discussed in his own work of 1822.

It has to be remarked that until 2012 [53] the determination of the correct form for contact interactions in $n$th-gradient continua was not obtained. Piola did not have at his disposal the mathematical tools from differential geometry, that were developed also, together with Gauss and Riemann, by the Italian school of Piola's scientific lineage, i.e. Francesco Brioschi (1824-1897), Eugenio Beltrami (1835-1900), Gregorio Ricci Curbastro (1853-1925) and Tullio Levi-Civita (1873-1941). However, Piola could prove, for a generic $n$th-gradient continuum, the following theorem, starting from the Principle of Virtual Velocities:

**Balance of forces and moment of forces (Piola):** *if a deformable nth-gradient continuum body is in an equilibrium configuration, then the resultant and resultant moment of applied external forces vanish.*

Piola, following d'Alembert and Lagrange, defines resultant forces and resultant moment of forces as the vectors needed to represent the work of a system of forces in a rigid virtual motion. Therefore, these concepts are mathematical abstract constructions that can be used to calculate equilibrium configurations. Resultant forces and resultant moment of forces are mathematically defined in order to allow for the characterization of equilibrium configurations and do not correspond to any directly measurable physical quantity. Remark that, while for first gradient continua Cauchy has proven that this necessary condition is also sufficient, in general, for a subclass of second gradient continua [144] and for all $n$th-gradient continua with $n \geq 3$ balance of forces and moment of forces select a set of configurations greater than the set of equilibrium configurations. In fact, in higher order continua contact interactions are not limited to forces and couples.

One can assume that the equilibrium necessary conditions, given by resultant forces balance and resultant moment of forces balance, represent all Euler-Lagrange conditions of the Total Potential Energy functional for first-gradient continua only, and as such, in this case, directly provide the governing equations in *strong* form; it is clear, however, that in a numerical approximation setting it is always convenient to reformulate such conditions in *weak* form, and for this purpose the use of a variational principle is preferable because it lends itself directly to providing the governing equations in an easily discretizable form.

## *Confusing a necessary condition with the fundamental Principle: the materialization of forces, i.e. auxiliary mathematical concepts*

As we have seen in the previous section, forces, and, in particular, contact forces, are a mathematical artifice introduced in order to deduce some consequences of the fundamental postulate of Mechanics, the Principle of Virtual Work, and, as such, they are of no use outside this context. Contact forces are, then, a mathematical invention, developed in the centuries to find some logical consequences of the above-mentioned Principle. The genesis of the concept of force and all the misunderstandings to which it was subject deserve an in-depth analysis, which is beyond the scope of this work (but see [100]). It is remarkable, however, that Archimedes did introduce in his *On the floating bodies* the concept of pressure and that the first textbooks in modern Mechanics (those, already cited, by d'Alembert and Lagrange) did apply the Principle of Virtual Work to deduce the equations of equilibrium and motion of perfect fluids.

The main idea that leads to the definition of resultant forces and resultant moment of forces can be traced back, in modern Continuum Mechanics literature, at least to Gabrio Piola. It is unfortunate that the complete works by Euler had not been published in an English translation until the second half of the twentieth century. The enormous corpus of the works by Euler, all written in Latin, may include some applications of the Principle of Virtual Work, or of the Total Potential Energy Minimum Principle, leading to the definition of resultant forces and resultant moments of forces, as the tradition in Mechanical literature attributes to Euler the introduction of these necessary conditions for equilibrium. We could not find any textbook clarifying this point and the original works by Euler are not easily accessible: however, see [19], Euler did deduce the equations of Elastica by using the Total Potential Energy Minimum Principle. It is remarkable that Truesdell wrote more than 400 pages in the series of Springer volumes gathering Euler Opus, without translating a single word of Euler's text.[11]

In order to characterize equilibrium configurations by using the Principle of Virtual Work, one can consider for every involved body those rigid virtual displacements that are allowed by applied constraints. In absence of applied constraints, therefore when we have a free body, the work done by externally applied loads on rigid displacements can be represented as linear functionals on the pair of vectors composed by translation and rotation velocities. By Riesz (1880-1956) representation theorem, these linear functionals are uniquely determined by two vectors when using the inner product for calculating the represented functionals images. The vector whose inner product with translation virtual velocity gives the virtual work done is called *resultant force* of the applied loads, while the corresponding vector giving the virtual work in correspondence with virtual angular velocity is called *resultant moment of forces* of the applied loads. It is therefore clear that the concept of force is generated while developing a mathematical theory to be used for deducing logical consequences

[11] Leonhardi Euleri Opera Omnia: Opera mechanica et astronomica. The rational mechanics of flexible or elastic bodies, 1638-1788 : introduction to Leonhardi Euleri opera omnia, vol. X et XI seriei secundae / C. Truesdell, Volume 10. Springer, Zurich, 1960.

from the basic principles of Mechanics. Albeit it is of great importance, it does not correspond to any directly measurable physical quantity and is a pure *superstructure* of use in mathematical reasonings: exactly as it happened for Eudoxus' spheres or Apollonius' epicycles. When a free body is in equilibrium, then the total work done on rigid displacements from equilibrium configurations must vanish, and, as a consequence, the resultant force and resultant moment of forces must vanish. It has to be remarked that also when the body is deformable such *necessary* conditions must be verified, whatever may be said by some scholars of Truesdellian orthodoxy. Moreover, by introducing Lagrange multipliers, one can add to active forces also reactive forces, in presence of constraints. Including reacting forces in the set of applied external loads, one gets the validity of fundamental balance equations for Mechanics (i.e. resultant forces equal to zero and resultant moment of forces equal to zero) also in the case of deformable constrained bodies.

When trying to calculate equilibrium configurations using analytical methods, fundamental balance equations are the most useful tool to be used in calculations. However, when using numerical computational tools one must resort either to the Total Potential Energy Minimum Principle or to the Principle of Virtual Work.

The most recent materialization of abstract mathematical concept can be observed when, within the framework of Truesdell's presentation of Continuum Mechanics, one finds statements attributing to the concept of forces a physical reality and when one reads that *the laws of balance of forces and moments of forces are based on physical evidence*. It is as if one could measure a functional defined in a Sobolev space and could get information about it based on physical intuitions.

Just as Eudoxus' model made sense and clearly served a certain purpose even though it had no pretension of being a *pictorial* description of physical reality, so the concept of force has a very precise reason for existing in the context of the theory where it was formulated: d'Alembert, Lagrange, Piola used this purely mathematical object to formulate a theorem by means of which the equilibrium equations of a mechanical system could be derived. The aim of this formulation was in fact to obtain a way of characterizing equilibrium, similarly as Eudoxus' aim was to find a way to describe the motions of the planets, including retrograde motions. When, however, the scientific epistemological consciousness decays, then there are those extemporary scholars who fail to understand the difference between models and reality, and, in the total resulting confusion, it happens that objects of secondary importance, such as the homocentric spheres for Eudoxus or the forces for Continuum Mechanics, become preponderant.

The confusion becomes total when Cauchy tries to deduce the equilibrium conditions in Continuum Mechanics by postulating the existence of the stress vector, in order to calculate resultant forces and moments of forces on sub-bodies of deformable bodies. Cauchy devised the ideal *Cauchy's cut* and corresponding contact interactions: he supposed to remove a part of a body and to replace this part with *an equivalent* system of forces, which are able to maintain the body in equilibrium. In fact, there may be a number of mathematical abstract concepts, in a model, which have the sole purpose of bringing together the various observable pieces of a theory, but having no relevance from an observational point of view. But the quantities at the basis of the model must be measurable: in the case of Mechanics, these measurable quantities are the kinematical ones. One remarks here that it is impossible to imagine an experiment able to measure Cauchy's stress vector. Cauchy started from some *ad hoc* assumptions, like the so-called *Cauchy postulate*, which is not a postulate, with the same logical status as the Principle of Virtual Work, but, on the contrary, a constitutive assumption [51, 52, 53, 45, 127, 59, 123, 130, 14]. Then he continued by postulating balance of forces and of moments of forces, proving the existence of a stress tensor by means of which he wrote a particular form of the Principle of Virtual Work that, for him, became a theorem. Therefore, in Truesdellian orthodoxy one finds oxymora like: *the theorem of the Principle of Virtual Work*. For continua whose deformation energy depends only on the first gradient of displacement, one could believe that d'Alembertian postulation of Cauchy postulation are two equivalent points of view. However, if one lists the higher number of basic assumptions needed to develop Cauchy postulations, when compared with those used in d'Alembertian one, then he will conclude, by using Occam razor, that the latter is much preferable. As expected, the Principle of Virtual Work postulation allows for easier generalization of the proposed models [51, 52, 53, 45], while keeping the Cauchy postulation renders nearly impossible any generalization if not adding a long series of *ad hoc* further assumptions [25, 64, 65, 67, 66, 101]. It is clear that, if one has postulated a Principle of Virtual Work and finds a series of Euler-Lagrange conditions that are logical consequence of the postulated principle, she/he will manage to repostulate the same mechanical model based on a list of balance laws, one for each independent equation obtained from the original Principle of Virtual Work. It has to be investigated if Cauchy was aware in 1823 of the results by Piola (made public in 1822) about the nature of contact forces in first gradient continua.

At this point, some questions arise spontaneously: what is a force and how to measure it? An extremely common answer, but accepted by many with a total lack of critical spirit, is to say that *a force is what is measured using a dynamometer*. However, even the most naive scholar knows that a dynamometer measures displacements and that the value of the *measured* force provided by such an instrument is obtained by applying theoretical concepts, i.e. Hooke's law. And this, after all, simply shows that force, albeit being a very important concept, is really a merely theoretical artifice, without any direct observational meaning. No direct experimental evidence concerns forces.

In conclusion, we can say that Cauchy, Navier and Poisson decided to postulate (instead of the mathematically too difficult Principle of Virtual Work) the balance of force and moments of forces, at the cost of losing the possibility of generalizing their model: contrarily to what done by Piola, they thus only considered continua whose deformation energy depends on the first gradient of displacement. To make their postulate convincing, they then materialized the concept of force by trying to convince themselves that this mathematical concept (vector used to express a variation of energy) is intuitive and physically understandable. We believe that it is absurd that this materialization process[12] occurred in the modern age and we absolutely agree with d'Alembert (1743):

> *I have completely banned the forces associated with the body in motion, dark and metaphysical beings, capable of doing nothing more than spreading darkness over a science that is clear in itself.*

## 1.7 Conclusive remarks

In this chapter we have started a discussion about some aspects of the sociological phenomena involved in transmission and re-elaboration of scientific theories. The study of the modalities of transmission of an original theory and the involved transformation induced in the transmission process is crucial. In this context, the most problematic feature is represented by the determination of the *first* or *original* sources of scientific theories. We are not interested in a personalistic research of the *first genius* who formulated a certain theory: the formulation of scientific models is, indeed, a choral endeavor where single contributions are like small bricks in a large building. Certainly, there are lucky bricklayers who manage to contribute by building a keystone: Einstein (1879-1955) did formulate, supposedly with the help of his wife Mileva Marić (1875-1948), the basic ideas of General Relativity. However, he himself admitted that without Levi-Civita and Ricci absolute calculus he would never had the possibility to even write his celebrated equations. Instead, we are interested in the study of the logical process that leads to the formulation of novel and predictive theories as we believe that this study may teach us how to invent such newer theories.

A very frustrating, and often even denied, phenomenon that is systematically observed in the History of Science regards the distortion of scientific models from their original form due to the decline that periodically afflicts human societies. In fact, many historians imagine the History of Science as the accumulation of knowledge with a permanent increase of scientific understanding and technological capacity, albeit admitting that the rate of this increase has been varying in different periods. Instead, it is our belief that, unfortunately and dangerously, human technological capacity and scientific understanding of reality may experience regression.

We have argued about our thesis by presenting two clarifying examples of how, during two of these decline periods, ideas, on which the human scientific progress had been founded, were misunderstood and how even simple logical concepts could be hard to understand for scholars during the decline ages. The more common epistemological regression phenomenon which can be observed in this context consists in the complete inability to recognize the difference between a model and the object that is described by this model. This loss of clarity of thought brings, in the end, to a complete detachment of the so-called intellectuals with the world reality

[12] Perhaps the similar materialization process occurred to Eudoxus' model is more understandable, considering the enormous regression of scientific knowledge that led to the Middle Ages.

and technology: this detachment produces the widespread belief that theoreticians are absolutely unable to respond to demands from practical needs. Such a belief seems to ignore, just to name few Greek scientists, that Archimedes or Archytas of Tarentum did prove, with their lives, that theoretical knowledge and technology are indissolubly linked.

As a first example we have reported the case of the materialization of the Eudoxus' planetary motion model. As we have frequently remarked along the chapter, this model was not the best fruit of the Hellenistic Science, but it had strong possibilities to be pictorially represented. When the scientific knowledge of Roman society became insufficient to fully understand Aristarchus' model together with the techniques introduced by Apollonius, the Eudoxus' model, simpler from the point of view of mathematical detail, was chosen and used as a faithful representation of reality. Unfortunately, we see many modern engineers to make similar choices, with consequences that are wrongly used to discredit Scientific Engineering.

We want to stress that, in that so dark (from a knowledge point of view) historical period starting with the Roman domination on the Hellenistic colonies in Sicily, at least the study of Euclidean geometry was preserved, probably simply following a well-rooted educational tradition. As Euclidean geometry was necessary to understand and manage the few scientific elements that were left in place after the collapse of Hellenistic societies, it can be conjectured that the persistence of Euclidean geometry teaching has been probably the main reason why the early humanists were still able to interpret, or at least perceive in its importance, a very complex scientific corpus such as Archimedes'.

Eudoxus' model, which may seem naive today (albeit some flat Earth groups still believe that it is too complicated), at the time when it was introduced probably represented a conceptual revolution comparable to the formulation of the theory of caloric. Instead, the scientific advances introduced by Aristarchus' model, sticking to the same metaphor, could be considered as the advances induced by the invention of Fourier's theory of heat. With the regression of scientific awareness, due to its possibility to be pictorially represented, Eudoxus' model ended up being taken as a part of reality.

The second example we have focused on is the materialization of the abstract mathematical concept of force. In this case, as we have repeatedly remarked, the substitution process observed is more sophisticated and serious and, from certain points of view, more difficult to explain than the similar one occurred to Eudoxus' model. The materialization of force did manage to persuade many scholars that a purely mathematical object had, instead, a physical reality: this mental process seems more misleading than the materialization of rails along which planets are running. In fact, at least planet have a physical reality.

We wish to emphasize that we do not intend here to pursue a modernist attitude according to which what happened to Eudoxus's model would be less serious just because it happened about two thousand years ago. The reader will certainly be in no doubt that we are convinced that the scientific advancement of the Hellenistic age was equal, if not in some aspects superior, to that occurred in the 18th century. The less dangerous decline of Hellenistic Science is, probably, due to the fact that this decline occurred in correspondence to a socio-cultural-political reversal of enormous momentum, such as the decline of Hellenistic states and the establishment and fall of the Roman Empire. Of course, as we have repeatedly suggested, one can associate the decline of society to its scientific decline (recall the discussion about Roman aqueducts), albeit there is a time delay between the loss of scientific knowledge and the subsequent technological collapse.

The case of the materialization of the concept of force, on the other hand, is much more alarming. In fact, this confusion has occurred in a time period when scientific culture continues to develop and still manages to induce remarkable technological developments. The fact that this avant-garde science continues to develop on sometimes an extremely confused conceptual basis leads us to reflect on the possible aberrations it could produce (and it is not certain that the aberration process is not already at an advanced stage). This pessimistic view can be counterbalanced by another consideration: in present times, most likely, we have a number of active living scientists which is greater than the cumulated number of scientists who ever lived on our Earth. In fact, one can consider that the quality of this group of scientists is not as homogeneous as it was during, for instance, the flourishing of Hellenistic Science or Illuminism. Therefore, it could be that we are observing simultaneously the rise of some scientific societies in some disciplines and countries together with the decline of other scientific societies and disciplines. Therefore, the net advancement of scientific knowledge is the result of a dynamic process where declining effects are counterbalanced by development effects. In conclusion, one can say that, until the number of scientific groups that are capable to base technological advancement on solid scientific grounds is great enough, we may hope that Dark Ages kind of decline can still be prevented.

One might say: actually, why bother with the fact that an epistemological misconception leads to the materialization of the concept of force? In the end, the concept of force is something that is used within the model anyway! No one today (apart from possibly the flat Earth groups, if they were able to understand it) could ever believe that Eudoxus' model is reality, that is, that the planets are stuck on spheres hinged to rotate rigidly relative to each other. Instead, it is commonplace, even among scientists, to believe that forces are something observable, despite so much evidence to the contrary. Let us consider a derived quantity, such as velocity: one does not need profound scientific knowledge to agree that it is not possible to measure velocity directly. Velocity can be estimated only by measuring space and time intervals, and only then one can derive an estimate of velocity from its kinematic definition. Similarly, in order to give a meaning to the concept of force, it is necessary to introduce a mathematical model: as clearly stated by d'Alembert, forces are mathematical concepts derived from basic postulates. In fact, the previously cited excerption by d'Alembert, concerning the obscurity of the concept of force, is completed by the following words[13]:

---

[13] The translation from French of this excerpt is by the authors of this Chapter. This is also true, in general, unless otherwise specified, for all other translations from Italian, French and Latin presented in this Chapter.

> "[...] *I must warn [the reader] that, in order to avoid circumlocutions, I often used the obscure term "force", and some other terms that are commonly employed when treating the Motion of Bodies; but I have never demanded to attach to this term other ideas than the ones resulting by the Principles that I have established, both in this Preface and in the first Part of this Treatise"* [d'Alembert, Traité de dynamique, 1743]

Forces are introduced in Mechanics as those vectors by means of which we can calculate virtual work using their inner product with virtual velocities: a very abstract definition indeed! Piola's theorem establishes a necessary condition for equilibrium, i.e the condition imposing that the resultant force and resultant moment of forces are both vanishing in an equilibrium configuration: a very abstract property indeed!

The so-called direct measures of forces are not direct at all. One measures other quantities and via the used theory the searched value of the force which she/he feels the need to talk about are determined. In a dynamometer one measures the deformation of, for example, a spring, then by further theoretical hypotheses (for example, one assumes that the measured deformation is purely elastic) and, finally, through the model (in the chosen example, Hooke's law) the *value* of the force can be calculated. If one presents the argument in this way, it is finally evident that force is a mere abstract object belonging to the used mathematical model.

Looking closely at the ontological misconception concerning the armillary spheres and the concept of force, the question naturally arises: why does the confusion between reality and model happen so often and remain so widespread?

As it will become clear from reading the following chapters, scientific progress has often been held hostage by power groups who, for political and power related reasons (or for mere ignorance), have blocked the development of certain ideas in favor of others. The fact that Newton's equations are common knowledge in the scientific world, while Euler-Lagrange's equations are still seen as an unnecessary mathematical complication, gives us a really clear indication of how and why scientific progress in Continuum Mechanics has taken the unfortunate path that we have described.

The present chapter intentionally serves as a (long) introduction to this Volume, whose ultimate aim is to analyze the influence of modalities of sources transmission on the development of scientific theories. As it will become clearer to the reader once she/he will be engaged in the various themes we have chosen to develop in this Volume, the role of sources transmission phenomena in the development of Science is central. This fact is certainly obvious in its positive aspects: Einstein could not have written the equations of Special Relativity without Poincaré (1854-1912) or those of General Relativity without Levi-Civita's and Ricci's contributions to differential geometry. However, the sources transmission modalities play an important role also in contexts involving less *noble* scientifically and much less humanly edifying actions. This circumstance emerges very clearly, for instance, from Heiberg's text of the Prolegomena of his critical edition of the Archimedean opus, and it has strongly prompted us to offer an English translation of its more relevant passages, which otherwise remained readable only in Latin. Heiberg demonstrates with philological methods how the process of transmission of a work is by no means simple and, indeed, is conditioned by a myriad of successive modifications and alterations. An aspect that Heiberg underlines in the text of the Prolegomena, and that we are sure will strike the reader, consists in the philological deduction, that Heiberg only suggests but that Marshall Clagett [36] (and other modern scholars [56, 99]) clearly demonstrates, of the fact that in his presumed translation into Latin of Archimedes' work Niccolò Tartaglia (1499/1500-1557) heavily used an earlier translation, due to William of Moerbeke (1215/35-1286 AD), without ever mentioning him. This is very striking because it is not unique in the History of Science: periodically, someone appropriates the results of others, probably relying on the scarcity of available sources and on linguistic barriers. For example, the only copy of Moerbeke's translation, which even seems to be autograph, has been longly lost and was only found in 1885 by the German classicist Valentin Rose (1829-1916) bound to other texts. Sometimes, some sources are only available in a certain language different from the current *lingua franca* and this circumstance, unfortunately, constitutes a insurmountable barrier for a large part of the scientific community: indeed, it is as if somebody were exploiting purposely linguistic barriers for hiding the true origins of sources. We have mentioned in this chapter, and will discuss in detail in a later chapter, the sad fate of Gabrio Piola's works, forgotten for about a century and a half, just because they were written in Italian. Forgotten or willingly ignored? The reader is invited to consider that the libraries of the most important universities in the world contained copies of Piola's works, and that even theories such as Peridynamics, which Piola introduced in the 19th century, were rediscovered in the 21th century.

We believe that the studies and analyses we presented in this work can be useful for those who want to seriously approach the study of Science, not stopping at the external appearance and the universally accepted version of its development. We hope that in the future it will no longer be possible for some people to deliberately steal the work of others (sometimes even without understanding it, and therefore distorting it), hiding or destroying the name of the true authors. Today we know, although the official version still struggles to recognize them, of the invaluable contributions of Archytas, William of Moerbecke, Piola and many others. We are certain that the memory of many other scholars has been completely destroyed. It is to them that we want to dedicate our work.

## Appendix: A Literary support to our theses

Albeit we tried to argue carefully about our point of view for the necessary revisitation of the History of Mechanics, we are aware that many criticisms may be attracted by the content of this chapter. In fact, in order to avoid to be considered inappropriate, many scholars preferred to insinuate some of our previous statements by using the artifice of hiding them in literary works, sometimes in the field of Science Fiction. Our attention has been particularly attracted by the masterpiece of Alfred Bester: *The stars, my destination*. We quote here some of the most relevant excerptions, in the sense we have specified, of this work.

*BETWEEN MARS AND JUPITER is spread the broad belt of the asteroids. Of the thousands, known and unknown, most unique to the Freak Century was the Sargasso Asteroid, a tiny planet manufactured of natural rock and wreckage salvaged by its inhabitants in the course of two hundred years.*

*They were savages, the only savages of the twenty-fourth century; descendants of a research team of scientists that had been lost and marooned in the asteroid belt two centuries before when their ship had failed. By the time their descendants were rediscovered they had built up a world and a culture of their own, and preferred to remain in space, salvaging and spoiling, and practicing a barbaric travesty of the scientific method they remembered from their forebears. They called themselves The Scientific People. The world promptly forgot them.*

*S.S. "Nomad" looped through space, neither on a course for Jupiter nor the far stars, but drifting across the asteroid belt in the slow spiral of a dying animalcule. It passed within a mile of the Sargasso Asteroid, and it was immediately captured by The Scientific People to be incorporated into their little planet. They found Foyle.*

*He awoke once while he was being carried in triumph on a litter through the natural and artificial passages within the scavenger asteroid.* [...]

*A crowd around the litter was howling triumphantly. "Quant Suff!" they shouted. A woman's chorus began an excited bleating: Ammonium bromide gr .11/2 Potassium bromide gr .3 Sodium bromide gr .2 Citric acid quant. suff. "Quant Suff!" The Scientific People roared. "Quant Suff!" Foyle fainted.* [...]

*The distant sun blazed through; the air was hot and moist. Foyle gazed around dimly. A devil face peered at him. Cheeks, chin, nose, and eyelids were hideously tattooed like an ancient Maori mask. Across the brow was tattooed JOSEPH. The "0" in JOSEPH had a tiny arrow thrust up from the right shoulder, turning it into the symbol of Mars, used by scientists to designate male sex.*

*"We are the Scientific Race," Joseph said. "I am Joseph; these are my people." He gestured. Foyle gazed at the grinning crowd surrounding his litter. All faces were tattooed into devil masks; all brows had names blazoned across them.* [...]

*"You are the first to arrive alive in fifty years. You are a puissant man. Very. Arrival of the fittest is the doctrine of Holy Darwin. Most scientific."*

*"Quant Suff!" the crowd bellowed.*

*Joseph seized Foyle's elbow in the manner of a physician taking a pulse. His devil mouth counted solemnly up to ninety-eight.*

*"Your pulse. Ninety-eight-point-six," Joseph said, producing a thermometer and shaking it reverently. "Most scientific."*

*"Quant Suff!" came the chorus. Joseph proffered an Erlenmeyer flask. It was labeled: Lung, Cat, c. s., hematoxylin & eosin. "Vitamin?" Joseph inquired. When Foyle did not respond, Joseph removed a large pill from the flask, placed it in the bowl of a pipe, and lit it. He puffed once and then gestured. Three girls appeared before Foyle. Their faces were hideously tattooed. Across each brow was a name: JOAN and MOIRA and POLLX. The "0" of each name had a tiny cross at the base.*

*"Choose." Joseph said. "The Scientific People practice Natural Selection. Be scientific in your choice. Be genetic."*

[Alfred Bester, "The stars my destination"]

## Acknowledgments

Mario Spagnuolo was supported by P.O.R. SARDEGNA F.S.E. 2014-2020 - Asse III "Istruzione e Formazione, Obiettivo Tematico: 10, Obiettivo Specifico: 10.5, Azione dell'accordo di Partenariato:10.5.12" Avviso di chiamata per il finanziamento di Progetti di ricerca - Anno 2017.

## References

1. Abali, B.E., Yang, H.: Parameter determination of metamaterials in generalized mechanics as a result of computational homogenization. In: D. Indeitsev, A. Krivtsov A. (Eds.) Advanced Problems in Mechanics. APM 2019. Lecture Notes in Mechanical Engineering, pp. 22–31. Springer, Cham, Switzerland (2019)
2. Abdoul-Anziz, H., Seppecher, P., Bellis, C.: Homogenization of frame lattices leading to second gradient models coupling classical strain and strain-gradient terms. Mathematics and Mechanics of Solids **24**(12), 3976–3999 (2019)
3. Alibert, J.J., Seppecher, P., dell'Isola, F.: Truss modular beams with deformation energy depending on higher displacement gradients. Mathematics and Mechanics of Solids **8**(1), 51–73 (2003)
4. Altenbach, H., Bîrsan, M., Eremeyev, V.A.: Cosserat-type rods. In: H. Altenbach, V. A. Eremeyev (Eds.) Generalized Continua from the Theory to Engineering Applications, pp. 179–248. Springer, Vienna (2013)
5. Altenbach, H., Eremeyev, V.: Strain rate tensors and constitutive equations of inelastic micropolar materials. International Journal of Plasticity **63**, 3–17 (2014)
6. Altenbach, H., Eremeyev, V.A.: On the theories of plates based on the Cosserat approach. In: G. A. Maugin, A. V. Metrikine (Eds.) Mechanics of generalized continua, pp. 27–35. Springer, New York (2010)
7. Altenbach, H., Eremeyev, V.A.: Cosserat media. In: H. Altenbach and V. A. Eremeyev (Eds). Generalized Continua from the Theory to Engineering Applications, pp. 65–130. Springer, Vienna (2013)
8. Altenbach, H., Eremeyev, V.A.: Cosserat-type shells. In: Altenbach H, Eremeyev V A (Eds). Generalized Continua from the Theory to Engineering Applications, pp. 131–178. Springer, Vienna (2013)
9. Altenbach, H., Forest, S.: Generalized continua as models for classical and advanced materials. Springer, Cham, Switzerland (2016)
10. Altenbach, J., Altenbach, H., Eremeyev, V.A.: On generalized Cosserat-type theories of plates and shells: a short review and bibliography. Archive of Applied Mechanics **80**(1), 73–92 (2010)
11. Aminpour, H., Rizzi, N.: On the modelling of carbon nano tubes as generalized continua. In: H. Altenbach, S. Forest (Eds.) Generalized Continua as Models for Classical and Advanced Materials, pp. 15–35. Springer, Cham, Switzerland (2016)
12. Andreaus, U., dell'Isola, F., Giorgio, I., Placidi, L., Lekszycki, T., Rizzi, N.L.: Numerical simulations of classical problems in two-dimensional (non) linear second gradient elasticity. International Journal of Engineering Science **108**, 34–50 (2016)
13. Auffray, N.: On the algebraic structure of isotropic generalized elasticity theories. Mathematics and Mechanics of Solids **20**(5), 565–581 (2015)
14. Auffray, N., dell'Isola, F., Eremeyev, V.A., Madeo, A., Rosi, G.: Analytical continuum mechanics à la hamilton–piola least action principle for second gradient continua and capillary fluids. Mathematics and Mechanics of Solids **20**(4), 375–417 (2015)

15. Barchiesi, E., dell'Isola, F., Hild, F., Seppecher, P.: Two-dimensional continua capable of large elastic extension in two independent directions: asymptotic homogenization, numerical simulations and experimental evidence. Mechanics Research Communications **103**, 103466 (2020)
16. Barchiesi, E., Eugster, S.R., dell'Isola, F., Hild, F.: Large in-plane elastic deformations of bi-pantographic fabrics: asymptotic homogenization and experimental validation. Mathematics and Mechanics of Solids **25**(3), 739–767 (2020)
17. Barchiesi, E., Ganzosch, G., Liebold, C., Placidi, L., Grygoruk, R., Müller, W.H.: Out-of-plane buckling of pantographic fabrics in displacement-controlled shear tests: experimental results and model validation. Continuum Mechanics and Thermodynamics **31**(1), 33–45 (2019)
18. Barchiesi, E., Khakalo, S.: Variational asymptotic homogenization of beam-like square lattice structures. Mathematics and Mechanics of Solids **24**(10), 3295–3318 (2019)
19. Benvenuto, E.: An introduction to the history of structural mechanics: Part I: Statics and resistance of solids. Springer, New York (2012)
20. Bersani, A., dell'Isola, F., Seppecher, P.: Lagrange multipliers in infinite dimensional spaces, examples of application. In: H. Altenbach, A. Öchsner (Eds.) Encyclopedia of Continuum Mechanics. Springer, Berlin (2019)
21. Bloch, A.: A new approach to the dynamics of systems with gyroscopic coupling terms. Philosophical Magazine **35 (series 7)**(244), 315–334 (1944)
22. Bloch, A.: Electromechanical analogies and their use for the analysis of mechanical and electromechanical systems. Journal of the Institution of Electrical Engineers-Part I: General **92**(52), 157–169 (1945)
23. Boutin, C., Giorgio, I., Placidi, L., et al.: Linear pantographic sheets: asymptotic micro-macro models identification. Mathematics and Mechanics of Complex Systems **5**(2), 127–162 (2017)
24. Cailletaud, G., Forest, S., Jeulin, D., Feyel, F., Galliet, I., Mounoury, V., Quilici, S.: Some elements of microstructural mechanics. Computational Materials Science **27**(3), 351–374 (2003)
25. Capriz, G.: Continua with microstructure, vol. 35. Springer, New York (2013)
26. Carcaterra, A., dell'Isola, F., Esposito, R., Pulvirenti, M.: Macroscopic description of microscopically strongly inhomogenous systems: A mathematical basis for the synthesis of higher gradients metamaterials. Archive for Rational Mechanics and Analysis **218**(3), 1239–1262 (2015)
27. Carroll, M.: Must elastic materials be hyperelastic? Mathematics and Mechanics of Solids **14**(4), 369–376 (2009)
28. Cazzani, A., Atluri, S.: Four-noded mixed finite elements, using unsymmetric stresses, for linear analysis of membranes. Computational Mechanics **11**(4), 229–251 (1993)
29. Cazzani, A., Lovadina, C.: On some mixed finite element methods for plane membrane problems. Computational Mechanics **20**(6), 560–572 (1997)
30. Cazzani, A., Malagù, M., Turco, E.: Isogeometric analysis of plane-curved beams. Mathematics and Mechanics of Solids **21**(5), 562–577 (2016)
31. Cazzani, A., Malagù, M., Turco, E., Stochino, F.: Constitutive models for strongly curved beams in the frame of isogeometric analysis. Mathematics and Mechanics of Solids **21**(2), 182–209 (2016)
32. Cazzani, A., Serra, M., Stochino, F., Turco, E.: A refined assumed strain finite element model for statics and dynamics of laminated plates. Continuum Mechanics and Thermodynamics **32**(3), 665–692 (2020)
33. Cazzani, A., Stochino, F., Turco, E.: An analytical assessment of finite element and isogeometric analyses of the whole spectrum of Timoshenko beams. ZAMM-Journal of Applied Mathematics and Mechanics/Zeitschrift für Angewandte Mathematik und Mechanik **96**(10), 1220–1244 (2016)
34. Chalmers, A.F.: What is this thing called science? Hackett Publishing Company, Indianapolis (2013)
35. Clagett, M. (ed.): Critical problems in the history of science. University of Wisconsin Press, Madison (1969)

36. Clagett, M.: William of Moerbeke: translator of Archimedes. Proceedings of the American Philosophical Society **126**(5), 356–366 (1982)
37. Clagett, M.: Greek science in antiquity. Dover, Mineola, NY (2001)
38. Cosserat, E., Cosserat, F.: Théorie des corps déformables. A. Hermann et fils, Paris (1909)
39. Cuomo, M., Contrafatto, L., Greco, L.: A variational model based on isogeometric interpolation for the analysis of cracked bodies. International Journal of Engineering Science **80**, 173–188 (2014)
40. dell'Isola, F., Andreaus, U., Placidi, L.: At the origins and in the vanguard of peridynamics, non-local and higher-gradient continuum mechanics: an underestimated and still topical contribution of Gabrio Piola. Mathematics and Mechanics of Solids **20**(8), 887–928 (2015)
41. dell'Isola, F., Barchiesi, E., Misra, A.: Naive model theory: Its applications to the theory of metamaterials design. In: F. dell'Isola and D. Steigmann (Eds.), Discrete and Continuum Models for Complex Metamaterials, pp. 141–196. Cambridge University Press, Cambridge (2020)
42. dell'Isola, F., Corte, A.D., Giorgio, I.: Higher-gradient continua: The legacy of Piola, Mindlin, Sedov and Toupin and some future research perspectives. Mathematics and Mechanics of Solids **22**(4), 852–872 (2017)
43. dell'Isola, F., Della Corte, A., Esposito, R., Russo, L.: Some cases of unrecognized transmission of scientific knowledge: from antiquity to Gabrio Piola's peridynamics and generalized continuum theories. In: H. Altenbach , S. Forest (Eds.) Generalized Continua as Models for Classical and Advanced Materials, pp. 77–128. Springer, Cham, Switzerland (2016)
44. dell'Isola, F., Giorgio, I., Pawlikowski, M., Rizzi, N.L.: Large deformations of planar extensible beams and pantographic lattices: heuristic homogenization, experimental and numerical examples of equilibrium. Proceedings of the Royal Society A: Mathematical, Physical and Engineering Sciences **472**(2185), 20150790 (2016)
45. dell'Isola, F., Madeo, A., Seppecher, P.: Cauchy tetrahedron argument applied to higher contact interactions. Archive for Rational Mechanics and Analysis **219**(3), 1305–1341 (2016)
46. dell'Isola, F., Maier, G., Perego, U., et al.: The complete works of Gabrio Piola: Volume I. Springer, Berlin (2014)
47. dell'Isola, F., Maier, G., Perego, U., et al.: The Complete Works of Gabrio Piola: Volume II. Springer, Cham, Switzerland (2019)
48. dell'Isola, F., Placidi, L.: Variational principles are a powerful tool also for formulating field theories. In: F. dell'Isola, S. Gavrilyuk (Eds.) Variational models and methods in solid and fluid mechanics, pp. 1–15. Springer, Vienna (2011)
49. dell'Isola, F., Rosa, L.: Saint Venant problem in linear piezoelectricity. In: V. V. Varadan, J. Chandra (Eds.) Smart Structures and Materials 1996: Mathematics and Control in Smart Structures - Proceedings of SPIE, vol. 2715, pp. 399–409 (1996)
50. dell'Isola, F., Sciarra, G., Vidoli, S.: Generalized Hooke's law for isotropic second gradient materials. Proceedings of the Royal Society A: Mathematical, Physical and Engineering Sciences **465**(2107), 2177–2196 (2009)
51. dell'Isola, F., Seppecher, P.: The relationship between edge contact forces, double forces and interstitial working allowed by the principle of virtual power. Comptes rendus de l'Académie des sciences. Série 2, Mécanique, physique, chimie, astronomie **321**(8), 303–308 (1995)
52. dell'Isola, F., Seppecher, P.: Edge contact forces and quasi-balanced power. Meccanica **32**(1), 33–52 (1997)
53. dell'Isola, F., Seppecher, P., Madeo, A.: How contact interactions may depend on the shape of Cauchy cuts in Nth gradient continua: approach "à la d'Alembert". Zeitschrift für Angewandte Mathematik und Physik **63**(6), 1119–1141 (2012)
54. dell'Isola, F., Seppecher, P., Placidi, L., Barchiesi, E., Misra, A.: Least action and virtual work principles for the formulation of generalized continuum models. In: F. dell'Isola, D. J. Steigmann (Eds.) Discrete and Continuum Models for Complex Metamaterials, p. 327. Cambridge University Press (2020)
55. dell'Isola, F., Steigmann, D.J.: Discrete and Continuum Models for Complex Metamaterials. Cambridge University Press, Cambridge (2020)

56. Dijksterhuis, E.J.: Archimedes. Princeton University Press, Princeton (2014)
57. Epstein, M., Smelser, R.: An appreciation and discussion of Paul Germain's «The method of virtual power in the mechanics of continuous media, I: Second-gradient theory». Mathematics and Mechanics of Complex Systems **8**(2), 191–199 (2020)
58. Eremeyev, V.A.: On Non-holonomic Boundary Conditions within the Nonlinear Cosserat Continuum. In: B. E. Abali, H. Altenbach, F. dell'Isola, V. A. Eremeyev, A. Öchsner (Eds.) New Achievements in Continuum Mechanics and Thermodynamics, pp. 93–104. Springer, Cham, Switzerland (2019)
59. Eremeyev, V.A., Altenbach, H.: Equilibrium of a second-gradient fluid and an elastic solid with surface stresses. Meccanica **49**(11), 2635–2643 (2014)
60. Eremeyev, V.A., dell'Isola, F., Boutin, C., Steigmann, D.: Linear pantographic sheets: existence and uniqueness of weak solutions. Journal of Elasticity **132**(2), 175–196 (2018)
61. Eremeyev, V.A., Ganghoffer, J.F., Konopińska-Zmysłowska, V., Uglov, N.S.: Flexoelectricity and apparent piezoelectricity of a pantographic micro-bar. International Journal of Engineering Science **149**, 103213 (2020)
62. Eremeyev, V.A., Lebedev, L.P., Altenbach, H.: Foundations of micropolar mechanics. Springer, Heidelberg (2012)
63. Eremeyev, V.A., Pietraszkiewicz, W.: Material symmetry group and consistently reduced constitutive equations for the elastic Cosserat continuum. In: H. Altenbach, S. Forest, A. Krivtsov (Eds.) Generalized Continua as Models for Materials. Springer, Berlin (2013)
64. Eringen, A.C.: Linear theory of micropolar elasticity. Journal of Mathematics and Mechanics pp. 909–923 (1966)
65. Eringen, A.C.: Linear theory of micropolar viscoelasticity. International Journal of Engineering Science **5**(2), 191–204 (1967)
66. Eringen, A.C.: Theory of micropolar plates. Zeitschrift für Angewandte Mathematik und Physik ZAMP **18**(1), 12–30 (1967)
67. Eringen, A.C.: Theory of micropolar elasticity. In: Microcontinuum field theories, vol. 1, pp. 101–248. Springer, New York (1999)
68. Eugster, S., dell'Isola, F., Steigmann, D.: Continuum theory for mechanical metamaterials with a cubic lattice substructure. Mathematics and Mechanics of Complex Systems **7**(1), 75–98 (2019)
69. Eugster, S.R., dell'Isola, F.: Exegesis of the Introduction and Sect. I from "Fundamentals of the Mechanics of Continua" by E. Hellinger. ZAMM-Journal of Applied Mathematics and Mechanics/Zeitschrift für Angewandte Mathematik und Mechanik **97**(4), 477–506 (2017)
70. Eugster, S.R., dell'Isola, F.: Exegesis of Sect. II and III. A from "Fundamentals of the Mechanics of Continua" by E. Hellinger. ZAMM-Journal of Applied Mathematics and Mechanics/Zeitschrift für Angewandte Mathematik und Mechanik **98**(1), 31–68 (2018)
71. Eugster, S.R., dell'Isola, F.: Exegesis of Sect. III. B from "Fundamentals of the Mechanics of Continua" by E. Hellinger. ZAMM-Journal of Applied Mathematics and Mechanics/Zeitschrift für Angewandte Mathematik und Mechanik **98**(1), 69–105 (2018)
72. Eugster, S.R., Glocker, Ch.: On the notion of stress in classical continuum mechanics. Mathematics and Mechanics of Complex Systems **5**(3-4), 299–338 (2017)
73. Evans, K.E.: Auxetic polymers: a new range of materials. Endeavour **15**(4), 170–174 (1991)
74. Evans, K.E., Alderson, A.: Auxetic materials: functional materials and structures from lateral thinking! Advanced Materials **12**(9), 617–628 (2000)
75. Feynman, R.P., Leighton, R.B., Sands, M.: The Feynman lectures on physics; vol. I. American Journal of Physics **33**(9), 750–752 (1965)
76. Forest, S.: Strain gradient elasticity from capillarity to the mechanics of nano-objects. In: A. Bertram, S. Forest (Eds.) Mechanics of Strain Gradient Materials, pp. 37–70. Springer, Cham, Switzerland (2020)
77. Franciosi, P.: Laminate system schemes for effective property estimates of architectured composites with co-(dis) continuous phases. Mechanics Research Communications **45**, 70–76 (2012)

78. Franciosi, P., Charles, Y.: Effective property estimates for n-phase composites with from all to none co-continuous phases. International Journal of Solids and Structures **96**, 110–125 (2016)
79. Franciosi, P., El Omri, A.: Effective properties of fiber and platelet systems and related phase arrangements in n-phase heterogenous media. Mechanics Research Communications **38**(1), 38–44 (2011)
80. Gallavotti, G.: Quasi periodic motions from Hipparchus to Kolmogorov. Atti della Accademia Nazionale dei Lincei. Classe di Scienze Fisiche, Matematiche e Naturali. Rendiconti Lincei. Matematica e Applicazioni **12**(2), 125–152 (2001)
81. George, D., Spingarn, C., Dissaux, C., Nierenberger, M., Rahman, R.A., Rémond, Y.: Examples of multiscale and multiphysics numerical modeling of biological tissues. Bio-Medical Materials and Engineering **28**(s1), S15–S27 (2017)
82. Germain, P.: La méthode des puissances virtuelles en mécanique des milieux continus, premiere partie: théorie du second gradient. Journal de Mécanique **12**(2), 235–274 (1973)
83. Germain, P.: The method of virtual power in continuum mechanics. Part 2: Microstructure. SIAM Journal on Applied Mathematics **25**(3), 556–575 (1973)
84. Germain, P.: Functional concepts in continuum mechanics. Meccanica **33**(5), 433–444 (1998)
85. Germain, P.: My discovery of mechanics. In: G.A. Maugin, R. Druot, F. Sidoroff (Eds.) Continuum Thermomechanics, P. Germain's Anniversary Volume, pp. 1–24. Kluver, New York (2000)
86. Germain, P.: The method of virtual power in the mechanics of continuous media, I: Second-gradient theory. Mathematics and Mechanics of Complex Systems **8**(2), 153–190 (2020)
87. Giorgio, I., Ciallella, A., Scerrato, D.: A study about the impact of the topological arrangement of fibers on fiber-reinforced composites: Some guidelines aiming at the development of new ultra-stiff and ultra-soft metamaterials. International Journal of Solids and Structures **203**, 73–83 (2020)
88. Giorgio, I., Rizzi, N.L., Andreaus, U., Steigmann, D.J.: A two-dimensional continuum model of pantographic sheets moving in a 3D space and accounting for the offset and relative rotations of the fibers. Mathematics and Mechanics of Complex Systems **7**(4), 311–325 (2019)
89. Greco, L.: An iso-parametric $G^1$-conforming finite element for the nonlinear analysis of Kirchhoff rod. Part I: the 2D case. Continuum Mechanics and Thermodynamics **32**(5), 1473–1496 (2020)
90. Greco, L., Cuomo, M.: An isogeometric implicit $G^1$ mixed finite element for Kirchhoff space rods. Computer Methods in Applied Mechanics and Engineering **298**, 325–349 (2016)
91. Greco, L., Cuomo, M., Contrafatto, L.: A reconstructed local B formulation for isogeometric Kirchhoff–Love shells. Computer Methods in Applied Mechanics and Engineering **332**, 462–487 (2018)
92. Greco, L., Cuomo, M., Contrafatto, L., Gazzo, S.: An efficient blended mixed B-spline formulation for removing membrane locking in plane curved Kirchhoff rods. Computer Methods in Applied Mechanics and Engineering **324**, 476–511 (2017)
93. Gurtin, M., Toupin, R.: A uniqueness theorem for the displacement boundary-value problem of linear elastodynamics. Quarterly of Applied Mathematics **23**(1), 79–81 (1965)
94. Gurtin, M.E.: A note on the principle of minimum potential energy for linear anisotropic elastic solids. Quarterly of Applied Mathematics **20**(4), 379–382 (1963)
95. Halliday, D., Resnick, R., Walker, J.: Fundamentals of physics. Wiley, Hoboken, NJ (2013)
96. Hardy, G.: A Mathematician's Apology. Cambridge University Press, Cambridge (1992)
97. Heath, T.L.: The works of Archimedes. Dover, Mineola, NY (2002)
98. Hesch, C., Schuß, S., Dittmann, M., Eugster, S., Favino, M., Krause, R.: Variational space–time elements for large-scale systems. Computer Methods in Applied Mechanics and Engineering **326**, 541–572 (2017)
99. Høyrup, J.: Archimedes: Reception in the Renaissance. In: M. Sgarbi (Ed.) Encyclopedia of Renaissance Philosophy. Springer, Cham, Switzerland (2019)
100. Jammer, M.: Concepts of force: A study in the foundations of dynamics. Harvard University Press, Cambridge, MA (1957)

101. Kafadar, C., Eringen, A.C.: Micropolar media—i the classical theory. International Journal of Engineering Science **9**(3), 271–305 (1971)
102. Khakalo, S., Balobanov, V., Niiranen, J.: Modelling size-dependent bending, buckling and vibrations of 2D triangular lattices by strain gradient elasticity models: applications to sandwich beams and auxetics. International Journal of Engineering Science **127**, 33–52 (2018)
103. Khakalo, S., Niiranen, J.: Form II of Mindlin's second strain gradient theory of elasticity with a simplification: For materials and structures from nano-to macro-scales. European Journal of Mechanics-A/Solids **71**, 292–319 (2018)
104. Khakalo, S., Niiranen, J.: Lattice structures as thermoelastic strain gradient metamaterials: Evidence from full-field simulations and applications to functionally step-wise-graded beams. Composites Part B: Engineering **177**, 107224 (2019)
105. Khakalo, S., Niiranen, J.: Anisotropic strain gradient thermoelasticity for cellular structures: Plate models, homogenization and isogeometric analysis. Journal of the Mechanics and Physics of Solids **134**, 103728 (2020)
106. Kron, G.: Numerical solution of ordinary and partial differential equations by means of equivalent circuits. Journal of Applied Physics **16**(3), 172–186 (1945)
107. Lagrange, J.L.: Traité de la résolution des équations numériques de tous les degrés: avec des notes sur plusieurs points de la théorie des équations algébriques. Courcier, Paris (1808)
108. Lagrange, J.L.: Mécanique analytique, vol. 1. Mallet-Bachelier, Paris (1853)
109. Laudato, M., Ciallella, A.: Perspectives in generalized continua. In: B. Abali, I. Giorgio (Eds.) Developments and Novel Approaches in Biomechanics and Metamaterials, pp. 1–13. Springer, Cham, Switzerland (2020)
110. Laudato, M., Manzari, L., Barchiesi, E., Di Cosmo, F., Göransson, P.: First experimental observation of the dynamical behavior of a pantographic metamaterial. Mechanics Research Communications **94**, 125–127 (2018)
111. Le Rond d'Alembert, J.B.: Traité de dynamique. David l'aîné, Paris (1743)
112. Mablekos, E., Weidmann, H.: Realization of negative impedance inverter using Kron's mesh primitive-branch approach. Journal of the Franklin Institute **286**(1), 59–75 (1968)
113. Makvandi, R., Reiher, J.C., Bertram, A., Juhre, D.: Isogeometric analysis of first and second strain gradient elasticity. Computational Mechanics **61**(3), 351–363 (2018)
114. Maugin, G.A.: A historical perspective of generalized continuum mechanics. In: Altenbach H., Maugin G., Erofeev V. (eds) Mechanics of Generalized Continua. Advanced Structured Materials, vol 7, pp. 3–19. Springer, Berlin, Heidelberg (2011)
115. Maxwell, J.C.: On a dynamical top, for exhibiting the phenomena of the motion of a system of invariable form about a fixed point, with some suggestions as to the Earth's motion. Transactions of the Royal Society of Edinburgh **21**(4), 559–570 (1857)
116. Mindlin, R.D.: Second gradient of strain and surface-tension in linear elasticity. International Journal of Solids and Structures **1**(4), 417–438 (1965)
117. Misra, A., Placidi, L., Scerrato, D.: A review of presentations and discussions of the workshop Computational Mechanics of Generalized Continua and Applications to Materials with Microstructure that was held in Catania 29–31 October 2015. Mathematics and Mechanics of Solids **22**(9), 1891–1904 (2017)
118. Misra, A., Poorsolhjouy, P.: Granular micromechanics model for damage and plasticity of cementitious materials based upon thermomechanics. Mathematics and Mechanics of Solids **25**(10), 1778–1803 (2020)
119. Pideri, C., Seppecher, P.: A second gradient material resulting from the homogenization of an heterogeneous linear elastic medium. Continuum Mechanics and Thermodynamics **9**(5), 241–257 (1997)
120. Pietraszkiewicz, W., Eremeyev, V.: On vectorially parameterized natural strain measures of the non-linear Cosserat continuum. International Journal of Solids and Structures **46**(11-12), 2477–2480 (2009)
121. Pingaro, M., Reccia, E., Trovalusci, P.: Homogenization of random porous materials with low-order virtual elements. ASCE-ASME Journal of Risk and Uncertainty in Engineering Systems, Part B: Mechanical Engineering **5**(3), 030905–1–10 (2019)

122. Piola, G.: Memoria intorno alle equazioni fondamentali del movimento di corpi qualsivogliono considerati secondo la naturale loro forma e costituzione. Memorie di Matematica e di Fisica della Società Italiana delle Scienze residente in Modena **24 Parte I (Matematica)**, 1–186 (1848)
123. Placidi, L.: A variational approach for a nonlinear 1-dimensional second gradient continuum damage model. Continuum Mechanics and Thermodynamics **27**(4-5), 623–638 (2015)
124. Placidi, L., Barchiesi, E., Misra, A.: A strain gradient variational approach to damage: a comparison with damage gradient models and numerical results. Mathematics and Mechanics of Complex Systems **6**(2), 77–100 (2018)
125. Placidi, L., dell'Isola, F., Barchiesi, E.: Heuristic homogenization of Euler and pantographic beams. In: C. Picu, J. F. Ganghoffer (Eds.) Mechanics of Fibrous Materials and Applications, pp. 123–155. Springer, Cham, Switzerland (2020)
126. Placidi, L., Misra, A., Barchiesi, E.: Simulation results for damage with evolving microstructure and growing strain gradient moduli. Continuum Mechanics and Thermodynamics **31**(4), 1143–1163 (2019)
127. Placidi, L., Rosi, G., Giorgio, I., Madeo, A.: Reflection and transmission of plane waves at surfaces carrying material properties and embedded in second-gradient materials. Mathematics and Mechanics of Solids **19**(5), 555–578 (2014)
128. Rahali, Y., Giorgio, I., Ganghoffer, J., dell'Isola, F.: Homogenization à la Piola produces second gradient continuum models for linear pantographic lattices. International Journal of Engineering Science **97**, 148–172 (2015)
129. Reccia, E., De Bellis, M.L., Trovalusci, P., Masiani, R.: Sensitivity to material contrast in homogenization of random particle composites as micropolar continua. Composites Part B: Engineering **136**, 39–45 (2018)
130. Rosi, G., Giorgio, I., Eremeyev, V.A.: Propagation of linear compression waves through plane interfacial layers and mass adsorption in second gradient fluids. ZAMM-Journal of Applied Mathematics and Mechanics/Zeitschrift für Angewandte Mathematik und Mechanik **93**(12), 914–927 (2013)
131. Russell, B.: The problems of philosophy. Oxford University Press, Oxford (2001)
132. Sarton, G.: A history of science: Ancient science through the golden age of Greece. Harvard University Press, Cambridge, MA (1952)
133. Scerrato, D., Giorgio, I.: Equilibrium of two-dimensional cycloidal pantographic metamaterials in three-dimensional deformations. Symmetry **11**(12), 1523 (2019)
134. Schiaparelli, G.V.: Le sfere omocentriche di Eudosso, di Callippo e di Aristotele. Hoepli, Milano (1875)
135. Sedov, L.: Models of continuous media with internal degrees of freedom. Journal of Applied Mathematics and Mechanics **32**(5), 803–819 (1968)
136. Sedov, L., Parkus, H.: Irreversible aspects of continuum mechanics and transfer of physical characteristics in moving fluids. Springer, New York (1968)
137. Sedov, L.I.: Mathematical methods for constructing new models of continuous media. Russian Mathematical Surveys **20**(5), 123–182 (1965)
138. Seppecher, P.: Etude des conditions aux limites en théorie du second gradient: cas de la capillarité. Comptes rendus de l'Académie des sciences. Série 2, Mécanique, Physique, Chimie, Sciences de l'univers, Sciences de la Terre **309**(6), 497–502 (1989)
139. Seppecher, P.: Second-gradient theory: application to Cahn-Hilliard fluids. In: G. A. Maugin, R. Drouot, F. Sidoroff (Eds.) Continuum Thermomechanics, pp. 379–388. Springer, Dordrecht (2000)
140. Seppecher, P., Alibert, J.J., dell'Isola, F.: Linear elastic trusses leading to continua with exotic mechanical interactions. Journal of Physics: Conference Series **319**(1), 012018 (2011)
141. Spagnuolo, M.: Circuit analogies in the search for new metamaterials: Phenomenology of a mechanical diode. In: H. Altenbach, V. A. Eremeyev, I. Pavlov, A. Porubov (Eds.) Nonlinear Wave Dynamics of Materials and Structures, pp. 411–422. Springer, Cham, Switzerland (2020)

142. Spagnuolo, M., Scerrato, D.: The mechanical diode: on the tracks of James Maxwell employing mechanical–electrical analogies in the design of metamaterials. In: B. E. Abali, I. Giorgio (Eds.) Developments and Novel Approaches in Biomechanics and Metamaterials, pp. 459–469. Springer, Cham, Switzerland (2020)
143. Torabi, J., Niiranen, J.: Microarchitecture-dependent nonlinear bending analysis for cellular plates with prismatic corrugated cores via an anisotropic strain gradient plate theory of first-order shear deformation. Engineering Structures **236**, 112117 (2021)
144. Toupin, R.A.: Theories of elasticity with couple-stress. Archive for Rational Mechanics and Analysis **17**, 85–112 (1964)
145. Tran, L.V., Niiranen, J.: A geometrically nonlinear Euler–Bernoulli beam model within strain gradient elasticity with isogeometric analysis and lattice structure applications. Mathematics and Mechanics of Complex Systems **8**(4), 345–371 (2020)
146. Truesdell, C., Noll, W.: The non-linear field theories of mechanics. In: S. Flügge (ed.) The non-linear field theories of mechanics, *Encyclopedia of Physics*, vol. III/3. Springer (1965)
147. Turco, E.: Tools for the numerical solution of inverse problems in structural mechanics: review and research perspectives. European Journal of Environmental and Civil Engineering **21**(5), 509–554 (2017)
148. Turco, E., dell'Isola, F., Cazzani, A., Rizzi, N.L.: Hencky-type discrete model for pantographic structures: numerical comparison with second gradient continuum models. Zeitschrift für Angewandte Mathematik und Physik **67**(85) (2016)
149. Turco, E., dell'Isola, F., Rizzi, N.L., Grygoruk, R., Müller, W.H., Liebold, C.: Fiber rupture in sheared planar pantographic sheets: Numerical and experimental evidence. Mechanics Research Communications **76**, 86–90 (2016)
150. Turco, E., Golaszewski, M., Cazzani, A., Rizzi, N.L.: Large deformations induced in planar pantographic sheets by loads applied on fibers: experimental validation of a discrete Lagrangian model. Mechanics Research Communications **76**, 51–56 (2016)
151. Turco, E., Misra, A., Pawlikowski, M., dell'Isola, F., Hild, F.: Enhanced Piola–Hencky discrete models for pantographic sheets with pivots without deformation energy: numerics and experiments. International Journal of Solids and Structures **147**, 94–109 (2018)
152. Turco, E., Rizzi, N.L.: Pantographic structures presenting statistically distributed defects: numerical investigations of the effects on deformation fields. Mechanics Research Communications **77**, 65–69 (2016)
153. Yang, H., Ganzosch, G., Giorgio, I., Abali, B.E.: Material characterization and computations of a polymeric metamaterial with a pantographic substructure. Zeitschrift für Angewandte Mathematik und Physik **69**(4), 105 (2018)
154. Yang, H., Müller, W.H.: Computation and experimental comparison of the deformation behavior of pantographic structures with different micro-geometry under shear and torsion. Journal of Theoretical and Applied Mechanics **57**(2), 421–434 (2019)

# Chapter 2
# Translation of Heiberg's Prolegomena

Mario Spagnuolo, Francesco dell'Isola, Beatrice Gerber and Antonio M. Cazzani

## 2.1 Translators' preface

In this chapter, we present the translation of the main excerpts of Heiberg's Prolegomena to his Archimedes Edition. This text was originally written in Latin [Heiberg, J. L. (1910). Archimedis Opera omnia cum commentariis Eutocii: Vol. 1-3. In aedibus BG Teubneri.] and contains the evidence of interesting phenomena in the transmission of ancient scientific texts. Considering the nature of the present work, which is rather interested in problems concerning the transmission and degradation of scientific knowledge through the centuries, the following translation concerns mainly the pages in which issues related to this type of problem are addressed. We have, however, omitted the translation of extremely technical parts of the philologist's work, which are beyond the scope of the present work. The original page numbering is preserved and indicated by margin notes. Accordingly, the footnotes from the original text are within the running text. The footnotes of this chapter are used to annotate the translation in order to clarify certain passages.

Mario Spagnuolo, e-mail: mario.spagnuolo@unica.it
DICAAR, Università degli Studi di Cagliari, Cagliari, Italy.
International Research Center on Mathematics and Mechanics of Complex Systems, University of L'Aquila, L'Aquila, Italy.
Francesco dell'Isola, e-mail: francesco.dellisola.aquila@gmail.com
DICEAA and International Research Center on Mathematics and Mechanics of Complex Systems, University of L'Aquila, L'Aquila, Italy.
Beatrice Gerber, e-mail: batrice_gerber@hotmail.com
Antonio M. Cazzani, e-mail: antonio.cazzani@unica.it
DICAAR, Università degli Studi di Cagliari, Cagliari, Italy.



F. dell'Isola et al. (eds.), *Evaluation of Scientific Sources in Mechanics*,
Advanced Structured Materials 152, https://doi.org/10.1007/978-3-030-80550-0_2

## 2.2 Prolegomena. On the Archimedean codices

VII

# PROLEGOMENA.
# On the Archimedean codices

by

**J. L. Heiberg**

When two years ago I searched for Archimedes' codices to properly analyze them, and the last part of *Archimedes' Questions* (*Quaestiones Archimedeae, Chapter VI, Copenhagen 1879*) is the most valuable about this analysis, I could not escape what the critic of the Torelliana Edition[1] (*Jenaer Literaturzeitung 1795 p. 610 sq.*) had already figured out, namely that the Florentine codex was the most excellent of all, connected with a very close link to the very ancient codex of Giorgio Valla[2]. Having no information about the age of this codex, except that Bandinius[3] had attributed it to the 13th century, I was inevitably persuaded by it to establish that the Florentine codex is itself as old as that of Valla, from which the Parisian codices B and C were copied. But difficulties remained both in the text of individual passages and especially in explaining how that codex had finally reached the Laurentian library; even that letter that was the premise of the first book *On the Sphere and Cylinder* seemed to be better preserved in codex B (*Quaest. Arch. p. 130*). So, I tried to explain these difficulties as well as I could (*Quaest. Arch. p. 132 sq.*). In fact, I was already beginning to doubt Bandinus' judgment even then. And after I had diligently examined and compared the Florentine codex by myself, I was persuaded that this codex in no way could be the same as Valla's codex, but rather his antigraph, copied with such diligence that the copyist reproduced the form even of the letters often with pedantic solicitude. So, I decided to correct this whole piece.

---

[1] Giuseppe Torelli (Verona, 1721-1781) is credited with a plentiful and meticulous edition of the Works of Archimedes in Greek and Latin, published in Oxford in 1792 by A. Robertson. Torelli's most important book is *Archimedis quae supersunt omnia, cum Eutocii Ascalonitae commentariis, ex recensione Josephii Torelli Veronensi cum nova versione latina (accedunt versiones variantes ex codd. Mediceo et Parisiensibus)*, "a typographically splendid work" [A. Frajese, 1974. Archimede: Opera (Italian). Turin: Unione tipografico-editrice torinese.]; Torelli also conducted a first, careful chronological reorganization of the Archimedean works. For more information about Torelli, one can refer to [Bagni, G. T. (1997). Un'intuizione dell'infinitesimo attuale: De nihilo geometrico (1758) di Giuseppe Torelli.].

[2] Giorgio Valla (1447-1500) was an Italian Humanist, presumably a relative of Lorenzo. He teached rhetoric in Pavia (1466-1477) and then in Genoa. Finally, in 1481, he moved to Venice. In 1496, he was arrested for politics, but he was later reintegrated into the professorship. He translated in Latin Greek philosophical and scientific works (Ptolemy, Galen, Alexander of Aphrodisia, Aristotle). He also collected many ancient codices and among them the codex **A** of Archimedes' works.

[3] Angelo Maria Bandini (1726-1803) was an Italian religious, librarian and art collector, as well as a cleric, scholar and bibliophile. Nowadays, he is best known for his work as a librarian. In fact, as their director, he elevated the Biblioteca Marucelliana and the Biblioteca Medicea to very high cultural heights. For these libraries, he compiled a monumental catalog, to which Heiberg refers several times in his *Prolegomena*.

Let's start by describing the Florentine codex. VIII

Well, the Florentine codex of the *Laurentian Medicean library plut. XXVII, 4* is made of parchment, it is written on thick parchments without any trace of strings, and it is very well preserved. It consists of 179 very large pages, which a recent hand has rather negligently marked with numbers; a previous hand had marked at the bottom right of the back page of each tenth sheet, the numbers of the fascicles and the first words of the following page. The codex quite clearly, though not very elegantly, has been written with many abbreviations; accents and spirits are omitted rather frequently; these, where they are present, have the square ⊢ or └ shape, very rarely the curved one, as we use today; here and there both spirit and accent have been placed on the same syllable, but never joined with the same trait. The first page has been written with all the accents with a rather faded ink, the title and the initial A are red. The following topics are contained in this codex:

(i) *On the Sphere and Cylinder* I-II, (ii) *Measurement of a circle*, (iii) *On Conoids and Spheroids*, (iv) *On Spirals*, (v) *On the equilibrium of planes*, (vi) *The Sand-reckoner*, (vii) *Quadrature of the Parabola*, (viii) Eutocius'[4] comments in the books II *On the Sphere and Cylinder*, in the booklet *Measurement of a circle* and in the books II *On the equilibrium of planes*, Heron's[5] excerptions about the measurements. At the end of the book the title is always repeated; moreover, at the end of the book *Quadrature of the Parabola* it also has:

εὐτυχοίης λέον γεώμετρα
πολλοὺς εἰς λυκάβαντας ἴοις πολὺ φίλτατε μούσαις[6]

and at the end of Eutocius' comments in the books about the sphere and the cylinder

Εὐτυκίου πινυτοῦ γλυκερὸς πόνος, ὅν ποτ᾽ ἐκεῖνος
γράψεν τοῖς φθονεροῖς πολλάκι μεμψάμενος[7]

The mathematical figures were always drawn with the same hand as the rest of the codex; in his work the copyist used ruler and compass, he refrained from using tools with which to draw conical sections and spirals; the reason for which he drew the latter so roughly, as if they were arcs of circumferences, is that he depicted them

---

[4] Eutocio of Ascalona (about 480-540?) was a Byzantine mathematician. Of his production, only few comments remain today on the books *On the Conics* of Apollonius of Perga and some works by Archimedes. These comments, especially those on the works of Archimedes, offer a very eloquent insight into Greek scientific development.

[5] Heron of Alexandria (I-III century A.D.) was an ancient Greek scientist, very skilled in the study of Mechanics (among his most famous inventions there is the *Eolipila*, the first steam engine known). He was a teacher of technical disciplines at the *Museum of Alexandria*. Heron carefully studied the works of Euclid and Archimedes. A strong conviction of his was the need for a complete preparation, made of theory and practice. Heron's masterpiece is the treatise on *Mechanics*, in which he systematizes the theoretical and practical aspects of Mechanics.

[6] "*May you have a good fate, O Leo the Geometrician, and for many years you may proceed widely, beloved of the Muses.*" Leo the Mathematician (790-869) – also known as the Geometrician – was a Byzantine philosopher and mathematician. In the 9th century, he was responsible for producing three codices containing the works of Archimedes in Constantinople: codex **A**, codex **𝔅** and codex **C**. The Valla's codex is recognized to be codex A. The vicissitudes of these three codices will be discussed in a further article.

[7] By the wise Eutocius gentle work, that once that famous man wrote, strongly blaming the envious.

without any help in a very poor and extremely negligent way. At the end of the codex neither

IX the word τὲλος [end] nor any other sign, showing that the code is finished, has been added; the last page is empty and is used as a cover. Cfr. *Bandinii catalogus II p. 14.*

I demonstrated that the Parisian codices 2360 (B) and 2361 (C) were copied from the ancient codex of Valla (Quest. Arch. p. 124 sq.) and, in this point, I will briefly recall the reasons.

In the Parisian codex B at the edge of page 120 the copyist has noted this: *These things are copied from that illustrious very ancient antigraph, former property of Giorgio Valla and that later became of the renowned sovereign Alberto Pio of Carpi*[8]*; this antigraph was, as we have said, very ancient and it had a great lack of clarity and not quantifiable because of the errors; moreover innumerable excerpts are not explained in any place. [...]*[9]

So, the B codex was copied from Valla's one. Also, in codex C we found this short preamble by George d'Armagnac[10]: "Do not be offended, diligent reader, to see this Author

X without any recommendation for himself or any preface: so the condition of the first page in the old exemplar, from which it was copied, had been worn and consumed by old age, that not even the name of Archimedes could be recognized, nor at that time was there anything else left in Rome by which this πρόσωπον [front page] could be restored. Every sign of both accent and spirit was missing everywhere; in the remaining parts it is intact and clear except for the second page of the last sheet of the book on measurements ἥρωνος [by Heron], was entirely deleted such as the name of Archimedes. However, in order for Gallia to rejoice at such a recommendation from the Author, I preferred in any way that a copy of it be produced for you [reader] at my expense, albeit that to mathematical lovers my guilt appears more negligent in this part." And at the edge of the codex, the copyist wrote this note: *Here the German Christopherus Auverus ended the treatise on the first day of 1544, at the expense of the very respected bishop of Rodez George d'Armagnac who was ambassador to [the Pope] Paul III in the Holy Church diocese in Rome for Francis King of the Celts, who was highly praised and held in the highest esteem.* From this Greek text it is clear that the codex C of Rome was copied in 1544 by Christoferus Auverus

[8] Alberto Pio, Prince of Carpi (1475-1531) was an Italian Renaissance prince and had a prominent role in the humanistic panorama. His education was directed by the humanist Giovanni Pico della Mirandola and the famous founder of Aldine Press in Venice, Aldo Manuzio. He played a leading role in the politics of relations between the Papacy and the Kingdom of France.

[9] In the original, this text is in Greek. Examples of errors and misunderstandings that the copyist highlights follow and are omitted here.

[10] Georges d'Armagnac (1501-1585), French Catholic cardinal and archbishop, was bishop of Rodez in 1530 and then was nominated ambassador to Venice (1536) and Rome (1540) and finally state councillor to Francis I. He was an important Maecenas for men of letters who put in communication with Francis I.

at the expense of George d'Armagnac. Already Guillaume Philandrier[11], who was, according to the letters, with George d'Armagnac and followed him both in Venice in 1541 and then in Rome, has in Vitruvius' edition (Lugdunum [Lion] 1552 and reprinted in 1586) these things on p. 357: "I had written these things, since, with the benevolence of Cardinal Pius Rodolphus of Carpi[12], I was given the ease of visiting and writing, while my Maecenas [i.e. George d'Armagnac] edited the volume on the sphere and the cylinder with the commentary of Eutocius, as an ornament in the future of the very august and well-stocked Library that You [Francis the First] established at the Fontainebleau. This volume had belonged to Georgius Valla and in it, because of the property of Doric language and the omission of spirits and accents, which would have produced any difficulty in reading, there are then notes of syllables and expressions, which are often not even recognized by the Greeks". So, there is hardly

any doubt that George d'Armagnac edited that codex of Valla, of which he made a copy for Philandrier, transcribing it to donate this antigraph to the Fontainebleau library. Consequently, it follows that codex C was also copied from Valla's codex, which is also confirmed by itself, since he arrived in the Paris library from the Fontainebleau one. XI

The Parisian paper codex 2360, once Medicean, contains the same works of Archimedes and the comments of Eutocius in the same order. After the book *Quadrature of the Parabola* presents the same verses. Some argue that it was written by Philandrier and hence attribute it to the 16th century; but this cannot be at all related to Philandrier's words attached above; in fact he says that he transcribed only from Valla's codex, i.e. he took note from it of those things that could be useful to him in the commentary to Vitruvius, and that if he had not copied everything, he would have remembered only the books *On the Sphere and Cylinder*; it seems that he only consulted these books and did not examine the others. Moreover, it appears from that inscription of the bookshelf, that the archetype codex, when codex B was copied from it, belonged to Albertus Pius; in fact, if already at that time it had been given to Rodolphus Pius, the name of him, not of Albertus, would have placed or certainly would have mentioned itself, too. For this reason, since Giorgius Valla died in 1499, and Albertus Pius in 1531, it must be concluded that codex B was written between these years.

Paris codex 2361 (C) is made of Fontainebleau paper and has the same works of Archimedes and Eutocius in the same order, as well as Heron's measurements, such as the Florentine codex. From this codex, F. Hultsch[13] has published the Heron's ones: *Heron, reliq. p. 188–207*. After the Heron's measurements, furthermore, two

[11] Guillaume Philandrier (1505-1563) was a French humanist and a friend of François Rabelais. He was also a cleric of Rodez. He became secretary of Bishop Georges d'Armagnac, whom he accompanied to Venice and Rome. He was involved in humanistic studies especially concerning Quintilian and Vitruvius.

[12] Rodolfo Pio di Carpi (1500-1564), cardinal and Italian archbishop, was Alberto's nephew.

[13] Friedrich Otto Hultsch (1833-1906) was a German classical philologist and a historian of ancient mathematics. He studied and then lectured in Leipzig.

fragments follow περὶ σταθμῶν[1]) [on weights] and περὶ μέτρων [on measures], which he published from this codex (*Metrolog. Script. 83–84. I p. 267–272; cfr. praef. p. XVII*). All these same things are also found in the same order in the Florentine codex, except that the last fragment περὶ μέτρων which is a little longer in this last one.

1) In Hultsch it is reported περὶ ταλάντων, but in the codex there is περὶ σταθμων, according to Charles Graux, and in the same way it is reported in the Florentine codex.

XII Even so, it is likely that the Florentine codex and codices B and C are related, and this is proved by other very reliable documents. First of all, we have seen that the codex of Valla was mutilated at the beginning, and for this reason the codex C is lacking of the greater part of the cover letter preceding the first book *On the Sphere and Cylinder*. In the Florentine codex too, the first page is in a different handwriting and with many lacunas, and in B the same passage is equally corrupted (cfr. vol. I p. 3 not.); the copyist added: Ἀρχιμήδους τοῦ περὶ σφαίρας καὶ κυλίνδρου τὸ προοίμιον λείπει· ἡ πρώτη γὰρ σελὶς τοῦ ἀντιγράφου ’φανῆς ἦν, ὡς ὁρᾷς[14]. Furthermore, in vol. III p. 4, 18 there is a lacuna between σύγκειται and τῇ ΑΒΓΔ in all codices; the copyist of codex B wrote: ἓν ὅλον σελίδιον ἢ καὶ δύο λείπει [at least one page or even two are missing]. Finally, the numbers of the propositions of the book περὶ κωνοειδέων [on conoids] are completely confused and reversed in the same way in all codices (cfr. Quaest. Arch. p. 123 sq.). Therefore, since the Florentine codex cannot have been copied by the Parisian codices, it remains established that Florentine codex is either the codex of Valla or derives from it. And I will explain why now I think this is true.

From the aforementioned note by George d'Armagnac taken from codex C, it becomes clear that the first and the last page of the codex of Valla indeed included written text but it could not be read because of their age and incomprehensibility. But in the Florentine codex, the first page is completely empty at the beginning, and eventually the omitted part of the cover letter has been added by another hand[1]), and there is neither any trace of an older script that has been obliterated, nor has anything ever been written in the lacunas. And the last part of Heron's fragments has been written brilliantly and plainly, like the rest of the codex, and after that nothing was written. I have said above that the last fragment in the Florentine codex is a little

1) Nicolaus Anziani, librarian of the Laurentian library, believes that the first page was written by the same hand as the rest of the codex. But the handwriting seems to me different, albeit not so much, and H. Menge and Charles Graux, who ascribe this part to XVI century, agree with me. For sure this fact is of interest, because in this part almost all accents and spiritus [rough or smooth breathing] are lacking.

XIII longer; therefore, the copyist could read [the text of] this codex a bit larger in the archetype than what the copyist of codex C could either read about fifty years later or believed to increase the value of the work. The fact that he was unable to read the archetype of the Florentine codex in its final part leads us to suppose that in

---

[14] *The preface of the Archimedes' book "On the Sphere and Cylinder" is incomplete: in fact, the first page of the antigraph was lacking, as you can see.*

the Florentine codex nor [the word] τέλος [end] nor any other sign with which the copyists have the habit to mark the completion of the work is found.

Moreover, that cover letter contains a disagreement with codex B, which cannot have been copied in this part from codex F; see firstly *vol. I p. 4, 6*: [...][15]; and in all these passages the writing of codex F is so clear that there is no doubt about this matter. Therefore, it had to be stated that that lacuna in codex F has finally been filled out later, when codex B was copied from it. This can be a casual independent event; but it is, nevertheless, verisimilar that the copyist of codex B, who in so many words deviates from the sources from which he is transcribing, then wrote that he had taken the preface from another copy, especially because he wrote this preface: Ἀρχιμήδους τοῦ περὶ σφαίρας καὶ κυλίνδρου τὸ προοίμιον λέιπει, ὡς ὁρᾷις [the preface of the book *On the Sphere and Cylinder* by Archimedes is incomplete [1]), as you can see]. I do not believe that these words could be taken by any other manuscript than that antigraph, which he has followed throughout the book.

Finally, certain passages among those that Valla translated from his codex into Latin are found, where Valla obviously did not have a corrupted writing of the Florentine codex but a good one of the codices B and C before his eyes:

[...][16]

1) I.e. it is incomplete (not: it is lacking), as it appears from the added words ὡς ὁρᾷις; in fact, the preface is written by the same hand as the rest. Cfr. Quaest. Archim. p. 121 not.

Having Valla himself barely reconstructed these passages by conjecture, what will XIV
be understood by those who have enlightened the errors repeated by him, from evidences in Valla and from the union of the B C codices, it follows that in these passages the text in Valla's codex was different and also more correct, as it is in the Florentine codex (*de III p. 76, 26; 124, 22 u. infra*)[17].

Even from those abbreviations, which are said to be obvious in Valla's codex by the copyist of codex B, a very strong evidence can be obtained, from which one can prove what we have proposed. There, in fact, (p. IX) instead of the syllable -οις this abbreviation ⁊ is indicated; but in the Florentine codex this [abbreviation] of this form is not found anywhere, instead a vertical and round ∂, which is a more recent form of this abbreviation (*O. Lehmann: Die Tachygraphischen Abkürzungen der Griechischen Handschriften (1880) p. 70–71*).

And although generally this is the nature of those passages, in which B C codices offer better writing than F code, so that it cannot be said that the copyists of B C codices could not have corrected a manifest and easy error on their own initiative in the same way, however, not only those better versions, which are frequent enough, can be found quite easily, if we establish that B C codices were derived from the same source as F codex, but there are also certain passages where the correction was more difficult and not such that two not too much cultured copyists seemed to have come across it by chance; such as in codex F at *I p. 6, 11* there is τοτε αξιωμα, but

[15] It follows a list of discrepancies between the two codices.

[16] It follows a list of comparisons between codices B C and F.

[17] The Translator's Note in footnote 29 on page 84 refers to the previous sentence.

codices B and C report the right form (τά τε ἀξιώματα); at *I p. 8, 11*: in codices B and C it can be read τομέα δὲ στερεὸν καλῶ, instead in codex F it appears that in a first stage there was only τομε; later, the same hand, which added the cover letter to the first page, filled the lacuna except that στερεόν is omitted leaving a small lacuna.

Finally, we must remember (in fact, this way of demonstrating especially in this codex is misleading and uncertain) that the form itself of the letters sometimes indicates a more recent origin. With regard to this, I, first of all, rely upon the judgment of Charles Graux[18], who, having diligently examined a photographic image, from which the table added to volume II was printed, of our codex so judged: *what seems to me the most probable thing at*

XV *the moment is that the [codex] Laurentino in question is the product of a 15th century copyist who had a ninth or tenth century manuscript as a model and tried to imitate it scrupulously even in the form of the letters*[19]. So, I found elsewhere the same statement that I also had deduced myself having progressed in other ways. But on the contrary, Guilelmus Gardthausen[20], to whom I had passed on the same image, did not think that this statement was to be rejected, although he himself preferred to attribute the codex to the 11th century from the form itself of the letters, unless documentary evidences could be obtained from another place to confirm that opinion; which I seem to have done here. The most important impediment, as both of those expert palaeographers pointed out, is the form of the letter φ, as it appears to have been written in a single line (see table, lines 1, 3, 4, 5, 6, etc.), which, apart from the ɸ form, does not appear before the 15th century (Gardthausen: Griech. Paleogr. p. 208)[21]. Moreover, this fact is also favorable to this opinion, namely that the writing, which at the beginning of the codex is very diligent and very sharp, becomes rather negligent towards the end and offers a less ancient appearance, and the accents, which are generally omitted, here and there, and indeed especially towards the end, are more frequent, just as the custom of the copyist seems to have meanwhile overcome the will to transmit the antigraph reliably. Finally, also the parchments reveal both by typology and by species a more recent time.

After examining all these things, we must admit that the Florentine codex is not the Valla's codex, but the latter is the common source of F B C codes. Then about our F codex itself we must acknowledge that Angelo Poliziano[22] wrote in Venice in 1491 to Lorenzo de' Medici (u. *Fabronius: vita Laurentii II p. 285*): *in Venice, I found some books by mathematicians Archimedes and Heron, which we lack ... and other*

---

[18] Charles Graux (1852-1882) was a French scholar in the field of classical and humanistic disciplines. He published important critical editions of works by Xenophon and Plutarch and pioneering, descriptive catalogs of the medieval copies of ancient Greek texts preserved in the libraries of Spain and Denmark. *Source Wikipaedia.*

[19] This text is in French, in the original Heiberg's essay.

[20] In the opinion of the translators, here Heiberg refers to Victor Emil Gardthausen (1843-1925), ancient historian, palaeographer, librarian.

[21] This piece on the form of the letter $\phi$ shows that Valla's codex should not be the Florentine codex.

[22] Angelo Ambrogini, known as Poliziano (1454-1494), was an Italian poet, humanist and philologist. He is generally considered the greatest of the Italian poets of the 15th century. He was also the author of works in Latin, Greek, and achieved extensive philological expertise.

*good things. So much so that Papa Janni*[23] *has something to write about for a piece*[24]. Since Giorgio Valla from 1486 to 1499 was teaching in Venice (Neue Jahrb. Suppl. XII p. 377), and Valla's codex, as we have recognized above, besides Archimedes also contained some fragments of Heron, we can hardly doubt that Angelo Poliziano

dealt with describing Valla's codex. But this means that Joannes Rhosos[1]) would XVI
have copied Archimedes' codex, copy that was not actually done. In fact, the codex F could not have been written by him, as it seems clear having compared some other codices[2]) written by him, which are very numerous, with our photographic table.

So, the F code was copied from Valla's own codex in 1491 or shortly after, while the B C codices were copied not much earlier. It therefore remains to be understood which of these three quasi-sister versions should be believed to be more correct than the others.

Then, I have demonstrated with documentary evidences, which I believe are quite robust in (*Quaest. Arch. p. 128–30*), that the B code has been copied by an educated copyist, who would have corrected many errors, many others would have tentatively corrected rather badly, and many errors are occurring on almost every page [...][25].

But we can clearly demonstrate that the F codex has been copied more accurately than the others. There are in fact certain passages in which we know by the interpretation of Valla himself[26] that in his codex there were the same foolish errors that were in the codex F, but that they were corrected in BC:[27]

1) In fact, he is "Papa Janni", as N. Anziani advised me. About Joannes Rhosos Cretean presbyter, very active copyist, it can be read Gardthausen p. 326 sq.

2) In Florence, I compared Laurentian codices XXXII, 6; LV, 9; LXXXI, 23; LXXXVI, 18, with the F codex itself.

[...][28] XVII

From these passages, one can understand how great and how pedantic the diligence of the copyist of codex F was; in fact, although this is hardly credible, it is clear that he could not correct these silly mistakes, as instead the copyists of B C codices did, because he wanted to return the archetype with the utmost fidelity. Therefore, it is clear that even where B C agree against the authority of the F codex, we must judge very cautiously about the scripture of the archetype from the latter, and that the Florentine codex also now must be considered the main source.

---

[23] Papa Janni, a Greek copyist, following the opinion of T. C. Dandolo, *"Firenze sino alla caduta della Repubblica: studii", Milano, Ubicini 1843 (p. 381).* In the following note, Heiberg identifies Papa Janni as John Rhosos.

[24] From Poliziano's Italian, in the original Heiberg's essay.

[25] Here Heiberg gives examples of such errors.

[26] Giorgio Valla, selected pieces from *"De expetendis et fugiendis rebus opus", Venice 1501.*

[27] In footnote 4), Heiberg cites his work where this point is fully discussed. Together with Henri Lebègue, he finds many errors in the Giuseppe Torelli's edition of the Archimedes' works *Archimedis quae supersunt omnia, cum Eutocii Ascalonitae commentariis, ex recensione Josephii Torelli Veronensi cum nova versione latina (accedunt versiones variantes ex codd. Mediceo et Parisiensibus).*

[28] Here Heiberg reports various examples of errors.

In short, the interpretation of Valla[29] coincides with F completely except in a few passages that I reported above on p. XIII. But also from other hints, we can conclude how much the codex F expresses an accurate image of the archetype. In fact, first of all most of the letters show a form far older than the 15th century, as can be understood from our photographic table, which represents a page taken from the first part of the codex (*I p. 156, 10–160, 11 editionis*). And the usage of abbreviations, the omission of accents and spiritus, their square form where they are present,

XVIII all these things suggest that the codex is ancient and [these circumstances] agree with [the hypothesis that] the transcription of Valla's codex was done by the copyist of the codex B in a very accurate way. Then, the mutilated part of the cover letter placed before the first book on the sphere and on the cylinder (I p. 2–6, 6), which was consumed in the first page of the archetype for usage and age, also in codex F is placed on the first page neither more or less [...][30].

Then, the copyist of the codex F with the same diligence, with which he imitated the form of the letters, he seems to have followed also the features of the antigraph, exactly as he made so that there was a correspondence page by page and even line by line.[1])

Thus, it may be possible that with the richness of these three codices, first of all the Florentine one, we recompose a certain [faithful] image of that archetype

1) See also what Jordan Hermes wrote about the Marciano codex 247 copied with similar diligence from the Marciano codex 246, XIV p. 264 sq.

XIX once owned by Giorgio Valla. Undoubtedly it [Valla's codex] was written in the 9th or 10th century, as Charles Graux[1]) conjectures from the clues given by the ancient form of the letters observed in F, moreover it was very similar to the Oxonian[31] codex of Euclid (*Bodleian. d'Orville ms. X, 1. Inf. 2. 30*; examples of it have been published in Paleographical Society tab. 65–66) with all the habits then in use in the summaries. It was copied quite diligently from an exemplar [written by] some man who was not inexperienced in mathematics; and in fact, those scholia found in almost all books are not added in the margin, [and are] especially very often present in in the books *On the Sphere and Cylinder*[2]). Those things that show a greater knowledge of mathematics than those that were known at the time could be improved by this copyist himself of the codex. It [Valla's codex] was provided with mostly excellently and diligently copied figures, but it was often wrong in the letters affixed to the figures and also in the figures and words themselves of Archimedes, which fact the copyist of the codex B (p. IX) noticed; examples from the codex F can be collected *Quaest. Archim. p. 125 sq.*, and several can be added. [...][32]

---

[29] See the sentence preceeding footnote 17 on page 81 .

[30] From here on various examples of comparisons of F codex and B C codices, where, in particular, the discrepancies with Torelli's edition are reported. This technical analysis is necessary for supporting Heiberg's conclusions.

[31] From Oxford.

[32] A list of this kind of errors then is included in Heiberg's text to support the last statement. In particular, Heiberg remarks that the text of demonstrations is corrupted by repetition and homoteleutic words.

1) In this regard the Valla's codex is called παλαιότατος [the most ancient] by the copyist of the codex B (u. supra p. IX), what he could not state about the Florentine codex itself.

2) However, it seems to me at least likely that these additions do not all come from the same man. I consider more ancient the additions that, being most of them in Doric, are inserted in the books on conics and spirals; the books on the sphere and the cylinder and on the measure of the circle seem to have been translated into a more standard [Greek] language and at the same time polluted by numerous additions only in a more recent period.

[...][33] XX

This codex therefore, once Giorgio Valla (1499) died, passed to Alberto Pio di Carpi who also acquired the other codices of Valla and probably his entire library, as it is clear from the inscription of a certain Scorialensis[34] codex (Miller: Catalogue de mss. grecs Escur. p. 454): *Donato*

*Bonturello*[35] *copied from the antigraph, which, previously owned by Giorgio Valla* XXI
*(and in fact he copied by his own hand), later became of the illustrious sovereign Alberto Pio of Carpi* [the Greek text continues in the narration of the passages of the Valla's codex]. About the destiny of the library of Alberto [Pio di Carpi], Stefano Borgia (*Anecdot. litterar. Romae 1773 ff. I p. 81*) passes the following information on. Alberto himself gave it as a gift to Agostino Steuco Eugubino, whose brother Fabio gave part of it to Cardinal Marcello Cervinio. It came from him by testament to Cardinal Guglielmo Sirleto, and since he died it was bought by Cardinal Ascanio Colonna. Then, through the hands of many, it finally was acquired by the Vatican. But it appears clearly that our codex had a different fate[1]); we see in fact that in 1544 it was owned by Rodolfo Pio, son of Alberto's brother, and in the catalogue of Sirleto's library (*Miller p. 323–324*) there is no mention of Archimedes' codex. Thus, perhaps because of its singular antiquity, it was kept in the Pious family. But we do not know where it would later arrive, nor one knows whether it was lost or happened to be in some library in Italy, which seems to me quite likely.

By now we see what relationship there may be between the other critical works and the Valla's and Florentine codices.

First, therefore, it appears that Pope Niccolò V had a codex of Archimedes, which he had taken care to have translated into Latin. In fact, Cardinal Nicola Cusano writes to him as follows (Opera p. 1004): "*You have in fact given me in these previous days the Geometry of the great Archimedes presented to you in Greek and thanks to your*

[33] Before this text here, the page continues with a list of errors and writing conventions including: the repetition of the same words, the exchange of apparently similar abbreviations, the omission of abbreviations, the very frequent permutation of vowels and syllables, the frequent confusing replacement of the connective particles δέ [but] and δή [therefore] in mathematical demonstrations, the presence of some abbreviations omitting terminal letters, the frequent short notations for double consonants. Finally, Heiberg remarks that the copyist of Valla's codex pertinaciously repeats the same errors many times in all his manuscripts.

[34] Scorialensis – The library of the Royal site of San Lorenzo de El Escorial.

[35] Donato Bernardino (Bonturello), 1483-1543, lived and worked at the court of the Pious in Carpi. He was preceptor of Rodolfo Pio and transcribed various Greek codices.

*care translated into Latin*". Who was Nicola's translator, we find out from the preface of the Basel edition (fol. 2 verso)[36]:

1) Also, Ambr. Morandus in its life of Steuchio (Steuchii opera. Venet. 1591, I praef. fol. 4 verso) does not tell that Alberto had given all the books to Steuchio. In fact, his words are: Fabio brother of Steuco, who had received it as a gift from Alberto Pio prince of Carpi, gave this magnificent library in great part to Marcello Cervino.

XXII "*He (Regiomontanus*[37]*) freely accepting his first invitation in Italy as soon as he reached a great fame of his name, saw many Greek books stolen during the defeat of Constantinople and copied out many by his own hands . Then, among others, [Regiomontanus] diligently copied Archimedes' books, which had been given to him by some friends, on the sphere and the cylinder, on the measure of the circle and on other things not only useful but also necessary for mankind, as it is clear from reading these books, that Iacopo Cremonese*[38]*, because the demand of Niccolò V Roman Pontiff, had already since long translated into Latin. [Iacopo Cremonese was] a man at those times worthy of double honor, being learned in Greek, since he seemed [to Pope Niccolò V] the only one who could absolve this work helped by [his] practice of languages. [Regiomontanus placed] many additions to the margins in Greek (since he also had a copy of the codices in Greek), where Cremonese's translation seemed him either to have been translated rather difficultly or to have been somehow corrupted in the copy*"[39]. This translation by Iacopo Cremonese was recovered in the Basel edition, as its same title alludes to: "*Works ... by Archimedes, which undoubtedly are better than all the others, once already donated to Latinity and now published for the first time*". For this reason, it is therefore legitimate to speculate about the codex of Pope Niccolò V. Right at once it seems clear that it was closely linked to Valla's codex; in fact that same lacuna vol. III p. 4, 18, which originated in that place as attested by the copyist of the codex B (see *supra p. XII*), since one or more sheets of Valla's codex had been lost (nor could it happen that that copyist hesitated in understanding such an easy thing, since the latter himself had in his hands Valla's codex), this lacuna, I said, was at this point in Niccolò V's codex. In fact, in Iacopo Cremonese's translation on p. 2 it is written: "*unam autem lineam in plano quocunque modo connexam quamvis sive ex rectis pluribus connectatur [sive ex curvis sive ex rectis pluribus connectatur ex ea connexione postulat appellari]*"[40]

---

[36] Heiberg refers to the *editio princeps* published at Basel in 1544 by Thomas Gechauff Venatorius.

[37] Regiomontanus, pseudonym of Johannes Müller of Königsberg (1436-1476), was a German mathematician, astronomer and astrologer.

[38] Iacopo da San Cassiano, also known as Iacobus Cremonensis, (between 1395 and 1413-1453/1454), was an Italian humanist and mathematician. He translated into Latin the corpus of the writings of Archimedes.

[39] This preface excerpt seems written by using a Latin style which is anticipating Baroquism, so that we have split into three parts a single sentence which occupied in the original text sixteen lines.

[40] Here Heiberg reports a truncated and reconstructed sentence in Cremonese's translation at the end of which Cremonese writes "*Here a page in the Greek exemplar is lacking.*". A possible translation of the reconstructed sentence is "*whereas a single line in the plane connected in any way either by numerous lines [or by curves or by numerous lines is connected, he postulates that it is named after that connection].*"

From here it necessarily follows that Niccolò's codex either was Valla's codex itself, a fact which had been suspected in *Quaest.*

*Arch. p. 139–140*, or had been copied from that.[1]) This last statement now seems XXIII
to me verisimilar, since those words which in the previous text[41] were closed in parentheses, with which I do not know who tried to fill the lacuna, had preserved traces of the Greek text origin; in fact, "*eam ex ea connexione*" seems to be a word by word translation of the Greek words τὴν ἐκ τῆς συνάψεως. And for this reason, the copyist of Pope Niccolò's codex added these words to fill the lacuna in Valla's codex up to a certain point.

A copy of Iacopo Cremonese's translation was copied by Regiomontanus with his hand, this circumstance is referred by Venatorius[42] (u. supra p. XXII) and by Regiomontanus himself in presence of Gassendi (Opera V p. 469: the translation is by Iacopo Cremonese, but corrected in some passages), and this copy is still in the library of the city of Nuremberg. Heinrich Menge was the first to warn about this fact (Neue Jahrb. f. Philologie 1880 p. 110). Later, myself examined this codex, having the Nuremberg Senate allowed without any restraint that it was sent to me to Copenhagen. The codex is made of paper centur. V, 15, and contains Archimedes' books and Eutocius' comments in this order: Archimedes' books *On the Sphere and Cylinder* I-II, *Measurement of a circle*, *On Conoids and Spheorids*, *On Spirals*, *On the equilibrium of planes* I-II, *Quadrature of the Parabola*, the *Sand-reckoner*, Eutocius' comments on the books *On the Sphere and Cylinder*, *Measurement of a circle*, *On the equilibrium of planes*. Greek words were noted on the margins in many places, just as we noticed that Venatorius had said (supra p. XXII). On the first page we read "*I belong to Thomas Venatorius*" and on the last page it is written by Venatorius "*Ioannes de Monte Regio was born in the year 1436 on the 6th of June at 4 and 40 minutes ... in the afternoon. Moreover, Regiomontanus dies in the year 1476 almost on the 8th of July*". In addition to Greek words Regiomontanus

1) And the translation is clearly, in the discrepancy of the writing, in accordance with the Florentine Codex. When this accordance does not happen, the circumstance is to be attributed either to the copyist of Niccolò's codex, who, as we shall see, does not abstain from interpolation, or to Regiomontanus. As an example of the accordance is the fact that the dittography of the codex F III p. 172, 17, which B C had corrected, was also before Cremonese's eyes p. 36, 7.

---

[41] See the sentence in Latin on page XXII.

[42] Thomas Gechauff, also known as Venatorius (1488-1551) is a Protestant geometer and theologian. Friend of the imperial adviser Pirckheimer, he edited the works of Archimedes.

XXIV not infrequently added on the margin corrections of Latin words, which were all accepted by Venatorius, which from this same codex generated the Latin translation added to its edition (Basel 1544).

From those passages that Regiomontanus annotated from his Greek codex, I suspect, as I proposed in *Quaest. Archim. p. 138*, it is strongly confirmed that Regiomontanus, friend of Bessarione[43], used our Venetian codex, which was once Bessarione's (u. infra). In fact, since it is not possible to conclude anything about Bessarione's codex from the many passages extracted from it by Regiomontanus, since the words used there are the same of all or many of our codices, in no passage, however, the errors proper to this class of codices, of which the Venetian codex is the main one, are also recurring in Regiomontanus' edition. And I will report these errors:

[...][44]

And sometimes Regiomontanus alludes with [the following] clear words to the exemplars of Bessarione, as he annotated with regard to the book On the equilibrium of planes II, 8: "*so it is written in the exemplar of the Cardinal [Bessarione] and it is likely that it was translated from Greek. But it was done badly*"; with regard to the book On the equilibrium of planes I, 15 [Regiomontanus writes]: "*it is bad. See the exemplar in both Latin and Greek by "dominus Nicenus". See also the ancient exemplar that is with Magister Paulus*" 1)). In fact,

1) In these passages, my collection of sources has been partly confirmed and partly corrected with benevolence by Frommannus, prefect of the Germanic Museum of Nuremberg, as I solicited him.

XXV the "dominus Nicenus" is Bessarione, who appears to have obtained the archbishopric of Nicea in 1436. His exemplar, of which Regiomontanus speaks, is in Latin and is preserved in Venice; it is the Latin codex CCCXXVII of the 15th century ("*once of Bessarione*", cf. *Marci bibliotheca codd. Mss. paraeside L. Theupolo. Venet. 1741 p. 140*), which contains Archimedes' books *On the Sphere and Cylinder* I-II with the presentation of Eutocius, *Measurement of a circle* with the presentation of Eutocius, *On Conoids and Spheroids*, *On Spirals*, *On the equilibrium of planes* with Eutocius, *Quadrature of the Parabola*, the *Sand-reckoner*. Although the name of the translator is omitted in said codex, it cannot be doubted that it is the apograph of the translation of Iacopo Cremonese; in fact, an order of books like this is found only in the translation of Iacopo Cremonese; in all other sources the booklet *The Sand-reckoner* is placed before the book *Quadrature of the Parabola*. One might therefore have believed that Regiomontanus had copied this same codex; in reality it seems that he used another copy, of which he then brought together the apograph with Bessarione's Latin codex. In fact, in the book *Quadrature of the Parabola* 14 words "*sicut autem ba ad bf, ita mensula de ad spacium q. spacium igitur q spacio r maius est. Nam hoc ostensum est*", which are missing in the Greek codices, in the Basel translation on p. 149 are instead enclosed in brackets; Regiomontanus also enclosed

[43] Bessarion, born perhaps Basil (1403-1472), was a Byzantine cardinal, humanist, and philosopher.

[44] Here Heiberg reports the comparison between the texts of the Regiomontanus edition, of the Venetian codex and of the Torelliana edition.

them in brackets and then in the margin wrote "*there is a void*"[1]) and just below "*in the exemplar of the dominus [Nicenus] there was an addition*", something that was undoubtedly taken from the Latin codex, not from the Greek one of Bessarione, as well as "*exemplar of Cardinal B*" in the above note (in *plan. aequil. II, 8*). The same thing can be deduced from the note in pl. aeq. I, 15. But above, on p. XXII, we see that Regiomontanus, according to the testimony of Venatorius, had a copy of several Greek codices, and this is confirmed by that annotation we have recalled on p. XXIV. A subsequent part of it [he reports]: "*see also an ancient exemplar belonging to magister Paulus*" is certainly written by Regiomontanus' hand, but with a different kind of ink

1) I take this from the Greek codices.

and very clearly added quite a bit later. With the same ink, with which this addition XXVI
was made, other numbers of the propositions, beginning with proposition 18, were added to the book on spiral lines. Here in the margin Regiomontanus annotated: "*those notes of the propositions from the new*[1]*) Greek exemplar*". [...][45] Since all these [substitutions in the numbering] completely match the numbers of the Florentine codex (with regard to the other codices, here we know nothing for sure; in the published ones the series of numbers has been corrected, except in the Cremonese translation [where] the number 19 is omitted), as will be clarified by our annotations, it is legitimate to suspect that that old exemplar of Magister Paulus was the same codex of Valla. And if this conjecture is true, from here we obtain a new evidence about the destiny of Valla's codex, that is, that before it came into Valla's possession, it belonged to "magister Paulus". He can hardly be any other than Paulus (Albertini) Veneto, born around 1430, died in 1475, a monk not unknown at that time, who also in a commemorative medal is called "M.", that is magister (Tiraboschi: Storia della letterat. Ital. VI p. 288 sq.).

Now we go to the codices in worse conditions. These codices are:

Codex Veneto Marciano CCCV on parchment, XV century (V) containing the same books of the Florentine codex and in the same order; after Archimedes and Eutocius it follows the same fragment of Heron (see Morellius: Biblioth. Manuscr. I p. 186). On the first page we read κτῆμα βησσαρίωνος καρδηναλέως [property of Cardinal Bessarione], and Bessarione corrected the most serious errors here and there, but numerous corrections seem to have been made by the copyist himself. Near the scholio περὶ ἑλίκ. 10 Bessarione added: σῆ. τοῦτο σχόλιόν

1) This means without doubt "from the Greek codex that I examined later"; in fact, it cannot be referred to the age of the codex

ἐστι εἰς τό ι θ(εώρημα) ὡραῖον πάνυ. In Eratosthenes' epigram the same verse XXVII
is different, above he wrote στίχοι ἡρωελεγεῖοι, then in volume III p. 114, 3 he corrected συνημῶν in συνηβῶν. The figures were perhaps added later by Bessarione himself.

The Parisian codex 2359, made of paper, once Medici, dates back to the 16th century (A); it contains the same works of Archimedes and Eutocius as contained

[45] In the following sentences Heiberg lists the propositions with the new numbering.

in the Venetian codex. This codex was written by the hand of two copyists, one of whom, starting from sheet 33, according to the testimony of Charles Graux is Nicolaus Murmuris, who around 1541-42 copied many codices in Venice.

The Parisian codex 2362, made of paper, [from the Library of] Fontainebleau, dating back to the 16th century (D), contains the same texts as codex A.

If we first rightly established that the Florentine codex was written precisely in 1491, it cannot happen that the Venetian codex, as I had previously believed, was copied from it, since Bessarione had already died in 1472 and the Venetian codex had arrived with all his Library in the Marciana Library in 1468. So, one must believe that the enormous similarity between these codices was forcefully caused by the fact that both codices are derived from the same archetype, that ancient codex by Giorgio Valla, so that from here it would be confirmed the reliability of the copyist of codex F in returning the antigraph not in a mediocre way. [...][46]

XXVIII [...][47]

So at least it can be questioned whether the F codex also in these parts of the text translated the archetype more reliably than all the other codices, although it may happen that the writing of Valla's codex was certainly authentic. Yet there are circumstances that confirm that the Venetian codex was not copied from the F codex itself. First of all it must be remembered that the fragment of Heron or rather of Epiphanus, added to it also in the Venetian codex as in the Parisian codex C (p. XI), is slightly shorter than in codex F (but nevertheless four words longer than in codex C); for this reason the copyist of codex V was not even able to elaborate an explanation to remove the difficulties in this part of the deteriorated archetype as much as the one who copied codex F. In fact, if the Venetian codex had been copied from codex F, it is not clear why the last part, equally easy to read as the others, was not transposed into the Venetian codex. [...][48]

XXIX So, we must conclude that the V codex does not derive from the codex F, but from its antigraph.

I already proved in *Quaest. Archim. p. 133* that the Parisian codices A and D, due to their many common lacunae, derive from code V. It remains questionable whether codex D was copied directly from codex V, which fact I now consider plausible, or from codex A; see *H. Menge: Neue Jahrbücher 1880 p. 111–112; Quaest. Archim. p. 137*. A diligent comparison of the V, A and D codices will settle this matter, if this question is worth of work. The following fact can certainly be considered sure, namely that codex A is the apograph of codex V, and it is of absolutely no value that the Parisian codex D was copied from codex V or from codex A.

In what follows it is presented the discussion about the codices of Tartaglia.

---

46 Here Heiberg reports some examples of the versions given by the Venetian codex and the F codex.

47 The examples started on the previous page continue here.

48 Here Heiberg reports a series of examples in order to show that the V codex does not derive from the F codex but directly from the antigraph.

Niccolò Tartaglia, from Brescia, a very famous mathematician († 1557) published in Venetia in the year 1543 the Latin translation of many works of Archimedes (*On the centers of gravity* or *aequerepentibus* I-II, *Quadrature of the Parabola*, *Measurement of a circle* and the *De insidentibus aquae* I[49]), in whose preface he wrote on the second page[50]: "*When by some chance certain books written in Greek of that very famous Philosopher Archimedes came into my hands, books which were broken and barely legible, I tried hard and carefully to the aim that the readable parts could be translated into our language, aim which was quite difficult. In fact, because of their age and of the negligence of those who held these books, I would like you to know that several mistakes have had to be corrected. Having seen the titles of these books and having examined all the work of this Philosopher, I knew that he had been considered by great and constant fame to be very outstanding, but now he is believed even more famous. That is why (as I have said) I looked greedily through these books, I went[1]) through them systematically and finally examined everything very carefully,*

1) [In the original text *procurri*]. Typographical error; it stands for: *percurri*. In general, Tartaglia's book is full of such errors, as can be seen from my note to the books περὶ ὀχουμένων. To amuse, I'm just going to add two examples here: fol. $2^{v}$: *valde aequerepentibus* stands for: *vel de aeq.*; fol. 19: *de centrum gravitatis vel duplationis aequerepentibus*. Below it says "oppositis" instead of "appositis".

*but when I found many distorted passages and certain inappropriate figures, which had nothing to do with it, I was almost forced to stop my plan. But I was consumed with an incredible desire to examine this work in order to purge it largely of errors, and I thought it worthy of seeing the light once I had adapted myself the figures and the words which seem to be particularly contradictory, and mainly I considered this part [of the work] to be worthy of seeing the light, part which I had made clear with words and examples as well as I could, instead the whole work, which will be done by me in a short time (I hope), will be wholly free of mistakes.*" Then, in 1565, two books *On the equilibrium of planes* from the notes of Tartaglia were published by the Venetian book seller Curzio Troiano (from the preface: *for this reason, because I [Curzio Troiano] still have with me the Archimedes' book "De insidentibus aquae" as made known by Niccolò himself and, as well as he [Tartaglia] could improve the manuscript, amended it from the mistakes of the copyist and, finally, made it vivid with his lucubrations, it seems to me to defraud all scholars of its possession until I have published all the things I still have of this ingenious man*). Both books are preceded by a separate title page in this edition and show the signature of the typographer and the same preface is preceding both books. The first book has been completely formatted as [done] by Tartaglia himself, except that the beginning and the end of the pages are not always the same because of larger figures (nevertheless, both books are bound together with four and a half pages); moreover, in the Tartaglia's edition the title is: Archimedes' book "*De insidentibus aquae*" (in the end: it is explicitly XXX

[49] In Greek it is περὶ ὀχουμένων – On floating bodies

[50] Heiberg reports this excerpt from the preface also to have the possibility of statuing that Tartaglia did not know Latin language well enough.

written *de insidentibus aquae liber*), instead for Curzio: Archimedes' first book "De insidentibus aquae" (in the end: it is explicitly written *de insidentibus aquae liber primus*). Nothing is changed in the book itself except for some further mistakes made by the typesetter.

Because Tartaglia has added book 1 "De insidentibus aquae" to the other books that he claims to have taken from the Greek codex without any reference to a new source, you could rightly believe that he found it in that old codex as well. But that is why we doubt this statement. In fact, at first, although the translation of the first book On floating bodies περὶ ὀχουμένων, if you look at it at the beginning, is similar to the translation of the rest of the books both in its whole style and in the reproduction of single words in Greek and Latin, the

XXXI Greek codex, nevertheless, seems to add something of its own; firstly, the Greek syntax, whose in the remaining books hardly a trace is discovered, is very accurately observed here, such as the genitive after a comparative instead of the ablative and similar things. 1) Moreover, it is clear that Tartaglia had in his hands no Greek codex eight years later; in fact, in the book "*Reasonings about his troubled invention*" [*Ragionamenti sopra la sua travagliata inventione*] (Venice 1551) it is written[51]: "*Since Your Lordship, o magnificent lord Count, in the last few days reasoned with me about the work of Archimedes Syracusan published myself and especially about that part, which is entitled "de insidentibus acquae", that part whose [Your Lordship] told me he was very eager to find and see the Greek original, from which that part was translated. For this reason, I understood that Your Lordship was looking for this original because of the darkness of the text, which is amplified in the aforementioned Latin translation. In order to avoid Your Lordship this effort to search for the Greek original (which perhaps you would find darker and more incorrect than the aforesaid Latin translation) I have clarified and detailed this part in this first reasoning of mine.*" Although it cannot be concluded with certainty that Tartaglia himself never had in his hands a Greek codex of this book, we cannot be surprised that this codex is so completely forgotten, after such a small period of years, that Tartaglia himself could not indicate to his patron, who seemed to ask for it very eagerly, the source of his translation. Moreover, the words: *which perhaps you would find darker and more incorrect*[52] would be rather surprising, if he himself had taken his translation from this Codex. The whole passage seems to me to be explainable quite simply if we have conjectured

1) In fact, it can be stated without any doubt that this and the other books edited by Tartaglia have been translated from Greek and not from the Arabic language, as someone might believe, since in sheet 11 we read "Archimenidis", a form of the name propagated by the Arabs through many medieval Latin translations. But the language and form of these books is such that it is to be believed that they necessarily flowed from a Greek source, as Tartaglia says.

---

[51] We completely agree with Bernardino Baldi [Cronica de matematici: overo Epitome dell' istoria delle vite loro, Urbino, Angelo Antonio Monticelli, 1707] where he states that the use of the Italian language by Tartaglia "*move a riso talhora chi legge le cose sue*" that is "*causes sometimes amusement in those who read his works*". We find Tartaglia's Italian to be rather dialectal also when compared to his contemporary authors.

[52] Italian text in the original: *qual forsi piu oscuro e incorretto lo ritrovaria.*

that Tartaglia never had, at least this part of, the Greek codex in his hands. It is XXXII
also worth mentioning that Tartaglia speaks about this book in such a way that he makes it clear that he obtains his Greek text from his own source and from others different place (*l'original greco, dove che tal parte era stata tradotta* [Italian text]). Moreover, to this evidence it is added the witnessing of Fr. Commandino, who was a man extremely educated in Greek mathematics and tirelessly searched the codices of this kind. Not so many years later (Bologna 1565), in the preface to the edition of the books *On floating bodies* περὶ ὀχουμένων on the second page, he writes the following: "*When, in fact, the Greek codex of Archimedes had not yet come to light, not only the translator who gave it to Latinity was wrong in many places, but also the codex itself has been corrupted and is lacking [parts] for the old age, as also the translator confesses.*" It is evident that these words, which I have taken from Commandino's preface above, refer to Tartaglia and to the transcribed copy of the codex used by Tartaglia, and we wonder why Commandino did not name Tartaglia. Moreover, it can be concluded from the aforementioned passage that Commandino believed that the books *On floating bodies* could be found in the same codex together with the other works. But even if this fact were true, one can conclude that Tartaglia did not know the Greek codex even from hearsay. By all these facts, I would be leaning rather towards believing that it was Tartaglia himself that translated the remaining books into Latin from that very ancient and broken Greek codex, but the first book of the "*De insidentibus aquae*", as well as the second, were offered to him already translated from Greek into Latin I do not know in which way.[1]) If this opinion is true, we must evidently suspect that that codex, which Tartaglia speaks about, was the codex of Valla, which was roughly copied at about the same time (1544) by the writer of codex C (supra p. X). In fact, it is not enough verisimilar,

1) Regarding the origin and authoritativeness of the fragment published by Angelo Mai, it is a difficult question to judge. At least it is certain that the translation of Tartaglia was not made from this form of the text, but from a much better one. I would consider that fragment to be insignificant and that the one who included propositions alone without demonstration, apart from the first one, did try to translate backwards into Greek the Latin translation of some learned man of the Middle Ages, the same thing which was attempted by Rivalto himself. Thurot's opinion seems identical.

that there were two codices of Archimedes in Italy at the same time, both of which XXXIII
were corrupted by age and partly difficult to read[1]), and both having been completely lost in our time[2]), especially because between Valla's codex and that of Tartaglia the commonality and affinity of the spelling mistakes is great. I will give some examples of this commonality and affinity:

[...][53]

1) Even though it appears that Tartaglia, in the books to be chosen to be published first, preferred the shorter one, it is verisimilar that the lacuna at the beginning of Book I on the sphere and the cylinder also dissuaded him from publishing these books, since he himself says he published those parts that could be read with the least effort.

2) Regarding the corrupted figures in Valla's codex, information that Tartaglia has transmitted about his codex (p. XXXI), even the copyist of the codex B complains about it (see supra p. IX).

53 Examples of the affinities between the codex of Valla and the codex of Tartaglia as reconstructed by Heiberg.

XXXIV [...]

Also, in Tartaglia, as well as in our codices, the propositions 1-2 of the book I *On the equilibrium of planes* read without numbers, so that the numbers of the propositions throughout the book are by two less.

In the books *Quadrature of the Parabola* and *Measurement of a circle*, Tartaglia caught another translation, although he does not even mention this fact with a single word (this fact creates the belief that he may have used a different source from the one used for the other books *On floating bodies*, even though he does not mention this circumstance clearly). In fact, the books *Quadrature of the Parabola* and *Measurement of a circle* had already been published in 1503 by Luca Gaurico[54] in Latin (Tetragonism[55] i.e. the quadrature of the circle discovered by Campano [Giovanni Campano da Novara], Archimedes of Syracuse and Boetius, who were very gifted in mathematics. In the preface [one finds]: Luca Gaurico of Gifuni from the Kingdom of Naples hails the scholars of mathematics: [datum in] given in the University of Padua [Almo Studio Patavino] 1503, 15 Kal. of August (July, the 17th). In the end of the book [one finds]: printed in Venice by Ioan. Bapt. Sessa a.D. 1503 August, the 28th.), and Tartaglia seized this translation literally keeping both the extremely inept mistakes and the perverse punctuation. He filled very few gaps and changed[1]) partly the figures and letters of the figures. From this source, Tartaglia seized the title on the 19th page: *The Tetragonism of Archimedes*, which title in the translation of Gaurico

1) Only, to my knowledge, Mazzucchelli recalled this fact: Notizie istoriche intorno alla vita, alle invenzioni ed agli scritti di Archimede siracusano [Historical facts on the life, the inventions and the works of Archimedes of Syracuse], Brescia 1737. 4. p. 95. After I came across this passage so long consigned to an unworthy oblivion, I myself became aware of a specimen of this very rare book, owned by the Royal Library in Copenhagen.

XXXV is the common title of both books; hence, the titles of the individual books in both Gaurico and Tartaglia are as follows: Incipit Archimenidis ("Archimedis" in Tartaglia) quadraturae parabolae, et: Archimedis Syracusani liber. What code he had followed, Gaurico does not specify; in the preface on the second page, he only writes the following: since the proof of Campano[56] and Archimedes about the tetragonism of the circle has come into our hands, I do not believe that it should be concealed. Therefore, it is not even clear whether he himself had translated these books from Greek or found a Latin translation. However, despite this, it is clear that his translation, which is by far worse than that of Iacobus of Cremona and follows so carefully the Greek [version] that he often deviates not only from the use of the Latin language but from any sense, is derived from a codex that is very similar to the codex of Valla, or exactly from it. In fact, not only in its archetype, as in the F codex and undoubtedly in Valla's codex, the propositions of the book on the quadrature of

[54] Luca Gaurico (1475-1558) was an Italian Catholic bishop and astrologer.

[55] Whose title was in Latin "Tetragonismus id est circuli quadratura"

[56] Campano di Novara, or Giovanni Campano (1220-1296), was an Italian mathematician, astronomer and astrologer. One of the most important scientists and mathematicians of the 13th century.

the parabola had absolutely no numbers, indicating that he very often divided the chapters very badly by connecting two and split one into two or more (Tartaglia has added the numbers of the chapters and has mostly followed the divisions of Gaurico, who only marked them in capital letters); but also most errors of Valla's codex are repeated here, such as:

[...][57] XXXVI –XLIV

[...] XLV

| Codex Vallae | Gauricus | Tartalea | Cremonensis |
|---|---|---|---|
| II p. 310, 26: ποτι τὸ Ζ· ὥστε μεῖζόν ἐστι τὸ Ζ τοῦ Κ. | quam ad spatium z quam spatium k. | quam ad spatium z. ergo spatium z maior est quam spatium k. | quam ad f. quare f spacium ipso k maius existit. |

In the last listed comparison, in Gaurico for a mistake of the typograph it is corrupted the sentence: "quare maius est spatium z" due to the homeoteleuton[58]; Tartaglia filled the lacuna clearly without having the Greek code in his hands (in fact the order of the words is different), and in this he made a mistake (maior[59]) shameful even for a child. For this reason, it must be deduced that even in the other corrections to Gaurico's errors Tartaglia did not use his codices, but he relied on his intuition[1]).

Now, we go back to Tartaglia's codices. If, as it was done above on p. XXXII, we have suspected that the codex "damaged and barely legible", of which Tartaglia speaks, was the same as Valla's codex, then it must be admitted that Tartaglia is the first to complain about the nature of the codex too much, as it appears by comparison with the Parisian codex C, which was correctly copied at about the same time by the same archetype; but perhaps Tartaglia was poorly versed in reading Greek codices. Thereupon, Tartaglia for his codex did not use the care and conscientiousness as he ought to have done, since in the book on the equilibrium of plane surfaces II, 9, instead of a presentation of a true [translation], he caught a manipulated and shortened Eutocius' paraphrase, as if it was originated from a true Archimedes' text (Quaest., Arch., p. 97)[2]). Finally, Tartaglia corrected and changed many text passages, for doing that he seems to have used some other codex which he only mentioned clearly in one place; in fact,

[57] Here there starts a list of errors repeated in the book edited by Gaurico and in Valla's codex. This list continues for many pages, between page XXXVI and page XLIV. We omit this list.

[58] The homoteleuton (in the original text, this word is in Greek) is the repetition of the ending of consecutive words. The sentence is replaced by "quam ad spatium z quam spatium k". The Cremonensis' translation is the most faithful to the Greek original.

[59] Here Heiberg quickly refers to the previously reproduced comparative table. Tartaglia uses the adjective "maior" i.e. "greater than" without inflecting it, an error which is by illiterates. In his work Cronica [Cronica de matematici: overo Epitome dell' istoria delle vite loro, Urbino, Angelo Antonio Monticelli, 1707], Bernardino Baldi writes "*Attese nondimeno così poco alla bontà della lingua, che move a riso talhora chi legge le cose sue*" (He studied so little to learn languages, that sometimes he provokes laughter to those who read his works).

1) In addition, Tartaglia at the beginning of the book *Quadrature of the Parabola* for two times in place of "Archimenides" (in Gaurico) wrote "Archimedes", and in place of "mathematicam" II p. 294, 11; 298, 2 "mecanicam".

2) Perhaps the faithful presentation in Valla's codex seemed to Tartaglia more difficult to read; what certainly results is that this codex gave him no reason to confuse Eutocius' presentation with the real one.

XLVI in the book *On the equilibrium of planes* II, 9 after the recovered demonstration of Eutocius on page $16^{v}$ it is added: "In another Greek copy one could read in this way". But before we carefully investigate this codex, we must say some few things about the Greek codex of Nuremberg cent. V app. 12 fol., whose existence I knew at first due to Menge (Neue Jahrb. 1880 p. 110); but after that I knew about it, I managed to bring it with me to Copenhagen. It is a codex of paper written in the 16th century, containing the same writings of Archimedes and Eutocius as codex F and in the same order. Venatorius, who corrected many mistakes with his own hand, partly on the margin, partly with strips of paper glued to the edge, used this codex to produce the main edition; moreover, he has improved many things to smartly change the script of the codex by scraping, adding lines or extending; for this reason, the corrections of this kind were very difficult to find and very often they took away the older writing altogether; some of them, however, seem to have been made by an older hand and perhaps the first hand who wrote the codex. In between, Venatorius has also written down notes that the typographer must read, so that it appears that this codex itself was in the hands of the typographers. In order to know the nature of these comments, an example should suffice; I p. 22–24, where the propositions in codex N$^{a}$ are divided in the same way as in codex F, he wrote on a strip of paper: "ὁμοίως [similar] paragraph. φανερὸν δέ [clear indeed] paragraph. δεικτέον δέ [to be proven indeed] demonstration of a new proposition. The note of the number ϛ should be prefixed", and this has been done in the edition Basil. p. 4. This codex N$^{a}$ was written in Rome or has certainly come to Nuremberg from Rome to Bilibaldus Pirckheymerus (ed. Basil. On the second page of Preface one reads: *Bilibaldus Pirckheymerus, whom you have experienced, as long as he lived, could not be difficultly admitted to be the most scholarly amongst the scholars, as he was a man of outstanding talent, having received a Greek writing copy of our Archimedes in Rome from some friend after a long expectation, he let the codex live as a guest of honor in his house*).

The codex N$^{a}$ shows to be of the same class as the other codices for the lacuna at the beginning of the first book *On the Sphere and Cylinder* and there is also a match of all the most serious errors altogether (Cfr. Quest. Arch. p. 138).

XLVII –LXXXIX *On these pages a technical philological analysis of many Archimedean codices is presented. This analysis seems to prove without doubt that Tartaglia did not master the subject to be a reliable editor and translator. Here Heiberg gives evidences that are used by Clagett [Clagett, M. The Use of the Moerbeke Translations of Archimedes in the Works of Johannes de Muris Isis, Vol. 43, No. 3 (Sep., 1952), pp. 236–242, p. 237] to state that: "Heiberg further observed that the Moerbeke translations of the Dimensio circuli and the De quadratura parabolae were published by Gauricus in Venice, 1503. These same two translations and those of the De centris gravium and Book I of the De insidentibus aquae were published by Tartaglia in*

*Venice in 1543. Tartaglia leaves us with the distinct and erroneous impression that the translation is original with him. We can note finally the publication in 1565 of the Moerbeke translation of the whole of the De insidentibus aquae by Trojanus Curtius in Venice and by Commandino in Bologna". Moreover, also adding his own philological analysis to the enormous work by Heiberg, Clagett [Vol III Chapt. 4 Sect. 2 p. 540] concludes that: "The central thrust of my argument on Tartaglia's role in Archimedean studies is that Tartaglia's knowledge of the works of Archimedes primarily derived from the translations of William of Moerbeke, although he certainly was familiar with the various extracts that appeared in Giorgio Valla's "De expetendis et fugiendis rebus" and no doubt also with the Cremonensis translations either in the Venice manuscript or in the edition princeps of Basel, 1544."*

## Acknowledgments

Mario Spagnuolo was supported by P.O.R. SARDEGNA F.S.E. 2014-2020 - Asse III "Istruzione e Formazione, Obiettivo Tematico: 10, Obiettivo Specifico: 10.5, Azione dell'accordo di Partenariato:10.5.12" Avviso di chiamata per il finanziamento di Progetti di ricerca - Anno 2017.

# Chapter 3
# Hellinger's 1913 encyclopedia article on the fundamentals of the mechanics of continua

Simon R. Eugster

## 3.1 Translator's preface

Ernst Hellinger (*September 30, 1883 - †March 28, 1950), who was born in Striegau, formerly Germany, enjoyed his scientific education at the Universities of Heidelberg, Breslau and Göttingen, [109]. As a pupil of D. Hilbert, he received his doctoral degree from the University of Göttingen in 1907. After two more years in Göttingen, as assistant of Hilbert, he moved to Marburg where he accepted a position as "Privatdozent". During that time, Hellinger had written his masterpiece on the foundations of continuum mechanics and finished in 1913 his fundamental review article *Die Allgemeinen Ansätze der Mechanik der Kontinua*, which appeared in the "Encyklopädie der mathematischen Wissenschaften mit Einschluss ihrer Anwendungen", Bd. IV-4, Hft. 5. As it used to be, Hellinger wrote the article in German. More than a hundred years had to pass, until recently, the author of this chapter together with F. dell'Isola have published an exegetic series about Hellinger's encyclopedia article, see [59, 61, 62, 60]. Besides a complete annotated translation into English, these articles give a critical analysis about the advancement of science. Then, due to the establishment of English as the upcoming scientific language and due to the refusal of a variational formulation of continuum mechanics in the subsequent period, Hellinger's contribution to the foundations of continuum mechanics had been ignored for decades and had almost fallen into oblivion. Solely, the Hellinger-Reissner principle has made its way directly into theoretical and numerical mechanics.

The work of Hellinger covers an incredible amount of still contemporary topics in modern continuum mechanics and testifies how advanced theoretical mechanics was at the beginning of the twentieth century. Even though Hellinger focused on the fundamentals of continuum mechanics, he presented within the very same variational framework the physics of optics, electrodynamics, thermodynamics and the theory of relativity. Accordingly, Hellinger's paper can be understood as a contribution to

Simon R. Eugster
University of Stuttgart, Stuttgart, Germany, e-mail: eugster@inm.uni-stuttgart.de



F. dell'Isola et al. (eds.), *Evaluation of Scientific Sources in Mechanics*,
Advanced Structured Materials 152, https://doi.org/10.1007/978-3-030-80550-0_3

continuum physics in general. In the same spirit, one has to consider the contribution given by [111].

With a side-by-side translation of Hellinger's article, also non-German speaking people should get the possibility to enjoy the reading of a crystal-clear and still topical article whose content has some enlightening parts. To make the reading as authentic as possible, the typesetting remains as close as possible to the original, the old-fashioned mathematical notation is left untouched and the original page-numbering is kept. The comparison of the translation with the original allows the reader to directly review the quality of the translation[1]. Despite the risk of some translated sentences being a bit clunky, a word-by-word translation has been carried out to avoid any implicit interpretation of the original text. Any additional word that has no correspondence in the German text is included in square brackets as follows: [xxxx].

The reader will immediately observe almost everywhere the antiquated mathematical notation, which can be viewed as an an intermediate notation towards index notation established by Ricci and Levi-Civita for tensor algebra and tensor calculus. However, the reader will also notice that Hellinger internalized the tensorial character of the introduced mathematical objects better than many contemporary scientists who are using the even more fashionable algebraic notation, which avoids the use of components but which requires the definitions of a myriad of operations. Moreover, the used notation is clarified at the very beginning for the vectorial equation

$$\bar{x} = x + \xi(x, y, z; \sigma) \qquad \begin{pmatrix} x, y, z \\ \xi, \eta, \zeta \end{pmatrix}$$

with the following explanation: «This signature and the analogous ones which follow, denote that besides the equation being written-out also those [equations] are valid, which arise by simultaneous cyclic permutation of $x, y, z$ and $\xi, \eta, \zeta$.» In index notation, one prefers denoting $x, y, z$ and $\xi, \eta, \zeta$ by $x_1, x_2, x_3$ and $\xi_1, \xi_2, \xi_3$, respectively, in order to write

$$\bar{x}_i = x_i + \xi_i(x_j; \sigma) \ .$$

In an algebraic notation, very commonly, lower case bold symbols are used to denote elements of the tree-dimensional Euclidean vector space resulting in

$$\bar{\mathbf{x}} = \mathbf{x} + \boldsymbol{\xi}(\mathbf{x}; \sigma) \ .$$

The internal virtual work appearing on p. 615 in Eq. (4) is another example for which we want to give a notational translation. In the expression

$$-\iiint\limits_{(V)} \sum_{\binom{x\,y\,z}{XYZ}} \left( X_x \frac{\partial \delta x}{\partial x} + X_y \frac{\partial \delta x}{\partial y} + X_z \frac{\partial \delta x}{\partial z} \right) dV$$

[1] Some translated excerpts of Hellinger's paper can also be found in Maugin [106].

Hellinger used the notation of Kirchhoff to denote the stress components of the stress tensor in the actual configuration, nowadays mostly referred to as the Cauchy stress and often denoted by $\boldsymbol{\sigma}$. Hence one can relate the components with respect to an orthonormal frame by

$$\sigma_{11} = X_x,\ \sigma_{12} = X_y,\ \ldots\ , \sigma_{33} = Z_z.$$

With the virtual displacement field $\delta\boldsymbol{x}$ defined over the current configuration, the components of its gradient $\nabla(\delta\boldsymbol{x})$ translate as

$$\frac{\partial \delta x_1}{\partial x_1} = \frac{\partial \delta x}{\partial x}, \ \frac{\partial \delta x_1}{\partial x_2} = \frac{\partial \delta x}{\partial y}, \ \ldots, \ \frac{\partial \delta x_3}{\partial x_3} = \frac{\partial \delta z}{\partial z}.$$

Using index notation together with Einstein's summation convention, which tacitly assumes summation over the range of the indices that appear twice in a term, we obtain the following compact form of the internal virtual work

$$-\iiint\limits_{(V)} \sum_{i,j=1}^{3} \sigma_{ij} \frac{\partial \delta x_i}{\partial x_j} dV = -\iiint\limits_{(V)} \sigma_{ij} \frac{\partial \delta x_i}{\partial x_j} dV = -\iiint\limits_{(V)} \sigma_{ij} \delta x_{i,j} dV.$$

Note, in the last equality, we have used the quite common notation to abbreviate partial derivatives. After a coordinate independent definition of the double contraction, often denoted by a colon, the internal virtual work can also be written as

$$-\iiint\limits_{(V)} \boldsymbol{\sigma} : \nabla(\delta\boldsymbol{x}) dV.$$

Every notation has its advantages and disadvantages and it is rather a matter of taste which notation one prefers. While Hellinger's notation doesn't require any education in tensor algebra and tensor calculus, the much more compact algebraic notation requires many conventions and can sometimes be even cryptic, if the operations are not introduced adequately. Nevertheless, notation is just notation and without the corresponding piece of prose the equations decompose into a collection of meaningless symbols. As a mathematical physicist, Hellinger mastered this difficulty brilliantly and presented in a unified way all field theories which had already been formulated at his times, assuming as fundamental paradigm for physics the concept of field. It is remarkable that the work of Hellinger seems to have given the starting point to the works of Paul Germain, Richard Toupin, and Leonid I. Sedov [39, 78, 77, 14, 38]. In a sense, the effort by Hellinger in framing continuum mechanics by using variational principles opened the way to the modern theory of metamaterials [37, 20, 142, 88]: in fact, the problem of synthesis of tailored metamaterials can be confronted more successfully when continuum models are introduced to describe the desired mechanical behavior [51].

Following the side-by-side translation, the chapter closes with a section in which several interesting parts of the work are analyzed and are set into context with more modern sources.

# 3.2 Die allgemeinen Ansätze der Mechanik der Kontinua

# IV 30. DIE ALLGEMEINEN ANSÄTZE DER MECHANIK DER KONTINUA.

Von

**E. Hellinger**

IN MARBURG A. L.

## Inhaltsübersicht.

# IV 30. FUNDAMENTALS OF THE MECHANICS OF CONTINUA.

By

**E. Hellinger**

IN MARBURG A. L.

---

## Contents.

---

## Litteratur.

Spezielle den vorliegenden Gegenstand betreffende Lehrbücher und Monographien liegen z. Z. in der Litteratur nicht vor. Von den wiederholt zu nennenden Werken seien hier folgende besonders zusammengestellt:

*A. v. Brill,* Vorlesungen zur Einführung in die Mechanik raumerfüllender Massen, Leipzig 1909.

*E. und F. Cosserat,* Théorie des corps déformables, Paris 1909. Ursprünglich als Appendix zur französischen Ausgabe von *O. D. Chwolson,* Traiteé de physique, t. II, Paris 1909 erschienen. Ein Auszug ist als Note an *P. Appell,* Traité de mécanique rationelle, t. III, 2. éd. (Paris 1909) beigefügt.

*P. Duhem,* Traité d'énergétique ou de thermodynamique générale, 3 vols, Paris 1911.

*G. Hamel,* Elementare Mechanik, Leipzig u. Berlin 1912.

*J. L. Lagrange,* Mécanique analytique, 4. éd. = Oeuvres complètes, Bd. 11 u. 12, (éd. par *G. Darboux*), Paris 1888/89.

*W. Voigt,* Kompendium der theoretischen Physik. Bd. 1 u. 2. Leipzig 1895/96.

Vgl. ausserdem die entsprechenden Abschnitte in den Lehrbüchern der Mechanik (s. die Übersichten in IV 1, *Voss*, IV 6 *Stäckel*, IV 11, *Heun*, IV 23, *Müller-Timpe*).

---

**1. Einleitung.** Das vorliegende Referat soll unter einheitlichem Gesichtspunkte einen zusammenfassenden Überblick über die verschiedenen *Formen* der Ansätze geben, durch die man in den einzelnen Gebieten der „Mechanik der Kontinua“ im weitesten Sinne, d. h. der Mechanik und Physik kontinuierlicher ausgedehnter Medien, den zeitlichen Ablauf oder auch den Gleichgewichtszustand der zu untersuchenden Vorgänge bestimmt; dabei wird immer nur an solche Kontinua gedacht, die nicht vermöge irgendwelcher einschränkender

---

## Literature.

Specific textboooks and monographs on the topic at hand are in the literature at the moment not available. From the repeatedly referred works, the following are listed in particular:

*A. v. Brill,* Vorlesungen zur Einführung in die Mechanik raumerfüllender Massen, Leipzig 1909.

*E. und F. Cosserat,* Théorie des corps déformables, Paris 1909. Appeared originally as appendix to the french edition of *O. D. Chwolson,* Traiteé de physique, t. II, Paris 1909. An extract is added as a note to *P. Appell,* Traité de mécanique rationelle, t. III, 2. éd. (Paris 1909) beigefügt.

*P. Duhem,* Traité d'énergétique ou de thermodynamique générale, 3 vols, Paris 1911.

*G. Hamel,* Elementare Mechanik, Leipzig u. Berlin 1912.

*J. L. Lagrange,* Mécanique analytique, 4. éd. = Oeuvres complètes, Bd. 11 u. 12, (éd. par *G. Darboux*), Paris 1888/89.

*W. Voigt,* Kompendium der theoretischen Physik. Bd. 1 u. 2. Leipzig 1895/96.

Cf. furthermore the corresponding sections in the textbooks of mechanics (for reviews see IV 1, *Voss*, IV 6 *Stäckel*, IV 11. *Heun*, IV 23. *Müller-Timpe*).

---

**1. Introduction.** With respect to a consistent point of view, the paper at hand shall give a recapitulatory overview on various *forms* of the axiomatic foundations, which in the particular fields of the "mechanics of continua" in the broadest sense, i. e. the mechanics and physics of continuously extended media, enable the determination of the time behavior or also the state of equilibrium of the analyzed processes; thereby these continua are kept in mind, which, due to any restricting

Bedingungen speziell endlich viele Freiheitsgrade besitzen. Die Möglichkeit, die Grundgleichungen verschiedener Disziplinen in analoge Formen zu bringen, hat man früh bemerkt: die „mechanischen“ Theorieen der Physik, die das physikalische Geschehen auf Bewegungserscheinungen der Materie zurückführen wollen, haben formal-mathematisch betrachtet geradezu den Inhalt, dass sie die Gleichungen der Physik als Sonderfälle der Gleichungen eines allgemeinen Systems bewegter Massen bzw. Massenpunkte erscheinen lassen; sie müssen also jene Analogien in Evidenz treten lassen.

Neben den eigentlichen mechanischen Theorien, die mehr oder weniger detaillierte Bilder vom Aufbau der Materie voran stellen, hat man zum Teil schon in den Anfängen, besonders aber seit der Mitte des 19. Jahrhunderts einen anderen an *J. L. Lagranges* analytischer Mechanik orientierten Weg eingeschlagen; so wie dort sämtliche zur Untersuchung kommenden Probleme wenigen sehr allgemeinen Prinzipien untergeordnet werden, so bemühte man sich die Grundansätze immer weiterer physikalischer Disziplinen in die Formen jener Prinzipien zu bringen, indem man die in ihnen auftretenden Grössen — Energie, Kräfte, usw. — von rein phänomenologischen Gesichtspunkten aus mit gewissen physikalischen Grössen identifizierte. Für Systeme mit endlich vielen Freiheitsgraden knüpft diese Entwicklung namentlich an die von *W. Thomson* (*Lord Kelvin*), *J. J. Thomson* und *H. v. Helmholtz* inaugurierten Untersuchungen über zyklische Systeme und deren Anwendungen und über die Reziprozitätssätze der Mechanik an.

Nun wendete bereits Lagrange seine Prinzipien direkt auf gewisse kontinuierliche Systeme (Flüssigkeiten, biegsame Fäden und Platten u. dgl.) an[1]); im Anschluss an die weitere Ausbildung dieser Ansätze, besonders durch die an *A. L. Cauchy* [2]) anknüpfende Entwicklung der Elastizitätstheorie, sowie unter der Einwirkung des Ausbaues anderer physikalischer, speziell optischer Theorien gewöhnte man sich immer mehr, auch ein kontinuierliches System als ein selbständiges Objekt der Mechanik (mit unendlich vielen Freiheitsgraden) zu betrachten, das sich seinerseits zwar in formaler Analogie zu der altbekannten Punktmechanik, aber doch völlig unabhängig von ihr behandeln lässt. Die so als selbständige Disziplin entwickelte „Mechanik des deformierbaren Kontinuums“ umfaßt den formalen Ansätzen nach neben der gewöhnlichen Elastizitätstheorie und Hydrodynamik sämtliche hier in

---

[1] Vgl. insbesondere 1. part., sect. IV, § II der „Mécanique analytique“.

[2] Entscheidend waren hier seine Untersuchungen über den Spannungsbegriff vom Jahre 1822 (Bull. de la Soc. philom. 1823, p. 9). Nähere Angaben s. in IV 23, Nr. **3**a, *Müller-Timpe*.

conditions, in particular are not reduced to continua with finitely many degrees of freedom. The possibility, to bring the fundamental equations of various theories into similar forms, has been noticed soon: the "mechanical" theories of physics, which try to explain the physical phenomena only by the motion of matter, contain from a formal-mathematical point of view the essence that the equations of physics appear as special cases of equations of a general system with moving masses or mass points; thus such [mechanical] theories must generate those analogies.

Besides the intrinsic mechanical theories, which put more or less detailed images of the constitution of matter at the basis [of their formulation], one has, partially in the beginning, but in a wider extent in the mid-19th-century, developed a new path following *J. L. Lagrange*'s analytical mechanics; indeed as in there all analyzed problems are based on a few very general principles, one has tried to bring the foundations of more and more physical disciplines into the form of those principles, by identifying the appearing quantities — energy, forces, and so on — with certain physical quantities [previously introduced] from a purely phenomenological point of view. For systems with finitely many degrees of freedom this development is presented in particular in the analysis inaugurated by *W. Thomson* (*Lord Kelvin*), *J. J. Thomson* and *H. v. Helmholtz* on cyclic systems with its applications and on the reciprocity theorems of mechanics.

Already Lagrange applied his principles directly to certain continuous systems (fluids, flexible wires and plates and similar ones)[1]); in connection with the further developments of these fundamentals, particularly with that one concerning the development of the theory of elasticity by following *A. L. Cauchy* [2]), as well as under the influence of the extension of other physical, especially optical theories, one got increasingly accustomed to consider a continuous system as an independent object of study in mechanics (with infinitely many degrees of freedom), which has to be in formal analogy to the well-known point mechanics, but which can be treated independently. This "mechanics of the deformable continuum" developed as an independent discipline, contains, due to the formal approaches [used], in addition to the common theory of elasticity and hydrodynamics all

---

[1] cf. especially 1. part., sect. IV, § II of "Mécanique analytique".

[2] Crucial were his analyses on the notion of stress from 1822 (Bull. de la Soc. philom. 1823, p. 9). For further details see IV 23, No. 3a, *Müller-Timpe*.

Betracht zu ziehende physikalischen Erscheinungen in kontinuierlich ausgedehnten Medien. Die Fortbildung dieser Betrachtungen wurde wesentlich beeinflusst durch die *Thermodynamik*, die prinzipiell das Gesamtgebiet der Physik zu umfassen strebt, und die dadurch, daß sie überall die Energiefunktion bzw. das Potential voranstellt, naturgemäss die Grundgleichungen der verschiedenen Einzelgebiete in analogen Formen liefert.

Alle diese Beziehungen sind in der mechanischen und physikalischen Literatur vielfach behandelt worden; vieles, was ausdrücklich nur in der Punktmechanik bzw. für Systeme von endlich vielen Freiheitsgraden ausgesprochen ist, lässt sich unmittelbar auf kontinuierliche Systeme ausdehnen. Es seien hier vorweg nur die Namen einiger Autoren genannt, die die hier in Betracht kommenden Beziehungen besonders berücksichtigt haben und die daher auch im folgenden häufig zur Geltung kommen: *W. Voigt* [3]), *P. Duhem* [4]) und *E.* und *F. Cosserat.*[5]) [6])

Der Zweck dieses Referates bedingt es, dass im folgenden das rein *formal*-mathematische Moment im Vordergrund steht: die Formulierung des *Ansatzes* der verschiedenen Probleme sowie ihre Zusammenfassung in eine einheitliche möglichst einfache und bequeme Formel. Sowohl die Untersuchung der *mechanischen* und *physikalischen* Bedeutung der Grössen und Gleichungen als auch die eigentlich *analytisch*-mathematische Theorie ist in den verschiedenen Referaten der Bände IV und V über die besonderen Disziplinen enthalten.

Als einheitliche *mathematische* Form, der sich die sämtlichen Einzelansätze am leichtesten einfügen, wird die des *Variationsprinzipes* verwendet. Allerdings genügt nicht die Gestalt, die in der eigentlichen Variationsrechnung in der Regel betrachtet wird, wo die unbekannten Funktionen so zu bestimmen sind, dass ein gewisses sie enthaltendes bestimmtes Integral einen Extremwert annimmt. Vielmehr handelt es sich hier vorzugsweise um diejenige Form, die die Variationsrechnung als notwendiges Kriterium des Extremums ergiebt, und in

---

[3] Neben vielen einzelnen Arbeiten besonders in seinem Kompendium der theoretischen Physik, 2 Bde., Leipzig 1895/96.

[4] In zahlreichen späterhin zu zitierenden Arbeiten; vgl. auch seinen Traité d'énergétique ou de thermodynamique générale, t. I. II, Paris 1911.

[5] Vgl. die als Appendix zur französischen Ausgabe von *O. D. Chwolson*, Traité de physique erschienene „théorie des corps déformables" (Paris 1909), von der ein Auszug der 2. Aufl. des 3. Bds. von *P. Appell* Traité de mécanique rationelle (Paris 1909) als Note beigefügt ist.

[6] Auch von Entwicklungen ähnlicher Art die *D. Hilbert* in einigen seiner Göttinger Vorlesungen gab, ist das folgende vielfach beeinflusst.

physical phenomena which are accounted for in [the theory of] continuously extended media. The advancement of this theory has been influenced significantly by *thermodynamics*, which aims to cover the entire field of physics, and by considering the energy function or rather the potential as the most fundamental concept, naturally yields the fundamental equations of various specific fields in similar forms.

All these equations have been treated in the mechanics and physics literature in many cases; much that has been stated explicitly in point mechanics or for systems with finitely many degrees of freedom, can immediately be extended to continuous systems. At a preeminent place, just the names of a few authors are mentioned, which have especially considered the relations treated here which often show to be useful in the following: *W. Voigt* [3]), *P. Duhem* [4]) and *E.* and *F. Cosserat.*[5]) [6])

The objective of this paper requires that the purely *formal*-mathematical aspect must have priority in what follows: The formulation of the *ansatz* of various problems as well as their collection to a unified and at most simple and convenient formula. Both the analysis of *mechanical* and *physical* interpretation of the quantities and equations as well as the essential *analytic*-mathematical theory of the particular disciplines are covered by various papers in the volumes IV and V.

As unifying *mathematical* form, in which all individual [methodological] fundamentals are included the easiest, the *variational principle* is applied. Although, the form commonly considered in the calculus of variations, in which the unknown function has to be determined such that a certain definite integral, containing the function, has an extremum, is not adequate. On the contrary, it concerns here particularly the form, which the calculus of variations yields as the necessary criterion of the extremum, and the form in

[3] Besides many individual works especially in his compendium of theoretical physics, 2 Bde., Leipzig 1895/96.

[4] In numerous works to cite later on; cf. also his Traité d'énergétique ou de thermodynamique générale, t. I. II, Paris 1911.

[5] Cf. "théorie des corps déformables" (Paris 1909), appearing as appendix to the french edition of *O. D. Chwolson*, Traité de physique, and which is added partially as a note to the 2. Edn. of the 3. vol. of *P. Appell* Traité de mécanique rationelle (Paris 1909).

[6] The following is influenced in many ways also from similar developements treated in some of the Göttinger lectures of *D. Hilbert*.

der man von altersher *das Prinzip der virtuellen Arbeit* ausdrückt: „Gegeben sind eine Reihe von Grössen $X, \dots, X_a, \dots$ in ihrer Abhängigkeit von den unbekannten Funktionen $x, \dots$ von $a, \dots, c$ und deren Ableitungen; diese Funktionen sollen der Bedingung genügen, dass ein bestimmtes Integral einer mit jenen $X, \dots, X_a, \dots$ als Koeffizienten gebildeten linearen Form der willkürlichen Funktionen $\delta x, \dots$ von $a, \dots c$ und ihrer Ableitungen

$$\int \cdots \int \left\{ X\delta x + \cdots + X_a \frac{\partial \delta x}{\partial a} + \cdots \right\} da \cdots dc$$

— oder eine Summe solcher Integrale — identische für alle (oder doch für alle gewissen Nebenbedingungen genügenden) $\delta x, \dots$ verschwindet."

Der Vorteil, den die Verwendung eines solchen Variationsprinzipes als Grundlage gegenüber anderen möglichen Formulierungen oder auch der direkten Inbetrachtnahme der Grundgleichungen gewährt, besteht ganz besonders darin, dass das Variationsprinzip im Stande ist, in *einer* Formel das Verhalten des betrachteten Mediums an allen Stellen und zu *jedem* Zeitpunkt zu bestimmen, speziell also auch neben den Gleichungen für das Innengebiet die *Randbedingungen* und die *Anfangsbedingungen* zu umfassen. Es ist zudem in seiner prägnanten Kürze in gewisser Hinsicht übersichtlicher als die Gleichungen und hat infolgedessen für die Behandlung neuer Gebiete, für die Aufstellung weiterer Verallgemeinerungen u. dgl. wesentliche *heuristische* Bedeutung; diese wird besonders betont durch die innige Beziehung des Variationsprinzipes zur Thermodynamik, durch deren Anspruch auf Allgemeingültigkeit es für die Begründung physikalischer Theorien beweisenden Wert bekommt. Auch bei der Durchführung von *Koordinatentransformationen* ist das Variationsprinzip gegenüber den expliziten Gleichungen im Vorteil; es lässt vielfach die *invariantentheoretische Natur* des betrachteten Problems, die Frage nach den Transformationsgruppen, die es ungeändert lassen, bequemer erkennen, ohne daß es der Einführung einer besonderen Symbolik bedarf. —

Nach einigen einleitenden Erörterungen über den Begriff des Kontinuums und seine Kinematik werden in dem *ersten* Abschnitt des Referates die Grundansätze der *Statik*, im *zweiten* die der *Kinetik* behandelt, jedesmal ohne Rücksicht darauf, was für Kraftwirkungen im einzelnen es sind, die das Kontinuum beeinflussen. Die Natur dieser Kraftwirkungen, speziell ihre Abhängigkeit von der Lage und Bewegung des Kontinuums (*Dynamik*) wird im *dritten* Abschnitt erörtert, wobei dann die einzelnen Disziplinen einzuordnen sind; hierbei kommen schliesslich auch in einer kurzen Skizze einerseits die Beziehungen zu den Ansätzen der *Thermodynamik*, andererseits das Verhalten der ein-which *the principle of virtual work* is expressed of old: "Given a series of quantities $X, \ldots, X_a, \ldots$ depending on the unknown functions $x, \ldots$ on $a, \ldots, c$ and on the derivatives thereof; these functions shall satisfy the condition, that a definite integral of a linear form on the arbitrary functions $\delta x, \ldots$ of $a, \ldots c$ and their derivatives composed with those [functions] $X, \ldots, X_a, \ldots$ as coefficients

$$\int \cdots \int \left\{ X \delta x + \cdots + X_a \frac{\partial \delta x}{\partial a} + \cdots \right\} da \cdots dc$$

— or the sum of such integrals — vanishes identically for all (or however for all constraint satisfying) $\delta x, \ldots$.

The advantage, which the application of such a variational principle as foundation allows versus other possible formulations or also the direct consideration of the fundamental equations, consists especially therein that the variational principle is capable to determine the behavior of the considered media in all points and to *every* instant of time in a *single* formula, in particular to contain besides the equations for the interior also the *boundary conditions* and the *initial conditions*. Furthermore from a certain point of view, it is in its concise brevity clearer as the equations and consequently it is [more suitable] for the treatment of new fields, for the formulation of further generalizations and it is therefore of essential *heuristic* relevance; this [circumstance] is emphasized especially through the profound relation of the variational principle to thermodynamics, as through requirement of generality it gains evident value for the foundations of physical theories. Also for the evaluation of *coordinate transformations* the variational principle is in advantage versus the explicit equations; in many cases it is possible to identify the *invariance* of the considered problem, i. e. the question of the transformation group letting the problem unchanged, easier and without the requirement to introduce a specific symbolism. —

Subsequent to some introductory discussions about the notion of a continuum and the kinematics thereof, in the *first* section of the paper the foundations of *statics*, in the *second* the foundations of *kinetics* are treated, each time without any consideration of what classes of force effects influence the continuum in particular. The nature of these force effects, especially their dependence on the position and the motion of the continuum (*dynamics*) is discussed in the *third* section, whereat the individual disciplines are classified; in this connection, on the one hand the relation to the methods of thermodynamics, on the other hand the behavior of the zelnen Wirkungsgesetze gegen Transformationen der Raum-und Zeitkoordinaten und damit auch die Auffassungen der *Relativitätstheorie der Elektrodynamik* zur Geltung.

**2. Der Begriff des Kontinuums.**

**2a. Das Kontinuum und seine Deformation.** Das allgemeine *dreidimensional ausgedehnte kontinuierliche Medium*, auf das sich die folgenden Betrachtungen beziehen, bedeutet — unter Abstraktion von allen individuellen Eigenschaften der Materie — eine Gesamtheit von materiellen Teilchen, die erstens voneinander *unterscheidbar* sein und zweitens den Raum bzw. einen stetig begrenzten Raumteil *stetig ausfüllen* sollen. Die erste Eigenschaft kommt darin zum Ausdruck, dass jedes Teilchen durch Angabe dreier Parameterwerte $a, b, c$ identifiziert wird, derart, dass verschiedene Teilchen in jedem Zustand, in dem man das Kontinuum etwa betrachtet, stets verschiedene Lagen haben; der von stetigen geschlossenen Flächen $S_0$ begrenzte Variabilitätsbereich $V_0$ dieser $a, b, c$ charakterisiert das Quantum der Materie, das in Betracht gezogen wird. Die zweite Forderung besagt, dass die Lagen aller Teilchen einen von stetigen geschlossenen Flächen $S$ begrenzten Raumteil $V$ erfüllen. Bestimmt man die Lage eines Teilchens durch seine kartesischen Koordinaten, so wird ein solcher Zustand analytisch gegeben durch drei Funktionen von $a, b, c$

$$x = x(a, b, c), \quad y = y(a, b, c), \quad z = z(a, b, c), \tag{1}$$

die $V_0$ auf $V$ abbilden, und deren Funktionaldeterminante

$$\Delta = \frac{\partial(x, y, z)}{\partial(a, b, c)} \tag{2}$$

innerhalb $V_0$ von Null verschieden, etwa positiv, ist. Man kann für $a, b, c$ die Koordinaten einer fest gewählten Ausgangslage nehmen; dann sind $x - a$, $y - b$, $z - c$ die Komponenten der Verschiebung, die jedes Teilchen beim Übergang zur Lage (1) erleidet, und die Funktionen (1) werden *stetige* Funktionen von $a, b, c$, sofern man die übliche Annahme macht, dass ursprünglich benachbarte Teile stets benachbart bleiben. Wir werden darüber hinaus stets voraussetzen, dass die Funktionen (1) hinreichend viele stetige Differentialquotienten nach ihren Argumenten haben; nur an einzelnen Punkten, Linien oder Flächen sollen Unterbrechungen dieser Stetigkeit stattfinden können (vgl. IV 1, Nr. **9**, *Voss*). Die gleichen Voraussetzungen werden wir im allgemeinen stillschweigend für die weiterhin auftretenden, physikalische Vorgänge darstellenden Funktionen zu machen haben.

Jedes Funktionensystem (1) beschreibt vollständig einen bestimmten *Deformationszustand* des Kontinuums; im allgemeinen gilt *jeder* De-individual constitutive laws with respect to transformations of space-time-coordinates and thereby the concept of the *theory of relativity of electrodynamics* eventually are profitably presented together in a short outline.

**2. The notion of a continuum.**

**2a. The continuum and its deformation.** The general *three-dimensional extended continuous medium*, on which the following presentation relates to, stands — under abstraction of all more specific properties of matter — for an aggregate of material particles, which first of all are *distinguishable* from each other and second which *continuously occupy* the space or rather a continuously bounded part of the space. The first property finds its expression [by assuming] that every particle is identified by the specification of three variable values $a, b, c$ such that different particles always have different positions in every state in which the continuum can be considered by any chance; [and that] the domain of variability $V_0$ of these $a, b, c$, bounded by the continuous and closed surfaces $S_0$, characterizes the portion of matter, which is taken into consideration. The second requirement states, that the positions of all particles occupy a part of the space $V$ bounded by a continuous and closed surface $S$. Determining the position of a particle by cartesian coordinates, analytically such a state is given by three functions of $a, b, c$,

(1) $$x = x(a, b, c), \quad y = y(a, b, c), \quad z = z(a, b, c),$$

mapping $V_0$ to $V$, and by their Jacobian

(2) $$\Delta = \frac{\partial(x, y, z)}{\partial(a, b, c)}$$

being different from zero within $V_0$, for instance positive. For $a, b, c$ one can take the coordinates of a fixed chosen initial position; then $x - a$, $y - b$, $z - c$ are the components of the displacement, which every particle undergoes by shifting them to the position (1), and the functions (1) become *continuous* functions of $a, b, c$, as long as the common assumptions are taken, that initially neighboring particles always remain neighboring. Moreover, we will always assume the functions (1) to have sufficiently many continuous difference quotients with respect to their arguments; only at individual points, lines or surfaces, discontinuities may occur (cf. IV 1, No. **9**, *Voss*). In general, we will have to make the same assumptions tacitly for the upcoming functions which describe physical processes.

Every system of functions (1) describes entirely a certain *state of deformation* of the cotinuum; in general *every* formationszustand, d. h. jedes Funktionentripel (1), das nur den soeben charakterisierten Stetigkeitsvoraussetzungen genügt, als zulässig; Be-schränkungen in der Art der möglichen Funktionen werden besondere Eigenschaften spezieller Medien zum Ausdruck bringen. In jedem Falle bestimmen die partiellen Ableitungen der Funktionen (1) in bekannter Weise die Verschiebungen, Verdrehungen und Formänderungen, die jedes sehr kleine Quantum (Volumelement) bei der Deformation erleidet (vgl. IV 14, Nr. **16**, *Abraham*).

Die Grundlage für die Untersuchung der Gleichgewichtsverhältnisse irgendeines Deformationszustandes (1) erhalten wir, wenn wir ihn mit einer sog. *unendlichkleinen virtuellen Verrückung* überlagern, die *virtuell* heisst, insofern sie willkürlich zu dem reell stattfindenden Deformationszustand hinzutritt.[7]) Um diesen Begriff in mathematisch präziser Form zu erhalten, ohne die bequeme übliche Bezeichnung und Verwendung der „unendlichkleinen" Grössen aufzugeben, betrachte man zunächst eine der Deformation (1) übergelagerte noch von einem Parameter $\sigma$ abhängige und mit $\sigma = 0$ verschwindende Deformation, die das ursprünglich an der Stelle $(x, y, z)$ befindliche Teilchen an die Stelle

$$\bar{x} = x + \xi(x, y, z; \sigma) \qquad \begin{pmatrix} x, y, z \\ \xi, \eta, \zeta \end{pmatrix}^{8)}$$

überführt; dabei sind $\xi, \eta, \zeta$ gegebene Funktionen von $x, y, z$ und von dem Parameter $\sigma$, der in einem (beliebig kleinen) $\sigma = 0$ umgebenden Bereich variieren kann. Vermöge (1) kann man auch unter Elimination von $x, y, z$ die so entstehenden neuen Deformationen in der alten Gestalt schreiben:

$$(3) \qquad \bar{x} = \bar{x}(a, b, c; \sigma), \quad \text{wo} \quad \bar{x}(a, b, c; 0) = x \qquad (x, y, z).$$

Ist $f$ irgendein von den Deformationsfunktionen (1) und ihren Ableitungen abhängiger Ausdruck, so bezeichnen wir allgemein als seine „Variation" den Ausdruck

$$\delta f(x, \ldots, x_a, \ldots) = \left\{ \frac{\partial}{\partial \sigma} f(\bar{x}, \ldots, \bar{x}_a, \ldots) \right\}_{\sigma=0}, \quad \text{wo } x_a = \frac{\partial x}{\partial a}, \ldots;$$

[7] So in Übereinstimmung mit der Terminologie von Voss (IV 1, Nr. **30**), die auch in den Lehrbüchern vielfach üblich ist. Andere (z. B. *Voigt*, Kompendium I, p. 27) sprechen von „virtuellen" Verrückungen erst dann, wenn die sonst beliebigen Verrückungen mit den für das System etwa bestehenden Bedingungen verträglich sind; *C. Neumann* (Ber. Ges. Wiss. Leipzig 31 (1879), p. 53 ff.) hat gelegentlich den Vorschlag von *Gauss* aufgenommen, dann von *fakultativen* Verrückungen zu reden.

[8] Diese Signatur und die analogen in der Folge bedeuten, dass neben der angeschriebenen Gleichung auch diejenigen gelten, die durch gleichzeitige zyklische Vertauschung von $x, y, z$ und $\xi, \eta, \zeta$ aus ihr entstehen.

state of deformation, i. e. every triple of functions (1), which satisfies the just characterized continuity assumptions, is admissible; Restrictions on the kind of possible functions express particular properties of special media. In any case, the partial derivatives of the functions (1) assign in the well-known manner the displacements, the rotations and the shape change, which each very little portion (volume element) undergoes during its deformation (cf. IV 14, No. **16**, *Abraham*).

We obtain the basis for the analysis of the equilibrium conditions of an arbitrary state of deformation (1), by superimposing it with a so-called *infinitesimal virtual displacement*, called *virtual*, since it is added arbitrarily to the actually occurring state of deformation.[7]) To obtain this notion in a mathematical rigorous way, without dropping the convenient and common expression and application of "infinitesimal" quantities, one considers at first a deformation, depending on a parameter $\sigma$, superimposed on the deformation (1) which vanishes for $\sigma = 0$ and shifts the particle being at the initial position $(x, y, z)$ to the position

$$\bar{x} = x + \xi(x, y, z; \sigma) \qquad \begin{pmatrix} x, y, z \\ \xi, \eta, \zeta \end{pmatrix};\ ^{8)}$$

thereby $\xi, \eta, \zeta$ are known functions of $x, y, z$ and of the parameter $\sigma$, which varies in an (arbitrary small) surrounding of $\sigma = 0$. Using (1) for the elimination of $x, y, z$, one can also write these new arising deformations in the old form:

$$\text{(3)} \qquad \bar{x} = \bar{x}(a, b, c; \sigma), \quad \text{where} \quad \bar{x}(a, b, c; 0) = x \qquad (x, y, z).$$

Let $f$ be any expression depending on the deformation functions (1) and their derivatives, then we generally denote its "variation" by the expression

$$\delta f(x, \ldots, x_a, \ldots) = \left\{ \frac{\partial}{\partial \sigma} f(\bar{x}, \ldots, \bar{x}_a, \ldots) \right\}_{\sigma=0}, \quad \text{with } x_a = \frac{\partial x}{\partial a}, \ldots;$$

---

[7] Thus in coincidence with the terminology of Voss (IV 1, No. **30**), which is also often common in textbooks. Others (e. g. *Voigt*, Kompendium I, p. 27) speak of "virtual" displacements only when the otherwise arbitrary displacements are admissible with respect to any constraints of the system; *C. Neumann* (Ber. Ges. Wiss. Leipzig 31 (1879), p. 53 ff.) occasionally has adopted the suggestion of *Gauss*, to speak then of *optional* displacements.

[8] This signature and the analogous ones which follow, denote that besides the equation being written-out also those [equations] are valid, which arise by simultaneous cyclic permutation of $x, y, z$ and $\xi, \eta, \zeta$.

dabei bleiben während der Differentiation $a, b, c$ konstant; die Operation $\delta$ ist daher mit der Differentiation nach $a, b, c$ vertauschbar:

$$\delta\frac{\partial f}{\partial a} = \frac{\partial \delta f}{\partial a}.$$

Verschwinden die 3 Funktionen

$$\left(\frac{\partial \bar{x}}{\partial \sigma}\right)_{\sigma=0} = \left(\frac{\partial \xi}{\partial \sigma}\right)_{\sigma=0} = \delta x(x, y, z) \qquad (x, y, z),$$

die vermittels (1) als Funktionen von $x, y, z$ angesehen werden können, nicht identisch in $x, y, z$, so kann man unter den üblichen Stetigkeitspostulaten setzen

$$\bar{x} = x + \sigma\delta x(x, y, z) \qquad (x, y, z), \tag{3'}$$

falls $\sigma$ so klein gewählt ist, dass $\sigma^2$ hinreichend klein gegenüber $\sigma$ wird; die so gegebene unendlich kleine virtuelle Verrückung des Kontinuums ist also bis auf den Faktor $\sigma$ durch die 3 Funktionen $\delta x, \delta y, \delta z$ bestimmt. Man kann diese Verrückung unmittelbar dem Begriff der in der Kinematik elastischer Medien betrachteten „unendlichkleinen Deformationen" einordnen (vgl. IV 14, Nr. **18**, *Abraham*) und findet insbesondere, dass die durch sie bedingte *„virtuelle Formänderung"* jedes Volumelements durch die 6 Grössen

$$\frac{\partial \delta x}{\partial x}, \frac{\partial \delta y}{\partial y}, \frac{\partial \delta z}{\partial z}, \frac{\partial \delta y}{\partial z} + \frac{\partial \delta z}{\partial y}, \frac{\partial \delta z}{\partial x} + \frac{\partial \delta x}{\partial z}, \frac{\partial \delta x}{\partial y} + \frac{\partial \delta y}{\partial x}, \tag{4}$$

ihre *„virtuelle Rotation"* durch

$$\frac{1}{2}\left(\frac{\partial \delta z}{\partial y} - \frac{\partial \delta y}{\partial z}\right), \frac{1}{2}\left(\frac{\partial \delta x}{\partial z} - \frac{\partial \delta z}{\partial x}\right), \frac{1}{2}\left(\frac{\partial \delta y}{\partial x} - \frac{\partial \delta x}{\partial y}\right) \tag{4'}$$

bestimmt wird — jedesmal abgesehen von dem Faktor $\sigma$.

Eine *Bewegung des Kontinuums* wird als eine vom Zeitparameter $t$ abhängige Folge von Deformationszuständen aufgefasst und demgemäss durch die drei nun noch von $t$ abhängigen Deformationsfunktionen

$$x = x(a, b, c; t), \quad y = y(a, b, c; t), \quad z = z(a, b, c; t) \tag{5}$$

dargestellt, die als Funktionen aller vier Variablen im notwendigen Umfange stetig und differenzierbar sind; bei festem $a, b, c$ stellt (5) die Bahn eines bestimmten Teilchens dar.

Ganz wie oben betrachtet man dann, indem man in die Formeln nur die Variable $t$ hineinnimmt, neben der Bewegung (5) noch die für $\sigma = 0$ in (5) übergehende Schar von Bewegungen

$$\bar{x} = \bar{x}(a, b, c; t; \sigma) = x + \sigma\delta x(x, y, z; t) \qquad (x, y, z) \tag{6}$$

für kleine Werte des Parameters $\sigma$ und bezeichnet $\delta x, \delta y, \delta z$ als Bestimmungsstücke dieser der Bewegung (5) übergelagerten *virtuellen Verrückung*.

thereby $a, b, c$ remain constant during the differentiation; thus, the operation $\delta$ commutes with the differentiation with respect to $a, b, c$:

$$\delta \frac{\partial f}{\partial a} = \frac{\partial \delta f}{\partial a}.$$

When the 3 functions

$$\left(\frac{\partial \bar{x}}{\partial \sigma}\right)_{\sigma=0} = \left(\frac{\partial \xi}{\partial \sigma}\right)_{\sigma=0} = \delta x(x, y, z) \qquad (x, y, z),$$

which, due to (1), can be seen as functions of $x, y, z$ do not vanish identically in $x, y, z$, then one can set according to the common continuity postulates

$$\text{(3}'\text{)} \qquad \bar{x} = x + \sigma \delta x(x, y, z) \qquad (x, y, z),$$

provided $\sigma$ is so small, that $\sigma^2$ becomes sufficiently small with respect to $\sigma$; up to the factor $\sigma$, the infinitesimal virtual displacement of the continuum as given is determined by the 3 functions $\delta x, \delta y, \delta z$. One can immediately bring these displacements into line with the notion of "infinitesimal deformations" considered in the kinematics of elastic media (cf. IV 14, No. **18**, *Abraham*) and finds particularly, that the "*virtual shape change*" of each volume element is determined by the 6 quantities

$$\text{(4)} \qquad \frac{\partial \delta x}{\partial x}, \frac{\partial \delta y}{\partial y}, \frac{\partial \delta z}{\partial z}, \frac{\partial \delta y}{\partial z} + \frac{\partial \delta z}{\partial y}, \frac{\partial \delta z}{\partial x} + \frac{\partial \delta x}{\partial z}, \frac{\partial \delta x}{\partial y} + \frac{\partial \delta y}{\partial x},$$

and its "*virtual rotation*" is determined by

$$\text{(4}'\text{)} \qquad \frac{1}{2}\left(\frac{\partial \delta z}{\partial y} - \frac{\partial \delta y}{\partial z}\right), \frac{1}{2}\left(\frac{\partial \delta x}{\partial z} - \frac{\partial \delta z}{\partial x}\right), \frac{1}{2}\left(\frac{\partial \delta y}{\partial x} - \frac{\partial \delta x}{\partial y}\right)$$

caused [by the virtual displacement] — in each case apart from the factor $\sigma$.

A *motion of the continuum* is considered as a sequence of states of deformations depending on a time parameter $t$ and is hence represented by the three [following] deformation functions which are now also depending on $t$

$$\text{(5)} \qquad x = x(a, b, c; t), \quad y = y(a, b, c; t), \quad z = z(a, b, c; t)\ ,$$

which as functions of all four variables are sufficiently continuous and differentiable; for fixed $a, b, c$, (5) represents the trajectory of a certain particle.

As above by taking the variable $t$ into the formulas, one considers then besides the motion (5) also the family of motions

$$\text{(6)} \qquad \bar{x} = \bar{x}(a, b, c; t; \sigma) = x + \sigma \delta x(x, y, z; t) \qquad (x, y, z)$$

holding for small values of the parameter $\sigma$ and implying (5) for $\sigma = 0$ and [one] denotes $\delta x, \delta y, \delta z$ as the characteristic quantities of the *virtual displacements* being superimposed on the motion (5).

**2b. Adjunktion physikalischer Parameter, insbesondere Dichte und Orientierung.** Jede physikalische Eigenschaft eines Mediums wird durch eine oder mehrere Funktionen von $a, b, c, t$ beschrieben, die zu den Deformationsfunktionen hinzutreten.

Von einer solchen Eigenschaft wird im folgenden allgemein Gebrauch gemacht werden: der Existenz einer *unveränderlichen Masse* $m$ für jedes Quantum $V_0'$ des Mediums, die sich als über $V_0'$ erstrecktes Integral einer für das Medium charakteristischen *Dichtefunktion* $\varrho_0 = \varrho_0(a, b, c)$ ausdrückt. Durch Übergang zur deformierten Lage (1) ergibt sich als *wirkliche Massendichte* $\varrho$ der Verteilung des Mediums

$$\varrho = \frac{\varrho_0}{\Delta}, \tag{7}$$

und die Masse im Teil $V'$ von $V$ ist

$$m = \iiint\limits_{(V')} \varrho \, dx \, dy \, dz = \iiint\limits_{(V_0')} \varrho_0 \, da \, db \, dc.$$

Veränderungen der Lage des Kontinuums legen an sich bezüglich des Verhaltens eines solchen adjungierten physikalischen Parameters noch nichts fest; man lässt indessen stets die Masse eines jeden Quantums, d. h. die Funktion $\varrho_0(a, b, c)$ bei einer virtuellen Verrückung ungeändert und ersetzt daher die Dichte $\varrho$ derart durch

$$\bar{\varrho} = \bar{\varrho}(x, y, z; \sigma) = \varrho + \sigma \delta \varrho(x, y, z), \tag{8}$$

dass (entsprechend der Kontinuitätsbedingungen, vgl. IV 15, Nr. **7**, p. 59 f. *A. E. H. Love*):

$$\delta \varrho_0 = \delta(\varrho \Delta) = 0 \quad \text{oder} \quad \delta \varrho + \varrho \frac{\partial(\delta x)}{\partial x} + \varrho \frac{\partial(\delta y)}{\partial y} + \varrho \frac{\partial(\delta z)}{\partial z} = 0. \tag{8'}$$

Entsprechendes soll bei einer Bewegung gelten, d. h. $\varrho_0(a, b, c)$ soll von $t$ unabhängig und $\varrho$ alsdann durch (7) gegeben sein.

Noch von einer hierhin gehörigen Begriffsbildung wird häufig Gebrauch zu machen sein, der Annahme nämlich, *dass für jedes Teilchen des Kontinuums die verschiedenen von ihm ausgehenden Richtungen charakteristisch verschiedene Bedeutung besitzen, und dass daher die Angabe seiner Orientierung wesentlich zur Beschreibung der Situation des Kontinuums gehört.* Solche Vorstellungen sind in der Molekulartheorie entstanden, indem man sich Körper von kristallinischer Struktur als Moleküle dachte, und bereits *S. D. Poisson* [9]) hat sie zur Gewinnung einer besseren Molekulartheorie der Elastizität zu verwenden versucht. Neuerdings haben *E.* und *F. Cosserat* [10]) ohne Heranziehung von Mole-

---

[9] Paris, Mém. de l'Acad. 18 (1842), p. 3, sowie in einigen vorangehenden Arbeiten; vgl. die ausführlichen Angaben in IV 23, Nr. **4c**, p. 39 (*Müller-Timpe*).

[10] Paris C. R. 145 (1907), p. 1409; 146 (1908), p. 68. Eine zusammen-

**2b. Introduction of physical parameters, in particular density and orientation.** Every physical property of a medium is described by one or more functions of $a, b, c, t$, which [may need to be] added to the deformation functions.

Subsequently, it will be generally made use of the following property: the existence of a *fixed mass* $m$ for any portion $V_0'$ of the medium, expressed by the integral over the domain $V_0'$ with the integrand being a *density function* $\varrho_0 = \varrho_0(a, b, c)$ which is characteristic for the medium. By transition to the deformed position (1) the *actual mass density* $\varrho$ of the distribution of the medium appears as

$$\varrho = \frac{\varrho_0}{\Delta}, \tag{7}$$

and the mass within the part $V'$ of $V$ is

$$m = \iiint\limits_{(V')} \varrho\, dx\, dy\, dz = \iiint\limits_{(V_0')} \varrho_0\, da\, db\, dc.$$

Changes in the position of the continuum determine nothing with regard to the behavior of such an introduced physical parameter; the mass of any portion, i. e. the function $\varrho_0(a, b, c)$ is in the meanwhile left unchanged for virtual displacements and one exchanges therefore the density $\varrho$ by

$$\bar{\varrho} = \bar{\varrho}(x, y, z; \sigma) = \varrho + \sigma\delta\varrho(x, y, z), \tag{8}$$

such that (analogous to the continuity conditions, cf. IV 15, No. **7**, p. 59 f. *A. E. H. Love*):

$$\delta\varrho_0 = \delta(\varrho\Delta) = 0 \quad \text{or} \quad \delta\varrho + \varrho\frac{\partial(\delta x)}{\partial x} + \varrho\frac{\partial(\delta y)}{\partial y} + \varrho\frac{\partial(\delta z)}{\partial z} = 0. \tag{8'}$$

The same shall hold for a motion, i. e. $\varrho_0(a, b, c)$ shall be independent of $t$ and $\varrho$ is consequently determined by (7).

Additionally, it will be frequently made use of a conceptualization which must be presented here, namely the assumption, *that for any particle of the continuum different directions radiating from these particles may present characteristically different behavior, and that therefore the specification of its orientation is essential in the description of the state of the continuum.* Such perceptions have been developed in the molecular theories, in which the bodies of crystalline structure are thought of as molecules, and already *S. D. Poisson* [9]) has tried to use [such perceptions] to arrive at a better molecular theory of elasticity. Recently, without referring to molecular perceptions, *E.* and *F. Cosserat* [10])

---

[9] Paris, Mém. de l'Acad. 18 (1842), p. 3, as well as some preceding works; cf. the detailed citations in IV 23, No. **4**c, p. 39 (*Müller-Timpe*).

[10] Paris C. R. 145 (1907), p. 1409; 146 (1908), p. 68. They have given a sum-

kularvorstellungen solche in jedem Teilchen mit einer bestimmten Orientierung behafteten Kontinua weitgehend behandelt.

In allgemeinster Weise kann dieser Begriff der orientierten Teilchen des Kontinuums analytisch formuliert werden[11]), indem man sich jedem Teilchen $a, b, c$ des Kontinuums ein *rechtwinkliges Axenkreuz* angeheftet denkt, dessen 3 Axen jeweils die Richtungskosinus $\alpha_i, \beta_i, \gamma_i$ $(i = 1, 2, 3)$ haben; drei unabhängige Parameter $\lambda, \mu, \nu$ (z. B. die Eulerschen Winkel), die die Orientierung eines solchen Dreikants in bezug auf das $x-y-z$-Koordinatensystem bestimmen, müssen neben den Funktionen (1) als Funktionen von $a, b, c$ bekannt sein:

$$\lambda = \lambda(a, b, c), \quad \mu = \mu(a, b, c), \quad \nu = \nu(a, b, c), \tag{9}$$

um den Zustand eines solchen Mediums völlig zu beschreiben.

Mit jeder virtuellen Verrückung des Kontinuums wird man jetzt eine *virtuelle Drehung* dieser Dreikante verbinden, indem man eine von einem Parameter $\sigma$ abhängige und für $\sigma = 0$ verschwindende Schar von Drehungen aus der Lage (9) heraus zugrunde legt und $\lambda, \mu, \nu$ unter Beschränkung auf hinreichend kleine Werte von $\sigma$ durch

$$\bar{\lambda} = \bar{\lambda}(a, b, c; \sigma) = \lambda + \sigma\delta\lambda(a, b, c) \qquad (\lambda, \mu, \nu) \tag{10}$$

ersetzt. Dabei kann man übrigens sowohl $\lambda, \mu, \nu$ als $\delta\lambda, \delta\mu, \delta\nu$ stets entweder als Funktionen von $a, b, c$ oder mit Hilfe von (1) als solche von $x, y, z$ auffassen. Die Variationen $\delta\alpha_1, \ldots, \delta\gamma_3$ der Richtungskosinus der 3 Axen selbst sind lineare homogene Funktionen der $\delta\lambda, \delta\mu, \delta\nu$ die man aus den expliziten Ausdrücken von $\alpha_1, \ldots, \gamma_3$ durch Differentiation nach $\sigma$ erhält; die Komponenten $\delta\pi, \delta\kappa, \delta\varrho$ der Winkelgeschwindigkeit der virtuellen Drehung nach den 3 Axen, die mit $\delta\alpha_1, \ldots, \delta\gamma_3$ durch die Formeln

$$\delta\pi = \beta_1\delta\gamma_1 + \beta_2\delta\gamma_2 + \beta_3\delta\gamma_3 = -(\gamma_1\delta\beta_1 + \gamma_2\delta\beta_2 + \gamma_3\delta\beta_3) \begin{pmatrix}\pi, \kappa, \varrho\\ \alpha, \beta, \gamma\end{pmatrix}, \tag{11}$$

$$\delta\alpha_i = \gamma_i\delta\kappa - \beta_i\delta\varrho \qquad \left(i = 1, 2, 3;\ \begin{matrix}\alpha, \beta, \gamma\\ \pi, \kappa, \varrho\end{matrix}\right) \tag{11'}$$

zusammenhängen und die übrigens, im Gegensatz zu dem bisherigen Gebrauch des Zeichens $\delta$, nicht Variationen bestimmter Funktionen von $a, b, c$ sind, werden also gleichfalls lineare homogene Funktionen von $\delta\lambda, \delta\mu, \delta\nu$, wir setzen

$$\delta\lambda = l_1\delta\pi + m_1\delta\kappa + n_1\delta\varrho \qquad \begin{pmatrix}\lambda, \mu, \nu\\ 1, 2, 3\end{pmatrix}. \tag{12}$$

---

fassende Darstellung haben sie in ihrer „théorie des corps déform."[5]) gegeben. Vgl. auch IV 11, II. Teil, *K. Heun.*

[11] Vgl. eine Bemerkung von *P. Duhem*, Ann. Éc. Norm. (3) 10 (1893), p. 206.

have extensively treated continua in which any particle is given a certain orientation.

In the most general way such a notion of oriented particles of the continuum can be formulated analytically[11]), by thinking that every particle $a, b, c$ of the continuum is endowed with an attached *orthonormal triad*, whose 3 axes have the directional cosines $\alpha_i, \beta_i, \gamma_i$ $(i = 1, 2, 3)$; besides the functions (1), three independent parameters $\lambda, \mu, \nu$ (e. g. Euler angles) must be given as functions of $a, b, c$ to determine the orientation of such a triad with respect to the $x$-$y$-$z$-coordinate system:

$$\text{(9)} \qquad \lambda = \lambda(a, b, c), \quad \mu = \mu(a, b, c), \quad \nu = \nu(a, b, c),$$

in order to describe the state of such a medium completely.

With every virtual displacement of the continuum a *virtual rotation* comes along by taking a family of rotations with respect to the orientation of the triads (9) depending on a parameter $\sigma$ which vanishes for $\sigma = 0$ and by exchanging $\lambda, \mu, \nu$ for sufficiently small values of $\sigma$ by

$$\text{(10)} \qquad \bar{\lambda} = \bar{\lambda}(a, b, c; \sigma) = \lambda + \sigma\delta\lambda(a, b, c) \qquad (\lambda, \mu, \nu).$$

Thereby, incidentally one can consider both $\lambda, \mu, \nu$ and $\delta\lambda, \delta\mu, \delta\nu$ always either as functions of $a, b, c$, or with the help of (1) as functions of $x, y, z$. The variations $\delta\alpha_1, \ldots, \delta\gamma_3$ of the directional cosines of the three axes are themselves linear homogeneous functions of $\delta\lambda, \delta\mu, \delta\nu$ being obtained by differentiation of the explicit expression of $\alpha_1, \ldots, \gamma_3$ with respect to $\sigma$; the components $\delta\pi, \delta\kappa, \delta\varrho$ of the angular velocity of the virtual rotation with respect to the 3 axes, which are connected to $\delta\alpha_1, \ldots, \delta\gamma_3$ by the formulas

$$\text{(11)} \qquad \delta\pi = \beta_1\delta\gamma_1 + \beta_2\delta\gamma_2 + \beta_3\delta\gamma_3 = -(\gamma_1\delta\beta_1 + \gamma_2\delta\beta_2 + \gamma_3\delta\beta_3) \begin{pmatrix} \pi, \kappa, \varrho \\ \alpha, \beta, \gamma \end{pmatrix},$$

$$\text{(11}'\text{)} \qquad \delta\alpha_i = \gamma_i\delta\kappa - \beta_i\delta\varrho \qquad \left(i = 1, 2, 3; \; \begin{matrix} \alpha, \beta, \gamma \\ \pi, \kappa, \varrho \end{matrix}\right)$$

and which by the way, in contrast to the former application of the symbol $\delta$, are not variations of a particular function of $a, b, c$, are in the same way linear homogeneous functions of $\delta\lambda, \delta\mu, \delta\nu$, and we set

$$\text{(12)} \qquad \delta\lambda = l_1\delta\pi + m_1\delta\kappa + n_1\delta\varrho \qquad \begin{pmatrix} \lambda, \mu, \nu \\ 1, 2, 3 \end{pmatrix}.$$

marizing version in their "théorie des corps déform."[5]). Cf. also IV 11, II. Teil, *K. Heun*.

[11] Cf. a note of *P. Duhem*, Ann. Éc. Norm. (3) 10 (1893), p. 206.

Daher bestimmen auch $\delta\pi, \delta\kappa, \delta\varrho$ (als Funktionen von $a, b, c$ oder $x, y, z$ gegeben) die virtuelle Verdrehung des Kontinuums.[12])

Alle diese Formeln lassen sich durch Aufnahme des Zeitparameters $t$ sofort auf den Fall der *Bewegung* ausdehnen.

**2c. Zwei- und eindimensionale Kontinua.** Durch Unterdrückung eines bzw. zweier der drei Parameter $a, b, c$ erhält man endlich unmittelbar auch die Ansätze zur Behandlung *zwei- und eindimensionaler Kontinua*, die *im dreidimensionalen Raume* gelegen sind.[13]) Ihre Lage in jedem Zustand wird gegeben durch

$$(13) \qquad x = x(a, b) \quad \text{bzw.} \quad x = x(a) \qquad (x, y, z);$$

die Parameter variieren in einem Bereich $S_0$ bzw. $C_0$ der $a$-$b$-Ebene bzw. der $a$-Axe, welcher durch (13) auf eine Fläche $S$ bzw. eine Kurve $C$ abgebildet wird. Auch hier kann man jedem Teilchen ein Axenkreuz aus *drei* zueinander senkrechten Richtungen zugeordnet denken [14]), das durch die Funktionen bestimmt wird

$$(14) \qquad \lambda = \lambda(a, b) \quad \text{bzw.} \quad \lambda = \lambda(a) \qquad (\lambda, \mu, \nu).$$

## I. Die Grundansätze der Statik.

### 3. Das Prinzip der virtuellen Verrückungen.

**3a. Kräfte und Spannungen.** Um auf diesem kinematischen Schema die dynamischen Eigenschaften des Kontinuums aufzubauen, knüpfen wir an den *Arbeitsbegriff* an. Der Gesamtheit der auf das Kontinuum infolge seines gegenwärtigen Deformationszustandes, infolge seiner Lage im Raume oder infolge irgendwelcher äusserer Umstände wirkenden Kräfte und Spannungen aller möglichen Arten — zunächst als Ganzes ohne Rücksicht auf ihren Ursprung betrachtet — ist gemeinsam, dass sie bei jeder virtuellen Verrückung eine *„virtuelle Arbeit"* $\delta A$ leisten; diese sehen wir als primär an und bestimmen sie folgendermassen: *$\delta A$ sei als lineare homogene Funktion der Gesamtheit der Werte der Verrückungskomponenten $\delta x, \delta y, \delta z$ innerhalb des Kontinuums gegeben und sei eine skalare, von der Wahl des Koordinatensystems unab-*

[12] Es sind die bekannten kinematischen Methoden der Flächentheorie (vgl. etwa III D 3, Nr. **10**, *R. v. Lilienthal* und *G. Darboux*, Leçons sur la théorie générale des surfaces), die *E.* und *F. Cosserat* hier zur Anwendung bringen (s. die ausführliche Darstellung in den „corps déform." ).

[13] In gewissem Sinne sind diese Probleme einfacher als die dreidimensionale Medien betreffenden; in der Tat gehören einzelne von ihnen zu den am frühesten eingehend behandelten Aufgaben der Mechanik der Kontinua (vgl. IV, 6, Nr. **22—24**, *P. Stäckel* und IV 11, Nr. **19, 20**, *K. Heun*).

[14] Vgl. die bei den in 10) zitierten Pariser Noten von *E.* u. *F. Cosserat* und Cap. II, III ihrer „corps déform."

Thus $\delta\pi, \delta\kappa, \delta\varrho$ (given as functions of $a, b, c$ or $x, y, z$) determine the virtual rotation of the continuum.[12])

By adding the time parameter $t$, all these formulas can immediately be extended to the case of a *motion*.

**2c. Two- and one-dimensional continuua.** By suppressing one or two of the three parameters $a, b, c$, one obtains immediately the basis for the treatment of *two- and one-dimensional continua*, which are embedded *in the three-dimensional space*.[13]) The position in every state is given by

$$(13) \qquad x = x(a, b) \quad \text{or} \quad x = x(a) \qquad (x, y, z);$$

the parameters vary in the domains $S_0$ and $C_0$ of the $a$-$b$-plane and the $a$-axis, respectively, which are mapped by (13) onto a surface $S$ and a curve $C$, respectively. Here, too, one can think of every particle with an attached triad consisting of *three* orthogonal directions[14]), which is determined by the functions

$$(14) \qquad \lambda = \lambda(a, b) \quad \text{or} \quad \lambda = \lambda(a) \qquad (\lambda, \mu, \nu).$$

## I. The foundations of statics.

### 3. The principle of virtual displacements.

**3a. Forces and stresses.** To build the dynamic properties of the continuum on this kinematic scheme, we take up the *notion of work*. The collection of forces and stresses of any kind which act on the continuum due to the current state of deformation, due the position in space or due to any external circumstances — for the moment in its entirety without considering its cause — they have in common, that for any virtual displacement they expend a *"virtual work"* $\delta A$; we see this [virtual work] as primitive and determine it as follows: *Let $\delta A$ be a linear homogeneous function of the entirety of values of the displacement components $\delta x, \delta y, \delta z$ within the continuum and let it be a scalar quantity, independent of the choice of the coordinate system.*

---

[12] There are the known kinematic methods of the geometry of surfaces (cf. for instance III D 3, No. **10**, *R. v. Lilienthal* and *G. Darboux*, Leçons sur la théorie générale des surfaces), which are applied here by *E.* and *F. Cosserat* (see for the detailed version in "corps déform.").

[13] In a certain manner these problems are easier than the ones for three-dimensional media; In fact, a few of them belong to the earliest problems which have been treated thoroughly in the mechanics of continua; (cf. IV, 6, No. **22—24**, *P. Stäckel* and IV 11, No. **19, 20**, *K. Heun*).

[14] Cf. the notes of Paris referred to in 10) of *E.* and *F. Cosserat* and Cap. II, III of their "corps déform."

*hängige Grösse.* Die Koeffizienten, mit denen die Einzelwerte von $\delta x, \ldots$ in $\delta A$ eingehen, sind die Bestimmungsstücke der einzelnen wirkenden Kraftsysteme; dass sie von den virtuellen Verrückungen selbst unabhängig sind (d. h. die Linearität des $\delta A$), bringt die Annahme zum Ausdruck, dass diese Verrückungen ihrer Kleinheit halber die auf jedes Teilchen ausgeübten Kraftwirkungen nicht modifizieren.

Um die sämtlichen Ansätze der Mechanik der Kontinua zu umfassen, ist es nicht notwendig, von dem allgemeinsten Ausdruck der beschriebenen Art für $\delta A$ auszugehen, der aus einer Summe von linearen Funktionen der Werte von $\delta x, \delta y, \delta z$ und ihren Ableitungen an irgendwelchen einzelnen Stellen des Kontinuums sowie von Linien-, Flächen- und Raumintegralen solcher Ausdrücke bestehen würde. Wir betrachten vielmehr zunächst einen Ausdruck — den wir später noch erweitern werden —, der aus einem über das ganze Gebiet $V$ des Kontinuums erstreckten Raumintegral sowie einem über seine Oberfläche $S$ erstreckten Flächenintegral besteht und dabei in dem ersteren noch eine Linearform der 9 Ableitungen der $\delta x, \delta y, \delta z$ nach $x, y, z$ enthält[15]):

$$(1)\qquad \begin{aligned} \delta A = &\iiint\limits_{(V)} \varrho(X\delta x + Y\delta y + Z\delta z)dV && = \delta A_1 \\ &- \iiint\limits_{(V)} \Big(X_x \frac{\partial \delta x}{\partial x} + X_y \frac{\partial \delta x}{\partial y} + \cdots + Z_z \frac{\partial \delta z}{\partial z}\Big)dV && + \delta A_2 \\ &+ \iint\limits_{(S)} (\overline{X}\delta x + \overline{Y}\delta y + \overline{Z}\delta z)dS && + \delta A_3 \end{aligned}$$

Die 15 hier auftretenden und sogleich näher zu diskutierenden Koeffizienten der Verrückungsgrössen sollen nun für jede Deformation des betrachteten Mediums bestimmte *überall endliche und nebst ihren Ableitungen, event. mit Ausnahme einzelner Flächen, stetige Funktionen* von

[15] Solche Ansätze für die virtuelle Arbeit sind als naheliegende Verallgemeinerungen der Formeln der Punktmechanik bei vielen speziellen Problemen früh entwickelt worden. Fast selbstverständlich war die Form der Summanden $\delta A_1$, $\delta A_3$, die ja nur das Summenzeichen der Punktmechanik durch das Integral ersetzen (vgl. etwa *Lagrange, Méc. an.*, 1. part, IV, 11); aber auch Terme von der Form $\delta A_2$ nur sehr spezialisiert, hat *Lagrange* schon z. B. bei der Behandlung der ausdehnbaren Fadens und der kompressiblen Flüssigkeit benutzt, Terme nämlich, die der Variation der Länge bzw. der Variation der Dichte proportional sind (s. *Méc. an.*, 1. part, V, 42; VIII,1). Darüber hinaus ist die Ausbildung des allgemeinen Ansatzes (5) jedenfalls durch die Auffassung der virtuellen Arbeit als Variation eines „Potentiales“ (s. Nr. **7**) angeregt worden, wie sie *C. L. Navier* in die Elastizitätstheorie einführte (s. IV 23, Nr. **5**, *Müller-Timpe*).

The coefficients, which enter $\delta A$ together with the values of $\delta x, \ldots$, are the characteristic quantities of the individual acting force systems; the independence of these components of the virtual displacements (i. e. the linearity of $\delta A$), expresses the assumption that these displacements, due to their smallness, do not modify the force effects exerted on every particle.

In order to include all fundamental equations of the mechanics of continua, it is not necessary to begin with the most general expression of the described form of $\delta A$, which would consist of a sum of linear functions of values of $\delta x, \delta y, \delta z$ and their derivatives at certain locations of the continuum as well as line, surface and volume integrals of such expressions. Instead, we consider at first an expression — which we will extend later on —, which consists of a volume integral over the whole domain $V$ of the continuum as well as a surface integral of its surface $S$ and thereby the former includes in addition a linear form of the 9 derivatives of $\delta x, \delta y, \delta z$ with respect to $x, y, z$[15]):

$$
\text{(1)}\quad
\begin{aligned}
\delta A = & \iiint\limits_{(V)} \varrho(X\delta x + Y\delta y + Z\delta z)dV && = \delta A_1 \\
& - \iiint\limits_{(V)} \left(X_x \frac{\partial \delta x}{\partial x} + X_y \frac{\partial \delta x}{\partial y} + \cdots + Z_z \frac{\partial \delta z}{\partial z}\right)dV && + \delta A_2 \\
& + \iint\limits_{(S)} (\overline{X}\delta x + \overline{Y}\delta y + \overline{Z}\delta z)dS && + \delta A_3
\end{aligned}
$$

The 15 coefficients of the displacement quantities, which appear in here and which are going to be discussed immediately in more detail, shall be, for any deformation of the considered medium, definite functions of $x, y, z$ or $a, b, c$ being *along with their derivatives everywhere bounded and, possibly with exceptions at individual surfaces, continuous*;

15 Such fundamental equations for the virtual work have early been developed as obvious generalization to the formulas of point mechanics for many special problems. Almost naturally was the form of the summands $\delta A_1$, $\delta A_3$, which replaces merely the sigma sign from point mechanics with the integral (cf. for instance *Lagrange, Méc. an.*, 1. part, IV, 11); but also terms of the form $\delta A_2$ only in very special form have been used by *Lagrange* e. g. for the treatment of the extensible wire and the compressible fluid, namely terms, which are proportional to the variation of the length or density, respectively (see *Méc. an.*, 1. part, V, 42; VIII,1). Moreover, the development of the generalized approach (5) has been initiated by the opinion to consider the virtual work as variation of a "potential" (see No. 7), as introduced by *C. L. Navier* in the theory of elasticity (see IV 23, No. **5**, *Müller-Timpe*).

$x, y, z$ oder $a, b, c$ sein; dann ist der anschauliche Sinn des Ansatzes (1), dass lediglich im allgemeinen stetig über den Raum sowie über einzelne Oberflächen verteilte *Kräfte* und stetig verteilte *Spannungen* berücksichtigt werden.

Zunächst sind nämlich der erste und letzte Summand von $\delta A$ den bekannten Arbeitsausdrücken der Punktmechanik ganz analog gebaut, nur dass als Faktor die Masse eines Volumelementes $\varrho dV$ bzw. das Flächenelement $dS$ auftritt; daher sind $X, Y, Z$ als Komponenten der auf die Masseneinheit des Mediums und $\overline{X}, \overline{Y}, \overline{Z}$ als Komponenten der auf die Flächeneinheit der Oberfläche berechneten an der betr. Stelle wirkenden Kraft zu deuten. Da $\delta x, \delta y, \delta z$ Axenprojektionen eines polaren Vektors sind, und da $\delta A$ als Skalar bei Koordinatentransformationen invariant bleibt, substituieren sich diese Kraftkomponenten bei Änderungen des rechtwinkligen Koordinatensystemes wie $\delta x, \delta y, \delta z$[16]): *diese Kräfte sind polare Vektoren.*

Eigentlich für die Mechanik der Kontinua charakteristisch ist der Summand $\delta A_2$. Die 9 Koeffizienten $X_x, X_y, \ldots, Z_z$ — in der bekannten *Kirchhoff*schen[17]) Bezeichnung —, die die Einwirkung der einzelnen Bestimmungsstücke der virtuellen Deformation auf die Arbeitsleistung messen, wird man als die *Komponenten des Spannungszustandes (stress)* an der betr. Stelle deuten, berechnet nach seiner Wirkung auf die Volumeneinheit. Ihr Verhalten bei Koordinatentransformationen ergiebt sich aus der Bemerkung, dass die 9 Ableitungen $\frac{\partial \delta x}{\partial x}, \ldots, \frac{\partial \delta z}{\partial z}$ von Vektorkomponenten sich bei orthogonalen Koordinatentransformationen wie die 9 Produkte aus den Komponenten zweier Vektoren (eine sog. *Dyade*[18]))

$$X_1 \cdot X_2, \quad X_1 \cdot Y_2, \quad \ldots, \quad Z_1 \cdot Z_2$$

[16] Vgl. IV 14, Nr. **2**, *Abraham*.

[17] J. f. Math. 56 (1858) = *G. Kirchhoff* Ges. Abhandl. (Leipzig 1882), p. 287.

[18] Die hiermit angedeutete Definition der Dyade als Komplex von Grössen mit bestimmtem Verhalten gegenüber den rechtwinkligen Koordinatentransformationen („Hauptgruppe" der räumlichen Änderungen), die durchaus im Kreise der *F. Klein*schen Auffassung der Geometrie, Vektoranalysis usw. liegt (vgl. insbesondere Zeitschr. f. Math. Phys. 47 (1902), p. 237 und Math. Ann. 62 (1906), p. 419, die Darstellung in IV 14, *Abraham* sowie *F. Klein*, Elementarmath. v. höh. Standp. aus, Bd. 2, 2. Aufl., Leipzig 1913, p. 90 ff., p. 534) scheint bisher noch nicht zur Grundlage einer selbständigen Darstellung gemacht zu sein. Der Name „dyadics" stammt von *J. W. Gibbs* (s. *Gibbs* und *Wilson*, Vektor Analysis, New York 1901, p. 260 ff.), der sie aus sog. linearen Vektorfunktionen entstehen lässt; von hier aus sind sie auch in die deutsche Litteratur übergegangen (vgl. IV 11, Nr. **1**c, *K. Heun*). Fasst man eine Dyade als Matrix von $3 \cdot 3$ Elementen auf, so ist der Dyadenkalkul in dem *Cayley*schen Matricenkalkul enthalten (vgl. über diesen I A 4, Nr. **10**[19]), *Study*).

in that case, the concrete meaning of the ansatz (1) is that we merely consider *forces*, which are in general continuously distributed on spatial domains as well as on individual surfaces, and [that we only take] continuously distributed *stresses* [into account].

To begin with, the first and the last summand of $\delta A$ are similar to the familiar work expressions of point mechanics, besides the appearance of the mass of a volume element $\varrho dV$ and the surface element $dS$ as factor, respectively; thus $X, Y, Z$ and $\overline{X}, \overline{Y}, \overline{Z}$ are to be interpreted as components of forces per unit mass of the medium and per unit area, respectively, acting at their corresponding position. Since $\delta x, \delta y, \delta z$ are the components of a polar vector, and since $\delta A$ remains as scalar invariant under coordinate transformations, for a change of the orthogonal coordinate system these force components transform like $\delta x, \delta y, \delta z$[16]): *these forces are polar vectors.*

Rather characteristic for the mechanics of continua is the summand $\delta A_2$. The 9 coefficients $X_x, X_y, \ldots, Z_z$ — in the familiar notation of *Kirchhoff*[17]) —, which measure the influence of the individual characteristic quantities of the virtual deformation on the expended work, can be interpreted as *components of the stress state* at the corresponding position, computed by their action per unit volume. Their behavior under coordinate transformation follows from the remark, that the 9 derivatives $\frac{\partial \delta x}{\partial x}, \ldots, \frac{\partial \delta z}{\partial z}$ of vector components transform under an orthogonal coordinate transformation in the same way as the 9 products of the components of two vectors (a so called *dyad*[18]))

$$X_1 \cdot X_2, \quad X_1 \cdot Y_2, \quad \ldots, \quad Z_1 \cdot Z_2$$

[16] Cf. IV 14, No. **2**, *Abraham.*

[17] J. f. Math. 56 (1858) = *G. Kirchhoff* Ges. Abhandl. (Leipzig 1882), p. 287.

[18] The herewith indicated definition of the dyad as complex of quantities with a particular behavior with respect to an orthogonal coordinate transformation ("basic group" of spatial transformations), which lies definitively within the notion of *F. Klein*'s geometry, vector analysis and more (cf. in particular Zeitschr. f. Math. Phys. 47 (1902), p. 237 and Math. Ann. 62 (1906), p. 419, the presentation in IV 14, *Abraham* as well as *F. Klein*, Elementarmath. v. höh. Standp. aus, Bd. 2, 2. Aufl., Leipzig 1913, p. 90 ff., p. 534) seems hitherto not to be the basis of an independent presentation. The name "dyadics" originates from *J. W. Gibbs* (see *Gibbs* and *Wilson*, Vektor Analysis, New York 1901, p. 260 ff.), who gives rise to them starting with so called linear vector functions; from here they have been transmitted to the German literature (cf. IV 11, No. **1**c, *K. Heun*). If one considers a dyad as a matrix of 3 · 3 elements, then the dyadic calculus is included within *Cayley*'s matrix calculus (cf. for this I A 4, No. **10**[19]), *Study*).

verhalten, während das bilineare Aggregat $X_x \cdot \frac{\partial \delta x}{\partial x} + \cdots$ invariant bleibt; daher müssen sich die Spannungskomponenten selbst wiederum wie Dyadenkomponenten transformieren, so dass man von einer *Spannungsdyade* spricht. Man kann diese, wie jede Dyade, zerspalten in einen (symmetrischen) Bestandteil von 6 Komponenten (ein *Tensortripel*[19]))

$$X_x, Y_y, Z_z, \tfrac{1}{2}(Y_z + Z_y), \tfrac{1}{2}(Z_x + X_z), \tfrac{1}{2}(X_y + Y_x) \tag{2}$$

und einen (schiefsymmetrischen) Bestandteil von 3 Komponenten

$$Z_y - Y_z,\ X_z - Z_x,\ Y_x - X_y, \tag{2'}$$

der einen *axialen Vektor* darstellt. Diese Zerlegung entspricht der in Nr. **2** angegebenen Hervorhebung zweier gesonderter Bestandteile (4), (4′) der virtuellen Deformation des Kontinuums, und ist aus ihr direkt zu entnehmen, wenn man den Integranden von $\delta A_2$ so zerlegt:

$$\sum_{\binom{x\,y\,z}{X\,Y\,Z}} \left\{ X_x \frac{\partial \delta x}{\partial x} + \tfrac{1}{2}(Y_z + Z_y)\left(\frac{\partial \delta y}{\partial z} + \frac{\partial \delta z}{\partial y}\right) + (Z_y - Y_z)\tfrac{1}{2}\left(\frac{\partial \delta z}{\partial y} - \frac{\partial \delta y}{\partial z}\right)\right\}.^{20)}$$

(Vgl. die Entwicklung in IV 14, Nr. **19**, *Abraham.*)

Insbesondere folgt hieraus, dass die 6 Grössen (2) denjenigen Teil des Spannungszustandes bestimmen, der bei einer unendlichkleinen reinen Formänderung des Kontinuums Arbeit leistet, also die *eigentlichen elastischen Wirkungen*, der Vektor (2′) aber denjenigen, der bei einer virtuellen Drehung der Volumenelemente, auch ohne Formänderung, in Betracht kommt, also die durch den Spannungszustand bedingten *Drehmomente*. Aus dem negativen Vorzeichen in (1) ergibt sich weiter, dass bei positivem $X_x$ positive Arbeit bei negativem $\frac{\partial \delta x}{\partial x}$ geleistet wird, dass also *Druck positiv* gemessen ist.

Um endlich die Bedeutung der Spannungskomponenten als *Flächen-kräfte* aus dem Ansatz (1) zu gewinnen[21]), denke man sich den Teil der virtuellen Arbeit berechnet, den die Spannungen innerhalb eines von der geschlossenen Fläche $S_1$ begrenzten Teilbereiches

[19] In der Bezeichnung von *W. Voigt*; vgl. darüber IV 14, Nr. **17** *M. Abraham*.

[20] Die Indizes am Summenzeichen und die analogen in der Folge bedeuten, daß die zu summierenden Ausdrucke durch gleichzeitige zyklische Vertauschung von $x, y, z$ und $X, Y, Z$ entstehen

[21] Das Folgende enthält die Überlegungen, die man seit *C. L. Navier* und *G. Green* macht, um aus dem Ansatz des elastischen Potentiales die Grundgleichungen nebst ihrer anschaulichen Bedeutung zu erhalten; man vergleiche das historische Referat in IV 23, Nr. **5** (*Müller-Timpe*) sowie z. B. die Darstellung in *H. v. Helmholtz*, Vorles. über theoret. Phys. II (Leipzig 1902), § 23.

while the bilinear aggregate $X_x \cdot \frac{\partial \delta x}{\partial x} + \cdots$ remains invariant; hence the stress components must also transform like the components of a dyad, with the result that one speaks of a *stress dyadic*. One can compose it, as any dyad, into a (symmetric) part with 6 components (a *tensor triple*[19]))

$$(2) \qquad X_x, Y_y, Z_z, \tfrac{1}{2}(Y_z + Z_y), \tfrac{1}{2}(Z_x + X_z), \tfrac{1}{2}(X_y + Y_x)$$

and a (skew-symmetric) part with 3 components

$$(2') \qquad Z_y - Y_z,\ X_z - Z_x,\ Y_x - X_y,$$

representing an *axial vector*. This decomposition corresponds to the two separate parts (4), (4′) of the virtual deformation of the continuum considered in No. **2**, and is obtained directly by decomposing the integrand of $\delta A_2$ as follows:

$$\sum_{\binom{x\,y\,z}{X\,Y\,Z}} \left\{ X_x \frac{\partial \delta x}{\partial x} + \tfrac{1}{2}(Y_z + Z_y)\left(\frac{\partial \delta y}{\partial z} + \frac{\partial \delta z}{\partial y}\right) + (Z_y - Y_z)\tfrac{1}{2}\left(\frac{\partial \delta z}{\partial y} - \frac{\partial \delta y}{\partial z}\right)\right\}.^{20)}$$

(Cf. the derivation in IV 14, No. **19**, *Abraham.*)

In particular it follows, that the 6 quantities (2) determine this part of the stress state, which expends work for an infinitesimally pure shape change of the continuum, i. e. the *actual elastic action*, the vector (2′) on the other hand determines that part, which can be considered for a virtual rotation of the volume element, also without shape change, i. e. *torques* induced by the stress state. From the negative sign of (1) it follows furthermore, that for positive $X_x$ and negative $\frac{\partial \delta x}{\partial x}$ positive work is expended, such that *pressure* is consequently measured *positive*.

To obtain from the ansatz (1) finally the interpretation of the stress components as *surface forces*[21]), one considers the virtual work contribution expended by a subdomain $V_1$ bounded by a closed surface $S_1$,

[19] In the notation of *W. Voigt*; cf. in addition IV 14, No. **17** *M. Abraham*.

[20] The indices of the sigma sign and similar ones in the following denote that the expressions to be summed arise by simultaneous cyclic permutation of $x$, $y$, $z$ and $X, Y, Z$

[21] The following includes the ideas, which are set since *C. L. Navier* and *G. Green*, to obtain from the ansatz concerning the elastic potential the fundamental equations in addition with its intuitive explanation; one should compare the historical presentation in IV 23, No. **5** (*Müller-Timpe*) as well as e. g. the presentation in *H. v. Helmholtz*, Vorles. über theoret. Phys. II (Leipzig 1902), § 23.

$V_1$ des Kontinuums leisten, d. i. das über $V_1$ erstreckte Teilintegral von $\delta A_2$; sind die Spannungskomponenten innerhalb $V_1$ ausnahmslos stetig, so geht dies durch partielle Integration (Anwendung des „Gaussschen Satzes“ , s. IV 14, p. 12) über in

$$\iiint\limits_{(V_1)} \sum_{\binom{x\,y\,z}{X\,Y\,Z}} \left(\frac{\partial X_x}{\partial x} + \frac{\partial X_y}{\partial y} + \frac{\partial X_z}{\partial z}\right)\delta x \cdot dV$$

$$+ \iint\limits_{(S_1)} \sum_{\binom{x\,y\,z}{X\,Y\,Z}} (X_x \cos nx + X_y \cos ny + X_z \cos nz)\delta x \cdot dS_1,$$

wo $n$ die nach $V_1$ hin gewendete Normalenrichtung der Fläche $S_1$ an der Stelle des Elementes $dS_1$ bedeutet. Durch Vergleich mit (1) folgt also, dass der Spannungszustand in $V_1$ die gleiche virtuelle Arbeit leistet, d. h. gerade so wirkt, als ob neben Volumenkräften in $V_1$ auf das Flächenelement $dS_1$ von $S_1$ pro Flächeneinheit berechnet die Kraft

$$X_n = X_x \cos nx + X_y \cos ny + X_z \cos nz, \qquad (X, Y, Z) \tag{3}$$

wirkt. Dieses *Cauchy*sche „Drucktheorem“ liefert dann bekanntlich durch Spezialisierung der Richtung von $n$ unmittelbar die Bedeutung der 9 Komponenten (vgl. IV 23, Nr. **3**a, *Müller-Timpe*).

**3b. Aufstellung des Prinzips der virtuellen Verrückungen.** Auf Grund dieser Begriffsbildungen lässt sich das *Prinzip der virtuellen Verrückungen*, das die Statik der diskontinuierlichen mechanischen Systeme beherrscht[22]), unmittelbar auf die Mechanik der Kontinua übertragen: *In einem bestimmten Deformationszustand ist ein kontinuierliches Medium, in dem gewisse Volumen- und Oberflächenkräfte* $X, \ldots$ *und* $\overline{X}, \ldots$ *und ein gewisser Spannungszustand* $X_x, \ldots$ *bestehen, dann und nur dann im Gleichgewicht, wenn die gesamte virtuelle Arbeit dieser Kräfte und Spannungen für jede virtuelle Verrückung, die mit den dem Kontinuum etwa auferlegten Nebenbedingungen verträglich ist, verschwindet:*

$$\iiint\limits_{(V)} \left\{\varrho \sum_{\binom{x\,y\,z}{X\,Y\,Z}} X\delta x - \sum_{\binom{x\,y\,z}{X\,Y\,Z}} \left(X_x \frac{\partial \delta x}{\partial x} + X_y \frac{\partial \delta x}{\partial y} + X_z \frac{\partial \delta x}{\partial z}\right)\right\} dV \tag{4}$$

$$+ \iint\limits_{(S)} \sum_{\binom{x\,y\,z}{X\,Y\,Z}} \overline{X}\delta x \cdot dS = 0.$$

Diese Übertragung hat tatsächlich bereits *J. L. Lagrange* [23]) vollzogen, nachdem er das *Bernoulli*sche Prinzip der virtuellen Verrückungen

[22] Vgl. IV 1, Nr. **30**, *Voss*.

[23] Mécan. anal., 1. part., sect. IV. § II, sowie bei einer Reihe spezieller Probleme in sect. V—VIII.

which is the integral over the part $V_1$ of $\delta A_2$; for continuous stress components within $V_1$, [this virtual work contribution] is transformed further by integration by parts (using the "Theorem of Gauss", s. IV 14, p. 12), to

$$\iiint\limits_{(V_1)} \sum_{\binom{x\,y\,z}{XYZ}} \left(\frac{\partial X_x}{\partial x} + \frac{\partial X_y}{\partial y} + \frac{\partial X_z}{\partial z}\right)\delta x \cdot dV$$
$$+ \iint\limits_{(S_1)} \sum_{\binom{x\,y\,z}{XYZ}} (X_x \cos nx + X_y \cos ny + X_z \cos nz)\delta x \cdot dS_1,$$

where $n$ denotes the normal of the surface $S_1$ at the position of the element $dS_1$ pointing in direction of $V_1$. By comparison with (1) it follows consequently, that the stress state in $V_1$ expends the same virtual work, i. e. acts equally as if besides volume forces in $V_1$ the force per area

$$X_n = X_x \cos nx + X_y \cos ny + X_z \cos nz, \qquad (X, Y, Z) \tag{3}$$

were acting on the surface element $dS_1$ of $S_1$. This "pressure theorem" of *Cauchy* provides then, by specializing the directions of $n$, immediately the interpretation of the 9 components (cf. IV 23, No. **3**a, *Müller-Timpe*).

**3b. Formulation of the principle of virtual displacements.** Due to these conceptualizations the *principle of virtual displacements*, dominating the statics of discrete mechanical systems[22]), can be adopted immediately for the mechanics of continua: *A continuous medium in a particular state of deformation, for certain volume and surface forces* $X, \ldots$ *and* $\overline{X}, \ldots$, *respectively, and for a certain stress state* $X_x, \ldots$, *is in equilibrium if and only if the total virtual work of these forces and stresses vanish for every virtual displacement, which is admissible with respect to the possibly imposed constraints of the continuum:*

$$\iiint\limits_{(V)} \left\{\varrho \sum_{\binom{x\,y\,z}{XYZ}} X\delta x - \sum_{\binom{x\,y\,z}{XYZ}} \left(X_x \frac{\partial \delta x}{\partial x} + X_y \frac{\partial \delta x}{\partial y} + X_z \frac{\partial \delta x}{\partial z}\right)\right\} dV$$
$$+ \iint\limits_{(S)} \sum_{\binom{x\,y\,z}{XYZ}} \overline{X}\delta x \cdot dS = 0. \tag{4}$$

This adoption has been implemented in fact already by *J. L. Lagrange* [23]), after postulating *Bernoulli*'s principle of virtual displacements

[22] Cf. IV 1, No. **30**, *Voss*.

[23] Mécan. anal., 1. part., sect. IV. § II, as well as for a series of particular problems in sect. V—VIII.

zur Grundlage seiner analytischen Mechanik gemacht hatte; für ihn ist eine selbstverständliche Folge der Gültigkeit dieses Prinzips in der Punktmechanik seine Anwendbarkeit auf die ihm zugänglichen Probleme der Mechanik der Kontinua, wo immer er den Arbeitsausdruck durch einen Grenzübergang von diskontinuierlichen Systemen aus oder durch direkte Intuition aufzustellen vermag. Man hat seither auch auf den weiteren der Behandlung erschlossenen Gebieten der Mechanik der Kontinua das Prinzip der virtuellen Verrückungen zur Geltung gebracht und hat sich dabei häufig, wie *Lagrange*, auf die Vorstellung gestützt, dass man das Kontinuum durch Systeme von endlichvielen Massenpunkten, und gleichzeitig alle physikalischen Vorgänge im Kontinuum durch entsprechende Vorgänge in diesen approximierenden Systemen annähern kann; freilich scheint eine axiomatische Präzisierung dieses Zusammenhanges, die vor allem die zur Umwandlung jener Analogisierungen in strenge Deduktionen notwendigen Stetigkeitsforderungen zu postulieren hätte, bisher nicht gegeben worden zu sein. Man mag es daher inzwischen vorziehen, für die Mechanik der Kontinua das eingangs formulierte Prinzip selbst als *oberstes Axiom* an die Spitze zu stellen (vgl. IV I, p. 72, *Voss*); man wird diesen Standpunkt um so lieber einnehmen, wenn man die Vorstellung kontinuierlich ausgedehnter Medien für naturgemässer hält als die abstrahierten „Massenpunkte" der Punktmechanik.[24]) Die Gewissheit der Richtigkeit dieses Axioms liegt einerseits darin begründet, dass ein solcher Ansatz unsern allgemeinen physikalischen Anschauungen und Denkgewohnheiten entspricht, vor allem aber darin, dass er anpassungsfähig genug ist, um die Erfahrungstatsachen hinreichend gut darzustellen.

**3c. Anwendung auf stetig deformierbare Kontinua.** Die bekannten formalen Operationen der Variationsrechnung gestatten es in jedem Falle leicht, das Prinzip der virtuellen Verrückungen in eine Anzahl von Gleichungen zwischen den Kräften und Spannungen umzusetzen.[25]) Betrachten wir zunächst nur als typisch das in keiner Weise durch Nebenbedingungen beschränkte *beliebig stetig deformierbare Medium*, so muß die Bedingung (4) für jedes System stetiger Funktionen $\delta x, \delta y, \delta z$ erfüllt sein. Die Umformung von (4) durch

[24] Diese Anschauung hat neuerdings besonders *G. Hamel* (Math. Ann. 66 (1908), p. 350 und Jahresb. d. Math.-Ver. 18 (1909), p. 357; vgl. auch sein Lehrbuch „Elementare Mechanik", Leipzig 1912) vertreten; er giebt dort eine vollständige Axiomatik der Mechanik der Kontinua, die das eine Grundprinzip, wie es hier benutzt ist, in eine Reihe unabhängiger Sätze auflöst

[25] So ist schon *Lagrange* in der Méc. an. bei den dort behandelten Probleme vorgegangen; s. Anm. 23.

as foundation of his analytical mechanics; for him the natural consequences of the validity of this principle of point mechanics is its applicability to problems of the mechanics of continua accessible for himself, whenever he is able to obtain the work expression by a limit process of discrete systems or by direct intuition. Since then one has also applied the principle of virtual displacements to further fields of the mechanics of continua, and has, like *Lagrange*, often based oneself on the perception, to be able to approximate the continuum by a system of finitely many mass points and that at the same time all physical processes in the continuum can be approximated by corresponding processes in these approximated systems; however it does not seem that such an axiomatic clarification of this connection have already been given, [a clarification] concerning the transformation of those intuitions into rigorous deductions would particularly have to postulate the necessary continuity requirements. Thus one may prefer in the meantime for the mechanics of continua, to choose the principle formulated at the beginning as the *highest axiom* (cf. IV I, p. 72, *Voss*); one prefers to take up this position anyway, when one considers the concept of continuously distributed media as more natural than the abstract "mass points" of point mechanics.[24]) The certainty of the validity of this axiom is justified that such an ansatz corresponds with our general physical intuition and habitual ways of thinking, but in particular therein, that it is adaptive to represent the empirical facts sufficiently enough.

**3c. Application to continuously deformable bodies.** The established formal operations of the calculus of variations enable easily, to transform the principle of virtual displacements into a number of equations between forces and stresses.[25]) Consider at first only the *aribtrarily continuously deformable medium* which is typically not at all constrained, then the condition (4) must be satisfied for every system of continuous functions $\delta x, \delta y, \delta z$.

---

[24] This perception has recently been represented in particular by *G. Hamel* (Math. Ann. 66 (1908), p. 350 and Jahresb. d. Math.-Ver. 18 (1909), p. 357; cf. also his textbook "Elementare Mechanik", Leipzig 1912); therein he introduces a complete axiomatic system of the mechanics of continua, in which the fundamental principle, used here, follows from a sequence of independent theorems.

[25] Already*Lagrange* proceeded in the Méc. an. in this way to treat the problems therein; see remark 23.

partielle Integration ergiebt dann, falls Kräfte, Spannungen und deren partielle Ableitungen überall in $V$ stetig sind, die Gleichungen:

1) an jeder Stelle des Bereiches $V$

(5a) $$\frac{\partial X_x}{\partial x} + \frac{\partial X_y}{\partial y} + \frac{\partial X_z}{\partial z} + \varrho X = 0 \qquad (X, Y, Z),$$

2) an jeder Stelle der Oberfläche $S$ mit der *äusseren* Normalenrichtung $n$

(5b) $$X_x \cos nx + X_y \cos ny + X_z \cos nz = \overline{X} \qquad (X, Y, Z).$$

Damit sind die sog. „Spannungsgleichungen" nebst den zugehörigen Oberflächenbedingungen gewonnen, die die *notwendigen und hinreichenden Bedingungen dafür geben, dass ein bestimmtes in einer gewissen Lage auf ein frei deformierbares Kontinuum wirkendes Kraft- und Spannungssystem im Gleichgewicht ist.*[26]) Freilich genügen diese Bedingungen keineswegs, um die Spannungs- und Kraftkomponenten zu *bestimmen*: dazu müssen noch die erst später zu behandelnden Relationen hinzutreten, die die Abhängigkeit der Kräfte und Spannungen von der wirklich stattfindenden Deformation des Kontinuums oder von irgendwelchen äusseren Ursachen zum Ausdruck bringen (vgl. IV 6, Nr. **26**, *Stäckel* und IV 23, **3**b, *Müller-Timpe*).

In (4), (5) sind die unabhängigen Variablen Koordinaten im *deformierten* Zustand des Kontinuums, und auch Kraft- und Spannungskomponenten finden ihre anschauliche Bedeutung als Wirkungen auf Massen- bzw. Flächeneinheiten des Mediums im *deformierten* Zustand. Demgegenüber verwendet man seit *S. D. Poisson* [27]) vielfach auch die $a, b, c$, aufgefasst als Koordinaten in der Ausgangslage des Mediums als unabhängige Variable; das führt zwar auf Kraftkomponenten von weniger unmittelbarer physikalischer Bedeutung, ist aber analytisch für viele Zwecke bequemer. Es wird nämlich, wenn

(6) $$k \cdot dS_0 = dS$$

---

[26] Die Gleichungen gehen auf *A. L. Cauchy* zurück, Exerc. de math. 2 (1827) = Oeuvres 7, sér. II, p. 141. Vgl. die weiteren Angaben hierüber in IV 23, Nr. **3**b, *Müller-Timpe*.

[27] Paris Mém. de l'Acad. 8 (1829), p. 387; J. éc. polyt. 20 (1831), p. 54. Dieser Unterschied ist vielfach übersehen worden, da er bei der Betrachtung unendlichkleiner Deformationen von einem spannungslosen Ruhezustande aus tatsächlich verschwindet; so ist er erst in der Entwicklung der Elastizitätstheorie endlicher Deformationen recht zur Geltung gekommen (vgl. unten Nr. **7** und **9**).

For forces, stresses and partial derivatives being continuous everywhere in $V$, the transformation of (4) yields by integration by parts the equations:

1) for every point of the domain $V$

$$\frac{\partial X_x}{\partial x} + \frac{\partial X_y}{\partial y} + \frac{\partial X_z}{\partial z} + \varrho X = 0 \qquad (X, Y, Z), \tag{5a}$$

2) for every point on the surface $S$ with *outward pointing* normal direction $n$

$$X_x \cos nx + X_y \cos ny + X_z \cos nz = \overline{X} \qquad (X, Y, Z). \tag{5b}$$

Thereby the so-called "equations of stress" together with the corresponding surface conditions are obtained, which give *the necessary and sufficient conditions that a particular force and stress system, acting on a freely deformable continuum in a certain position, is in equilibrium.*[26]) Certainly these conditions are not enough, to *determine* the stress and force components: To this we must add the relations which will be treated later on and which express the dependence of forces and stresses on the actual deformation of the continuum or on any external causes (cf. IV 6, No. **26**, *Stäckel* and IV 23, **3**b, *Müller-Timpe*).

In (4), (5) the independent variables are the coordinates of the *deformed* state of the continuum, and also force and stress components have their descriptive meaning as effect per unit mass or surface of the medium in the deformed state. On the other hand, since *S. D. Poisson* [27]) one often refers to $a, b, c$, being the coordinates of the initial position of the medium as independent variables; indeed, this leads to force components of physical interpretation less immediate, but it is for many cases analytically more convenient. Setting

$$k \cdot dS_0 = dS \tag{6}$$

[26] These equations can be traced back to *A. L. Cauchy*, Exerc. de math. 2 (1827) = Oeuvres 7, sér. II, p. 141. Cf. the further references about this in IV 23, No. **3**b, *Müller-Timpe*.

[27] Paris Mém. de l'Acad. 8 (1829), p. 387; J. éc. polyt. 20 (1831), p. 54. This difference has been frequently overlooked, since it vanishes in fact for the consideration of infinitesimal deformations from a stress free state of equilibrium; so it has shown to be useful only when the development of the theory of elasticity of finite deformations (cf. below No. **7** and **9**) [was achieved].

gesetzt und Nr. **2**, (7) berücksichtigt wird:

$$(7)\quad \delta A = \iiint\limits_{(V_0)} \Big[\varrho_0 \sum_{(xyz)} X\delta x - \sum_{\binom{x\,y\,z}{XYZ}} \Big(X_a \frac{\partial \delta x}{\partial a} + X_b \frac{\partial \delta x}{\partial b} + X_c \frac{\partial \delta x}{\partial c}\Big)\Big] dV_0 + \iint\limits_{(S_0)} \sum_{\binom{x\,y\,z}{XYZ}} \overline{X} k \delta x \cdot dS_0,$$

wobei

$$(8)\qquad \Delta \cdot X_x = X_a \frac{\partial x}{\partial a} + X_b \frac{\partial x}{\partial b} + X_c \frac{\partial x}{\partial c} \qquad (X, Y, Z; x, y, z).$$

Daher sind, wie durch Auflösung und Vergleich mit (3) folgt, $X_a, Y_a, Z_a$ die Komponenten der Flächenkraft, die auf ein Element der Fläche $a$ = const. vermöge des Spannungszustandes in der nach der Seite wachsender $a$ hin gelegenen Materie wirkt, berechnet auf die Einheit der Fläche in der Ausgangslage im $a$-$b$-$c$-Raum.[28]) Aus (7) entsteht eine *neue Form der Gleichgewichtsbedingungen*[28]) genau so wie (5a), (5b) aus (4) entstehen:

$$(9\text{a})\qquad \frac{\partial X_a}{\partial a} + \frac{\partial X_b}{\partial b} + \frac{\partial X_c}{\partial c} + \varrho_0 X = 0 \quad \text{innerhalb } V_0 \qquad (X, Y, Z),$$

$$(9\text{b})\qquad X_a \cos n_0 a + X_b \cdot \cos n_0 b + X_c \cos n_0 c = k\overline{X} \quad \text{auf } S_0 \quad (X, Y, Z);$$

hierbei bedeutet $n_0$ die äussere Normalenrichtung des Flächenelementes $dS_0$ im $a$-$b$-$c$-Raum.

**3d. Beziehungen zur Mechanik starrer Körper.** Man kann die Gleichgewichtsbedingungen (5) noch in etwas anderer Weise aus dem Prinzip (4) herleiten und erhält dadurch den Zusammenhang mit dem nach dem Vorgange von *A. L. Cauchy* [29]) vielfach zu ihrer direkten Aufstellung benutzten *„Erstarrungsprinzip"*, dass jeder aus dem deformierten Kontinuum herausgeschnittene Teil unter der Einwirkung der in seinem Inneren angreifenden Volumkräfte und der an seiner Oberfläche angreifenden Kräfte (3) wie ein starrer Körper im Gleichgewicht sein muss. Zu diesem Ende braucht man nur gewisse *unstetige* Verrückungen zu betrachten, die freilich den Zusammenhang des stetig deformierbaren Kontinuums verletzen und für die $\delta A$ daher zunächst

---

[28] Vgl. IV 23, Nr. **6** *(Müller-Timpe)* und etwa die ausführliche Darstellung (die freilich Symmetrie der Spannungsdyade voraussetzt) bei *E.* und *F. Cosserat*; Ann. de Toulouse, X (1896), p. 146; die Schreibweise $X_a, X_b, \ldots$ erscheint konsequenter als die dort gebrauchte $A_x, B_x, \ldots$, da sie grosse Buchstaben für die Bezeichnung der Komponenten, die Indizes aber für die Charakterisierung des betrachteten Flächenelemente beibehält.

[29] Bull. soc. philomath. 1823, p. 9 und Exerc. de math. 2 (1827) = Oeuvres, sér. II, t. 7, p. 141; vgl. die Angaben in IV 6, Nr. **26**, *Stäckel* und IV 23, Nr. **3**b, *Müller-Timpe*.

and considering No. **2**, (7), we obtain namely:

$$(7)\quad \delta A = \iiint\limits_{(V_0)} \Big[\varrho_0 \sum_{(xyz)} X\delta x - \sum_{\binom{x\,y\,z}{XYZ}} \Big(X_a \frac{\partial \delta x}{\partial a} + X_b \frac{\partial \delta x}{\partial b} + X_c \frac{\partial \delta x}{\partial c}\Big)\Big] dV_0 + \iint\limits_{(S_0)} \sum_{\binom{x\,y\,z}{XYZ}} \overline{X} k \delta x \cdot dS_0,$$

where

$$(8)\qquad \Delta \cdot X_x = X_a \frac{\partial x}{\partial a} + X_b \frac{\partial x}{\partial b} + X_c \frac{\partial x}{\partial c} \qquad (X, Y, Z; x, y, z).$$

Therefore, by solving and comparing with (3) it follows, that $X_a, Y_a, Z_a$ are the components of the surface force, computed per unit area of the initial position in the $a$-$b$-$c$-space, which due to the stress state acts via an element of the surface $a$ = const. on the matter lying on this side for which $a$ is increasing.[28]) From (7) a *new form of the equilibrium conditions*[28]) arises, in the same manner as (5a), (5b) arise from (4):

$$(9a)\qquad \frac{\partial X_a}{\partial a} + \frac{\partial X_b}{\partial b} + \frac{\partial X_c}{\partial c} + \varrho_0 X = 0 \quad \text{in } V_0 \qquad (X, Y, Z),$$

$$(9b)\qquad X_a \cos n_0 a + X_b \cdot \cos n_0 b + X_c \cos n_0 c = k\overline{X} \quad \text{on } S_0 \quad (X, Y, Z);$$

hereby $n_0$ denotes the outward pointing normal direction of the surface element $dS_0$ in the $a$-$b$-$c$-space.

**3d. Relation to the mechanics of rigid bodies.** It is also possible to derive the equilibrium conditions (5) in a slightly different way starting with the principle (4) and thereby one obtains the connection to the "rigidifying principle", frequently used for the direct derivation of the equilibrium conditions according to the approach of *A. L. Cauchy* [29]), stating that every part cut out of the deformable continuum exposed to the volume forces applied within the part and the forces applied on the surface (3) must be in equilibrium like a rigid body. For this, one only has to consider certain *discontinuous* displacements, which certainly violate the connection of the continuously deformable continuum and for which $\delta A$ does

---

[28] Cf. IV 23, No. **6** *(Müller-Timpe)* and for instance the detailed presentation (which presumes certainly the symmetry of the stress dyad) of *E.* and *F. Cosserat*; Ann. de Toulouse, X (1896), p. 146; the notation $X_a, X_b, \ldots$ seems to be more consistent than $A_x, B_x, \ldots$, used there, since it remains the capital letters for the denotation of the components, but the indices for the characterization of the considered surface element.

[29] Bull. soc. philomath. 1823, p. 9 and Exerc. de math. 2 (1827) = Oeuvres, sér. II, t. 7, p. 141; cf. the references in IV 6, No. **26**, *Stäckel* and IV 23, No. **3**b, *Müller-Timpe*.

nicht zu verschwinden braucht: man kommt aber zum Ziele, wenn man sie durch eine Schar *stetiger* virtueller Verrückungen approximiert.

So werde eine Verrückung, die in einem Teilgebiet $V_1$ von $V$ mit der Grenzfläche $S_1$ konstante Werte $\delta x = \alpha$, $\delta y = \beta$, $\delta z = \gamma$ hat, außerhalb $V_1$ aber 0 ist (d. i. eine *Translation* des Bereiches $V_1$), durch stetige virtuelle Verrückungen angenähert, indem $V_1$ mit einem beliebig kleinen Gebiete $V_2$ umgeben wird, innerhalb dessen $\delta x, \delta y, \delta z$ von $\alpha, \beta, \gamma$ nach 0 stetig abfallen. Für eine solche virtuelle Verrückung folgt aus (4):

$$\iiint\limits_{(V_1)} \varrho(X\alpha + Y\beta + Z\gamma)dV_1 + \iint\limits_{(S_1)} (X_n\alpha + Y_n\beta + Z_n\gamma)dS_1$$
$$+ \iiint\limits_{(V_2)} \sum_{\binom{x\,y\,z}{X\,Y\,Z}} \Big(\varrho X + \frac{\partial X_x}{\partial x} + \frac{\partial X_y}{\partial y} + \frac{\partial X_z}{\partial z}\Big)\delta x \cdot dV_2 = 0,$$

wo $n$ die von $V_1$ fortzeigende Normale von $dS_1$ ist. Lässt man nun $V_2$ immer kleiner werden, so wird das zweite Integral beliebig klein, da die $X, X_x, \ldots$ und ihre Ableitungen endlich bleiben, und es ergeben sich, da $\alpha, \beta, \gamma$ beliebig ist, drei Gleichungen

$$\iiint\limits_{(V_1)} \varrho X dV_1 + \iint\limits_{(S_1)} X_n dS = 0 \quad (X, Y, Z). \tag{10}$$

Das sind genau die Gleichungen, die durch Anwendung des sog. *Schwerpunktsatzes* auf den im oben geschilderten Sinne starr gedachten und aus dem Kontinuum herausgeschnittenen Teil $V_1$ entstehen. Wegen der Willkür des Bereiches $V_1$ kann man dann bekanntlich aus (10) die Gleichungen (5a) gewinnen (vgl. IV 23, *Müller-Timpe*, p. 23).

Geht man in ähnlicher Weise von einer starren Drehung eines Teilbereiches $V_1$ mit den Komponenten $qz - ry$, $rx - pz$, $py - qx$ aus, so folgen drei Gleichungen:

$$\iiint\limits_{(V_1)} \{\varrho(Zy - Yz) + Y_z - Z_y\}dV_1 + \iint\limits_{(S_1)} (Z_n y - Y_n z)dS_1 = 0 \ (X, Y, Z). \tag{11}$$

Das stimmt nur dann mit dem auf $V_1$ als starren Körper angewandten *Flächensatz* überein, wenn man den Momenten der räumlich verteilten Kräfte $X, Y, Z$ und der Flächenkräfte $X_n, Y_n, Z_n$ noch ein direkt am Volumelement angreifendes Drehmoment entgegengesetzt gleich dem Vektorbestandteil (2′) der Spannungsdyade hinzurechnet. Postuliert man also den Flächensatz in der üblichen Form, dass die Summe der Momente der Volumen- und Flächenkräfte verschwindet, so folgt daraus unmittelbar die Symmetrie der Spannungsdyade.[30])

[30] Diese Forderung hat *G. Hamel*[24]) als „Boltzmannsches Axiom" unter seine Axiome der Mechanik der Volumenelemente aufgenommen.

not have to vanish at first: though one succeeds by approximating it with a family of *continuous* virtual displacements.

Hence, a displacement, which on a subset $V_1$ of $V$ with boundary $S_1$ has constant values $\delta x = \alpha$, $\delta y = \beta$, $\delta z = \gamma$, but is 0 outside of $V_1$ (which is a *translation* of the domain $V_1$), is approximated by continuous virtual displacements, by surrounding $V_1$ with an arbitrary small region $V_2$, in which $\delta x, \delta y, \delta z$ decrease continuously from $\alpha, \beta, \gamma$ to 0. For such a virtual displacement it follows from (4):

$$\iiint\limits_{(V_1)} \varrho(X\alpha + Y\beta + Z\gamma)dV_1 + \iint\limits_{(S_1)} (X_n\alpha + Y_n\beta + Z_n\gamma)dS_1 + \iiint\limits_{(V_2)} \sum_{\binom{x\,y\,z}{XYZ}} \left(\varrho X + \frac{\partial X_x}{\partial x} + \frac{\partial X_y}{\partial y} + \frac{\partial X_z}{\partial z}\right)\delta x \cdot dV_2 = 0,$$

where $n$ is the normal of $dS_1$ being outward pointing with respect to $V_1$. Letting $V_2$ become smaller and smaller, the second integral gets arbitrary small, since $X, X_x, \ldots$ and the derivatives thereof remain finite, and since $\alpha, \beta, \gamma$ are arbitrary, one obtains the three equations

(10) $$\iiint\limits_{(V_1)} \varrho X dV_1 + \iint\limits_{(S_1)} X_n dS = 0 \quad (X, Y, Z).$$

These are precisely the equations obtained by the application of the so-called *center-of-mass theorem* on the part $V_1$ being cut out of the continuum and being regarded as rigid in the above mentioned manner. Due to the arbitrariness of the domain $V_1$, as is generally known, one can gain from (10) the equations (5a). (cf. IV 23, *Müller-Timpe*, p. 23).

Proceeding on the assumption of a rigid rotation of the subset $V_1$ with components $qz - ry$, $rx - pz$, $py - qx$, consequently three equations follow:

(11) $$\iiint\limits_{(V_1)} \{\varrho(Zy - Yz) + Y_z - Z_y\}dV_1 + \iint\limits_{(S_1)} (Z_n y - Y_n z)dS_1 = 0 \; (X, Y, Z).$$

This coincides exactly with the *law of equal area* applied to $V_1$ if one adds to the moments of the spatially distribute forces $X, Y, Z$ and the surface forces $X_n, Y_n, Z_n$ an opposed torque exerted directly at the volume element corresponding to the vector components (2′) of the stress dyad. By postulating the law of equal areas in the usual form, i. e. the sum of the moments of the volume and surface forces vanishes, thereof the symmetry of the stress dyad follows immediately.[30])

[30] This requirement denoted as "Boltzmann axiom" has been included by *G. Hamel*[24]) into his axioms of the mechanics of volume elements.

In nahem Zusammenhange mit diesen Tatsachen steht eine andere Auffassung des Prinzipes der virtuellen Verrückungen, die von vornherein nur die eigentlichen *Kräfte*, die Massenkräfte $X, Y, Z$ und die Flächenkräfte $\overline{X}, \overline{Y}, \overline{Z}$, als gegeben betrachtet; es ist die folgende leichte Fortbildung der Formulierung von *G. Piola* [31]) : *Für das Gleichgewicht ist notwendig, dass die virtuelle Arbeit der angeführten Kräfte*

$$\iiint\limits_{(V)} (X\delta x + Y\delta y + Z\delta z)dV + \iint\limits_{(S)} (\overline{X}\delta x + \overline{Y}\delta y + \overline{Z}\delta z)dS$$

*verschwindet für alle rein translatorischen virtuellen Verrückungen des ganzen Bereiches $V$.* Drückt man diese Nebenbedingung für die Verrückungen, nämlich durch die 9 partiellen Differentialgleichungen aus:

$$\frac{\partial \delta x}{\partial x} = 0, \frac{\partial \delta x}{\partial y} = 0, \ldots, \frac{\partial \delta z}{\partial z} = 0,$$

so kann man nach dem bekannten Kalkül der Variationsrechnung 9 zugehörige Lagrangesche Faktoren $-X_x, -X_y, \ldots, -Z_z$ einführen und erhält dann genau die Gleichung (4) des alten Prinzips, wobei sich also die *Komponenten der Spannungsdyade als Lagrangesche Faktoren gewisser Starrheitsbedingungen* erweisen. Sie werden natürlich durch dieses Variationsprinzip nicht bestimmt, spielen vielmehr genau die gleiche Rolle wie die inneren Spannungen in den statisch unbestimmten Problemen der Mechanik starrer Körper.[32])

Stellt man die gleiche Forderung für *alle* starren Bewegungen von $V$ überhaupt (statt nur für die Translationen), so erhält man genau den in IV 23, p. 23 wiedergegebenen *Piola*schen Ansatz, der gemäss den 6 Nebenbedingungen nur 6 Lagrangesche Faktoren und damit eine symmetrische Spannungsdyade, liefert.

**3e. Zwei- und eindimensionale Kontinua im dreidimensionalen Raume.** Alle diese Ansätze lassen sich unmittelbar auch für die am Ende von Nr. **2** berührten zwei- und eindimensionalen Kontinua, die im dreidimensionalen Raume gelegen sind, aufstellen.[33]) Die einzige Modifikation ist, dass sich die Dimension der Integrationsgebiete ändert, und dass statt der Ableitungen der virtuellen Verrückungen nach den drei Raumkoordinaten diejenigen nach den zwei bzw. der einen Koordinate innerhalb des deformierten Mediums eingehen.

---

[31] Modena Mem. 24, parte 1 (1848), p. 1; vgl. IV 23, Nr. **3**b, *Müller-Timpe.*

[32] Vgl. auch IV 6, Nr. **26** (*Stäckel*) , p. 550 und IV 23, Nr. **3**b (*Müller-Timpe*), p. 24.

[33] Für eine Reihe besonderer Probleme finden sich auch diese Ansätze schon in *Lagrange*, Mécan. anal.; s. 1. part, sect IV, Nr. 25 ff.; sect. V, chap. III.

In close relationship to these facts is another notion of the principle of virtual displacements considering at first only the actual *forces*, i. e. the forces per unit mass $X, Y, Z$ and the surface forces $\overline{X}, \overline{Y}, \overline{Z}$, as given; it is the following slightly advanced formulation of *G. Piola* [31]): *For the equilibrium it is necessary that the virtual work of the applied forces*

$$\iiint\limits_{(V)} (X\delta x + Y\delta y + Z\delta z)dV + \iint\limits_{(S)} (\overline{X}\delta x + \overline{Y}\delta y + \overline{Z}\delta z)dS$$

*vanishes for all purely translational virtual displacements of the whole domain $V$.* Expressing these constraints of the displacements identical to the 9 partial differential equations:

$$\frac{\partial \delta x}{\partial x} = 0, \frac{\partial \delta x}{\partial y} = 0, \dots, \frac{\partial \delta z}{\partial z} = 0,$$

then, due to the well-known calculus of variations one can introduce 9 corresponding Lagrange multipliers $-X_x, -X_y, \dots, -Z_z$ and [one] obtains precisely equation (4) of the old principle, whereas the *components of the stress dyad* appear as *Lagrange multiplier of certain rigidity constraints*. Certainly, they are not determined by this variational principle and play in fact exactly the same role as the internal stresses of the statically indeterminate problems of rigid body mechanics.[32])

Assuming the same requirement for *all* rigid motions of $V$ at all (instead of mere translations), one obtains precisely *Piola*'s ansatz given in IV 23, p. 23, which provides due to the 6 constraints only 6 Lagrange multipliers and consequently a symmetric stress dyad.

**3e. Two- and one-dimensional continua in the three-dimensional space.** All these fundamentals can immediately be formulated also for two- and one-dimensional continua being embedded in the three-dimensional space, which have been mentioned at the end of No. **2**.[33]) The only modification is the change in the dimension of the domain of integration and that instead of the derivatives of the virtual displacements with respect to the three spatial directions, the derivatives with respect to the two or one coordinate within the deformed medium enter.

31 Modena Mem. 24, parte 1 (1848), p. 1; vgl. IV 23, No. **3**b, *Müller-Timpe*.

32 Cf. also IV 6, No. **26** (*Stäckel*) , p. 550 and IV 23, No. **3**b (*Müller-Timpe*), p. 24.

33 For a series of particular problems these fundamentals can already be found in *Lagrange*, Mécan. anal.; s. 1. part, sect IV, No. 25 ff.; sect. V, chap. III.

Betrachten wir im einzelnen zunächst ein *zweidimensionales Kontinuum*, das im deformierten Zustande ein einfach zusammenhängendes Flächenstück $S$ mit der Randkurve $C$ bildet; auf $S$ sei ein — der Einfachheit halber — *orthogonales* Parametersystem $u, v$ festgelegt, Längen- und Flächenelement sei durch

$$ds^2 = E du^2 + G dv^2, \; dS = h du dv, \; h = \sqrt{EG}$$

gegeben, und es bezeichne $\varrho$ die Flächendichte der Massenbelegung. Dann betrachten wir die virtuelle Arbeit:

$$\text{(12)} \qquad \delta A = \iint\limits_{(S)} \sum_{\binom{x\,y\,z}{X\,Y\,Z}} \left\{ \varrho X \delta x - \left( \frac{X_u}{\sqrt{E}} \frac{\partial \delta x}{\partial u} + \frac{X_v}{\sqrt{G}} \frac{\partial \delta x}{\partial v} \right) \right\} dS + \int\limits_{(C)} \sum_{\binom{x\,y\,z}{X\,Y\,Z}} \overline{X} \delta x ds.$$

Hier bedeuten $X, Y, Z, \overline{X}, \overline{Y}, \overline{Z}$ die Komponenten der an der Masseneinheit innerhalb $S$ bzw. an der Längeneinheit auf $C$ angreifenden Kraft, über die Grössen $X_u, \ldots$ aber lassen sich ganz analoge Aussagen entwickeln, wie oben über $X_x, \ldots$; sie bewirken einerseits gewisse an den auf $S$ gelegenen Massen angreifende Kräfte, andererseits einen innerhalb $S$ herrschenden Spannungszustand derart, dass auf jedes in $S$ gelegene Linienelement vermöge des Spannungszustandes auf einer Seite pro Längeneinheit eine Kraft

$$\text{(13)} \qquad X_\nu = X_u \cos(\nu, u) + X_v \cos(\nu, v)$$

wirkt; hierin bedeutet $n$ die innerhalb $S$ gelegene nach der betrachteten Seite hin weisende Normalenrichtung des Elementes.

Für ein Medium, das alle stetigen Verrückungen zulässt, kann man die Bedingung $\delta A = 0$ des Prinzips der virtuellen Verrückungen in 6 Gleichgewichtsbedingungen[34]) auflösen, indem man $\delta A$ durch die bekannten Methoden der partiellen Integration umformt:

$$\text{(14a)} \qquad \frac{1}{h} \left( \frac{\partial \sqrt{G} X_u}{\partial u} + \frac{\partial \sqrt{E} X_v}{\partial v} \right) + \varrho X = 0 \quad \text{auf } S \quad (X, Y, Z),$$

$$\text{(14b)} \qquad X_u \cos \nu u + X_v \cos \nu v = \overline{X} \quad \text{auf } C \quad (X, Y, Z),$$

hier bedeutet $\nu$ diejenige Richtung, die innerhalb der Fläche $S$ normal auf $C$ steht und von dem betrachteten Flächenstück abgewandt ist. — Auch diese Gleichungen kann man leicht auf die Anfangsparameter $a, b$ transformieren, wenn man von dem transformierten Aus-

[34] Die allgemeine Form dieser Gleichungen unter den verschiedensten Auffassungen geben *E.* und *F. Cosserat*, Corps déform., chap. III, übrigens sogleich für den Fall orientierter Teilchen (s. Nr. **4**b; vgl. auch IV 11, Nr. **20**, *K. Heun*). Über die seit *Lagranges* Ansätzen [33]) behandelten speziellen Probleme vgl. ausserdem IV 6, Nr. **24**, *Stäckel*.

At first, we consider in particular a *two-dimensional continuum*, which consists in the deformed state of a simply connected surface $S$ with boundary curve $C$; on $S$ let — for the sake of simplicity — $u, v$ be an *orthogonal* system of parameters and let the line and surface elements be

$$ds^2 = Edu^2 + Gdv^2, \; dS = hdudv, \; h = \sqrt{EG}$$

and $\varrho$ denotes the surface density of the mass distribution. Then we consider the virtual work [expression]:

$$\text{(12)} \qquad \delta A = \iint\limits_{(S)} \sum_{\binom{x\,y\,z}{XYZ}} \left\{ \varrho X \delta x - \left( \frac{X_u}{\sqrt{E}} \frac{\partial \delta x}{\partial u} + \frac{X_v}{\sqrt{G}} \frac{\partial \delta x}{\partial v} \right) \right\} dS + \int\limits_{(C)} \sum_{\binom{x\,y\,z}{XYZ}} \overline{X} \delta x ds.$$

Herein $X, Y, Z, \overline{X}, \overline{Y}, \overline{Z}$ denote the components of the applied forces in $S$ per unit mass and on $C$ per unit length, for the quantities $X_u, \ldots$ similar conclusions can be drawn as above for $X_x, \ldots$; on the one hand they result in certain forces exerted on the masses of $S$, on the other hand [they cause] a stress state within $S$ such that due to this stress state on one side of every line element lying on $S$ the force per unit length

$$\text{(13)} \qquad X_\nu = X_u \cos(\nu, u) + X_v \cos(\nu, v)$$

acts; herein $n^\dagger$ denotes the normal direction lying within $S$ and pointing in direction of the considered side of the element.

For a medium, which allows for all continuous displacements, one can solve the condition $\delta A = 0$ of the principle of virtual displacements with respect to 6 equilibrium conditions[34]), by transforming $\delta A$ using the familiar methods of integration by parts:

$$\text{(14a)} \qquad \frac{1}{h} \left( \frac{\partial \sqrt{G} X_u}{\partial u} + \frac{\partial \sqrt{E} X_v}{\partial v} \right) + \varrho X = 0 \quad \text{on } S \quad (X, Y, Z),$$

$$\text{(14b)} \qquad X_u \cos \nu u + X_v \cos \nu v = \overline{X} \quad \text{on } C \quad (X, Y, Z),$$

here $\nu$ denotes the direction which is within the surface $S$ and is normal to the curve $C$ pointing away from the considered surface. — Also these equations can easily be transformed with respect to the initial parameters $a, b$, when starting with the transformed

$^\dagger$ Most probably, it should be a $\nu$ – (TN)

34 The general form of these equations from different viewpoints are given by *E.* and *F. Cosserat*, Corps déform., chap. III, by the way directly for the case of oriented particles (see No. **4**b; cf. also IV 11, No. **20**, *K. Heun*). For the particular problems treated since *Lagrange*'s fundamental studies [33]) cf. also IV 6, No. **24**, *Stäckel*.

druck der virtuellen Arbeit

$$\delta A = \iint\limits_{(S_0)} \sum_{\binom{x\,y\,z}{XYZ}} \left\{\varrho_0 X - \left(X_a \frac{\partial \delta x}{\partial a} + X_b \frac{\partial \delta x}{\partial b}\right)\right\} da\,db + \int\limits_{(C_0)} \sum_{\binom{x\,y\,z}{XYZ}} \overline{X} \delta x \frac{ds}{ds_0} ds_0 \tag{15}$$

ausgeht, wobei

$$h \frac{\partial(u,v)}{\partial(a,b)} X_u = X_a \frac{\partial u}{\partial a} + X_b \frac{\partial u}{\partial b} \qquad (X, Y, Z; u, v); \tag{16}$$

durch Vergleich mit (13) ergiebt sich, dass $X_a, \ldots$ die vermöge des Spannungszustandes auf Linienelemente $a = \text{const.}$, $b = \text{const.}$ wirkenden Kräfte bedeuten, berechnet auf Längeneinheiten in der $a$-$b$-Ebene.

Ganz analog gestaltet sich alles bei *eindimensionalen Kontinuis.*[35]) Ist $s$ $(0 \leqq s \leqq l)$ die Bogenlänge auf der in deformierter Gestalt gebildeten Kurve, so hat man

$$\delta A = \int\limits_0^l \sum_{\binom{x\,y\,z}{XYZ}} \left\{\varrho X \delta x - X_s \frac{\partial \delta x}{\partial s}\right\} ds + \left[\sum_{\binom{x\,y\,z}{XYZ}} \overline{X} \delta x\right]_{s=0}^{s=l}, \tag{17}$$

wo die Bedeutung der einzelnen Größen sich ganz wie soeben ergiebt, und bei willkürlichen stetigen Variationen lauten die Gleichgewichtsbedingungen

$$\frac{dX_s}{ds} + \varrho X = 0 \quad \text{für} \quad 0 < s < l \quad (X, Y, Z) \tag{18a}$$

$$X_s = \overline{X} \quad \text{für} \quad s = 0,\ s = l \quad (X, Y, Z). \tag{18b}$$

Auch hier ist es mitunter zweckmässig, unter Benutzung der Formel

$$\delta A = \int\limits_0^{l_0} \sum_{\binom{x\,y\,z}{XYZ}} \left\{\varrho_0 X \delta x - X_a \frac{\partial \delta x}{\partial a}\right\} da + \left[\sum_{\binom{x\,y\,z}{XYZ}} \overline{X} \delta x\right]_{a=0}^{a=l_0}, \quad X_s \frac{ds}{da} = X_a \tag{19}$$

den Anfangsparameter $a$ als Unabhängige einzuführen.

### 4. Erweiterungen des Prinzipes der virtuellen Verrückungen.

**4a. Auftreten höherer Ableitungen der Verrückungen.** Man kann an dem in Nr. **3** formulierten Ansatz des Prinzipes der virtuellen Verrückungen noch eine Reihe von Erweiterungen anbringen, die es erst in weitestem Masse befähigen, alle in der Mechanik der Kontinua auftretenden Gesetze zu umfassen. Am nächsten liegt es, in die virtuelle Arbeit pro Volumenelement eine Linearform der 18 *zweiten Ableitungen* der virtuellen Verrückungen $\frac{\partial^2 \delta x}{\partial x^2}, \ldots$ aufzunehmen. In der Tat haben Probleme, bei denen es sich als nötig erwies, die

---

[35] Vgl. *E.* und *F. Cosserat*, Corps déformables, chap. II sowie IV 11, Nr. **19** (*K. Heun*) und IV 6, Nr. **23** (*P. Stäckel*).

expression of the virtual work

$$(15)\qquad \delta A = \iint\limits_{(S_0)} \sum_{\binom{x\,y\,z}{XYZ}} \left\{ \varrho_0 X - \left( X_a \frac{\partial \delta x}{\partial a} + X_b \frac{\partial \delta x}{\partial b} \right) \right\} da\, db + \int\limits_{(C_0)} \sum_{\binom{x\,y\,z}{XYZ}} \overline{X} \delta x \frac{ds}{ds_0} ds_0$$

where

$$(16)\qquad h \frac{\partial(u,v)}{\partial(a,b)} X_u = X_a \frac{\partial u}{\partial a} + X_b \frac{\partial u}{\partial b} \qquad (X, Y, Z; u, v);$$

a comparison with (13) results in the interpretation, that $X_a, \ldots$ denote the forces due to the stress state acting at the line elements $a = \text{const.}$, $b = \text{const.}$, computed with respect to the unit of length in the $a$-$b$-plane.

Everything is entirely analogous for *one-dimensional continua*.[35]) Let $s$ $(0 \leqq s \leqq l)$ be the arc length of the curve representing the deformed state, then one has

$$(17)\qquad \delta A = \int_0^l \sum_{\binom{x\,y\,z}{XYZ}} \left\{ \varrho X \delta x - X_s \frac{\partial \delta x}{\partial s} \right\} ds + \left[ \sum_{\binom{x\,y\,z}{XYZ}} \overline{X} \delta x \right]_{s=0}^{s=l},$$

in which the interpretation of each quantity is obtained as above, and for arbitrary continuous variations the equilibrium conditions are

$$(18a)\qquad \frac{dX_s}{ds} + \varrho X = 0 \quad \text{for} \quad 0 < s < l \quad (X, Y, Z)$$

$$(18b)\qquad X_s = \overline{X} \quad \text{for} \quad s = 0,\ s = l \quad (X, Y, Z).$$

By using the formula

$$(19)\qquad \delta A = \int_0^{l_0} \sum_{\binom{x\,y\,z}{XYZ}} \left\{ \varrho_0 X \delta x - X_a \frac{\partial \delta x}{\partial a} \right\} da + \left[ \sum_{\binom{x\,y\,z}{XYZ}} \overline{X} \delta x \right]_{a=0}^{a=l_0}, \quad X_s \frac{ds}{da} = X_a$$

it is also here convenient to introduce the initial parameter $a$ as independent quantity.

**4. Enhancement of the principle of virtual displacements.**

**4a. Appearance of higher order derivatives of displacements.** One can apply a number of enhancements to the ansatz of the principle of virtual displacements presented in No. **3**, which enable it to cover in the broadest sense all laws appearing in the mechanics of continua. The most obvious is to add to the virtual work per unit volume a linear form of the 18 *second derivatives* of the virtual displacements $\frac{\partial^2 \delta x}{\partial x^2}, \ldots$. In fact there have been problems, in which it was necessary

[35] Cf. *E.* and *F. Cosserat*, Corps déformables, chap. II as well as IV 11, No. **19** (*K. Heun*) and IV 6, No. **23** (*P. Stäckel*).

Energiefunktionen von den *zweiten* Ableitungen der Deformationsfunktionen abhängen zu lassen, auf hierhin gehörende Ausdrücke geführt; in erster Linie kommt dies für die ein- und zweidimensionalen Kontinua (Drähte und Platten) in Betracht.[36]

Eine eingehende Behandlung dieses Ansatzes vom allgemeinen Standpunkte aus scheint nicht vorzuliegen, und sie erübrigt sich durch die Bemerkung, dass man durch partielle Integration die neuen Zusatzglieder des Volumenintegrals auf Glieder zurückführen kann, die lediglich die *ersten* Ableitungen der $\delta x, \delta y, \delta z$ enthalten; die neuen Wirkungen im Innern des Körpers ordnen sich also dem alten Begriff der Spannungsdyade ein. Freilich tritt dabei ein *Oberflächenintegral* von der Gestalt

$$(1) \qquad \iint\limits_{(S)} \sum_{\binom{x\,y\,z}{X\,Y\,Z}} \left(\overline{X}_x \frac{\partial \delta x}{\partial x} + \overline{X}_y \frac{\partial \delta x}{\partial y} + \overline{X}_z \frac{\partial \delta x}{\partial z}\right) dS$$

neu hinzu, das einmal das Vorhandensein einer *Oberflächenspannung* beweist, wie sie in Nr. **3**e bei einem selbständig existierenden zweidimensionalen Kontinuum betrachtet wurde, darüber hinaus aber im allgemeinen noch in (12) von Nr. **3**e nicht enthaltene Terme besitzt, die von den Ableitungen der $\delta x, \dots$ *normal* zur Fläche abhängen. Diese neuen an der Oberfläche angreifenden Spannungswirkungen scheinen noch keine Anwendung gefunden zu haben, während jene anderen Glieder lediglich zu den alten Randbedingungen (5b) von Nr. **3** einen Beitrag von der Form der in (14a) auftretenden Glieder liefern und allenfalls an Grenzlinien oder Unstetigkeitslinien der Oberfläche noch Linienkräfte vom Typus (14b) ergeben.[37]

**4b. Medien mit orientierten Teilchen.** Dehnen wir ferner unsere Betrachtungen auf die in Nr. **2**b definierten Medien mit orientierten Teilchen aus, so muss die neue Annahme in Kraft treten, *dass auch bei jeder virtuellen Rotation des Kontinuums eine virtuelle Arbeit geleistet wird, die eine lineare homogene Funktion der Gesamtheit der Werte der Rotationskomponenten $\delta\pi, \delta\kappa, \delta\varrho$ ist* für die wir den Nr. **3**, (1) analogen Ansatz machen:

$$(2) \qquad \iiint\limits_{(V)} \varrho(L\delta\pi + M\delta\kappa + N\delta\varrho)dV + \iint\limits_{(S)} (\overline{L}\delta\pi + \overline{M}\delta\kappa + \overline{N}\delta\varrho)dS$$
$$- \iiint\limits_{(V)} \left(L_x \frac{\partial \delta\pi}{\partial x} + L_y \frac{\partial \delta\pi}{\partial y} + \dots + N_z \frac{\partial \delta\varrho}{\partial z}\right) dV.$$

[36] Vgl. die Erörterungen der Potentialansätze in Nr. **7**a, p. 645 sowie auch Nr. **8**a, p. 660.

[37] Vgl. unten Nr. **12**.

to let the energy function depend on the *second* derivatives, which have led to expressions belonging to here; primarily this comes into consideration for one- and two-dimensional continua (wires and plates).[36])

A thorough treatment of this [new] ansatz from a more general point of view seems not to be available [in the literature] and becomes unnecessary by remarking, that one can transform using integration by parts the new additional terms in the volume integral to terms which include merely the *first* derivatives of $\delta x, \delta y, \delta z$; hence, the new effects within the body are classified in the sense of the old notion of the stress dyad. Certainly, a new *surface integral* of the form

$$(1) \qquad \iint\limits_{(S)} \sum_{\binom{x\,y\,z}{X\,Y\,Z}} \left(\overline{X}_x \frac{\partial \delta x}{\partial x} + \overline{X}_y \frac{\partial \delta x}{\partial y} + \overline{X}_z \frac{\partial \delta x}{\partial z}\right) dS$$

appears, which at one point proves the existence of a *surface tension* as it has been considered in No. **3**e for an independently existing two-dimensional continuum, it may contain in addition expressions which are not included in (12) of No. **3**e [and] which depend on the derivatives of the $\delta x, \ldots$ *normal* to the surface. These new effects of stresses interactions applied at the surface, seem not to have found any application so far,while in contrast the other [remaining] terms simply contribute to the old boundary conditions (5b) of No. **3** in the same form as the terms appearing in (14a), and possibly lead to line distributed forces in the sense of (14b) at interfaces or lines of discontinuities.[37])

**4b. Media with oriented particles.** When we enhance our consideration furthermore to the media with oriented particles defined in No. **2**b, then a new assumption must become valid, *that, also for every virtual rotation of the continuum, virtual work is expended being a linear homogeneous function of the totality of values of the rotational components* $\delta\pi, \delta\kappa, \delta\varrho$ for which we make the ansatz analogous to No. **3**, (1):

$$(2) \qquad \iiint\limits_{(V)} \varrho(L\delta\pi + M\delta\kappa + N\delta\varrho)dV + \iint\limits_{(S)} (\overline{L}\delta\pi + \overline{M}\delta\kappa + \overline{N}\delta\varrho)dS$$
$$- \iiint\limits_{(V)} \left(L_x \frac{\partial \delta\pi}{\partial x} + L_y \frac{\partial \delta\pi}{\partial y} + \cdots + N_z \frac{\partial \delta\varrho}{\partial z}\right) dV.$$

36 Cf. the discussions about the potential-based approaches in No. **7**a, p. 645 as well as No. **8**a, p. 660.

37 Cf. below No. **12**.

Hieran kann man völlig analoge Erörterungen wie in Nr. **3**a schliessen, wobei man als selbstverständlich wieder die Voraussetzung der Endlichkeit und Stetigkeit der 15 neu auftretenden Koeffizienten übernimmt. Zunächst stellen $L, M, N$ bzw. $\overline{L}, \overline{M}, \overline{N}$ die Komponenten je eines *axialen Vektors* dar, der als das auf eine Stelle innerhalb des Körpers (pro Masseneinheit) bzw. auf eine Stelle der Oberfläche (pro Flächeneinheit berechnete) *Drehmoment* aufzufassen ist; denn in der Tat haben wir hier eine Kraftwirkung von genau der in der Mechanik starrer Körper so bezeichneten Art. Die 9 Grössen $L_x, \ldots, N_z$ weiterhin verhalten sich bei Koordinatentransformationen wie die Komponenten einer Dyade mit der Modifikation, dass sie bei Spiegelungen das Vorzeichen wechseln[38]); ihre Bedeutung kann man darin finden, dass

$$(3) \qquad L_n = L_x \cos nx + L_y \cos ny + L_z \cos nz \qquad (L, M, Z)$$

die Komponenten des Drehmoments darstellt, das auf ein Flächenelement durch die auf der Seite der positiven Normalenrichtung $n$ gelegene Materie ausgeübt wird, berechnet auf die Flächeneinheit.

Wir übernehmen nun das Prinzip der virtuellen Verrückungen für das neue Kontinuum in der erweiterten Form, *dass in der durch die 6 Funktionen* Nr. **2**, (1) *und* (9) *beschriebenen Gleichgewichtslage die durch* (2) *ergänzte virtuelle Arbeit für jedes zulässige System virtueller Verrückungen verschwinden soll.* Für das völlig frei stetig deformierbare Kontinuum, für das auch die Axenkreuze unabhängig voneinander und von der Grösse der Verrückungen drehbar sind, sind dann $\delta x, \ldots, \delta\pi, \ldots$ 6 völlig willkürliche stetige Funktionen, und durch Wiederholung der Überlegungen von Nr. **3**c findet man, dass die dort aufgestellten Bedingungen (5) ungeändert bleiben und nur durch folgende zuerst von *W. Voigt*[39]) aufgestellten und neuerdings in dem *Cosserat*schen Werke[40]) ausführlich betrachteten 2 Gleichungs-

---

[38] Für Tensorkomponenten (d. h. bei einer symmetrischen Dyade) hat *W. Voigt* (vgl. Lehrbuch der Kristallphysik, Leipzig 1910, p. 132 ff.) das entsprechende Verhalten durch das Beiwort axial ausgedrückt, gegenüber polaren Tensoren, deren Komponenten bei Inversion ihr Vorzeichen nicht wechseln. Man vergleiche über diese Klassifikation auch die in 18) zitierte Litteratur.

[39] Gött. Abhandl. 34 (1887), p. 11, wo *Voigt* an die Poissonschen Vorstellungen[9]) anschliesst. Vgl. auch das Referat in *Voigts* Vortrag auf dem internat. Physiker-Kongress in Paris 1900 (Rapp. prés. au congr. T. I, p. 277 = Gött. Nachr., math.-phys. Kl. 1900, p. 117) und die von direkter Bezugnahme auf Molekularvorstellungen freie Darstellung in *Voigts* Kompendium I, p. 219 ff.

[40] *E.* u. *F. Cosserat*, Corps déform., chap. IV, inbes. p. 137. Vgl. auch IV 11, Nr. **21**, *K. Heun.*

Here one can an discuss similar arguments as in No. **3**a, where one naturally assumes again the requirements of the finiteness and the continuity of the 15 emerging coefficients. At first $L, M, N$ and $\overline{L}, \overline{M}, \overline{N}$ represent the components of an *axial vector*, which has to be understood as a *torque* at a point within the body (per unit of mass) or at a point on the surface (per unit of area), respectively; then in fact we have here a force effect of exactly the same kind as in rigid body mechanics. Under coordinate transformations, the quantities $L_x, \ldots, N_z$ still transform like the components of a dyad with the modification that the sign changes for reflections[38]); their interpretation can be found therein, that

$$(3) \qquad L_n = L_x \cos nx + L_y \cos ny + L_z \cos nz \qquad (L, M, Z)^{\dagger}$$

represents the components of the torque per unit area, which is exerted via a surface element on the matter being on the side of the positive normal direction $n$.

We now assume the principle of virtual displacements for the new continuum in enhanced form, *that in the equilibrium position described by the 6 functions* No. **2**, (1) *and* (9), *the virtual work augmented by* (2) *must vanish for every admissible set of virtual displacements*. Being assured that the continuously deformable continuum is completely free, for which the triads can be each other relatively rotated independently also of the magnitude of the displacements, then $\delta x, \ldots, \delta\pi, \ldots$ are 6 completely arbitrary continuous functions, and by repeating the considerations of No. **3**c one finds, that the conditions (5) formulated therein remain unchanged and that they have to be completed only by following two sets of three equations formulated first by *W. Voigt*[39]) and recently discussed in detail in the work of *Cosserat*[40]):

[38] For tensor components (i. e. for a symmetric dyad) *W. Voigt* (cf. Lehrbuch der Kristallphysik, Leipzig 1910, p. 132 ff.) has denoted the corresponding behavior using the adjective axial, in contrast to polar tensors, whose components do not change sign under inversion. About this classification one compares also the literature cited in 18).

$^{\dagger}$ It should rather be $(L, M, N)$ – (TN)

[39] Gött. Abhandl. 34 (1887), p. 11, where *Voigt* builds on the notions of Poisson.[9]). Cf. also the discussion in *Voigts* presentation at the international congress of physicists in Paris 1900 (Rapp. prés. au congr. T. I, p. 277 = Gött. Nachr., math.-phys. Kl. 1900, p. 117) and the exposition in *Voigts* Kompendium I, p. 219 ff, being free of any direct reference to molecular perceptions.

[40] *E.* and *F. Cosserat*, Corps déform., chap. IV, in particular p. 137. Cf. also IV 11, No. **21**, *K. Heun*.

tripel zu ergänzen sind[41]):

$$(4a)\qquad \frac{\partial L_x}{\partial x} + \frac{\partial L_y}{\partial y} + \frac{\partial L_z}{\partial z} + \varrho L = 0 \quad \text{in } V \qquad (L, M, N),$$

$$(4b)\qquad L_x \cos nx + L_y \cos ny + L_z \cos nz = \overline{L} \quad \text{auf } S \qquad (L, M, N).$$

Auch diese Gleichungen kann man wieder auf die Anfangsparameter $a, b, c$ transformieren, indem man die virtuelle Arbeit der inneren Flächenmomente auf die Form

$$(2')\qquad -\iiint\limits_{(V_0)} \Big( \sum_{\binom{\pi\,\kappa\,\varrho}{L\,M\,N}} L_a \frac{\partial \delta\pi}{\partial a} + L_b \frac{\partial \delta\pi}{\partial b} + L_c \frac{\partial \delta\pi}{\partial c} \Big) dV_0$$

transformiert, wo

$$(5)\qquad \Delta \cdot L_x = L_a \frac{\partial x}{\partial a} + L_b \frac{\partial x}{\partial b} + L_c \frac{\partial x}{\partial c} \qquad (L, M, N; x, y, z),$$

und wo $L_a, M_a, N_a$ das auf ein Element der Fläche $a$ = const wirkende, auf die Flächeneinheit im undeformierten Zustande berechnete Drehmoment bedeuten. An die Stelle von (4) treten dann neben Nr. **3**, (9) die Gleichungstripel[42]):

$$(6a)\qquad \frac{\partial L_a}{\partial a} + \frac{\partial L_b}{\partial b} + \frac{\partial L_c}{\partial c} + \varrho_0 L = 0 \quad \text{in } V \qquad (L, M, N),$$

$$(6b)\qquad L_a \cos n_0 a + L_b \cos n_0 b + L_c \cos n_0 c = k\overline{L} \quad \text{auf } S \qquad (L, M, N).$$

Auch hier kann man wieder den Zusammenhang mit den Gleichgewichtsbedingungen am starren Körper erhalten, indem man einmal von einer Translation, dann von einer Rotation eines aus $V$ herausgeschnittenen und starr gedachten Teilbereiches $V_1$ ausgeht, innerhalb dessen man sich nun auch die Axenkreuze starr mit dem Kontinuum verbunden, also parallel mit sich fortgeführt bzw. starr mitgedreht denkt; approximiert man diese unstetige Verrückung genau wie in Nr. **3**d durch stetige virtuelle Verrückungen, so findet man einmal ungeändert die Gleichungen Nr. **3** (10) des Schwerpunktssatzes wieder, dann aber an Stelle der Formeln (11) drei Gleichungen

$$(3.7)\qquad \iiint\limits_{(V_1)} \{\varrho(Zy - Yz + L) + Y_z - Z_y\}\, dV_1 + \iint\limits_{(S_1)} \{Z_n y - Y_n z + L_n\}\, dS_1 = 0 \qquad \binom{L, M, N}{X, Y, Z},$$

[41] Diese Gleichungen sind, abgesehen von der Festsetzung der Vorzeichen, noch insofern von denen von *Voigt* und *Cosserat* verschieden, als dort das gesamte auf ein Teilchen wirkende Drehmoment $\varrho L + Y_z - Z_y, \ldots$ mit einem Buchstaben bezeichnet ist.

[42] In etwas verschiedener Bezeichnung bei *E.* u. *F. Cosserat*, Corps déformables, p. 132.

(4a) $$\frac{\partial L_x}{\partial x} + \frac{\partial L_y}{\partial y} + \frac{\partial L_z}{\partial z} + \varrho L = 0 \quad \text{in } V \qquad (L, M, N),$$

(4b) $$L_x \cos nx + L_y \cos ny + L_z \cos nz = \overline{L} \quad \text{on } S \qquad (L, M, N).^{41)}$$

Also these equations can be transformed to be formulated with respect to the initial parameters $a, b, c$, by transforming the virtual work of the internal surface torques to the form

(2′) $$-\iiint\limits_{(V_0)} \Big( \sum_{\binom{\pi\kappa\varrho}{LMN}} L_a \frac{\partial \delta\pi}{\partial a} + L_b \frac{\partial \delta\pi}{\partial b} + L_c \frac{\partial \delta\pi}{\partial c} \Big) dV_0,$$

where

(5) $$\Delta \cdot L_x = L_a \frac{\partial x}{\partial a} + L_b \frac{\partial x}{\partial b} + L_c \frac{\partial x}{\partial c} \qquad (L, M, N; x, y, z),$$

and where $L_a, M_a, N_a$ denote the torque acting on an element of the surface $a =$ const, computed with respect to the unit of area in the undeformed state. The equations of (4) are then substituted besides No. **3**, (9) by the triple of equations[42]):

(6a) $$\frac{\partial L_a}{\partial a} + \frac{\partial L_b}{\partial b} + \frac{\partial L_c}{\partial c} + \varrho_0 L = 0 \quad \text{in } V \qquad (L, M, N),$$

(6b) $$L_a \cos n_0 a + L_b \cos n_0 b + L_c \cos n_0 c = k\overline{L} \quad \text{on } S \qquad (L, M, N).$$

Also here one can obtain a connection to the equilibrium conditions of the rigid body, by starting in one case with a translation, then with a rotation, of a subset $V_1$ cut out of $V$ thought of as being rigid, within which the triads are rigidly fixed with the continuum, considering them consequently to be carried along in parallel and rigidly, respectively; approximating these discontinuous displacements exactly as in No. **3**d by continuous displacements, one finds on the one hand the unchanged equations No. **3** (10) of the center-of-mass theorem, but then one finds instead of the formulas (11) three equations

(3.7) $$\iiint\limits_{(V_1)} \left\{\varrho(Zy - Yz + L) + Y_z - Z_y\right\} dV_1 + \iint\limits_{(S_1)} \{Z_n y - Y_n z + L_n\}\, dS_1 = 0 \qquad \binom{L, M, N}{X, Y, Z},$$

[41] Except for the assignment of the signs, these equations are insofar different from the ones of *Voigt* and *Cosserat*, because there the entire torque $\varrho L + Y_z - Z_y, \ldots$ acting on a particle is denoted by a single letter.

[42] In a slightly different notation in *E.* and *F. Cosserat*, Corps déformables, p. 132.

die den Flächensatz unter den jetzt stattfindenden Umständen ausdrücken. Aus diesen 6 Integralbedingungen, die für *jeden* Teilbereich $V_1$ gelten sollen, kann man wieder die Gleichgewichtsgleichungen (4) herleiten.[43])

Sind die Dreikante nicht mehr frei beweglich, so modifizieren sich die Gleichgewichtsbedingungen (4) und Nr. **3**, (5), da man dann die bei den Summanden (2) und Nr. **3**, (1) der virtuellen Arbeit nicht mehr gesondert behandeln darf. Es sei hier nur auf den Fall hingewiesen, dass die Axen des Dreikants fest mit dem Medium verbunden sind; dann wird eine jede virtuelle Verrückung eine Verdrehung der Dreikante mit den Grössen Nr. **2**, (4′) als Komponenten zur Folge haben, und daher werden insbesondere neue Glieder zu den Komponenten der Spannungsdyade additiv hinzutreten. Man hat das benutzt, um auch unter Verwendung einer symmetrischen Spannungsdyade ($X_y = Y_x, \ldots$) das Auftreten von Drehmomenten zu deuten.[44])

Bei *zwei- und eindimensionalen Medien* mit orientierten Teilchen (s. Nr. **2**c) ergiebt sich ganz analog, dass bei Verwendung der früheren Bezeichnungen zu der virtuellen Arbeit für die Fläche (Nr. **3**e, (12)) der Summand

$$(8)\qquad \iint\limits_{(S)} \sum_{\binom{\pi\,\kappa\,\varrho}{L N M}} \left\{ \varrho L \delta\pi - \left( \frac{L_u}{\sqrt{E}} \frac{\partial \delta\pi}{\partial u} + \frac{L_v}{\sqrt{G}} \frac{\partial \delta\pi}{\partial v} \right) \right\} dS + \int\limits_{(C)} \sum_{\binom{\pi\,\kappa\,\varrho}{L N M}} \bar{L} \delta\pi \, ds,$$

für die Kurve (Nr. **3**e, (17)) ein entsprechender

$$(9)\qquad \int_0^l \sum_{\binom{\pi\,\kappa\,\varrho}{L N M}} \left\{ \varrho L \delta\pi - L_s \frac{\partial \delta\pi}{\partial s} \right\} ds + \Big[ \sum_{\binom{\pi\,\kappa\,\varrho}{L N M}} \bar{L} \delta\pi \Big]_0^l$$

hinzutritt; demgemäss erhält man im ersten Falle neben Nr. **3**e, (14)) noch die Gleichgewichtsbedingungen[45])

$$(10)\qquad \begin{array}{ll} \frac{1}{h}\left( \frac{\partial \sqrt{G} L_u}{\partial u} + \frac{\partial \sqrt{E} L_v}{\partial v} \right) + \varrho L = 0 & \text{auf } S \\ L_u \cos \nu u + L_v \cos \nu v = \bar{L} & \text{auf } C \end{array} \qquad (L, M, N),$$

---

[43] So geht *Voigt*, Kompendium I, p. 219 vor.

[44] Siehe etwa *J. Larmor*, London math. Soc. Proc. 23 (1892), p. 127, Combébiac, Bull. soc. de math. 30 (1902), p. 108, 242.

[45] Vgl. *F.* und *E. Cosserat*, Corps déform., chap. III, sowie IV 11, Nr. **20**, (*K. Heun*).

which express the law of equal areas in the current context. From these 6 integral conditions, which have to be satisfied for *every* subset $V_1$, one can again derive the equilibrium conditions (4).[43])

When the triads are not any more free to move, then the equilibrium conditions (4) and No. **3**, (5) are modified, since the summands (2) and No. **3**, (1) of the virtual work cannot be treated separately anymore. We just want to mention the case when the axes of the triad are fixed to the medium; then for every virtual displacement this will imply a rotation of the triad with the magnitude No. **2**, (4′) as components, and thus in particular new terms will be added to the components of the stress dyad. One has used this to interpret the appearance of torques even when using a symmetric stress dyad ($X_y = Y_x, \dots$).[44])

For *two- and one-dimensional media* with oriented particles (see No. **2**c) it yields similarly, by the application of the earlier used notation, that to the virtual work of the surface (No. **3**e, (12)) the summand

$$(8) \qquad \iint\limits_{(S)} \sum_{\binom{\pi\kappa\varrho}{LNM}} \left\{ \varrho L \delta\pi - \left( \frac{L_u}{\sqrt{E}} \frac{\partial \delta\pi}{\partial u} + \frac{L_v}{\sqrt{G}} \frac{\partial \delta\pi}{\partial v} \right) \right\} dS + \int\limits_{(C)} \sum_{\binom{\pi\kappa\varrho}{LNM}} \bar{L} \delta\pi ds,$$

is added and that to the virtual work of the curve (No. **3**e, (17)) a corresponding [term]

$$(9) \qquad \int\limits_0^l \sum_{\binom{\pi\kappa\varrho}{LNM}} \left\{ \varrho L \delta\pi - L_s \frac{\partial \delta\pi}{\partial s} \right\} ds + \Bigg[ \sum_{\binom{\pi\kappa\varrho}{LNM}} \bar{L} \delta\pi \Bigg]_0^l$$

is added; accordingly, one obtains in the first case in addition to No. **3**e, (14), the equilibrium equations[45])

$$(10) \qquad \begin{aligned} &\frac{1}{h}\left( \frac{\partial \sqrt{G} L_u}{\partial u} + \frac{\partial \sqrt{E} L_v}{\partial v} \right) + \varrho L = 0 \quad \text{on } S \\ &L_u \cos \nu u + L_v \cos \nu v = \bar{L} \quad \text{on } C \end{aligned} \qquad (L, M, N),$$

[43] In this way *Voigt*, Kompendium I, p. 219 proceeds.

[44] See for instance *J. Larmor*, London math. Soc. Proc. 23 (1892), p. 127, Combébiac, Bull. soc. de math. 30 (1902), p. 108, 242.

[45] Cf. *F.* and *E. Cosserat*, Corps déform., chap. III, as well as IV 11, No. **20**, (*K. Heun*).

im zweiten Falle neben Nr. **3**e, (18)) noch diejenigen[46])

$$\begin{aligned} \frac{dL_s}{ds} + \varrho L &= 0 \quad \text{für } 0 < s < l \\ L_s &= \bar{L} \quad \text{für } \quad s = 0,\ s = l \end{aligned} \qquad (L, M, N). \tag{11}$$

Auch die Deutung der $L_u, \ldots$ als spezifische Drehmomente bezogen auf den deformierten Zustand ergiebt sich analog zu Nr. **3**e; sie hängen mit den entsprechenden auf den undeformierten Zustand bezogenen Grössen zusammen durch Gleichungen vom Typus

$$h\frac{\partial(u,v)}{\partial(a,b)}L_u = L_a\frac{\partial u}{\partial a} + L_b\frac{\partial u}{\partial b} \quad \text{bzw.} \quad L_s\frac{ds}{da} = L_a. \tag{12}$$

**4c. Auftreten von Nebenbedingungen.** Bisher wurde das Prinzip der virtuellen Verrückungen vorzugsweise auf solche Fälle angewandt, in denen das Kontinuum in jeder möglichen Weise stetig deformierbar war. In der Formulierung des Prinzipes sind aber unmittelbar auch solche Kontinua umfasst, deren *Beweglichkeit durch Bedingungen irgendwelcher Art beschränkt* ist, und tatsächlich betreffen gerade einige der ersten Probleme der Mechanik der Kontinua, die *Lagrange*[47]) behandelt hat, solche Fälle. Diese Bedingungen drücken sich in erster Linie durch *Gleichungen* für die die Deformation beschreibenden Funktionen (1), (9) von Nr. **2** aus, in welche übrigens neben den Funktionen selbst auch ihre Ableitungen nach $a, b, c$ eingehen können; typisch ist eine Gleichung

$$\omega(a,b,c;x,y,z;x_a,\ldots,z_a;\lambda,\mu,\nu;\lambda_a,\ldots,\nu_c)=0, \text{wo}\, x_a = \frac{\partial x}{\partial a}, \ldots \tag{13}$$

für *jeden* Punkt des Bereiches $V_0$, doch kann man ähnliche Gleichungen auch nur für Teilbereiche, Grenzflächen oder dgl. aussprechen. In jedem Falle werden dadurch die möglichen Deformationen bzw. die möglichen Verdrehungen der adjungierten Dreikante eingeschränkt, oder es werden auch bestimmte Beziehungen zwischen Verdrehung des Dreikants und Deformation gefordert (z. B. eine bestimmte Orientierung der Dreikante gegen den Raum oder gegen das Medium; vgl. oben S. 626); das Auftreten von $a, b, c$ in (13) besagt, dass die Art der Bedingung von Teilchen zu Teilchen wechseln kann. Setzt man nun in (13) die variierte Deformation Nr. **2**, (3) bzw. (10) ein, so

---

[46] Vgl. *F.* und *E. Cosserat*, Corps déform., chap. II, sowie IV 11, Nr. **19**, (*K. Heun*).

[47] Mécan. anal., 1. Part., Sect. V, Chap. III (unausdehnbarer Faden u. dgl.), Sect. VIII (inkompressible Flüssigkeit).

in the second case one obtains in addition to No. **3**e, (18)) the following[46])

$$(11)\qquad \begin{aligned} \frac{dL_s}{ds} + \varrho L &= 0 \quad \text{for } 0 < s < l \\ L_s &= \overline{L} \quad \text{for} \quad s = 0,\ s = l \end{aligned} \qquad (L, M, N).$$

Also the interpretation of $L_u, \dots$ as specific torques formulated with respect to the deformed state is obtained similarly to No. **3**e; they are connected to the corresponding quantities formulated with respect to the undeformed configuration with the equations of the kind

$$(12)\qquad h\frac{\partial(u, v)}{\partial(a, b)} L_u = L_a \frac{\partial u}{\partial a} + L_b \frac{\partial u}{\partial b} \quad \text{or} \quad L_s \frac{ds}{da} = L_a.$$

**4c. Appearance of constraints.** Hitherto the principle of virtual displacements has been applied particularly for cases in which the continuum was continuously deformable in all possible ways. In the formulation of the principle also such continua are immediately included whose *movability is constrained by restrictions of any kind*, and in fact just some of the first problems in the mechanics of continua, treated by *Lagrange* [47]), are cases of this kind. Primarily, these constraints are expressed by *equations* for the functions (1), (9) of No. **2** describing the deformation, in which besides the functions also their derivatives with respect to $a, b, c$ can enter; typically is an equation

$$(13)\qquad \omega(a, b, c; x, y, z; x_a, \dots, z_a; \lambda, \mu, \nu; \lambda_a, \dots, \nu_c) = 0, \text{ where } x_a = \frac{\partial x}{\partial a}, \dots$$

for *every* point of the domain $V_0$, but it is also possible to formulate similar equations for subsets, interfaces or similar ones. In any case thereby possible deformations or possible rotations of the adjoint triads are restricted, or particular relations between rotation of the triad and the deformation are demanded (e. g. a particular orientation of the triad with respect to the space or the medium; cf. above p. 626); The appearance of $a, b, c$ in (13) indicates, that the type of the condition can vary from particle to particle. Inserting the variation of the deformation No. **2**, (3) or (10) in (13), then

[46] Cf. *F.* and *E. Cosserat*, Corps déform., chap. II, as well as IV 11, No. **19**, (*K. Heun*).

[47] Mécan. anal., 1. Part., Sect. V, Chap. III (inextensible wire and similar ones), Sect. VIII (incompressible fluid).

ergiebt sich durch Differentiation nach $\sigma$

$$(14) \quad \delta\omega \equiv \sum_{(x\,y\,z)} \left(\frac{\partial\omega}{\partial x}\delta x + \frac{\partial\omega}{\partial x_a}\delta x_a + \frac{\partial\omega}{\partial x_b}\delta x_b + \frac{\partial\omega}{\partial x_c}\delta x_c\right)$$
$$+ \sum_{(\lambda\,\mu\,\nu)} \left(\frac{\partial\omega}{\partial\lambda}\delta\lambda + \frac{\partial\omega}{\partial\lambda_a}\delta\lambda_a + \frac{\partial\omega}{\partial\lambda_b}\delta\lambda_b + \frac{\partial\omega}{\partial\lambda_c}\delta\lambda_c\right) = 0,$$

und da nach Nr. **2**, S. 608 die $\delta x_a, \ldots$ mit den Ableitungen der $\delta x, \ldots$ übereinstimmen, liegt hier eine *lineare homogene Bedingung für die virtuellen Verrückungen* vor.

*Das Prinzip der virtuellen Verrückungen* fordert dann, *dass $\delta A$ für alle (14) genügenden Funktionen $\delta x, \ldots$ verschwindet,* und das kann man, wenn die Gleichungen (14) nicht zufällig die Elimination einer der Verrückungskomponenten gestatten, durch Einführung eines Lagrangeschen Faktors[48]) $\lambda$ in die Form

$$(15) \qquad \delta A + \iiint\limits_{(V)} \lambda\delta\omega dV = 0 \quad \text{für alle} \quad \delta x, \ldots$$

umsetzen, die genau der des ursprünglichen Prinzipes entspricht; an Stelle des Raumintegrales treten event., wenn (13) nur längs einzelner Flächen oder Kurven bestehen soll, oder das Kontinuum überhaupt nur eine Fläche oder Kurve erfüllt, Flächen- oder Kurvenintegrale. Über die Bedeutung des Faktors $\lambda$ als „Druck" wird später (Nr. **8**b, S. 662) noch zu sprechen sein.

Endlich ist noch der Möglichkeit zu gedenken, die gleichfalls aus der Mechanik diskreter Systeme wohlbekannt ist, dass *„einseitige" Ne-benbedingungen* auftreten, die die Form von *Ungleichungen* haben — sei es z. B., dass die Grenzfläche des Kontinuums in ihrer Beweglichkeit nur nach einer Seite hin eingeschränkt ist, sei es dass die Deformationsgrössen im Inneren gewissen Ungleichungen unterliegen (man denke etwa an Körper, die keine Kompression über eine gewisse Grenze hinaus gestatten, oder ähnliche Festsetzungen). Dann wird auch hier das Gleichgewicht gegeben durch die *Fourier*sche Formulierung[49]) des Prinzips der virtuellen Verrückungen, dass nämlich für jedes den Nebenbedingungen genügende System von virtuellen Verrückungen die virtuelle Arbeit negativ oder Null ist:

$$\delta A \leqq 0.$$

[48] Die Behandlung mehrdimensionaler Variationsprobleme mit dieser Methode wurde von Lagrange an den in 47) genannten Problemen das erste Mal entwickelt; vgl. II A8, p. 622, *Kneser*.

[49] Vgl. IV 1, Nr. **34**, *Voss*; die Formulierung bei *Gauss* (Principia generalia theoriae figurae fluidorum in statu aequilibrii, Gott. Comment. rec. 7 (1830) = Werke 5, p. 35, deutsch von *R. H. Weber* in Ostwald's Klassiker der exakten Wiss. Nr. 135, Leipzig 1903) berücksichtigt von vornherein die Ausdehnung auf Kontinua.

differentiation with respect to $\sigma$ yields

$$(14)\quad \delta\omega \equiv \sum_{(x\,y\,z)} \left(\frac{\partial\omega}{\partial x}\delta x + \frac{\partial\omega}{\partial x_a}\delta x_a + \frac{\partial\omega}{\partial x_b}\delta x_b + \frac{\partial\omega}{\partial x_c}\delta x_c\right) + \sum_{(\lambda\,\mu\,\nu)} \left(\frac{\partial\omega}{\partial\lambda}\delta\lambda + \frac{\partial\omega}{\partial\lambda_a}\delta\lambda_a + \frac{\partial\omega}{\partial\lambda_b}\delta\lambda_b + \frac{\partial\omega}{\partial\lambda_c}\delta\lambda_c\right) = 0,$$

and since due to No. **2**, p. 608 the $\delta x_a, \ldots$ coincide with the derivatives of $\delta x, \ldots,$ there is a *linear homogeneous condition for the virtual displacements.*

*The principle of virtual displacements* then claims, *that $\delta A$ vanishes for all functions $\delta x, \ldots$ which satisfy (14),* and this can be realized, when equation (14) does not allow accidentally the elimination of a displacement component, by the introduction of a Lagrange multiplier[48]) $\lambda$ in the form

$$(15)\qquad \delta A + \iiint_{(V)} \lambda\delta\omega dV = 0 \quad \text{for all} \quad \delta x, \ldots$$

which corresponds exactly with the original principle; instead of volume integrals possibly there appear surface or curve integrals, when (13) exists only along individual surfaces or curves, or when the continuum is in fact merely a surface or a curve. The denotation of the factor $\lambda$ as "pressure" will be addressed later on (No. **8**b, p. 662).

Finally, one should think of the possibility, likewise well-known from the mechanics of discrete systems, that *"unilateral" constraints* appear, which have the form of *inequalities* — let it be e. g., that the boundary of the continuum is restricted in its movability in one direction, let it be that the deformation quantities in the inside are subjected to certain inequalities (one can think of bodies, which do not allow any compression beyond a certain threshold, or similar conditions). Then also here, the equilibrium will be determined by *Fourier*'s formulation[49]) of the principle of virtual displacements, that for any system of virtual displacements, satisfying the constraints, the virtual work is negative or zero:

$$\delta A \leqq 0.$$

---

48 The treatment of multidimensional variational problems has been developed for the first time by Lagrange for the problems referred to in 47); cf. II A8, p. 622, *Kneser*.

49 Cf. IV 1, No. **34**, *Voss*; The formulation in *Gauss* (Principia generalia theoriae figurae fluidorum in statu aequilibrii, Gott. Comment. rec. 7 (1830) = Werke 5, p. 35, german of *R. H. Weber* in Ostwald's Klassiker der exakten Wiss. No. 135, Leipzig 1903) a priori considers the enhancement to continua.

## II. Die Grundansätze der Kinetik.

**5a. Die Bewegungsgleichungen des Kontinuums.** Aufgabe der Kinetik ist festzustellen, welche Bewegung in dem bisher betrachteten Kontinuum eintritt, wenn irgendwie in der Zeit gegebene Kraftwirkungen in ihm stattfinden, oder umgekehrt, welche Wirkungen zur Aufrechterhaltung einer bestimmten Bewegung notwendig sind. Dabei sind die Wirkungskomponenten wie in der Statik als Koeffizienten des Arbeitsausdruckes $\delta A$ gegeben gedacht, während die Art ihrer Abhängigkeit von den Bewegungsfunktionen zunächst offen bleibt. Wir befassen uns vorerst nur mit den gewöhnlichen in Nr. **3** betrachteten Medien. Der Übergang von der Statik zur Kinetik kann dann genau wie in der Mechanik der diskontinuierlichen Systeme mit Hilfe des *d'Alembertschen Prinzips* (vgl. IV 1, Nr. **36**, *Voss*) geschehen; seine Übertragung auf kontinuierliche Systeme bietet sich fast von selbst dar, wenn man sich wie in der Statik (S. 616) von dem Gedanken eines Grenzüberganges zum Kontinuum leiten lässt, bzw. direkt im Sinne der Analogie zwischen Punktsystemen und Kontinuis vorgeht. Von diesen Gesichtspunkten aus hat bereits *Lagrange* [50]) die von ihm behandelten Probleme der Hydrodynamik angefasst.

Demnach kann man ganz entsprechend der von *d'Alembert* [51]) selbst entwickelten Auffassung für die allgemeine Mechanik der Kontinua das folgende Prinzip aussprechen: *Betrachtet man die während der Bewegung in einem bestimmten Zeitmoment auf das Quantum $V_0$ des Mediums wirkenden Kräfte und Spannungen, so befinden sie sich im statischen Gleichgewicht im früheren Sinne, wofern man ihnen an jeder Stelle noch Kräfte („Trägheitskräfte") hinzufügt, deren Komponenten auf die Masseneinheit des Kontinuums berechnet den Komponenten der Beschleunigung entgegengesetzt gleich sind:*

$$-\frac{\partial^2 x}{\partial t^2} = -x'', \quad -\frac{\partial^2 y}{\partial t^2} = -y'', \quad -\frac{\partial^2 z}{\partial t^2} = -z''. \text{ [52])}$$

Erweist es sich auch vielfach als zweckmässig, diesen Satz als Axiom an die Spitze der Kinetik zu stellen, so bleibt die Frage offen, in welche unabhängigen Bestandteile man ihn zerlegen kann und wie weit diese von den statischen Axiomen unabhängig sind — eine Frage, die in genau der gleichen Bedeutung schon in der Mechanik der diskontinuierlichen Systeme auftritt. Es sei daher hier

---

[50] Vgl. insbesondere Méc. anal., 2. part., Sect. XI, § I.

[51] Traité de dynamique, Paris 1743. Vgl. IV 1 *Voss*, p. 77 [209]).

[52] Mit Akzenten werden im folgenden stets die Ableitungen der Bewegungsfunktionen (1) nach $t$ bei konstantem $a, b, c$ bezeichnet.

## II. The foundations of kinetics.

**5a. The equations of motion of the continuum.** It is the task of kinetics to determine, what motion arises within the continuum considered so far, when force effects occur in it somehow specified in time, or vice versa, which effects are required for the maintenance of a particular motion. As in statics, the effects are thereby thought as given by the coefficients of the expression of work $\delta A$, while the kind of their dependence on the function of motion remains open at first. For now, we address only ordinary media considered in No. **3**. The transition from statics to kinetics can occur as in the mechanics of discrete systems with the help of *d'Alembert's Principle* (cf. IV 1, No. **36**, *Voss*); its generalization to continuous systems is offering itself almost automatically, if one let lead oneself as in statics (S. 616) by the idea of a limit process to the continuum, or if one proceeds directly in the sense of the analogy between point systems and continua. From these points of view, already *Lagrange* [50]) has tackled those problems of hydrodynamics which he considered.

According to [this last analogy] one can state the following principle which is in full correspondence with the notion of the general mechanics of continua developed by *d'Alembert* [51]) himself: *Considering the forces and stresses which act during the motion at a particular instant of time on the quantum* $V_0$ *of the media, then they are in static equilibrium in the former sense, provided that one adds in every position additional forces ("forces of inertia"), whose components computed per unit of mass of the continuum are the same but opposite to the accelerations:*

$$-\frac{\partial^2 x}{\partial t^2} = -x'', \quad -\frac{\partial^2 y}{\partial t^2} = -y'', \quad -\frac{\partial^2 z}{\partial t^2} = -z''.\text{[52])}$$

Even if it frequently seems to be convenient to put this principle as an axiom on the top of kinetics, the question remains, in which independent constituent parts one can divide it and how much are these independent of the static axioms — a question, which appears in the very same sense already in the mechanics of discrete systems. Therefore it should be

---

[50] Cf. in particular Méc. anal., 2. part., Sect. XI, § I.

[51] Traité de dynamique, Paris 1743. Cf. IV 1 *Voss*, p. 77 [209]).

[52] In the following, apostrophes denote the derivatives of the functions of motion (1) with respect to $t$ for constant $a, b, c$.

nur kurz bemerkt, daß diess *d'Alembert*sche Prinzip einmal die wesentlich dem zweiten *Newton*schen Axiom äquivalente Tatsache enthält, dass die Beschleunigung eines frei gedachten Volumelementes gleich der Summe aller auf dasselbe wirkenden Kräfte ist, daß es aber andererseits — worauf *G. Hamel*[53]) nachdrücklich hingewiesen hat — eine von diesem ersten Bestandteil logisch durchaus unabhängige Aussage enthält: Wirken auf ein Kontinuum solche Kräfte, dass die für jedes Teilchen nach dem zweiten *Newton*schen Gesetz folgenden Beschleunigungen mit den kinematischen Bedingungen des Systemes verträglich sind, so treten diese Beschleunigungen auch wirklich ein.

Führt man nunmehr in das *d'Alembert*sche Prinzip das Prinzip der virtuellen Verrückungen als Gleichgewichtsbedingung ein, so erhält man das von *Lagrange*[54] ) als Grundformel der Dynamik benutzte *Variationsprinzip.* Man denke sich für jeden Moment $t$ die Bewegung wie in Nr. **2**, (6) überlagert mit einer unendlichkleinen virtuellen Verrückung, die mit den im Moment $t$ für das Kontinuum etwa bestehenden kinematischen Bedingungen verträglich ist; *dann muss die durch die Trägheitskräfte ergänzte virtuelle Arbeit stets verschwinden:*

$$(1) \qquad -\iiint\limits_{(V)} \varrho(x''\delta x + y''\delta y + z''\delta z)dV + \delta A = 0,$$

und dies für jeden Zeitpunkt $t$ des Bewegungsverlaufes. Im Falle eines beliebig stetig deformierbaren Körpers folgen hieraus unter den gleichen Annahmen wie in Nr. **3**c als Gleichungen der Bewegung für jeden Punkt des Kontinuums und zu jeder Zeit:

$$(2) \qquad \varrho x'' = \varrho X + \frac{\partial X_x}{\partial x} + \frac{\partial X_y}{\partial y} + \frac{\partial X_z}{\partial z} \qquad \begin{pmatrix} x, & y, & z \\ X, & Y, & Z \end{pmatrix},$$

während die Randbedingungen (5b) von Nr. **3** ungeändert für jeden Zeitmoment $t$ bestehen bleiben. Diese Gleichungen bestimmen die Bewegung wiederum erst dann, wenn der Zusammenhang der Kraft und Spannungskomponenten mit den Bewegungsfunktionen festgelegt ist.

Was nun die Berücksichtigung kinematischer Nebenbedingungen anlangt, so beziehen wir uns hier ausschliesslich auf den Fall sog. *holonomer* Bedingungen, die *keine zeitlichen* Ableitungen der Bewegungsfunktionen enthalten.[55]) Eine solche Bedingung unterscheidet sich von

---

[53] *G. Hamel*, Math. Ann. 66 (1908), p. 354; p. 386 ist die Unabhängigkeit für die Mechanik starrer Körper durch ein Beispiel gezeigt; vgl. auch die ausführlichere Darstellung in *Hamel*s Elementarer Mechanik, p. 306f.

[54] Méc. analyt., 2. part., Sect. II.

[55] Will man Probleme mit *nichtholonomen* Bedingungsgleichungen mit dem *d'Alembert*schen Prinzip behandeln, so muss man in der Mechanik der Kontinua wie bereits in der Punktmechanik davon absehen, dass die variierte

noticed here just shortly, how such [a formulation of] *d'Alembert*'s principle incorporates the statement being equivalent to *Newton*'s second axiom: that the acceleration of an imagined freely moving volume element is equal to the sum of all applied forces to this element; that this principle incorporates on the other hand — which has firmly been pointed out by *G. Hamel* [53]) — a statement being logically independent of this first constituent part: If there are forces acting on a continuum, such that the accelerations of every particle following *Newton*'s second law are admissible with respect to the systems kinematic constraints, then these accelerations really occur.

By introducing now the principle of virtual displacements as equilibrium condition into *d'Alembert*'s principle, one consequently obtains the *variational principle* used by *Lagrange* [54] ) as basic formula in dynamics. For every instant $t$, one thinks of the motion as in No. **2**, (6) superposed by an infinitesimal virtual displacement, which is admissible with respect to the kinematic constraints of the continuum arising at the instant of time $t$; *then the virtual work augmented by the inertia forces must vanish always:*

$$(1) \qquad -\iiint\limits_{(V)} \varrho(x''\delta x + y''\delta y + z''\delta z)dV + \delta A = 0,$$

and this for every instant of time $t$ of the path of motion. For the case of an arbitrarily continuously deformable body and under the same assumptions as in No. **3**c, one can deduce as equations of the motion for every point of the continuum at any instant of time [the following ones]:

$$(2) \qquad \varrho x'' = \varrho X + \frac{\partial X_x}{\partial x} + \frac{\partial X_y}{\partial y} + \frac{\partial X_z}{\partial z} \qquad \begin{pmatrix} x, & y, & z \\ X, & Y, & Z \end{pmatrix},$$

while the boundary conditions (5b) of No. **3** remain unchanged for every instant of time $t$. However, these equations only determine the motion, when the connection of the force and the stress components with the functions of motion is specified.

Concerning the consideration of kinematic constraints, we refer here exclusively to the case of so called *holonomic* constraints, which contain *no time* derivatives of the functions of motion.[55]) Such a constraint differs from

---

53 *G. Hamel*, Math. Ann. 66 (1908), p. 354; p. 386 the independence for the mechanics of rigid bodies is demonstrated in an example; cf. also the extensive presentation in *Hamel*'s Elementarer Mechanik, p. 306f.

54 Méc. analyt., 2. part., Sect. II.

55 If one wants to treat problems including *nonholonomic* constraints in the mechanics of continua using *d'Alembert*'s principle, then, as in point mechanics, one has to refrain from considering that the variation

der in Nr. **4**c betrachteten Form nur durch das explizite Auftreten von $t$:

$$\omega(a, b, c; x, y, z; x_a, \ldots, z_c; t) = 0. \tag{3}$$

Für die virtuellen Verrückungen kommt nun nur die Form dieser Bedingung zur Zeit $t$ in Betracht; die variierte Lage soll (für beliebig kleine $\sigma$) der Bedingung (3) für den betrachteten festen Wert von $t$ genügen, so daß durch Differentiation nach $\sigma$ („Variation der Bewegung bei festem t") folgt

$$\sum_{(xyz)} \frac{\partial \omega}{\partial x} \delta x + \sum_{(xyz,abc)} \frac{\partial \omega}{\partial x_a} \delta x_a = 0 \qquad \text{für jedes } t. \tag{3'}$$

Hieraus ergeben sich wie in Nr. **4**c angedeutet die Bewegungsgleichungen.

**5b. Übergang zu dem sog. Hamiltonschen Prinzip.** Man kann nun auch weiterhin ganz analog bekannten Entwicklungen der Punktmechanik das *d'Alembert*sche Prinzip in andere die Bewegung bestimmende Variationsprinzipe umformen; insbesondere handelt es sich hier darum, die von der Bewegung (den Trägheitskräften) herrührenden Glieder in die Variation eines einzigen für jeden Bewegungsvorgang bestimmten Ausdruckes überzuführen.

Grundlegend sind, wie bei *Lagrange* [56]), die Identitäten

$$x''\delta x = \frac{d}{dt}(x' \cdot \delta x) - \delta(\tfrac{1}{2}x'^2) \qquad (x, y, z),$$

die durch wiederholte Differentiation aus Nr. **2**, (6) nach den voneinander unabhängigen Veränderlichen $\sigma, t$ folgen. Trägt man das in (1) ein und berücksichtigt, daß die Operationssymbole $\frac{d}{dt}$ und $\delta$ ohne Rücksicht auf den Faktor $\varrho$ vor das Integral gezogen werden können, da nach Einführung der $a, b, c$ als Integrationsvariable sowohl der Integrationsbereich $V_0$ als auch der verbleibende Faktor ($\varrho_0$ von von $\sigma$ und $t$ unabhängig sind, so ergibt sich

$$-\frac{d}{dt} \iiint_{(V)} \varrho \sum_{(xyz)} x'\delta x \cdot dV + \delta T + \delta A = 0, \tag{4}$$

Bewegung für kleine $\sigma$ den Bedingungen genügt — vielmehr wird die Bedingungsgleichung für die Verrückungen rein *formal* durch Ersetzen der zeitlichen Differentiation durch die Operation $\delta$ gewonnen; vgl. unten p. 633. Man vergleiche hierzu IV 1, Nr. **37**, **38** (*Voss*) und die dort zitierte Litteratur, insbesondere *O. Hölder*, Gött. Nachr., math.-phys. Kl. 1896, p. 141–143, ferner die seither erschienenen Darstellungen von *G. Hamel*, Zeitschr. Math. Phys. 50 (1904), p. 1 und Math. Ann. 59 (1904), p. 416.

[56] Mécan. anal., 2. part., sect. IV, art. 3.

the form considered in No. **4**c only by the explicit appearance of $t$:

$$\omega(a, b, c; x, y, z; x_a, \ldots, z_c; t) = 0. \tag{3}$$

Now, for the virtual displacements this form of condition at time $t$ comes into question only; the varied position shall (for arbitrarily small $\sigma$) satisfy the condition (3) for the considered fixed value of $t$, such that by differentiating with respect to $\sigma$ ("variation of the motion for fixed t") it follows

$$\sum_{(xyz)} \frac{\partial \omega}{\partial x}\delta x + \sum_{(xyz,abc)} \frac{\partial \omega}{\partial x_a}\delta x_a = 0 \qquad \text{for every } t. \tag{3'}$$

As mentioned in No. **4**c the equations of motion arise out of this.

**5b. Transition to the so-called Hamilton's principle.** In a rather similar way to the well-known derivations in point mechanics, one can still transform *d'Alembert*'s principle into other variational principles which determine the motion; here the point concerns in particular the transformation of the terms coming from the motion (the inertia forces) into the variation of a single expression being determined for every motion.

Essential are, as in *Lagrange* [56]), the identities

$$x''\delta x = \frac{d}{dt}(x' \cdot \delta x) - \delta(\tfrac{1}{2}x'^2) \qquad (x, y, z),$$

which follow from repeated differentiation of (6) from No. **2** with respect to the independent variables $\sigma, t$. By substituting that [relation] into (1), we obtain by taking into account that regardless of the factor $\varrho$ the operation-symbols $\frac{d}{dt}$ and $\delta$ can be dragged in front of the integral, because after introducing $a, b, c$ as variables of integration, both the domain of integration $V_0$ and the remaining factor $\varrho_0$ are independent of $\sigma$ and $t$

$$-\frac{d}{dt}\iiint\limits_{(V)} \varrho \sum_{(xyz)} x'\delta x \cdot dV + \delta T + \delta A = 0, \tag{4}$$

of the motion for small $\sigma$ satisfies the constraints — moreover the constraint equations for the displacements are gained in a pure *formal* way by replacing the time differentiation by the operation $\delta$; cf. below p. 633. Hereto one shall compare IV 1, No. **37**, **38** (*Voss*) and the cited literature therein, in particular *O. Hölder*, Gött. Nachr., math.-phys. Kl. 1896, p. 141–143, furthermore the presentations appeared so far from *G. Hamel*, Zeitschr. Math. Phys. 50 (1904), p. 1 and Math. Ann. 59 (1904), p. 416.

[56] Mécan. anal., 2. part., sect. IV, art. 3.

wobei zur Abkürzung die gesamte *kinetische Energie*

$$(5)\qquad T = \frac{1}{2}\iiint\limits_{(V_0)} \varrho_0 \sum_{(xyz)} x'^2 dV_0 = \frac{1}{2}\iiint\limits_{(V)} \varrho \sum_{(xyz)} x'^2 dV$$

eingeführt ist. Gl. (4) ist die von *G. Hamel* [57]) und *K. Heun* [58]) unter dem Namen *Lagrangesche Zentralgleichung* als Grundlage der Mechanik der Systeme mit endlich vielen Freiheitsgraden verwendete Gleichung, die also auch in der Mechanik der Kontinua im gleichen Sinne gilt[59]), und die mit (1) völlig äquivalent ist: *die Bewegung erfolgt so, dass für jede mit den etwa stattfindenden Bedingungen verträgliche virtuelle Verrückung in jedem Moment die zeitliche Ableitung der virtuellen Arbeit des Impulses* $(x', y', z')$ *pro Masseneinheit gleich der Summe der Variation der kinetischen Energie und der virtuellen Arbeit sämtlicher Kraftwirkungen ist.*[60])

Betrachtet man nun die Bewegung im Zeitintervalle $t_0 \leqq t \leqq t_1$, so gilt (4) für jeden Moment, und durch Integration nach $t$ in den Grenzen $t_0, t_1$, ergiebt sich, wenn die virtuellen Verrückungen für die Momente $t = t_0, t_1$ gleich Null genommen werden, das sog. *Hamiltonsche Prinzip* [61]): *Lagert man über die Bewegung des Kontinuums irgendwelche mit den etwa stattfindenden Bedingungen verträgliche virtuelle Verrückungen, die für die Momente durchweg verschwinden, so verschwindet das von* $t_0$ *bis* $t_1$ *erstreckte Zeitintegral der Summe von virtueller Arbeit und Variation der kinetischen Energie:*

$$(6)\qquad \int\limits_{t_0}^{t_1} (\delta T + \delta A)\,dt = 0.$$

Da in (6) die virtuellen Verrückungen für jedes Zeitintervall will-

---

[57] *G. Hamel*, Zeitschr. Math. Phys. 50 (1904), p. 14.

[58] *K. Heun*, Lehrbuch der Mechanik, T. 1: Kinematik (Leipzig 1906), p. 92. Vgl. auch IV 11, Nr. **11**, *K. Heun*.

[59] Vgl. IV 11, Nr. **19** bis **21**, *K. Heun*.

[60] Variiert man auch den Zeitparameter $t$, so kann man ebenso die von *G. Hamel* (Math. Ann. 59 (1904), p. 423) und *K. Heun* [58]) als *allgemeine Zentralgleichung* bezeichnete Relation auf die Mechanik der Kontinua übertragen; vgl. IV 11, Nr. **19** bis **21**, *K. Heun*.

[61] Dies Prinzip wurde für einzelne Teilgebiete der Mechanik der Kontinua von verschiedenen Seiten sehr bald aufgestellt, nachdem man es einmal für die Punktmechanik besass (s. IV 1, Nr. **42**, *Voss*); man vergleiche ausser der später zu zitierenden Litteratur der Einzeldisziplinen *A. Walter*, Anwendung der Methode Hamiltons auf die Grundgleichungen der math. Theorie der Elastizität, Diss. Berlin 1868, sowie die zusammenfassenden Darstellungen in *Kirchhoffs* Mechanik, p. 117 f und *W. Voigts* Kompendium I, p. 227 ff.

where the abbreviation of the total *kinetic energy*

$$(5)\qquad T = \tfrac{1}{2} \iiint\limits_{(V_0)} \varrho_0 \sum_{(xyz)} x'^2 dV_0 = \tfrac{1}{2} \iiint\limits_{(V)} \varrho \sum_{(xyz)} x'^2 dV$$

has been introduced. Eq. (4) is the equation used by *G. Hamel*[57]) and *K. Heun*[58]) under the name *central equation of Lagrange* as the foundation of mechanics of systems with finitely many degrees of freedom, which holds in the same manner also for the mechanics of continua[59]), and is completely equivalent to (1): *the motion happens to be such that for every virtual displacement being admissible with respect to the possible constraints, the time derivative of the virtual work of the momentum* $(x', y', z')$ *per unit of mass is, at any instant, equal to the sum of the variation of the kinetic energy and the virtual work of all force effects.*[60])

Considering now the motion within the time interval $t_0 \leqq t \leqq t_1$, then (4) holds for every instant, and by integrating in $t$ with the boundaries $t_0, t_1$, it follows the so-called *Hamilton's Principle* when the virtual displacements are taken to be zero at the instants $t = t_0, t_1$[61]): *By superimposing to the motion of the continuum any virtual displacements which vanish without exception at the instants* [$t_0$ *and* $t_1$] *being admissible with respect to the possibly occurring constraints, then the time integral from* $t_0$ *to* $t_1$ *of the sum of the virtual work and the variation of the kinetic energy vanishes:*

$$(6)\qquad \int_{t_0}^{t_1} (\delta T + \delta A)dt = 0.$$

Since for every time interval in (6) the virtual displacements

[57] *G. Hamel*, Zeitschr. Math. Phys. 50 (1904), p. 14.

[58] *K. Heun*, Lehrbuch der Mechanik, T. 1: Kinematik (Leipzig 1906), p. 92. Cf. also IV 11, No. **11**, *K. Heun*.

[59] Cf. IV 11, No. **19** to **21**, *K. Heun*.

[60] By varying also the time-parameter $t$, one can also obtain the relation denoted by *G. Hamel* (Math. Ann. 59 (1904), p. 423) and *K. Heun*[58]) as *general central equation* to the mechanics of continua; cf. IV 11, No. **19** to **21**, *K. Heun*.

[61] This principle has been formulated very soon by different authors for individual branches of the mechanics of continua, after one has got it for point mechanics (see IV 1, No. **42**, *Voss*); one shall compare besides the literature of individual disciplines cited later on *A. Walter*, Anwendung der Methode Hamiltons auf die Grundgleichungen der math. Theorie der Elastizität, Diss. Berlin 1868, as well as the summarizing presentation in *Kirchhoff's* Mechanik, p. 117 f. and *W. Voigts* Kompendium I, p. 227 ff.

kürlich gewählt werden können, kann man ebenso leicht rückwärts aus (6) auf (4) oder (1) schliessen: *diese Prinzipe sind völlig äquivalent.*

Man kann nun weiterhin direkt aus diesen Sätzen das *Prinzip der kleinsten Wirkung* in seinen verschiedenen Formen für die Mechanik der Kontinua herleiten[62]); doch scheint es da — abgesehen von Fällen, die auf Systeme mit endlichvielen Freiheitsgraden zurückkommen — noch keine wesentliche Anwendung gefunden zu haben.

**5c. Das Prinzip des kleinsten Zwanges.** Man kann die Trägheitsglieder im *d'Alembert*schen Prinzip auch ohne die Integration nach der Zeit in die Variation eines für jeden Bewegungszustand bestimmten nur vom Zustand im Moment $t$ abhängigen Ausdruckes überführen, wobei freilich das Auftreten zweiter zeitlicher Ableitungen zugelassen werden muß. So entsteht das *Gauss*sche Prinzip des kleinsten Zwanges[63]), das *A. v. Brill* neuerdings zum Ausgangspunkt einer systematischen Behandlung der Mechanik der Kontinua gewählt hat.[64])

Um dies Prinzip zu gewinnen, entnehmen wir die virtuelle Verrückung einer Schar variierter Bewegungen Nr. **2**, (6) von folgender besonderer Art: Jedes Teilchen $a, b, c$ soll in dem betrachteten Zeitmoment $t$ dieselbe Lage und dieselbe Geschwindigkeit besitzen wie bei der wirklichen Bewegung, d. h. es soll für ebenjenen Wert $t$ gelten:

$$\delta x(a,b,c;t) = 0, \quad \delta x'(a,b,c;t) = 0 \quad (x,y,z), \tag{7}$$

während die Variationen $\delta x'', \delta y'', \delta z''$ der Beschleunigungen von Null verschieden sind. Diese drei Funktionen kann man nun in jedem Falle als Bestimmungsstücke der in (1) eingehenden Verrückung verwenden. Im Falle eines frei deformierbaren Kontinuums ist das evident. Besteht aber eine Bedingung der Form (3), so ergiebt sich durch zweimalige Differentiation nach $t$

$$\sum_{(xyz)} \frac{\partial \omega}{\partial x} x'' + \sum_{(xyz,abc)} \frac{\partial \omega}{\partial x_a} x_a'' + \cdots = 0,$$

[62] Beispielsweise die Betrachtungen von *O. Hölder*, Die Prinzipien von Hamilton und Maupertius (Gött. Nachr., math.-phys. Kl. 1896, p. 122 ff.) lassen sich unmittelbar auf Kontinua ausdehnen.

[63] *Gauss*' Werke V, p. 23 = Journal f. Math. 4 (1829). Die erste analytische Formulierung dieses von *Gauss* nur in Worten ausgesprochenen Prinzipes haben *R. Lipschitz*, Journ. f. Math. 82 (1877), p. 321 ff. und wenig später *J. W. Gibbs*, Amer. Journ. 2 (1879), p. 49 = Scientif. Pap. II (New-York 1906), p. 1 gegeben. Über die weitere Litteratur s. IV 1, Nr. **39**, *A. Voss*.

[64] *A. v. Brill*, Vorlesungen zur Einführung in die Mechanik raumerfüllender Massen, Leipzig 1909.

can be chosen arbitrarily, one can also easily imply the other way round that (4) or (1) follow from (6): *these principles are completely equivalent.*

Furthermore, one can now derive directly from these propositions the *principle of least action* in its various forms for the mechanics of continua[62]); nevertheless it seems — except for cases reducing to systems with finitely many degrees of freedom — that it has not found essential applications [up to now].

**5c. The principle of least constraint.** Also without the integration in time, one can transform the inertia terms in *d'Alembert*'s principle into the variation of an expression determined for every motion by the state at the instant [of time] $t$ only, in which certainly the appearance of time derivatives of second order must be allowed. In this way *Gauss's* principle of least constraint emerges[63]), which *A. v. Brill* has recently chosen as the starting point of a systematic treatment of the mechanics of continua.[64])

To obtain this principle, we take from the virtual displacements a familiy of varied motions No. **2**, (6) of the following special kind: For the considered instant of time $t$, every particle $a, b, c$ shall have the same position and the same velocity as the actual motion, i. e. it shall hold for that very value $t$:

$$(7) \qquad \delta x(a,b,c;t) = 0, \quad \delta x'(a,b,c;t) = 0 \quad (x,y,z),$$

while the variations $\delta x'', \delta y'', \delta z''$ of the accelerations are different from zero. One can now use these three functions as characteristic quantities of the displacements involved in (1). In the case of a freely deformable continuum this is evident. However, when there is a condition of the form (3), two times differentiation with respect to $t$ yields

$$\sum_{(xyz)} \frac{\partial \omega}{\partial x} x'' + \sum_{(xyz,abc)} \frac{\partial \omega}{\partial x_a} x_a'' + \cdots = 0,$$

62 For instance the considerations of *O. Hölder*, Die Prinzipien von Hamilton und Maupertius (Gött. Nachr., math.-phys. Kl. 1896, p. 122 ff.) can be extended immediately to continua.

63 *Gauss*' Werke V, p. 23 = Journal f. Math. 4 (1829). The first analytic formulation of this principle proposed by *Gauss* only verbally has been given by *R. Lipschitz*, Journ. f. Math. 82 (1877), p. 321 ff. and soon after by *J. W. Gibbs*, Amer. Journ. 2 (1879), p. 49 = Scientif. Pap. II (New-York 1906), p. 1. For further literature see IV 1, No. **39**, *A. Voss*.

64 *A. v. Brill*, Vorlesungen zur Einführung in die Mechanik raumerfüllender Massen, Leipzig 1909.

wo durch die Punkte bekannte Funktionen der $x, \ldots, x_a, \ldots,$ und ihrer ersten zeitlichen Ableitungen angedeutet sind. Durch Variation, d. h. Differentiation nach $\sigma$, folgt wegen (7) für den festen Moment $t$

$$\sum_{(xyz)} \frac{\partial \omega}{\partial x} \delta x'' + \sum_{(xyz,abc)} \frac{\partial \omega}{\partial x_a} \delta x''_a = 0,$$

und das ist tatsächlich genau die oben für die $\delta x$ aufgestellte Bedingung (3′). Die Einführung der Funktionen $\delta x'', \ldots$ in (1) ist daher gestattet und ergiebt nach leichter Umformung das folgende neue Prinzip[65]): *Variiert man die wirkliche Bewegung eines Kontinuums in einem bestimmten Moment so, dass Lage und Geschwindigkeit eines jeden Teilchens erhalten bleiben, aber die Beschleunigung den etwa stattfindenden Nebenbedingungen entsprechend geändert wird, so verschwindet stets die folgende Integralsumme:*

$$(8) \quad -\delta \iiint\limits_{(V)} \tfrac{1}{2} \varrho \sum_{(xyz)} x''^2 dV + \iiint\limits_{(V)} \Big( \varrho \sum_{(XYZ)} X \delta x'' - \sum_{(XYZ,xyz)} X_x \frac{\partial \delta x''}{\partial x} \Big) dV + \iint\limits_{(S)} \sum_{(XYZ)} \overline{X} \delta x'' dS = 0.$$

An Stelle der hier auftretenden Variation einer der „mittleren Beschleunigung“ entsprechenden Größe[66]) kann man auch das genaue Analogon des *Gaussschen Zwanges* einführen; denn ordnet man der variierten Bewegung die gleichen ungeänderten Kräfte zu, so kann man (8) schreiben

$$(8') \quad -\delta \Big\{ \iiint\limits_{(V)} \tfrac{1}{2} \varrho \sum_{\binom{xyz}{XYZ}} (x'' - X)^2 dV \Big\} - \iiint\limits_{(V)} \sum_{(XYZ,xyz)} X_x \frac{\partial \delta x''}{\partial x} dV + \iint\limits_{(S)} \sum_{(XYZ)} \overline{X} \delta x'' dS = 0.$$

Die wesentliche Bedeutung dieses Prinzips besteht wie in der Punktmechanik darin, dass es völlig ungeändert auch bei Systemen mit *nichtholonomen* Nebenbedingungen Geltung hat. Besteht etwa eine solche Bedingungsgleichung, in der neben den Bewegungsfunktionen und ihren räumlichen Ableitungen auch die ersten *zeitlichen* Differentialquotienten auftreten:

$$\omega(a, b, c; x, y, z; x_a, \ldots, z_c; x', y, z'; x'_a, \ldots, z'_c; t) = 0,$$

so erhält man durch einmalige Differentiation nach $t$ die Bedingung

[65] *Brill*, a. a. O., p. 61 ff.

[66] Sie ist zuerst von *P. Appell*, Paris C. R. 129 (1899), p. 317 und in einer Reihe weiterer Arbeiten (s. IV 1, Nr. **38**, *Voss*) in diesem Zusammenhange benutzt worden.

where with the points some known functions $x, \ldots, x_a, \ldots,$ and their first time derivatives are indicated. By the variation, i. e. differentiation with respect to $\sigma$, it follows due to (7) for the fixed instant $t$

$$\sum_{(xyz)} \frac{\partial \omega}{\partial x} \delta x'' + \sum_{(xyz,abc)} \frac{\partial \omega}{\partial x_a} \delta x_a'' = 0,$$

and this is in fact exactly condition (3′) for the $\delta x$ formulated above. Thus, the introduction of the functions $\delta x'', \ldots$ in (1) is allowed and after slight transformations the following new principle is obtained.[65]): *Varying the actual motion of a continuum in a certain instant in such a way that the position and the velocity of every particle are conserved, but the acceleration is changed agreeing with the possible constraints, then the following sum of integrals always vanishes:*

$$(8) \quad -\delta \iiint_{(V)} \tfrac{1}{2}\varrho \sum_{(xyz)} x''^2 dV + \iiint_{(V)} \Big(\varrho \sum_{(XYZ)} X\delta x'' - \sum_{(XYZ,xyz)} X_x \frac{\partial \delta x''}{\partial x}\Big) dV + \iint_{(S)} \sum_{(XYZ)} \overline{X} \delta x'' dS = 0.$$

Instead of the variation of a quantity corresponding to an "averaged acceleration" appearing here[66]) one can also introduce the exact analogy of *Gauss's constraint*; then by attributing to the varied motion the same unchanged forces, one can write (8) as

$$(8') \quad -\delta \Big\{ \iiint_{(V)} \tfrac{1}{2}\varrho \sum_{\binom{xyz}{XYZ}} (x'' - X)^2 dV \Big\} - \iiint_{(V)} \sum_{(XYZ,xyz)} X_x \frac{\partial \delta x''}{\partial x} dV + \iint_{(S)} \sum_{(XYZ)} \overline{X} \delta x'' dS = 0.$$

The significant relevance of this principle lies, as in point mechanics, in the fact that it is valid completely unchanged for systems with *nonholonomic* constraints. For instance, [when] such a constraint equation, in which besides the functions of motion and their spatial derivatives also their first differential quotient with respect to *time* appear:

$$\omega(a, b, c; x, y, z; x_a, \ldots, z_c; x', y, z'; x_a', \ldots, z_c'; t) = 0,$$

then one obtains by once differentiating with respect to $t$ the condition

[65] *Brill*, op. cit. p. 61 ff.

[66] It has been used in this context at first by *P. Appell*, Paris C. R. 129 (1899), p. 317 and in a series of further works (s. IV 1, No. **38**, *Voss*).

für die Werte der Beschleunigung im festen Momente $t$, und durch Variation (Differentiation nach $\sigma$) ergiebt sich wegen (7)

$$\sum_{(xyz)} \frac{\partial \omega}{\partial x'} \delta x'' + \sum_{(xyz,abc)} \frac{\partial \omega}{\partial x'_a} \delta x''_a = 0,$$

was nunmehr (8) als Nebenbedingung zuzufügen ist.

Ist $\omega$ speziell linear in den Geschwindigkeiten $x', \ldots, x'_a, \ldots$ so ist dies Resultat dem Wesen nach identisch mit der Form, in der man das *d'Alembert*sche Prinzip vielfach für nichtholonome Bedingungen ausspricht[55]), wobei an Stelle der dort nur formal eingeführten virtuellen Verrückungen eben die Variationen der Beschleunigung treten.

Ein weiterer Vorzug dieses Prinzips vor dem *d'Alembert*schen, der indessen in der Mechanik der Kontinua bisher kaum ausgenutzt zu sein scheint, besteht darin, dass es auch für die Behandlung kinetischer Probleme mit Ungleichungsnebenbedingungen die geeignete Grundlage bietet: man hat nur zu fordern, dass der Ausdruck (8) für alle nach den Nebenbedingungen im Moment $t$ bei fester Lage und Geschwindigkeit der einzelnen Teilchen zulässigen Variationen der Beschleunigung kleiner oder gleich Null ist — genau wie es für die Punktmechanik schon *Gauss* [67]) besonders betont hat.

**5d. Ansätze allgemeinerer Natur.** Von Ansätzen, die über die bisher umschriebenen gewissermassen klassischen Formen der Grundgleichungen der Kinetik hinausführen, ist an erster Stelle eine *Verallgemeinerung des Hamiltonschen Prinzips* zu nennen, die ganz ähnlich bereits in der Dynamik der Systeme mit endlichvielen Freiheitsgraden eine wichtige Rolle spielt[68]); sie besteht darin, daß man zur Bildung der kinetischen Energie $T$ eine allgemeinere Funktion der Geschwindigkeitskomponenten, insbesondere eine definite quadratische Form verwendet:[69])

$$T = \tfrac{1}{2} \iiint_{(V)} \mathfrak{T} dV, \quad \text{wo } \mathfrak{T} = \varrho_{11} x'^2 + 2\varrho_{12} x' y' + \cdots . \tag{9}$$

Alsdann folgen aus dem *Hamilton*schen Prinzip (6) Bewegungsgleichungen, die sich von (2) nur dadurch unterscheiden, dass an Stelle von $\varrho \cdot x'', \ldots$ tritt $\frac{d}{dt}\left(\frac{\partial \mathfrak{T}}{\partial x'}\right), \ldots$. Die 6 Koeffizienten $\varrho_{11}, \ldots$ sind ge-

[67] *Gauss*, Werke V, p. 27.

[68] Vgl. IV 12, *P. Stäckel*

[69] Diese Ansätze spielen in den älteren optischen Theorien *Lord Rayleighs* die entscheidende Rolle; s. bes. Phil. Magaz. (4) 41 (1871), p. 519 (vgl. V 21, Nr. **29**, *A. Wangerin*). Derselbe Ansatz bei *T. J. Bromwich*, Lond. math. Soc. Proc. 34 (1902), p. 307.

the values of the acceleration [have to satisfy] for the fixed instant $t$, and by calculating the variation (differentiation with respect to $\sigma$) [subject to conditions] (7) it follows

$$\sum_{(xyz)} \frac{\partial \omega}{\partial x'} \delta x'' + \sum_{(xyz,abc)} \frac{\partial \omega}{\partial x'_a} \delta x''_a = 0,$$

which henceforth has to be added to (8) as a constraint.

Especially, when $\omega$ is linear in the velocities $x', \ldots, x'_a, \ldots$, then this result is, by its nature, identical to the form in which one frequently states *d'Alembert's* principle for nonholonomic conditions[55]), whereas the variations of the acceleration substitute the virtual displacements formally introduced therein.

An additional advantage of this principle in contrast to *d'Alembert*'s [principle], which has hardly been used so far in the mechanics of continua, lies therein, that it provides a suitable basis for the treatment of kinetic problems with inequality constraints: one only has to ask expression (8) to be smaller or equal to zero for all variations of the acceleration which at an instant $t$ for a fixed position and velocity are admissible with respect to the constraints — exactly as it already has been stressed in particular by *Gauss*.[67])

**5d. Principles of general nature.** To mention principles, which go beyond the so far discussed classical forms of the fundamental equations of kinetics, one must cite a *generalization of Hamilton's principle*, which plays quite similarly an important role in the dynamics of systems with finitely many degrees of freedom[68]); [the generalization] lies therein, to use for the formation of the kinetic energy $T$ a more general function in the components of velocities, particularly a definite quadratic form:[69])

$$(9) \qquad T = \tfrac{1}{2} \iiint_{(V)} \mathfrak{T}\, dV, \text{ where } \mathfrak{T} = \varrho_{11} x'^2 + 2\varrho_{12} x' y' + \cdots .$$

Thereupon from *Hamilton*'s principle (6) there follows equations of motion, which differ from (2) only in the point, that $\varrho \cdot x'', \ldots$ is substituted by $\frac{d}{dt}\left(\frac{\partial \mathfrak{T}}{\partial x'}\right), \ldots$. The 6 coefficients $\varrho_{11}, \ldots$ are

[67] *Gauss*, Werke V, p. 27.

[68] Cf. IV 12, *P. Stäckel*

[69] These approaches play an essential role in older optical theories of *Lord Rayleigh*; see especially Phil. Magaz. (4) 41 (1871), p. 519 (cf. V 21, No. **29**, *A. Wangerin*). The same approach in *T. J. Bromwich*, Lond. math. Soc. Proc. 34 (1902), p. 307.

gebene Funktionen von $a, b, c$; sie bestimmen gemeinsam die Dichte (Trägheitsmasse) des Mediums, die sonach von der Richtung abhängig ist (*kinetische Anisotropie*).

Sehr viel weiter holt ein *anderer Ansatz* aus, der über die besondere Form der kinetischen Energie bzw. der vermöge der Bewegung auftretenden „Trägheitskräfte" ebensowenig eine Annahme macht, wie der Arbeitsausdruck in Nr. **3** bezüglich der Natur der Kräfte und Spannungen: Man erweitere das Prinzip der virtuellen Verrückungen durch die einer vierten unabhängigen Variablen — der Zeit $t$ — entsprechenden Operationen (Integration nach $t$ und Hinzufügung von Gliedern mit zeitlichen Ableitungen der $\delta x, \ldots$) und bezeichne als *virtuelle Arbeit des bewegten Kontinuums im Zeitintervall* $t_0, t_1$ bei einer virtuellen Verrückung der Bewegung das vierfache Integral, in dem $a, b, c, t$ als Unabhängige aufgefasst sind):[70])

$$(10) \qquad \int_{t_0}^{t_1} dt \iiint_{(V_0)} dV_0 \Bigg( \sum_{\binom{x\,y\,z}{XYZ}} \Big( \varrho_0 X \delta x + X_t \frac{\partial \delta x}{\partial t} \Big) - \sum_{\binom{xyz}{XYZ};\, a\,b\,c} X_a \frac{\partial \delta x}{\partial a} \Bigg).$$

Dabei sind die *Impulskomponenten* $X_t, \ldots$, die den Einfluss der Bewegung repräsentieren, ebenso wie Kraft- und Spannungsgrössen $X, \ldots, X_a, \ldots$ in ihrer Abhängigkeit von den Bewegungsfunktionen gemäss der speziellen Natur des Kontinuums gegeben zu denken; die bisher angenommenen geläufigen Trägheitskräfte erhält man, wie (5), (6) zeigt, für $X_t = \varrho_0 x'$, während (9) einem allgemeinen linearen Ansatz in $x', y', z'$ entspricht. Weiterhin können zu (10) wie in Nr. **3** noch analoge Integrale über den Rand des Integrationsgebietes im $a$-$b$-$c$-$t$-Raume hinzutreten. *Die Bewegung geht nun so vor sich, dass die virtuelle Arbeit* (10) *für jede unendlichkleine mit den etwa bestehenden Nebenbedingungen verträgliche virtuelle Verrückung verschwindet;* daraus kann man nach den bekannten Methoden leicht die Bewegungsgleichungen entnehmen — beispielsweise folgt für ein beliebig stetig deformierbares Kontinuum:

$$(11) \qquad \frac{dX_t}{dt} = \varrho_0 X + \frac{\partial X_x}{\partial a} + \frac{\partial X_y}{\partial b} + \frac{\partial X_z}{\partial c} \qquad (X, Y, Z),$$

und analog ergeben sich die Randbedingungen. Ganz wie in Nr. **3**c

[70] Für den Spezialfall, dass diese virtuelle Arbeit die Variation eines „Wirkungsintegrales" ist, sind diese Ansätze systematisch aufgestellt und verfolgt von *E.* u. *F. Cosserat*, Corps déformables, p. 156 ff. (vgl. Nr. **7**b). — In einer durch die Anforderungen der Relativitätstheorie modifizierten Form tritt derselbe Ansatz auf bei *H. Minkowski*, Grundgleichungen der elektromagnet. Vorgänge in bewegten Körpern, Gött. Nachr. 1908, p. 106 (vgl. Nr. **16**).

known functions of $a, b, c$; they determine together the density (mass of inertia) of the medium, which therefore depends on the direction (*kinetic anisotropy*).

Much further goes an *another postulation*, which makes on the particular form of the kinetic energy or the "inertia forces" due to the motion just as little assumptions, as the work expression in No. **3** on the nature of forces and stresses: One shall augment the principle of virtual displacements with a fourth independent variable — the time $t$ — [with] corresponding operations (integration in $t$ and addition of terms with time derivatives of $\delta x, \ldots$) and [one shall] denote the fourfold integral, in which $a, b, c, t$ are considered to be independent, as *virtual work of the moving continuum in the time interval* $t_0, t_1$:[70])

$$(10) \qquad \int\limits_{t_0}^{t_1} dt \iiint\limits_{(V_0)} dV_0 \left( \sum_{\binom{x\,y\,z}{X\,Y\,Z}} \left( \varrho_0 X \delta x + X_t \frac{\partial \delta x}{\partial t} \right) - \sum_{\left(\substack{xyz\\XYZ};\, a\,b\,c\right)} X_a \frac{\partial \delta x}{\partial a} \right).$$

Thereby, the *components of momentum* $X_t, \ldots$, representing the influence on the motion, shall be seen in the same way as the force and the stress quantities $X, \ldots$, $X_a, \ldots$ in their relation to the functions of motion according to the special nature of the continuum; one obtains the inertia forces commonly assumed so far, as (5), (6) show, for $X_t = \varrho_0 x'$, while (9) corresponds to a general linear ansatz in $x', y', z'$. In addition, there can be added to (10) as in No. **3** similar integrals over the boundary of the domain of integration in the $a$-$b$-$c$-$t$-space. *The motion now takes place in such a way, that the virtual work* (10) *vanishes for every infinitesimal virtual displacement being admissible with respect to the possible constraints;* According to the well known methods, one can easily extract out of this the equations of motion — for instance for an arbitrarily continuously deformable continuum [it] follows:

$$(11) \qquad \frac{dX_t}{dt} = \varrho_0 X + \frac{\partial X_x}{\partial a} + \frac{\partial X_y}{\partial b} + \frac{\partial X_z}{\partial c} \qquad (X, Y, Z),$$

and the boundary conditions are obtained analogously. Like in No. **3**c

[70] For the special case, that this virtual work is the variation of an "action integral", these approaches are formulated and pursued by *E.* and *F. Cosserat*, Corps déformables , p. 156 ff. (cf. No. **7**b). — In a form modified by the requirements of the theory of relativity the same approach appears in *H. Minkowski*, Grundgleichungen der elektromagnet. Vorgänge in bewegten Körpern, Gött. Nachr. 1908, p. 106 (cf. No. **16**).

kann man in (10), (11) $x, y, z$ statt $a, b, c$ als Unabhängige einführen[71]).

Ganz analog hat man die allgemeine Kinetik der in Nr. **4**b betrachteten *Medien mit orientierten Teilchen* auszubauen, wenn man diesen Teilchen einen Trägheitswiderstand gegen Winkelbeschleunigungen zuschreibt: Man hat, um hier sogleich den allgemeinsten Ausdruck zu formulieren, zu (10) nur das Nr. **4**, (2) analoge Integral[72])

$$(12) \qquad \int_{t_0}^{t_1} dt \iiint_{(V_0)} dV_0 \Big( \sum_{\left(\substack{LMN \\ \pi\kappa\varrho}\right)} \Big(\varrho_0 L \delta\pi + L_t \frac{\partial \delta\pi}{\partial t}\Big) - \sum_{\left(\substack{LMN \\ \pi\kappa\varrho}; \, abc\right)} L_a \frac{\partial \delta\pi}{\partial a} \Big)$$

hinzuzufügen, wo $L_t, \ldots$ den Impuls der inneren Rotation bestimmen, und kann hieraus wie in Nr. **4**b in jedem Falle die Bewegungsgleichungen, die bei freier Beweglichkeit der Dreikante ein zweites (11) analoges Tripel aufweisen, herleiten[73])

Alle diese Betrachtungen sind mit leichten Modifikationen auch auf die Dynamik zwei- und eindimensionaler Kontinua anwendbar.[74])

## III. Die Formen der Wirkungsgesetze.

### A. Formulierung der allgemeinen Typen.

**6. Die Typen der Abhängigkeit der Kraftwirkungen von den Deformationsgrössen.** Während in den bisherigen Erörterungen die Wirkungskomponenten — unter diesem Ausdruck seien der Kürze halber neben den Kräften und Spannungen aller Arten auch die Impulsgrössen von Nr. **5**d mitinbegriffen — nur formal als Koeffizienten des Ausdrucks der virtuellen Arbeit in Betracht kamen, ist nunmehr von ihrem Zusammenhang mit den Bestimmungsstücken der Deformation bzw. der Bewegung des Kontinuums Rechenschaft zu geben, der bestehen und bekannt sein muss, wenn anders die angegebenen Grundgleichungen überhaupt die Deformation bzw. Bewegung des Kontinuums bestimmen sollen. Er muss überdies die anschaulich evidente Tatsache zum Ausdruck bringen, dass in jedem Kontinuum durch Bewegungen und Deformationen gewisse Kraftwirkungen ausgelöst werden, und dass umgekehrt durch einwirkende Kräfte und Spannungen Bewegungen und Deformationen hervorgerufen werden. Dabei muss in

---

[71] Vgl. *E.* u. *F. Cosserat*, a. a. O., p. 187 ff.

[72] *E.* u. *F. Cosserat*, a. a. O., p. 156 ff., p. 167 ff.

[73] Vgl. auch IV 11, Nr. **21**c (*K. Heun*)

[74] *E.* u. *F. Cosserat*, a. a. O., p. 121. Die Ansätze der Kinetik ein- und zweidimensionaler Kontinua ordnen sich denen der Statik zwei- bzw. dreidimensionaler ein.

in (10), (11), one can introduce $x, y, z$ instead of $a, b, c$ as independent [variables][71]).

Completely analogously one has to extend the general kinetics of *media with oriented particles* considered in No. **4**b, if one associates to these particles a resistance of inertia against angular accelerations: In order to formulate readily the most general expression, one has to add to (10) only the integral being analogous to No. **4**, (2)[72])

$$(12) \qquad \int_{t_0}^{t_1} dt \iiint_{(V_0)} dV_0 \left( \sum_{\binom{LMN}{\pi\kappa\varrho}} \left( \varrho_0 L \delta\pi + L_t \frac{\partial \delta\pi}{\partial t} \right) - \sum_{\left(\substack{LMN \\ \pi\kappa\varrho}\, ;\, a\,b\,c\right)} L_a \frac{\partial \delta\pi}{\partial a} \right),$$

where $L_t, \ldots$ determines the momentum of the internal rotation, and [one] can herefrom derive as in No. **4**b in every case the equations of motion, which have for a free movability of the triad a second triple [of equations] analogous to (11).[73])

All these considerations are with slight modifications also applicable for the dynamics of two- and one-dimensional continua.[74])

## III. The forms of constitutive laws.

### A. Formulation of general classes.

**6. The classes with dependence of the force effects on the deformation quantities.** While in the previous discussions the effects — for the sake of brevity this expression includes besides forces and stresses of any kind also the momentum quantities of No. **5**d — have been considered in a mere formal way as coefficients of the virtual work expression, henceforth, [we] have to account for their connection with the characteristic quantities of the deformation or the motion of the continuum, which has to exist and has to be known, when after all the stated fundamental equations shall determine the deformation or the motion of the continuum. Moreover, [this connection] must express the clearly evident fact, that in every continuum due to motions and deformations certain force effects are induced, and that vice versa due to impressed forces and stresses motions and deformation are caused. Thereby

[71] Cf. *E.* and *F. Cosserat*, op. cit. p. 187 ff.

[72] *E.* and *F. Cosserat*, op. cit. p. 156 ff., p. 167 ff.

[73] Cf. also IV 11, No. **21**c (*K. Heun*)

[74] *E.* and *F. Cosserat*, op. cit. p. 121. The postulations regarding the kinetics of one- and two-dimensional continua can be based on those [used] in the statics of two- and three-dimensional [continua], respectively.

erster Linie der Unterschied zur Geltung kommen, ob die Kraftwirkungen *äussere* sind, d. h. in den Beziehungen des betrachteten Mediums zu ausserhalb gelegenen Medien oder Wirkungsquellen ihren Ursprung haben (Fernkräfte, Drucke an den Grenzflächen usw.), oder *innere*, d. h. auf der materiellen Konstitution des einzelnen Mediums und den gegenseitigen Einwirkungen seiner Teile beruhen. Die zuletzt genannten Wirkungen sind für den vorliegenden Zweck wesentlicher; insofern die gesuchten Gleichungen sie liefern, charakterisieren sie das eigentümliche dynamische Verhalten eines jeden Mediums innerhalb der allen Kontinuis gemeinsamen Formen und können daher geradezu als *Stoffgleichungen* bezeichnet werden.

Bei der Erörterung, wie diese Stoffgleichungen im allgemeinen beschaffen sind, genügt es, in erster Linie auf die in Nr. **3** behandelten Medien und die eigentlichen Spannungsgrössen $X_x, \ldots, Z_z$ und allenfalls auf die Kraftkomponenten $X, Y, Z$ Bezug zu nehmen. Danach lassen sich dann die entsprechenden allgemeinen Schemata für die in Nr. **4**, auftretenden Spannungsgrößen in weiterem Sinne und für die Impulskomponenten von Nr. **5**d leicht aufstellen — die bei diesen bisher tatsächlich angewendeten Ansätze ordnen sich übrigens den speziellen in Nr. **7**b,f zu besprechenden Abhängigkeitstypen unter. Die Werte der Spannungskomponenten $X_x, \ldots, Z_z$ die dem zur Zeit $t$ an der Stelle

$$x = x(a, b, c; t), \quad y = y(a, b, c; t), \quad z = z(a, b, c; t) \tag{1}$$

befindlichen Teilchen $a, b, c$ entsprechen, müssen durch die Stoffgleichungen für jede mögliche Bewegung des Kontinuums gegeben sein; sie werden also explizit dargestellt als irgendwie geartete von $a$, $b$, $c$, $t$ und den Funktionen (1) abhängige Ausdrücke, in die neben den Werten dieser Funktionen und ihrer örtlichen und zeitlichen Ableitungen an der Stelle $a, b, c, t$ möglicherweise auch ihre Werte an andern Stellen $\bar{a}$, $\bar{b}$, $\bar{c}$, $\bar{t}$ und überhaupt ihr Gesamtverlauf im Variabilitätsbereich ihrer vier Veränderlichen (Integrale u. dgl.) eingehen — also, symbolisch geschrieben, in der Form:

$$F(a, b, c, t; x(\bar{a}, \bar{b}, \bar{c}, \bar{t}), \ldots\ldots). \tag{2}$$

Geht man zu einem andern rechtwinkligen Koordinatensystem $x, y, z$ über, so sind diese neun Ausdrücke der Spannungskomponenten wie die Komponenten einer Dyade zu transformieren (und ebenso die Ausdrücke für $X, Y, Z$ wie Vektorkomponenten usw.); handelt es sich um innere Kraftwirkungen, so müssen zwischen den transformierten Komponenten und den neuen Koordinaten Gleichungen genau der alten Form bestehen.

primarily the [following] difference must be clarified, if the force effects are *external*, i. e. [the effects] have their cause in the relation to media and sources of effects located outside the considered medium (long-range forces, pressures at the boundary and such like), or *internal*, i. e. [the effects] are based on the material constitution of the particular medium and the mutual effects of the particles thereof. The last-named effects are for the objective at hand more essential; provided that the desired equations yield these [effects], they characterize the specific dynamic behavior of each one medium within the common classes of all continua and can consequently be denoted as *material laws*.

In the discussion, how these material laws are constituted in general, it is enough to refer primarily to the media treated in No. **3** and to the effective quantities of stress $X_x, \ldots, Z_z$ and if necessary to the force components $X, Y, Z$. Thereafter, the corresponding general schemes for the quantities of stress in the broader sense appearing in No. **4** and for the components of momentum of No. **5**d can be formulated easily — after all, the formulations [of the material laws] for these [effects] being effectively applied so far, can be deduced as special classes of dependence [which need] to be discussed in No. **7**b,f. The values of the stress components $X_x, \ldots, Z_z$ corresponding to the particle $a, b, c$ located at time $t$ at the position

$$(1) \qquad x = x(a,b,c;t), \quad y = y(a,b,c;t), \quad z = z(a,b,c;t),$$

must be given by the material laws for every possible motion of the continuum; hence [the values] are represented explicitly as expressions of any kind depending on $a$, $b$, $c$, $t$ and the functions (1). [These expressions] also include besides the values of the functions [(1)] and their spatial and time derivatives at the positions $a, b, c, t$ possibly values at other positions $\bar{a}, \bar{b}, \bar{c}, \bar{t}$ and in general the complete history in the domain of variability of the four variables (integrals and similar ones) — Hence, symbolically written in the form:

$$(2) \qquad F(a,b,c,t; x(\bar{a},\bar{b},\bar{c},\bar{t}), \ldots\ldots).$$

Changing over to another orthogonal coordinate system $x, y, z$, then these nine expressions of the stress components have to be transformed like the components of a dyad (and similarly the expressions for $X, Y, Z$ like vector components and so on); if it concerns internal force effects, then there must exist equations between the transformed components and the new coordinates [which are] exactly of the old form.

Wir betrachten nun der Reihe nach die einzelnen möglicherweise in den Spannungsausdrücken auftretenden Argumente; natürlich können die im folgenden einzeln diskutierten Wirkungen auch gleichzeitig stattfinden. In erster Linie ist da zu bemerken, daß explizites Vorkommen der Parameter $a, b, c$ auf *Inhomogenität* des Mediums, d. h. Verschiedenheit seiner Eigenschaften von Teilchen zu Teilchen, hindeutet; Auftreten des Zeitparameters $t$ bedeutet in ihrem zeitlichen Verlauf von vornherein bestimmte, d. h. ohne Rücksicht auf die wirklich stattfindende Bewegung gegebene *äussere Einwirkungen.*

Das eigentlich Wesentliche ist natürlich die Art des Eingehens der Funktionen (1) selbst[75]); betrachten wir zunächst den Fall, daß nur ihr Wertverlauf in beliebig kleiner Umgebung der Stelle $a, b, c, t$, d. h. die Werte der Funktionen und ihrer Ableitungen an dieser Stelle, in (2) auftreten, daß also (2) von der Form ist

$$(3) \qquad F(a, ..., t; x(a, ..., t), ...; x_a(a, ..., t), ..., x_t(a, ..., t); x_{aa}(a, ..., t), ...).$$

Das Vorkommen der Lokalwerte von $x, y, z$ selbst bedeutet Wirkungen, die von der wirklichen Lage der einzelnen Teilchen im Raum abhängen, wie es beispielsweise äussere gegebene Kraftfelder (Schwere oder dgl.) sind. Charakteristischer für die Kontinua sind indessen die Nahewirkungen, die sich im Auftreten von Spannungen infolge lokaler Deformationen äussern. Als Bestimmungsstücke der gesamten Deformation an einer Stelle (nicht bloss der reinen Formänderung der elementaren Elastizitätstheorie) betrachtet man bekanntlich in erster Linie die Werte der neun *ersten örtlichen Ableitungen* daselbst (vgl. IV 14, Nr. **16**); die in Rede stehende Wirkung kommt daher in expliziter Abhängigkeit der Spannungskomponenten von den Werten $x_a, \ldots, z_c$ an der Stelle $a, b, c, t$ zum Ausdruck. Die Art dieser Abhängigkeit muss hervortreten lassen, ob und welche einzelnen Bestandteile der Deformation alleinigen oder vorzugsweisen Einfluss auf die Spannung bzw. auf die einzelnen Bestandteile der Spannung besitzen, wie später bei der Behandlung der einzelnen Gebiete zur Geltung kommen wird.

Der Deformationszustand an einer Stelle wird genauer beschrieben, wenn man neben den ersten noch *höhere örtliche Ableitungen* der Funktionen (1) heranzieht, d. h. die Deformation in der Umgebung durch eine Transformation höheren Grades statt durch eine lineare approximiert; die Abhängigkeit der Spannungen von der Deformation wird also vollständiger wiedergegeben sein, wenn man auch diese höheren

[75] Die im folgenden zunächst anzuführenden Abhängigkeitstypen sind ihrer Form nach in der Regel zuerst in der Entwicklung der Elastizitätstheorie aufgetreten; nähere Angaben werden unter IIIB in den Nrn. **9**—**16** zu machen sein.

We consider now successively each argument possibly appearing in the expressions of stress; certainly, these effects which are discussed in the following individually can also appear simultaneously. Primarily, it has to be noted that the explicit appearance of the parameter $a, b, c$ indicates *inhomogeneity*, i. e. difference of the properties from particle to particle; [the] appearance of the time parameter $t$ indicates given *external excitations*, whose progress in time is a priori determined, irrespective of the actual occurring motion.

The bare essentials are certainly how the functions (1) themselves enter [in the functional dependence expressed by (2)][75]); to begin with, we consider the case that [for the functions (1)] only their behavior in an arbitrary small vicinity of the position $a, b, c, t$, i. e. the values of the functions and their derivatives at this position, appear in (2), hence that (2) is of the form

$$F(a, ..., t; x(a, ..., t), ...; x_a(a, ..., t), ..., x_t(a, ..., t); x_{aa}(a, ..., t), ...). \tag{3}$$

The occurrence of local values of $x, y, z$ themselves corresponds with effects, which depend on the actual position of the individual particles in space, as they are for example external given force fields (gravity or similar ones). More characteristic for continua are however short-range effects, which manifest themselves in the appearance of stresses due to local deformations. As characteristic quantities of the whole deformation at a position (not only the pure shape change of the elementary theory of elasticity), one considers, as is well known, primarily the values of the nine *first spatial derivatives* thereof (cf. IV 14, No. **16**); thus, the considered effect at hand expresses itself with an explicit dependence of the stress components on the values $x_a, \ldots, z_c$ at the position $a, b, c, t$. The type of these dependences must clarify, if and which individual elements of the deformation have exclusive or mainly influence on the stress or on the individual elements of the stress, as it will become clear later in the discussion of the particular fields.

The state of deformation at a position is described more precisely, if one uses besides the first also *higher spatial derivatives* of the functions (1), i. e. the deformation in the neighborhood is approximated by a transformation of higher order instead of a linear one; the dependence of the stresses on the deformation will be represented more completely, if one includes also these higher

[75] The classes of dependence, quoted in the following to begin with, have according to their form usually appeared at first in the development of the theory of elasticity; particulars are to be given in IIIB for the Nos. **9**—**16**.

Ableitungen in die Stoffgleichungen hineinnimmt. Tatsächlich hat man bisher nicht höhere als zweite Ableitungen in Betracht gezogen, und zwar wird das erst dann nötig, wenn der Zustand des Mediums *sehr rasch* mit dem Ort variiert; die Spannungen an einer Stelle hängen dann also auch von dem örtlichen Abfall der gewöhnlichen Deformationsgrössen 1. Ordnung ab.

Ebenso wie die Lokalwerte der örtlichen Ableitungen können in der Kinetik auch die Werte der *zeitlichen Ableitungen* der Funktionen (1) an der Stelle $a, b, c, t$ in (3) explizit eingehen; man hat da namentlich die Geschwindigkeitskomponenten $x', y', z'$ der Teilchen selbst und die „Änderungsgeschwindigkeiten" der Deformationsgrössen $x'_a, \ldots, z'_c$ — die, mit $dt$ multipliziert, auch als Komponenten der Deformation der Umgebung des Teilchens $a, b, c$ vermöge der von $t$ bis $t + dt$ vor sich gehenden Bewegung aufgefasst werden können[76]) — in Betracht gezogen. Diese Ansätze werden den Erscheinungen der *äusseren* und *inneren Reibung*, der *Zähigkeit* u. dgl. gerecht.

Bei allen Gesetzen vom Typus (3) ist die Frage von prinzipieller Bedeutung, wie diese Gleichungen sich bei einer Transformation der Richtungen der $a$-$b$-$c$-Parameterlinien durch diesen Punkt $a, b, c$ verhalten, während die $x$-$y$-$z$-Koordinaten ungeändert bleiben. Dadurch wird nämlich bestimmt, ob und welche verschiedenen Richtungen durch einen Punkt des Mediums für dessen Konstitution, soweit sie sich in den betrachteten Stoffgleichungen äussert, gleichbedeutend sind, d. h. es wird über *Isotropie oder Aeolotropie des Mediums* entschieden; unter besonderen Verhältnissen ist dieser Zusammenhang in der Kristallphysik eingehend studiert worden, wobei nur durch die Beschränkung auf unendlichkleine Deformationen der Unterschied zwischen den Transformationen der $a, b, c$ und der $x, y, z$ nicht zur Geltung kommt.[77])

Für den allgemeineren Fall, dass in die Stoffgleichungen (2) auch die Werte der Funktionen (1) an anderen Stellen und zu anderen Zeiten eingehen, ist der charakteristische Ansatz von hinreichender Allgemeinheit — zunächst für die *Statik* —, die Spannungskomponenten gleich Raumintegralen über das ganze Kontinuum zu setzen

$$(4) \qquad \iiint\limits_{(V_0)} f(a, \ldots; x, \ldots; x_a, \ldots; \bar{a}, \ldots; \bar{x}, \ldots; \bar{x}_a, \ldots)\, d\bar{a}\, d\bar{b}\, d\bar{c},$$

[76] *Stokes*, Cambridge Phil. Trans. 8 (1845) = Math. and Phys. Papers 1 (1880), p. 80; vgl. auch IV 15, Nr. **7**, *Love*.

[77] Vgl. etwa *F. Neumann*, Vorles. üb. d. Theorie der Elastizität (Leipzig 1885), p. 164; *W. Voigt*, Abh. Ges. d. Wiss. Göttingen 34 (1887), 36 (1890), Kompendium I, p. 128 ff. und p. 333, sowie besonders Lehrb. d. Krystallphysik (Leipzig 1910), § 286 ff., § 370 ff., § 414 ff., § 462.

derivatives in the material laws. In fact, one has considered so far [derivatives which are] not higher than second derivatives, this is namely required not until then, when the state of the medium varies *very quickly* in space; the stresses at a position then depends also on the spatial slope of the common deformation quantities of 1. order.

Equally to the local values of the spatial derivatives, in kinetics also the values of the *time derivatives* of the functions (1) at the position $a, b, c, t$ can enter in (3) explicitly; One has considered in particular the velocity components $x', y', z'$ of the particles themselves and the "velocities of change" of the deformation quantities $x'_a, \dots, z'_c$ — which, multiplied by $dt$, can also be interpreted as components of the deformation of the neighborhood of the particle $a, b, c$ due to the ongoing motion between $t$ and $t + dt$[76]). These basic approaches satisfy the phenomena of *external* and *internal friction*, i. e. *viscosity* and similar ones.

For all laws of the class (3) the question, how these equations behave under a transformation of the directions of the $a$-$b$-$c$ parameter lines through these points $a, b, c$, while the $x$-$y$-$z$-coordinates remain unchanged, is of fundamental evidence. Thereby it is determined namely, if and which different directions through a point of the medium are tantamount for its constitution, provided that it is expressed in the considered material laws, i. e. it is decided on *isotropy or aeolotropy of the medium*; for specific conditions this connection has been studied thoroughly in the physics of crystals, where due to the mere restriction to infinitesimal deformation the difference between transformations of $a, b, c$ and $x, y, c$ does not appear.[77])

For the more general case, that within the material laws (2) also the values of the functions (1) at different positions and for different times enter, the characteristic ansatz is of sufficient generality — at first for *statics* —, to identify the components of stress with volume integrals over the whole continuum

$$\text{(4)} \qquad \iiint\limits_{(V_0)} f(a, \dots; x, \dots; x_a, \dots; \bar{a}, \dots; \bar{x}, \dots; \bar{x}_a, \dots) d\bar{a}\, d\bar{b}\, d\bar{c},$$

[76] *Stokes*, Cambridge Phil. Trans. 8 (1845) = Math. and Phys. Papers 1 (1880), p. 80; cf. also IV 15, No. **7**, *Love*.

[77] Cf. for instance *F. Neumann*, Vorles. üb. d. Theorie der Elastizität (Leipzig 1885), p. 164; *W. Voigt*, Abh. Ges. d. Wiss. Göttingen 34 (1887), 36 (1890), Kompendium I, p. 128 ff. and p. 333, as well as in particular Lehrb. d. Krystallphysik (Leipzig 1910), § 286 ff., § 370 ff., § 414 ff., § 462.

deren Integranden gegebene Funktionen der Werte der Deformationsfunktionen (1) und ihrer Ableitungen für die Teilchen $a, b, c$ und $\bar{a}, \bar{b}, \bar{c}$ sind. Damit sind die Fernwirkungen innerhalb des Mediums umfasst: eine Wirkung an der Stelle $x, y, z$ entsteht infolge der Zustände an allen anderen Stellen des Kontinuums. Aber neben die aus der klassischen Mechanik bekannten von Massenteilchen zu Massenteilchen wirkenden durch solche Ansätze dargestellten *Kräfte* nach Art der Attraktionskräfte treten hier neu die von *P. Duhem*[78]) betrachteten Fernwirkungen („*influence*") auf, vermöge deren an jeder Stelle des Kontinuums sich superponierende Kräfte oder Spannungen durch die an allen andern Stellen des Kontinuums stattfindenden *Deformationen* ausgelöst werden.

In der *Kinetik* wird man diesen Ansatz noch dahin ausdehnen, dass man ein Zeitintegral über den gesamten Bewegungsverlauf; oder vielmehr — unsern allgemeinen Vorstellungen von Ursache und Wirkung entsprechend — über die Zeit *vor* dem betrachteten Moment $t$ hinzunimmt; der Integrand enthält dabei die Werte der Funktionen (1) sowie ihrer Ableitungen für die Momente $t$ und $\bar{t}$ $(-\infty < \bar{t} \leqq t)$:

$$(5) \qquad \int_{-\infty}^{t} d\bar{t} \iiint_{(V)} d\bar{a}\, d\bar{b}\, d\bar{c}\, f(a,\ldots,t;x,\ldots,x_a;\ldots,x_t;\ldots;\bar{a},\ldots,\bar{t};\bar{x},\ldots;\bar{x}_a;\ldots\bar{x}_t;\ldots).$$

Solche Ausdrücke für die Spannungskomponenten hat zuerst *L. Boltzmann*[79]) zur Formulierung der Erscheinung der elastischen Nachwirkung verwendet, bei der die in einem Moment stattfindenden Spannungen tatsächlich abhängen von allen Zuständen, die das Medium vorher durchlaufen hat. Neuerdings hat *V. Volterra*[80]) die Untersuchung der durch diese Integralansätze entstehenden Probleme aufgenommen, nachdem er in der Theorie der Integro-Differentialgleichungen ein neues Mittel zu ihrer analytischen Behandlung sich geschaffen hatte; er läßt übrigens in (5) auch mehrfache Integrationen nach der Zeit zu, wobei der Integrand von den Werten für mehr als zwei Zeitmomente abhängt. Für die sämtlichen hier umfaßten Probleme, bei denen die Wirkungen in einem Moment von der ganzen Vorgeschichte des Systems abhängen, nimmt er die von *E. Picard*[81]) eingeführte Bezeichnung „*hereditäre Mechanik*" auf.

[78] *P. Duhem*, J. de math. (4) 8 (1892), p. 311; Ann. de l'Éc. norm. (3) 10 (1893), p. 215, und 21 (1904), p. 117.

[79] Wien. Ber. 70 (1874), p. 275 = Pogg. Annalen, Ergänzungsbd. 7 (1876), p. 624 = Wissensch. Abh. I, p. 616.

[80] Die allgemeinen Ansätze sind in den Roma, Acc. Linc. Rend. (5) 18, 2 (1909), p. 295 und Acta math. 35 (1912), p. 295 enthalten.

[81] Riv. di Scienz. 1 (1907), p. 14.

whose integrands are given functions of the values of the deformation functions (1) and their derivatives for the particles $a, b, c$ and $\bar{a}, \bar{b}, \bar{c}$. Thereby long-range effects are included within the medium: an effect at the position $x, y, z$ emerges in consequence of the states at all other positions of the continuum. However, besides *forces* represented by basic approaches known from classical mechanics acting from mass particles to mass particles according to the class of forces of attraction, here it appears anew the long-range effects ("influence") considered by *P. Duhem*[78]), due to which at every position of the continuum superposed forces or stresses are caused by *deformations* taking place at all other positions of the continuum.

In the [field] of *kinetics* one will augment this ansatz such, that one adds a time integral over the whole motion; or rather — corresponding with our general notion of action and reaction — over the time *before* the considered instant $t$; thereby the integrand contains the values of the functions (1) as well as their derivatives at the instants $t$ and $\bar{t}$ $(-\infty < \bar{t} \leqq t)$:

$$(5) \qquad \int_{-\infty}^{t} d\bar{t} \iiint_{(V)} d\bar{a}\, d\bar{b}\, d\bar{c}\, f(a,\ldots,t;x,\ldots,x_a;\ldots,x_t;\ldots;\bar{a},\ldots,\bar{t};\bar{x},\ldots;\bar{x}_a;\ldots\bar{x}_t;\ldots).$$

Originally, *L. Boltzmann*[79]) has used such expressions for the stress components to formulate the phenomenon of elastic residual effects, for which the stresses occurring in one instant depend in fact on all states the medium has passed in advance. Recently, *V. Volterra*[80]) has taken up the studies of problems arising from these integral formulations, once he has created with the theory of integro-differential equations a new tool for the analytical treatment thereof; by the way, he also allows in (5) multiple integrations with respect to time, where the integrands depend on the values of more than two instants of time. For all herein contained problems, for which the effects at one instant depend on the whole previous history of the system, he takes up the name "hereditary mechanics" introduced by *E. Picard*[81]).

[78] *P. Duhem*, J. de math. (4) 8 (1892), p. 311; Ann. de l'Éc. norm. (3) 10 (1893), p. 215, and 21 (1904), p. 117.

[79] Wien. Ber. 70 (1874), p. 275 = Pogg. Annalen, Ergänzungsbd. 7 (1876), p. 624 = Wissensch. Abh. I, p. 616.

[80] The general fundamentals are included in Roma, Acc. Linc. Rend. (5) 18, 2 (1909), p. 295 and Acta math. 35 (1912), p. 295.

[81] Riv. di Scienz. 1 (1907), p. 14.

Beschränkt man sich auf analytische Funktionen, so kann man unter entsprechenden Konvergenzvoraussetzungen das Zeitintegral (5) durch eine Funktion aller (unendlich vielen) zeitlichen Ableitungen der Funktionen $x, \ldots, x_a, \ldots$ im Moment $t$ ersetzen, wie dies *W. Voigt*[82]) bei der Anwendung auf elastische Nachwirkung tut.

Alle diese Formen der Stoffgleichungen sind vorzugsweise in dem speziellen Falle behandelt worden, dass die Deformationen des Kontinuums „*unendlichklein*" sind (vgl. IV 14, Nr. **16**, *Abraham*). Die Funktionen (1) umfassen diesen Fall, wenn man $a, b, c$ als Raumkoordinaten des Teilchens in der Ausgangslage auffasst und (vgl. Nr. **2**a, p. 607) mit Hilfe eines auf beliebig kleine Werte beschränkten Parameters $\sigma$ setzt:

$$x(a, b, c; t) = a + \sigma \cdot u(a, b, c; t) + \cdots \qquad \begin{pmatrix} x, y, z \\ u, v, w \end{pmatrix}, \tag{6}$$

wobei durchweg die höheren Potenzen von $\sigma$ gegenüber den niederen zu vernachlässigen sind. Hängt nun eine Wirkungskomponente von diesen Bewegungsfunktionen durch ein Gesetz von der Form (3) ab, so hat man den Ausdruck $F$ durch die ersten Glieder seiner Entwicklung nach Potenzen von $\sigma$ vermöge (6) zu ersetzen; verschwinden die linearen Glieder in $\sigma$ nicht identisch, so erhält man daher an Stelle von (3) ein Gesetz der Form:

$$F + \sigma\{F_x \cdot u + \cdots + F_{x_a} \cdot u_a + \cdots + F_{x_t} \cdot u_t + \cdots + F_{x_{aa}} \cdot u_{aa} + \cdots\}. \tag{3'}$$

$F, F_x, \ldots$ sind die Werte der Funktion (3) und ihrer Ableitungen für $\sigma = 0$, also bekannte Funktionen von $a, b, c, t$; das Wirkungsgesetz ist nunmehr *linear* in den Lokalwerten der die unendlichkleine Deformation bestimmenden Funktionen $u(a, b, c; t), \ldots$, und ihrer Ableitungen — entsprechend dem *Hookeschen Gesetz* der Elastizitätstheorie (vgl. IV 23, Nr. **4**). Das von $\sigma$ freie Glied entspricht den Anfangskräften oder -spannungen, die in dem undeformierten Kontinuum möglicherweise herrschen können. Ebenso könnte man aber auch Spannungsgesetze betrachten, bei denen der Koeffizient von $\sigma^1$ verschwindet[83]); dann würden für unendlichkleine Deformationen die Spannungen mindestens *quadratisch* von den Deformationen abhängen — entgegen dem Hookeschen Gesetz, das sonach auch für unendlichkleine Deformationen nicht notwendig gelten muss.

---

[82] Kompendium I, p. 458; vgl. auch *Cl. Maxwell*, Scientif. Papers 2, p. 623.

[83] Hierauf hat bei der Diskussion über die Gültigkeit des *Hooke*schen Gesetzes besonders nachdrücklich *B. de Saint-Venant* hingewiesen; vgl. seine Bemerkungen in *Navier*, De la résistance des corps solides, 3e éd. (Paris 1864), p. 662 und *Clebsch*, Théorie de l'élasticité des corps solides (Paris 1883), p. 39.

By restricting oneself to analytic functions, then, under corresponding convergence requirements, one can replace the time integral (5) by a function of all (infinitely many) time derivatives of the functions $x, \ldots, x_a, \ldots$ at the instant of time $t$, as it is done by *W. Voigt*[82]) for the treatment of elastic residual effects.

All these forms of material laws have been treated mainly for the special case that the deformations of the continuum are "infinitesimally small" (cf. IV 14, No. **16**, *Abraham*). The functions (1) include this case, when one considers $a, b, c$ as spatial coordinates of the particle in the initial position and (cf. No. **2**a, p. 607) by setting [these functions] with the help of a parameter $\sigma$, restricted to arbitrary small values, to:

$$(6) \qquad x(a,b,c;t) = a + \sigma \cdot u(a,b,c;t) + \cdots \quad \begin{pmatrix} x, y, z \\ u, v, w \end{pmatrix},$$

where throughout the higher powers of $\sigma$ are neglected with respect to the lower ones. If an effect depends now on these functions of motion by a law of the form (3), then one has to replace the expression $F$ with the first terms of its expansion with respect to the powers of $\sigma$ due to (6); If the linear terms in $\sigma$ do not vanish identically, then one obtains as a result instead of (3) a law of the form:

$$(3') \qquad F + \sigma\{F_x \cdot u + \cdots + F_{x_a} \cdot u_a + \cdots + F_{x_t} \cdot u_t + \cdots + F_{x_{aa}} \cdot u_{aa} + \cdots\}.$$

$F, F_x, \ldots$ are the values of the function (3) and its derivatives for $\sigma = 0$, thus known functions of $a, b, c, t$; the material law is now *linear* in the local values of the functions $u(a,b,c;t), \ldots$ determining the infinitesimal deformations, and the derivatives [of these functions]—according to *Hooke's law* of the theory of elasticity (cf. IV 23, No. **4**). The term without $\sigma$ corresponds to initial forces or stresses, which can possibly exist in the undeformed continuum. Similarly, one could even consider stress laws, for which the coefficient of $\sigma^1$ vanishes[83]); then for infinitesimally small deformations, the stresses would depend on the deformations at least *quadratic* — contrary to Hooke's law, which consequently does not have to be valid necessarily even for infinitesimally small deformations.

---

[82] Kompendium I, p. 458; cf. as well *Cl. Maxwell*, Scientif. Papers 2, p. 623.

[83] With particular emphasis, *B. de Saint-Venant* has pointed this out in the discussion on the validity of *Hooke*'s law; cf. his remarks in *Navier*, De la résistance des corps solides, 3e éd. (Paris 1864), p. 662 and *Clebsch*, Théorie de l'élasticité des corps solides (Paris 1883), p. 39.

Haben die Stoffgleichungen die Integralform (4), (5), so ergiebt genau die gleiche Betrachtung Reduktion des Integranden auf eine lineare — möglicherweise freilich auch auf eine höhere — Funktion der Werte der Verschiebungen und ihrer Ableitungen an den Stellen $a, b, c, t$ und $\bar{a}, \bar{b}, \bar{c}, \bar{t}$; beispielsweise wird aus (4)

$$(4') \qquad \iiint\limits_{(V_0)} (f + \sigma\{f_x \cdot u + f_{\bar{x}} \cdot \bar{u} + \cdots + f_{x_a} \cdot u_a + f_{\bar{x}_a} \cdot \bar{u}_a + \cdots\}) d\bar{a}\, d\bar{b}\, d\bar{c},$$

und ähnlich vereinfacht sich im Falle zeitlicher Nachwirkung das Integral (5).

**7. Medien mit einer charakteristischen Zustandsfunktion.**

Besonders häufig werden in der Mechanik der Kontinua Medien mit solchen Wirkungen betrachtet, deren charakteristische Gleichungen sich auf *eine einzige* Funktion der Zustandsgrößen zurückführen lassen. Eine solche Reduktion entspringt, wenn wir zunächst von der Statik reden, vor allem aus der Annahme, *dass die in Betracht kommende virtuelle Arbeit für jede virtuelle Verschiebung bis aufs Vorzeichen gleich ist der Variation eines einzigen nur von dem jeweiligen Deformationszustande abhängigen skalaren Ausdruckes, des „Potentials" oder der „potentiellen Energie" der wirkenden Kräfte oder Spannungen*[84]); diese Annahme kann auf allgemeine thermodynamische Sätze zurückgeführt werden.

**7a. Das gewöhnliche Potential und seine nächsten Verallgemeinerungen.** Die einfachste Form dieses Potentials wird durch die Eigenschaft charakterisiert, *dass das Potential eines in Teile zerlegten Bereiches gleich der Summe der Potentiale* $\Phi^*$ *der Teilbereiche* $V^*$ *ist*[85]). Unter den naheliegenden Voraussetzungen, *dass* $\Phi^*$ *sich stetig mit der Grenzfläche von* $V^*$ *ändert, und dass der Quotient* $\Phi^* : V^*$ *gegen einen bestimmten Grenzwert* $\bar{\varphi}$ *konvergiert, wenn* $V^*$ *sich unbegrenzt um eine bestimmte Stelle* $x, y, z$ *zusammenzieht* — und dies gleichmässig im ganzen Bereich $V$ —, folgt leicht[86]), dass das Potential

[84] Für einfache Fälle hat schon *Lagrange* in der Méc. anal. eine solche Annahme aus der Mechanik diskreter Massen auf Kontinua übertragen (Prem. part., Sect. IV, Nr. 25) und sie speziell auf die Hydrostatik angewendet, indem er der virtuellen Arbeit einen der Variation der Volumdilatation proportionalen Term anfügt (1. part., sect VIII, Nr. 1); ihre weitere Ausbildung hat sie dann in der Elastizitätstheorie erfahren, und zwar hat *G. Green* (Cambr. Phil. Soc. Trans. 1838 = Math. Papers (London 1871), p. 245) zum erstenmal aus ihr die Grundgleichungen abgeleitet. Vgl. dazu IV 23, Nr. **5**b

[85] Diese Annahme ist schon seit der ersten direkten Einführung des elastischen Potentiales als selbstverständlich mehr oder weniger ausdrücklich verwendet worden. Eine ausführliche Darlegung giebt *P. Duhem*, Le potential thermodynamique et la pression hydrostatique, Ann. Éc. Norm. (3) 10 (1893), p. 183.

[86] Vgl. *P. Duhem*, l. c., p. 187 ff. Es liegt hier nur eine präzisere Formu-

If the material laws are of the integral form (4), (5), then the same considerations lead to a reduction of the integrand to a linear — certainly perhaps also to a higher [order] — function of the values of the displacements and their derivatives at the positions $a, b, c, t$ and $\bar{a}, \bar{b}, \bar{c}, \bar{t}$; for example, (4) becomes

$$(4') \qquad \iiint_{(V_0)} (f + \sigma\{f_x \cdot u + f_{\bar{x}} \cdot \bar{u} + \cdots + f_{x_a} \cdot u_a + f_{\bar{x}_a} \cdot \bar{u}_a + \cdots\}) d\bar{a}\, d\bar{b}\, d\bar{c},$$

and similarly the integral (5) simplifies for the case of temporal residual effects.

**7. Media with one characteristic state function.**

In the mechanics of continua particularly often media are considered whose characteristic equations can be reduced to *one single* function of the state variables. In case we talk at first about statics, such a reduction originates particularly from the assumption, *that the virtual work coming into question is, up to sign, for every virtual displacement equivalent to the variation of a single scalar expression depending only on the corresponding state of deformation, [which is] the "potential" or the "potential energy" of the acting forces and stresses*[84]); this assumption can be traced back to general theorems of thermodynamics.

**7a. The common potential and its closest generalizations.** The most simple form of this potential is characterized by the property *that the potential of a domain dissected into parts is equal to the sum of the potentials $\Phi^*$ of the [corresponding] subdomains $V^*$*[85]). Under the obvious assumptions *that $\Phi^*$ changes continuously with the boundary of $V^*$ and that the quotient $\Phi^* : V^*$ converges to a certain limit value $\bar{\varphi}$, when $V^*$ contracts around a certain point $x, y, z$ indefinitely* — and this regularly in the whole domain $V$ —, it follows easily[86]) that the potential

[84] For simple cases already *Lagrange* has interpreted in the Méc. anal. such an assumption from the mechanics of discrete masses for continua (Prem. part., Sect, IV, No. 25) and applied it particularly in hydrostatics, by adding to the virtual work a term being proportional to the variation of the volume dilatation (1. part., sect VIII, No. 1); [this assumption] has undergone a further development in the theory of elasticity, namely *G. Green* (Cambr. Phil. Soc. Trans. 1838 = Math. Papers (London 1871), p. 245) has derived from it the fundamental equations for the first time. Cf. thereto IV 23, No. 5b

[85] Already since the first direct introduction of the elastic potential, this assumption, as [being] natural, has been used more or less explicitly. A detailed explanation is given by *P. Duhem*, Le potential thermodynamique et la pression hydrostatique, Ann. Éc. Norm. (3) 10 (1893), p. 183.

[86] Cf. *P. Duhem*, l. c., p. 187 ff. It is here merely a precise form-

des gesamten Kontinuums $V$ (und ähnlich das eines jeden Teilbereiches) durch das über $V$ erstreckte Raumintegral der Ortsfunktion $\bar{\varphi}$ dargestellt wird:

$$(1) \qquad \Phi = \iiint\limits_{(V)} \bar{\varphi}\, dx\, dy\, dz = \iiint\limits_{(V_0)} \varphi\, da\, db\, dc, \quad \text{wo } \varphi = \bar{\varphi}\frac{\partial(x,y,z)}{\partial(a,b,c)}.$$

$\bar{\varphi}$ ist die auf die Volumeinheit des deformierten Kontinuums, $\varphi$ die auf die Volumeinheit des Ausgangszustandes bezogene *Energiedichte*; es sind *skalare Gössen*, die für jedes in einem bestimmten Deformationszustande betrachtete Kontinuum stetige oder doch abteilungsweise stetige Funktionen von $x, y, z$ bzw. $a, b, c$ sind. Die Beschaffenheit des Kontinuums unabhängig von der jeweils stattfindenden Deformation ist bestimmt, wenn $\varphi$ als Funktion des Gesamtverlaufes der Deformationsfunktionen gegeben ist; soll das Zerlegungsaxiom für jede Deformation gelten, so kann $\varphi$ nur die Werte der Funktionen und ihrer Ableitungen an der betrachteten Stelle explizit enthalten:

$$(2) \qquad \varphi = \varphi(a,b,c; x(a,b,c), \ldots; x_a(a,b,c), \ldots; x_{aa}(a,b,c), \ldots)$$

Handelt es sich um innere Kraftwirkungen, so muß diese Funktion gegenüber rechtwinkligen Koordinatentransformationen im $x$-$y$-$z$-Raume invariant sein.

Wir nehmen zunächst an, dass *nur die ersten Ableitungen auftreten.* Um den Zusammenhang mit den Wirkungskomponenten zu finden[87]), bilden wir das Potential für die variierte Deformation (Nr. **2**a, (3)); dann ergiebt sich als Variation von $\Phi$

$$\delta\Phi = \iiint\limits_{(V_0)} \sum_{(x\,y\,z)} \left(\frac{\partial\varphi}{\partial x}\delta x + \frac{\partial\varphi}{\partial x_a}\delta x_a + \frac{\partial\varphi}{\partial x_b}\delta x_b + \frac{\partial\varphi}{\partial x_c}\delta x_c\right) da\, db\, dc,$$

wobei in die Ableitungen von $\varphi$ die unvariierten Werte von $x, y, z$ und ihren Ableitungen einzusetzen sind. Aus der Identität

$$(3) \qquad \delta A = -\delta\Phi \quad \text{für alle} \quad \delta x, \delta y, \delta z$$

folgen für ein Medium, das alle stetigen virtuellen Verrückungen gestattet, durch Gleichsetzung der Koeffizienten der $\delta x, \ldots$ und ihrer Ab-

lierung des von altersher in der Mechanik üblichen Prozesses der Umwandlung von Funktionen eines Gebietes (wie Masse u. dgl.) in bestimmte Integrale vor. Übrigens braucht man die gleichmässige Konvergenz von $\Phi^* : V^*$ nur für eine bestimmte, $V$ im Limes erschöpfende Einteilung vorauszusetzen und kann ausserdem natürlich Unterbrechungen der Stetigkeit und gleichmässigen Konvergenz an einzelnen Flächen zulassen.

[87] Es kommt hier lediglich das in der Variationsrechnung übliche Verfahren zur Bildung der ersten Variation mehrfacher Integrale in Betracht, wie es Lagrange (Misc. Taur. 2 (1760/61) = Oeuvres 1, p. 353) zuerst ausgebildet und in der Méc. anal. vielfach angewendet hat.

of the whole continuum $V$ (and similarly that of any subdomain) is represented by the volume integral, ranging over $V$, of the spatial function $\bar{\varphi}$:

$$(1) \qquad \Phi = \iiint\limits_{(V)} \bar{\varphi}\, dx\, dy\, dz = \iiint\limits_{(V_0)} \varphi\, da\, db\, dc, \text{ where } \varphi = \bar{\varphi}\frac{\partial(x, y, z)}{\partial(a, b, c)}.$$

$\bar{\varphi}$ is the *energy density* per unit of volume of the deformed continuum, $\varphi$ is [the energy density] per unit of volume of the initial state; these are *scalar quantities*, which are for every continuum considered in a certain state of deformation continuous or yet piecewise continuous functions of $x, y, z$ and $a, b, c$, respectively. The nature of the continuum, independent of each of the occurring deformation, is determined when $\varphi$ is given as a function of the complete history of the deformation functions; if the dissection axiom shall hold for every deformation, then $\varphi$ can explicitly contain only the values of the functions and their derivatives [evaluated] at the considered position:

$$(2) \qquad \varphi = \varphi(a, b, c; x(a, b, c), \ldots; x_a(a, b, c), \ldots; x_{aa}(a, b, c), \ldots)$$

If it is about internal force effects, this function must be invariant with respect to orthogonal coordinate transformations in the $x$-$y$-$z$-space.

At first we consider, that *only the first derivatives appear*. To find the connection with the effects[87]), we compute the potential of the varied deformation (No. **2**a, (3)); Then the variation of $\Phi$ is obtained as

$$\delta\Phi = \iiint\limits_{(V_0)} \sum_{(x\,y\,z)} \left(\frac{\partial\varphi}{\partial x}\delta x + \frac{\partial\varphi}{\partial x_a}\delta x_a + \frac{\partial\varphi}{\partial x_b}\delta x_b + \frac{\partial\varphi}{\partial x_c}\delta x_c\right) da\, db\, dc,$$

where in the derivatives of $\varphi$ the unvaried values of $x, y, z$ and the derivatives thereof have to be inserted. From the identity

$$(3) \qquad \delta A = -\delta\Phi \quad \text{for all} \quad \delta x, \delta y, \delta z$$

[and] by equating the coefficients of $\delta x, \ldots$ and the derivatives thereof, for a medium, which allows for all continuous virtual displacements,

ulation of the process, common of old in mechanics, of the transformation of functions of a domain (as e. g. mass) into certain integrals. By the way, one needs to assume the uniform convergence of $\Phi^* : V^*$ only for a certain partition [which] in the limit tends to $V$ and can in addition certainly allow disconnections of the continuity and the uniform convergence at individual surfaces.

87 Here, merely the approach common in the calculus of variations for the computation of the first variation for multiple integrals comes into consideration, as Lagrange (Misc. Taur. 2 (1760/61) = Oeuvres 1, p. 353) originally has formulated it and [as he has] applied it in the Méc. anal. in many cases.

leitungen unmittelbar die Stoffgleichungen. Verwendet man für $\delta A$ etwa den Ansatz Nr. **3**c, (7), so wird[88])

(4) $$\varrho_0 X = -\frac{\partial\varphi}{\partial x}, \quad X_a = \frac{\partial\varphi}{\partial x_a} \qquad \begin{pmatrix} x, y, z \\ X, Y, Z \end{pmatrix}; a, b, c\Big);$$

geht man mittels (8) von Nr. **3**c und (1) zu den auf die deformierte Lage bezogenen Grössen über, so erhält man[89]):

(5) $$\begin{cases} \varrho X = -\frac{\partial\overline{\varphi}}{\partial x}, \\ X_x = \frac{\partial\overline{\varphi}}{\partial x_a}\cdot x_a + \frac{\partial\overline{\varphi}}{\partial x_b} x_b + \frac{\partial\overline{\varphi}}{\partial x_c} x_c + \overline{\varphi}, \\ X_y = \frac{\partial\overline{\varphi}}{\partial x_a} y_a + \frac{\partial\overline{\varphi}}{\partial x_b}\cdot y_b + \frac{\partial\overline{\varphi}}{\partial x_c}\cdot y_c, \\ X_z = \frac{\partial\overline{\varphi}}{\partial x_a} z_a + \frac{\partial\overline{\varphi}}{\partial x_b} z_b + \frac{\partial\overline{\varphi}}{\partial x_c} z_c. \end{cases} \qquad \begin{pmatrix} x & y & z \\ X & Y & Z \end{pmatrix}$$

Damit sind die sämtlichen in Nr. **5** betrachteten Stoffgleichungen auf die einzige Gleichung (2) zurückgeführt, die $\varphi$ bzw. $\overline{\varphi}$ als skalare Funktion des lokalen Deformationszustandes giebt.

Hängt die Energiedichte (2) *auch von den zweiten Ableitungen* $x_{aa}, x_{ab}, \ldots$ der Deformationsfunktionen ab — was wiederum nur bei sehr rascher Änderung des Zustandes mit dem Orte in Betracht kommt —, so werden in $\delta\Phi$ neue Glieder mit den zweiten Ableitungen der virtuellen Verrückungen $\delta x_{aa} = \frac{\delta^2 \delta x}{\delta a^2}, \ldots$ auftreten, und das kommt gerade auf die in Nr. **4**a besprochenen Zusatzglieder zu dem ursprünglichen Ausdruck der virtuellen Arbeit hinaus; alsdann hängen sowohl die Komponenten dieser neuen Wirkung, deren Ausdrücke durch $\varphi$ sich unmittelbar ergeben, wie die alten Spannungskomponenten, deren Ausdrücke leicht zu modifizieren sind, auch von den zweiten Ableitungen $x_{aa}, \ldots$ ab.

Ein spezieller Fall, der sich hier einordnet, sei besonders hervorgehoben: dass nämlich zu dem Potential (1) ein *Integral über die Oberfläche des Kontinuums* additiv hinzutritt:

(6) $$\Phi_1 = \iint_{(S)} \overline{\psi} dS = \iint_{(S_0)} \psi dS_0,$$

wobei die „Flächendichte" $\overline{\psi}$ bzw. $\psi$ das Potentials von den Werten der

[88] *G. Kirchhoff*, Sitzungsber. d. Akad. Wien, math.-nat., Kl. 9 (1852), p. 772.

[89] *J. Boussinesq*, Mém. prés. p. div. sav., Paris **20** (1872), note 3. p. 591. Hier ist nur $\varphi$ statt $\overline{\varphi}$ verwendet, aber, was das Wesentliche ist, es werden zum erstenmal die Komponenten $X_x, \ldots$ statt $X_a, \ldots$ bestimmt. Diese Formeln sind übrigens in der Elastizitätstheorie endlicher Deformationen wiederholt neu hergeleitet und ausgebildet worden.

the material laws follow immediately. If one uses for $\delta A$ for instance the Ansatz No. **3**c, (7), then[88])

$$(4) \qquad \varrho_0 X = -\frac{\partial \varphi}{\partial x}, \quad X_a = \frac{\partial \varphi}{\partial x_a} \qquad \begin{pmatrix} x, y, z \\ X, Y, Z \end{pmatrix}; a, b, c\Big);$$

changing over to the quantities related to the deformed position by use of (8) from No. **3**c and (1), then one obtains[89]):

$$(5) \qquad \begin{cases} \varrho X = -\frac{\partial \bar{\varphi}}{\partial x}, \\ X_x = \frac{\partial \bar{\varphi}}{\partial x_a} \cdot x_a + \frac{\partial \bar{\varphi}}{\partial x_b} x_b + \frac{\partial \bar{\varphi}}{\partial x_c} x_c + \bar{\varphi}, \\ X_y = \frac{\partial \bar{\varphi}}{\partial x_a} y_a + \frac{\partial \bar{\varphi}}{\partial x_b} \cdot y_b + \frac{\partial \bar{\varphi}}{\partial x_c} \cdot y_c, \\ X_z = \frac{\partial \bar{\varphi}}{\partial x_a} z_a + \frac{\partial \bar{\varphi}}{\partial x_b} z_b + \frac{\partial \bar{\varphi}}{\partial x_c} z_c. \end{cases} \qquad \begin{pmatrix} x & y & z \\ X & Y & Z \end{pmatrix}$$

Thereby, all of the considered material laws in No. **5** are reduced to the single equation (2), which gives $\varphi$ or $\bar{\varphi}$ as scalar functions of the local state of deformation.

If the energy density (2) depends *also on the second derivatives* $x_{aa}, x_{ab}, \ldots$ of the deformation functions — which in turn only comes into consideration for very quick changes in space of the state —, then new terms with second derivatives in the virtual displacements $\delta x_{aa} = \frac{\delta^2 \delta x}{\delta a^2}, \ldots$ will appear in $\delta\Phi$, and this results precisely in the in No. **4**a discussed additional terms to the original expression of the virtual work; thereupon both the components of this new effect, whose expressions emerge immediately from $\varphi$, and the old stress components, whose expressions have to be modified slightly, depend also on the second derivatives $x_{aa}, \ldots$.

A special case, which can be classified here, shall be emphasized especially: namely, that to the potential (1) an *integral over the surface of the continuum* can be added:

$$(6) \qquad \Phi_1 = \iint_{(S)} \bar{\psi} dS = \iint_{(S_0)} \psi dS_0,$$

where the "surface density" $\bar{\psi}$ or $\psi$ of the potential depends on the values at the surface $S$ of the

[88] *G. Kirchhoff*, Sitzungsber. d. Akad. Wien, math.-nat., Kl. 9 (1852), p. 772.

[89] *J. Boussinesq*, Mém. prés. p. div. sav., Paris **20** (1872), note 3. p. 591. Here only $\varphi$ instead of $\bar{\varphi}$ is used, but, what is essential, for the first time the components $X_x, \ldots$ instead of $X_a, \ldots$ are determined. By the way, in the theory of elasticity of finite deformations these formulas have been repeatedly derived and formulated anew.

Deformationsfunktionen und ihrer *ersten* Ableitungen an der Oberfläche $S$ abhängt; ein solches Potential kann analog dem obigen die Form (1) bestimmenden Axiom dadurch charakterisiert werden, dass sich $\Phi_1^* : S^*$ einem endlichen Werte $\bar{\psi}$ nähert, wenn sich das Oberflächenstück $S^*$ um eine Stelle zusammenzieht. Man kann (6) tatsächlich in ein Raumintegral über $V$ oder einen Teilraum umformen, wenn man zweite Ableitungen $x_{aa}, \ldots$ hinzunimmt. Übrigens kann man $\delta\Phi_1$ auch direkt bilden und bekommt dann für die virtuelle Arbeit unmittelbar ein Glied der in Nr. **4**a, (1) betrachteten Form.

Hängt $\bar{\psi}$ speziell nur von den Werten der Deformationsfunktionen $x, y, z$ selbst, nicht von ihren Ableitungen ab, so hat $\delta\Phi_1$ gerade die Form der Arbeit $\delta A_3$ der an der Oberfläche des Kontinuums angreifenden Druckkräfte (Nr. **3**, (1)), und zwar werden deren Komponenten

$$\bar{X} = \frac{\partial\bar{\psi}}{\partial x}, \quad \bar{Y} = \frac{\partial\bar{\psi}}{\partial y}, \quad \bar{Z} = \frac{\partial\bar{\psi}}{\partial z}. \tag{6a}$$

Man kann diese Potentialansätze leicht derart ausdehnen, dass sie Kraftwirkungen von der in Nr. **6** betrachteten allgemeineren Gestalt (4) liefern. Man braucht dazu nur, nach dem Vorgange von *P. Duhem*[90]), an die Stelle des Axioms von der additiven Zusammensetzung der Potentiale der Teilbereiche zum Gesamtpotential die Annahme zu setzen, *dass bei einer Zerlegung des Kontinuums in n Teilbereiche $V_1, \ldots, V_n$ das Potential $\Phi$ eine Doppelsumme*

$$\Phi = \sum_{i,k=1}^{n} \Phi_{ik}$$

*von $n^2$ Summanden wird, deren jeder $\Phi_{ik}$ nur von dem Zustand zweier Teilbereiche $V_i, V_k$ abhängt.* Unter Hinzunahme ähnlicher Stetigkeitsannahmen, wie oben angedeutet, wird dann $\Phi$ gleich einem sechsfachen, zweimal über $V$ bzw. $V_0$ ausgedehnten Integral, dessen Integrand von den Werten der Deformationsfunktionen und ihrer Ableitungen in zwei Argumentpunkten $a, b, c$ und $\bar{a}, \bar{b}, \bar{c}$ abhängt[91]):

$$\Phi = \iiint\limits_{(V_0)} \iiint\limits_{(V_0)} \varphi(a, \ldots; x, \ldots; x_a, \ldots; \bar{a}, \ldots; \bar{x}, \ldots; \bar{x}_a, \ldots)\, da \ldots d\bar{c} \tag{7}$$

(Speziell kann hierin, wenn $\varphi$ einen von der Stelle $\bar{a}, \bar{b}, \bar{c}$ unabhängigen Summanden aufweist, auch ein Summand der Form (1) inbegriffen sein.) Die Variation von $\Phi$ wird

$$\delta\Phi = \iiint\limits_{(V_0)} \iiint\limits_{(V_0)} \left\{ \sum_{(x\,y\,z)} \left( \frac{\partial\varphi}{\partial x}\delta x + \frac{\partial\varphi}{\partial\bar{x}}\delta\bar{x} \right) + \sum_{(x\,y\,z;a\,b\,c)} \left( \frac{\partial\varphi}{\partial x_a}\delta x_a + \frac{\partial\varphi}{\partial\bar{x}_a}\delta\bar{x}_a \right) \right\} da \ldots d\bar{c},$$

[90] *P. Duhem*, l. c., p. 188.

[91] *P. Duhem*, l. c., p. 205.

deformation functions and the *first* derivatives thereof; such a potential can be characterized analogously to the foregoing axiom, [which] determines the form (1), that $\Phi_1^* : S^*$ approaches a finite value $\bar{\psi}$, when the surface element $S^*$ contracts around a point. In fact, one can transform (6) into a volume integral over $V$, if one adds second derivatives $x_{aa}, \ldots$ . By the way, one can also compute $\delta\Phi_1$ directly and obtains then for the virtual work immediately a term of the form considered in No. **4**a, (1).

If $\psi$ depends in particular only on the values of the deformation functions $x, y, z$ themselves, [and] not on the derivatives thereof, then $\delta\Phi_1$ has precisely the form of the work $\delta A_3$ of the compressive forces applied at the surface of the continuum (No. **3**, (1)), and indeed their components become

$$\bar{X} = \frac{\partial\bar{\psi}}{\partial x}, \quad \bar{Y} = \frac{\partial\bar{\psi}}{\partial y}, \quad \bar{Z} = \frac{\partial\bar{\psi}}{\partial z}. \tag{6a}$$

One can easily extend such potential-based approaches in such a way, that they yield force effects of the more general form (4) considered in No. **6**. Thereto one only needs, according to the procedure of *P. Duhem*[90]), to substitute the axiom of the additive composition of the potential of the subdomains to the total potential by the assumption, *that for a dissection of the continuum into n subdomains* $V_1, \ldots, V_n$, *the potential* $\Phi$ *becomes a double sum*

$$\Phi = \sum_{i,k=1}^{n} \Phi_{ik}$$

*of* $n^2$ *summands, each of which* $\Phi_{ik}$ *depend only on the state of two subdomains* $V_i, V_k$. By the application of similar continuity assumptions as mentioned above, $\Phi$ becomes equal to a sixfold integral, [which is] twice over $V$ or $V_0$, whose integrand depends on the values of the deformation functions and the derivatives thereof in two points $a, b, c$ and $\bar{a}, \bar{b}, \bar{c}$ [91]):

$$\Phi = \iiint\limits_{(V_0)} \iiint\limits_{(V_0)} \varphi(a, \ldots; x, \ldots; x_a, \ldots; \bar{a}, \ldots; \bar{x}, \ldots; \bar{x}_a, \ldots) da \ldots d\bar{c} \tag{7}$$

(In particular, when $\varphi$ includes a summand independent of the point $\bar{a}, \bar{b}, \bar{c}$, a summand of the form (1) can be included herein.) The variation of $\Phi$ becomes

$$\delta\Phi = \iiint\limits_{(V_0)} \iiint\limits_{(V_0)} \Big\{ \sum_{(x\,y\,z)} \Big( \frac{\partial\varphi}{\partial x}\delta x + \frac{\partial\varphi}{\partial\bar{x}}\delta\bar{x} \Big) + \sum_{(x\,y\,z;a\,b\,c)} \Big( \frac{\partial\varphi}{\partial x_a}\delta x_a + \frac{\partial\varphi}{\partial\bar{x}_a}\delta\bar{x}_a \Big) \Big\} da \ldots d\bar{c},$$

[90] *P. Duhem*, l. c., p. 188.

[91] *P. Duhem*, l. c., p. 205.

und aus der Identität (3) folgen daher für ein Medium, das alle stetigen virtuellen Verrückungen gestattet, als Kraft- und Spannungskomponenten an der Stelle $a, b, c$:

$$\text{(8)}\qquad \begin{cases} \varrho_0 X = -\iiint\limits_{(V_0)} \frac{\partial(\varphi+\varphi_1)}{\partial x} d\bar{a}\, d\bar{b}\, d\bar{c}, \\ X_a \;=\; \iiint\limits_{(V_0)} \frac{\partial(\varphi+\varphi_1)}{\partial x_a} d\bar{a}\, d\bar{b}\, d\bar{c}; \end{cases} \qquad \begin{pmatrix} x, y, z \\ X, Y, Z \end{pmatrix}; a,\; b,\; c$$

dabei bedeutet $\varphi_1$ die aus $\varphi$ durch Vertauschung der überstrichenen und nicht überstrichenen Argumente entstehende Funktion. Wie oben ergeben sich hieraus sofort die Stoffgleichungen, die $X_x, \ldots$ mit Hilfe der einen Funktion $\varphi$ ausdrücken; *P. Duhem* hat dies unter speziellen, den Verhältnissen der reinen Elastizitätstheorie entsprechenden Annahmen ausführlich entwickelt.[92]) Ansätze von dieser Art sind es im Grunde, die bei dem Aufbau der Mechanik der Kontinua auf *Molekularvorstellungen* vielfach benutzt werden.[93]) Die Doppelsummen, die man da für die Potentiale von Molekülsystemen ansetzt, werden durch die Grenzübergänge gerade zu Integralen vom Typus (7), und die Aufgabe der Theorie ist es, solche Annahmen zu formulieren, daß sie sich bei richtiger Führung der Grenzübergänge in Potentiale der einfachen Formen (1) bzw. (6) transformieren; man vergleiche etwa die Darstellung von *H. Minkowski* in V 9, Nr. **14**.

Besonders hervorzuheben ist wieder die Gestaltung der Potentialansätze in dem Falle *„unendlichkleiner" Deformation des Kontinuums* (Nr. **6**, (6)). Die Ausdrücke der Kraft- und Spannungskomponenten werden nach (4), bei Vernachlässigung quadratischer Glieder in $\sigma$[94]),

$$\text{(9a)}\qquad \varrho_0 X = -\frac{1}{\sigma}\frac{\partial\tilde{\varphi}}{\partial u}, \quad X_a = -\frac{1}{\sigma}\frac{\partial\tilde{\varphi}}{\partial u_a}, \qquad \begin{pmatrix} u, v, w \\ X, Y, Z \end{pmatrix}; a,\; b,\; c$$

dabei bedeutet $\tilde{\varphi}$ diejenigen in $\sigma$ linearen und quadratischen Gliedern der Potenzentwicklung der Energiedichte $\varphi$, die von den in $\sigma$ linearen Gliedern der Reihe (6) von Nr. **6** herrühren:

$$\text{(9b)}\quad \begin{aligned} \tilde{\varphi} &= \varphi^0 + \sigma(\varphi^0_x u + \cdots + \varphi^0_{x_a} u_a + \cdots) \\ &+ \frac{\sigma^2}{2}(\varphi^0_{xx} u^2 + \varphi^0_{xy} uv + \cdots + \varphi^0_{xx_a} uu_a + \cdots + \varphi^0_{x_a x_a} u_a^2 + \varphi^0_{x_a x_b} u_a u_b + \cdots), \end{aligned}$$

[92] *P. Duhem* Ann. Éc. Norm., (3) 21 (1904), p. 117 ff. Auch separat: Récherche sur l'élasticité, Paris 1906.

[93] Z. B. in der *Navier*schen Theorie des elastischen Potentials (vgl. IV 23, Nr. **5**a, *Müller-Timpe*) und in der Theorie der Kapillarität von *P. S. Laplace* und *C. Fr. Gauss* (vgl. V 9, Nr. **13**, *Minkowski*).

[94] *H. Poincaré*, Leçons sur la théorie de l'Élasticité, Paris 1892, p. 54 ff.; *E.* u. *F. Cosserat*, Ann. de la Fac. des Sciences de Toulouse 10 (1896), p. J. 70 ff.

and from the identity (3) [it] therefore follows for a medium, which allows for all continuous virtual displacements, the force and stress components at the point $a, b, c$:

$$(8)\qquad \begin{cases} \varrho_0 X = -\underset{(V_0)}{\iiint} \frac{\partial(\varphi+\varphi_1)}{\partial x} d\bar{a}\, d\bar{b}\, d\bar{c}, \\ X_a \ = \underset{(V_0)}{\iiint} \frac{\partial(\varphi+\varphi_1)}{\partial x_a} d\bar{a}\, d\bar{b}\, d\bar{c}; \end{cases} \qquad \left(\begin{matrix} x, y, z \\ X, Y, Z \end{matrix}; a,\ b,\ c\right)$$

thereby $\varphi_1$ corresponds to the function arising from $\varphi$ by the permutation of the overlined and non-overlined arguments. As above the material laws, which express $X_x, \ldots$ by means of the one function $\varphi$, emerge directly out of this; *P. Duhem* has developed this [Ansatz] with respect to special assumptions corresponding with the circumstances of a pure theory of elasticity.[92]) Basically, there are approaches of this type, which are frequently used for the foundations of the mechanics of continua based on the *perception of molecules*.[93]) The double sums, which one formulates there for the potentials of the systems of molecules, become within the limit processes directly to integrals of type (7), and it is the question of the theory to formulate such assumptions, that for a correct guidance of the limit processes they transform into potentials of the simple forms (1) or (6); One confers for instance the presentation of *H. Minkowski* in V 9, No. **14**.

To be emphasized particularly is again the formulation of the potential-based approaches for the case of *"infinitesimal" deformation of the continuum* (No. **6**, (6)). By neglecting the quadratic terms in $\sigma$[94]), the expressions of force and stress components turn according to (4) into

$$(9a)\qquad \varrho_0 X = -\frac{1}{\sigma}\frac{\partial\tilde{\varphi}}{\partial u}, \quad X_a = -\frac{1}{\sigma}\frac{\partial\tilde{\varphi}}{\partial u_a}, \qquad \left(\begin{matrix} u, v, w \\ X, Y, Z \end{matrix}; a,\ b,\ c\right)$$

thereby $\tilde{\varphi}$ corresponds to those terms of the power series of the energy density $\varphi$ being linear and quadratic in $\sigma$, which arise from the terms linear in $\sigma$ of the series (6) of No. **6**:

$$(9b)\qquad \tilde{\varphi} = \varphi^0 + \sigma(\varphi^0_x u + \cdots + \varphi^0_{x_a} u_a + \cdots)$$
$$+ \frac{\sigma^2}{2}(\varphi^0_{xx} u^2 + \varphi^0_{xy} uv + \cdots + \varphi^0_{xx_a} uu_a + \cdots + \varphi^0_{x_a x_a} u_a^2 + \varphi^0_{x_a x_b} u_a u_b + \cdots),$$

92 *P. Duhem* Ann. Éc. Norm., (3) 21 (1904), p. 117 ff. Also separately: Récherche sur l'élasticité, Paris 1906.

93 E. g. in *Navier*'s theory of the elastic potential (cf. IV 23, No. **5**a, *Müller-Timpe*) and in the theory of capillarity of *P. S. Laplace* and *C. Fr. Gauss* (cf. V 9, No. **13**, *Minkowski*).

94 *H. Poincaré*, Leç ons sur la théorie de l'Élasticité, Paris 1892, p. 54 ff.; *E.* and *F. Cosserat*, Ann. de la Fac. des Sciences de Toulouse 10 (1896), p. J. 70 ff.

worin die mit der Marke 0 versehenen Ableitungen von $\varphi$ für $\sigma = 0$, d. h. für die Argumente $x = a, \dots, x_a = 1, x_b = 0, \dots$ zu nehmen sind. Die Ausdrücke (9a) haben in der Tat den in Nr. **6**, (3′) betrachteten Typus des *Hooke*schen Gesetzes; vorausgesetzt ist dabei natürlich, dass die $\tilde{\varphi}$ zusammensetzenden Glieder der Entwicklung von $\varphi$ nicht identisch verschwinden. Die auf die deformierte Lage des Kontinuums bezogenen Spannungskomponenten $X_x, \dots$ unterscheiden sich gemäß Nr. **3**c, (8) von den $X_a, \dots$ um folgende in $\sigma$ lineare Ausdrücke:

$$X_x - X_a = \sigma(-\varphi^0_{x_a}(v_b + w_c) + \varphi^0_{x_b} u_b + \varphi^0_{x_c} u_c), \dots, \tag{10}$$

und diese werden nur dann Null bzw. von der Grössenordnung $\sigma^2$ der sonst vernachlässigten Grössen, wenn die durch $\varphi^0_{x_a}$ gegebenen „Anfangsspannungen" vor der unendlichkleinen Deformation verschwinden.[95]) — Es bedarf danach keiner genaueren Ausführungen, wie man in ähnlicher Weise den allgemeineren *Duhem*schen Potentialansatz (7), für unendlichkleine Deformationen umzubilden hat.

**7b. Der Potentialansatz für Medien mit orientierten Teilchen.** Nach dem Vorgange von *E.* und *F. Cosserat*[96]) kann man diesen Potentialansatz auch auf die Kontinua ausdehnen, deren Teilchen mit einer bestimmten Orientierung behaftet sind; man braucht nur anzunehmen, dass die sonst wie in Nr. **7**a definierte Energiedichte $\varphi$ ausser von den bisher betrachteten Grössen auch von den die momentane Orientierung des Teilchens $a, b, c$ bestimmenden Parametern $\lambda, \mu, \nu$ (Nr **2**b, (9)) und deren (ersten) Ableitungen nach $a, b, c$ abhängt:

$$\varphi = \varphi(\lambda(a, b, c), \dots; \lambda_a(a, b, c), \dots, \nu_c(a, b, c)). \tag{11}$$

Eine virtuelle Drehung der einzelnen Teilchen Nr. **2** (10) liefert zur Variation des Potentials dann den folgenden Beitrag:

$$\delta\Phi = \iiint\limits_{(V_0)} \sum_{(\lambda\,\mu\,\nu)} \left(\frac{\partial\varphi}{\partial\lambda}\delta\lambda + \frac{\partial\varphi}{\partial\lambda_a}\delta\lambda_a + \frac{\partial\varphi}{\partial\lambda_b}\delta\lambda_b + \frac{\partial\varphi}{\partial\lambda_c}\delta\lambda_c\right) da\,db\,dc.$$

Führt man nun vermöge Nr. **2**, (11), (12) die Winkelgeschwindigkeiten $\delta\pi, \delta\kappa, \delta\varrho$ der virtuellen Verdrehung ein und beachtet, dass

$$\delta\lambda_a = \frac{\partial\delta\lambda}{\partial a} = \sum_{\binom{lmn}{\pi\kappa\varrho}} \left(\frac{\partial l_1}{\partial a}\delta\pi + l_1\frac{\partial\delta\pi}{\partial a}\right) \qquad \left(\begin{matrix}\lambda, \mu, \nu\\ 1, 2, 3\end{matrix}; a, b, c\right),$$

so ergeben sich durch Identifikation von $-\delta\Phi$ mit dem Arbeitsausdruck Nr. **4**, (2) bzw. (2′) die folgenden Formeln für die auf die

[95] *J. Boussinesq*, a. a. O.[89]), p. 598, *E. u. F. Cosserat*, l. c., p. J. 74 f.

[96] *E.* und *F. Cosserat*, „Corps déformables" [5]), chap. IV, p. 122 ff.

wherein the derivatives of $\varphi$ signed with the label 0 have to be evaluated for $\sigma = 0$, i. e. for the arguments $x = a, \ldots, x_a = 1, x_b = 0, \ldots$ . The expressions (9a) are in fact of the class of *Hooke*'s law considered in No. **6**, (3′); thereby [it] is naturally required, that the terms of the expansion of $\varphi$ constituting $\bar{\varphi}$ do not vanish identically. The stress components with respect to the deformed position of the continuum $X_x, \ldots$ differ according to No. **3**c, (8) from $X_a, \ldots$ by the following expressions linear in $\sigma$:

$$X_x - X_a = \sigma(-\varphi^0_{x_a}(v_b + w_c) + \varphi^0_{x_b} u_b + \varphi^0_{x_c} u_c), \ldots, \tag{10}$$

and these become only zero or of the order of magnitude $\sigma^2$ of the otherwise neglected quantities, when the "initial stresses" given by $\varphi^0_{x_a}$ vanish before the infinitesimal deformation.[95]) — Thereafter no more detailed presentation is required, how one reformulates in a similar way the more general potential-based approach of *Duhem* (7) for infinitesimal deformations.

**7b. The potential-based approach for media with oriented particles.**

According to the procedure of *E.* and *F. Cosserat*[96]) one can extend this potential-based approach also to continua, whose particles are endowed with a certain orientation; one only has to assume, that the energy density $\varphi$, usually defined as in No. **7**a, depends besides the so far considered quantities also on the parameters $\lambda, \mu, \nu$, [which] determine the actual orientation of the particle $a, b, c$ (No. **2**b, (9)), and the (first) derivatives with respect to $a, b, c$ of these [parameters]:

$$\varphi = \varphi(\lambda(a, b, c), \ldots; \lambda_a(a, b, c), \ldots, \nu_c(a, b, c)). \tag{11}$$

A virtual rotation of the individual particles No. **2** (10) then yields the following contribution to the variation of the potential:

$$\delta\Phi = \iiint\limits_{(V_0)} \sum_{(\lambda\,\mu\,\nu)} \left(\frac{\partial\varphi}{\partial\lambda}\delta\lambda + \frac{\partial\varphi}{\partial\lambda_a}\delta\lambda_a + \frac{\partial\varphi}{\partial\lambda_b}\delta\lambda_b + \frac{\partial\varphi}{\partial\lambda_c}\delta\lambda_c\right) da\, db\, dc.$$

If one introduces now due to No. **2**, (11), (12) the angular velocities $\delta\pi, \delta\kappa, \delta\varrho$ of the virtual rotation and by considering that

$$\delta\lambda_a = \frac{\partial\delta\lambda}{\partial a} = \sum_{\binom{lmn}{\pi\kappa\varrho}} \left(\frac{\partial l_1}{\partial a}\delta\pi + l_1\frac{\partial\delta\pi}{\partial a}\right) \qquad \left(\begin{matrix}\lambda, \mu, \nu \\ 1, 2, 3\end{matrix}; a, b, c\right),$$

then, by identification of $-\delta\Phi$ with No. **4**, (2) and (2′), respectively, the following formulas for the

[95] *J. Boussinesq*, op. cit.[89]), p. 598, *E. and F. Cosserat*, l. c., p. J. 74 f.

[96] *E.* and *F. Cosserat*, "Corps déformables" [5]), chap. IV, p. 122 ff.

Massen- und Flächenelemente wirkenden Drehmomente[97]):

$$(12)\qquad \begin{cases} \varrho_0 L = -\sum\limits_{\binom{\lambda\mu\nu}{123}} \left\{ \frac{\partial\varphi}{\partial\lambda}\cdot l_1 + \frac{\partial\varphi}{\partial\lambda_a}\frac{\partial l_1}{\partial a} + \frac{\partial\varphi}{\partial\lambda_b}\frac{\partial l_1}{\partial b} + \frac{\partial\varphi}{\partial\lambda_c}\frac{\partial l_1}{\partial c} \right\}, \\ L_a = \frac{\partial\varphi}{\partial\lambda_a} l_1 + \frac{\partial\varphi}{\partial\mu_a} l_2 + \frac{\partial\varphi}{\partial\nu_a} l_3. \end{cases} \left( \begin{matrix} L, M, N \\ l, m, n \end{matrix}; a, b, c \right)$$

Es ist vielfach zweckmäßig, in diese Formeln die den $\delta\pi, \delta\kappa, \delta\varrho$ analogen Winkelgeschwindigkeiten einzuführen, die bei der Überführung des zu einem Teilchen gehörigen Dreikantes in das eines Nachbarteilchens auftreten; wir betrachten speziell die in der Richtung der Parameterlinien $a, b, c$ benachbarten Teilchen, also die Winkelgeschwindigkeitskomponenten

$$(13)\qquad p_a = \sum_{(123)} \beta_1 \frac{\partial\gamma_1}{\partial a},\; q_a = \sum_{(123)} \gamma_1 \frac{\partial\alpha_1}{\partial a},\; r_a = \sum_{(123)} \alpha_1 \frac{\partial\beta_1}{\partial a} \qquad (a, b, c).$$

Dann hat man analog den Relationen (12) von Nr. **2**

$$\frac{\partial\lambda}{\partial a} = l_1 p_a + m_1 q_a + n_1 r_a \qquad \left( \begin{matrix} \lambda, \mu, \nu \\ 1, 2, 3 \end{matrix}; a,\ b,\ c \right),$$

und kann in dem Ausdruck (11) der Energiedichte die $\lambda_a, \dots, \nu_c$ durch die Winkelgeschwindigkeiten $p_a, \dots, r_c$ ersetzen:

$$(14)\qquad \varphi = \varphi(\lambda, \mu, \nu; p_a, p_b, \dots, r_c).$$

Bildet man aus diesem Ausdruck $\delta\Phi$ und berücksichtigt die aus (13) folgenden Relationen (das Analogon der sog. „Übergangsgleichungen“[98]) der Kinetik)

$$\delta p_a = \frac{\partial\delta\pi}{\partial a} + r_a \delta\kappa - q_a \delta\varrho \qquad \left( \begin{matrix} p, q, r \\ \pi, \kappa, \varrho \end{matrix}; a,\ b,\ c \right),$$

so ergiebt sich durch analoge Betrachtungen, wie sie zu (12) führten[97]):

$$(15)\qquad \begin{cases} \varrho_0 L = -\left\{ \frac{\partial\varphi}{\partial\lambda} l_1 + \frac{\partial\varphi}{\partial\mu} l_2 + \frac{\partial\varphi}{\partial\nu} l_3 + \sum\limits_{(abc)} \left( q_a \frac{\partial\varphi}{\partial r_a} - r_a \frac{\partial\varphi}{\partial q_a} \right) \right\} \left( \begin{matrix} L, M, N \\ l, m, n \end{matrix} \right) \\ L_a = \frac{\partial\varphi}{\partial p_a}. \qquad\qquad \left( \begin{matrix} L, M, N \\ p, q, r \end{matrix}; a,\ b,\ c \right) \end{cases}$$

Der Übergang zu den auf das deformierte Kontinuum bezogenen Drehmomentkomponenten $L_x, \dots, N_z$ ist mit Hilfe von Nr. **4**b, (5) leicht zu vollziehen.

---

[97] Diese Formeln finden sich in dem *Cosserat*schen Buche nicht explizit angegeben, da dort die unten ausgeführte Annahme eines „Euklidischen“ Potentiales an der Spitze steht; sie sind indessen in den Gleichungen von p. 132 ff. und 141 bzw. p. 130 ff. und 134 ff. enthalten; die Identifizierung geschieht am leichtesten von den p. 138 ff. angegebenen Formeln für die Arbeit aus.

[98] Sie gehen in diese über, wenn $a$ durch den Zeitparameter ersetzt wird; vgl. IV **6** (*P. Stäckel*), No. **30**, p. 584 f. und Anm.[417]) sowie IV 11 (*K. Heun*), No. **14**c.

torques acting at the mass and surface elements emerge[97]):

$$(12)\quad \begin{cases} \varrho_0 L = -\sum\limits_{\binom{\lambda\mu\nu}{123}} \left\{ \frac{\partial\varphi}{\partial\lambda} \cdot l_1 + \frac{\partial\varphi}{\partial\lambda_a}\frac{\partial l_1}{\partial a} + \frac{\partial\varphi}{\partial\lambda_b}\frac{\partial l_1}{\partial b} + \frac{\partial\varphi}{\partial\lambda_c}\frac{\partial l_1}{\partial c} \right\}, \\ L_a \;= \frac{\partial\varphi}{\partial\lambda_a} l_1 + \frac{\partial\varphi}{\partial\mu_a} l_2 + \frac{\partial\varphi}{\partial\nu_a} l_3. \end{cases} \left( \begin{matrix} L, M, N \\ l, m, n \end{matrix} ; a, b, c \right)$$

Often it is useful to introduce in these formulas the angular velocities analogously to the $\delta\pi, \delta\kappa, \delta\varrho$, which appear in the transition from one triad of a particle to the one of the neighboring particle; we consider especially the neighboring particles in direction of the parameter lines $a, b, c$, i. e. the components of the angular velocities

$$(13)\quad p_a = \sum_{(123)} \beta_1 \frac{\partial\gamma_1}{\partial a},\; q_a = \sum_{(123)} \gamma_1 \frac{\partial\alpha_1}{\partial a},\; r_a = \sum_{(123)} \alpha_1 \frac{\partial\beta_1}{\partial a} \quad (a, b, c).$$

Then one has analogously to the relation (12) of No. **2**

$$\frac{\partial\lambda}{\partial a} = l_1 p_a + m_1 q_a + n_1 r_a \qquad \left( \begin{matrix} \lambda, \mu, \nu \\ 1, 2, 3 \end{matrix} ;\; a,\; b,\; c \right),$$

and [one] can substitute the $\lambda_a, \ldots, \nu_c$ with the angular velocities $p_a, \ldots, r_c$ in the expression (11) of the energy density:

$$(14)\qquad \varphi = \varphi(\lambda, \mu, \nu; p_a, p_b, \ldots, r_c).$$

If one computes $\delta\Phi$ with this expression and by considering the relation following from (13) (the analogue to the so called "transition equations"[98]) of kinetics)

$$\delta p_a = \frac{\partial\delta\pi}{\partial a} + r_a \delta\kappa - q_a \delta\varrho \qquad \left( \begin{matrix} p, q, r \\ \pi, \kappa, \varrho \end{matrix} ;\; a,\; b,\; c \right),$$

then similar considerations which lead to (12)[97]) result in:

$$(15)\quad \begin{cases} \varrho_0 L = -\left\{ \frac{\partial\varphi}{\partial\lambda} l_1 + \frac{\partial\varphi}{\partial\mu} l_2 + \frac{\partial\varphi}{\partial\nu} l_3 + \sum\limits_{(abc)} \left( q_a \frac{\partial\varphi}{\partial r_a} - r_a \frac{\partial\varphi}{\partial q_a} \right) \right\} \left( \begin{matrix} L, M, N \\ l, m, n \end{matrix} \right) \\ L_a \;= \frac{\partial\varphi}{\partial p_a}. \qquad\qquad \left( \begin{matrix} L, M, N \\ p, q, r \end{matrix} ;\; a,\; b,\; c \right) \end{cases}$$

Using No. **4**b, (5), the transformation to the components of the torques with respect to the deformed continuum $L_x, \ldots, N_z$ can be carried out easily.

---

[97] These formulas cannot be found explicitly in the book of the *Cosserats*, since therein the assumption of a "Euclidean" potential, which is achieved below, forms the basis; however, they are contained in the equations of p. 132 ff. and 141 or p. 130 ff. and 134 ff.; The identification occurs easiest starting from the formulas for the work given on p. 138 ff.

[98] They change into these [equations], when $a$ is replaced by a time parameter; cf. IV **6** (*P. Stäckel*), No. **30**, p. 584 f. and remark[417]) as well as IV 11 (*K. Heun*), No. **14**c.

*E.* und *F. Cosserat* betrachten insbesondere die durch diesen Ansatz dargestellten *inneren Wirkungen* in einem Medium, bei denen $\varphi$ als Funktion der $x, \ldots$ und $\lambda, \ldots$ invariant gegen rechtwinklige Koordinatentransformationen im $x$-$y$-$z$-Raume ist, oder — was dasselbe bedeutet — bei denen jede Bewegung des mitsamt den adjungierten Dreikanten erstarrt gedachten Kontinuums das Potential ungeändert lässt; ein solches Potential nennen sie ein *euklidisches* (*action Euclidienne*). Um diese Klasse von Potentialen zu umschreiben, verwenden sie in jedem Punkte $x, y, z$ als (bewegliches) Bezugssystem die momentane Lage des dem gerade dort befindlichen Teilchen angehefteten Dreikantes; an Stelle der Komponenten $p_a, \ldots, r_c$ treten die Komponenten der gleichen Winkelgeschwindigkeiten in Bezug auf diese neuen Axen:

$$\text{(16a)} \qquad \mathfrak{p}_a = \alpha_1 p_a + \beta_1 q_a + \gamma_1 r_a = \alpha_3 \frac{\partial \alpha_2}{\partial \alpha} + \beta_3 \frac{\partial \beta_2}{\partial \alpha} + \gamma_3 \frac{\partial \gamma_2}{\partial \alpha} \left( \begin{matrix} \mathfrak{p}, \mathfrak{q}, \mathfrak{r} \\ 1, 2, 3 \end{matrix} ; a, b, c \right),$$

und in gleicher Weise mögen die 9 Deformationsgrössen $x_a, \ldots, z_c$ transformiert werden in:

$$\text{(16b)} \qquad \mathfrak{x}_a = \alpha_1 x_a + \beta_1 y_a + \gamma_1 z_a \qquad \left( \begin{matrix} \mathfrak{x}, \mathfrak{y}, \mathfrak{z} \\ 1, 2, 3 \end{matrix} ; a, b, c \right).$$

*Dann ist das allgemeinste euklidische Potential, das höchstens von den ersten Ableitungen der Deformationsfunktionen abhängt, eine willkürliche Funktion dieser 18 Grössen* $\mathfrak{p}_a, \ldots \mathfrak{z}_c$, die ausserdem noch explizit $a, b, c$ enthalten kann[99]):

$$\text{(17)} \qquad \varphi = \varphi(a, \ldots; \mathfrak{x}_a, \ldots, \mathfrak{z}_c; \mathfrak{p}_a, \ldots, \mathfrak{r}_c).$$

Zur Herleitung der Gleichgewichtsbedingungen für diesen Ansatz führt man auch die Komponenten der virtuellen Verrückung und Verdrehung nach den neuen beweglichen Axen ein:

$$\begin{aligned} \delta \mathfrak{x} &= \alpha_1 \delta x + \beta_1 \delta y + \gamma_1 \delta z, \\ \delta \mathfrak{i} &= \alpha_1 \delta \pi + \beta_1 \delta \kappa + \gamma_1 \delta \varrho; \end{aligned} \qquad \left( \begin{matrix} \mathfrak{x}, \mathfrak{y}, \mathfrak{z} \\ \mathfrak{i}, \mathfrak{j}, \mathfrak{k} \\ 1, 2, 3 \end{matrix} \right)$$

dann hat man die „Übergangsgleichungen"

$$\begin{aligned} \delta \mathfrak{x}_a &= \frac{\partial \delta \mathfrak{x}}{\partial a} + \mathfrak{q}_a \delta \mathfrak{z} - \mathfrak{r}_a \delta \mathfrak{y}_a \delta \mathfrak{k} - \mathfrak{z}_a \delta \mathfrak{j}, \\ \delta \mathfrak{p}_a &= \frac{\partial \delta \mathfrak{i}}{\partial a} + \mathfrak{q}_a \delta \mathfrak{k} - \mathfrak{r}_a \delta \mathfrak{j} \end{aligned} \qquad \left( \begin{matrix} \mathfrak{x}, \mathfrak{y}, \mathfrak{z} \\ \mathfrak{p}, \mathfrak{q}, \mathfrak{r}; \; a, b, c \\ \mathfrak{i}, \mathfrak{j}, \mathfrak{k} \end{matrix} \right)$$

und kann daher die Variation des mit (17) gebildeten Potentiales

$$\delta \Phi = \iiint\limits_{(V_0)} \sum_{\left( \begin{smallmatrix} \mathfrak{x}, \mathfrak{y}, \mathfrak{z} \\ \mathfrak{p}, \mathfrak{q}, \mathfrak{r} \end{smallmatrix} ; a, b, c \right)} \left( \frac{\partial \varphi}{\partial \mathfrak{x}_a} \delta \mathfrak{x}_a + \frac{\partial \varphi}{\partial \mathfrak{p}_a} \delta \mathfrak{p}_a \right) da \, db \, dc$$

---

[99] *E.* u. *F. Cosserat*, „Corps déformables", p. 127.

*E.* and *F. Cosserat* considered in particular the *internal actions* in a medium represented by this ansatz[.] For these [actions] $\varphi$ is, as a function of $x, \ldots$ and $\lambda, \ldots$, invariant with respect to orthogonal coordinate transformations in the $x$-$y$-$z$-space, or — what implies the same — these [actions for which] every motion of the continuum together with the adjoint triads being regarded as rigid leaves the potential unchanged; they call such a potential a *euclidean* one (*action Euclidienne*). To describe this class of potentials, they use in every point $x, y, z$ as (moving) frame of reference the actual position of the triad attached to the particle just located there; The components $p_a, \ldots, r_c$ are substituted by the components of the same angular velocities formulated with respect to these new axes:

$$\text{(16a)} \qquad \mathfrak{p}_a = \alpha_1 p_a + \beta_1 q_a + \gamma_1 r_a = \alpha_3 \frac{\partial \alpha_2}{\partial \alpha} + \beta_3 \frac{\partial \beta_2}{\partial \alpha} + \gamma_3 \frac{\partial \gamma_2}{\partial \alpha} \begin{pmatrix} \mathfrak{p}, \mathfrak{q}, \mathfrak{r} \\ 1, 2, 3 \end{pmatrix} ; a, b, c \Big),$$

and in a similar way the 9 deformation quantities $x_a, \ldots, z_c$ shall be transformed to:

$$\text{(16b)} \qquad \mathfrak{x}_a = \alpha_1 x_a + \beta_1 y_a + \gamma_1 z_a \qquad \left( \begin{matrix} \mathfrak{x}, \mathfrak{y}, \mathfrak{z} \\ 1, 2, 3 \end{matrix} ; a, b, c \right).$$

*Then the most general euclidean potential, which depends at most on the first derivatives of the deformation functions, is an arbitrary function of these 18 quantities* $\mathfrak{p}_a, \ldots \mathfrak{z}_c$, [functions] which moreover can explicitly contain $a, b, c$[99]):

$$\text{(17)} \qquad \varphi = \varphi(a, \ldots; \mathfrak{x}_a, \ldots, \mathfrak{z}_c; \mathfrak{p}_a, \ldots, \mathfrak{r}_c).$$

For the derivation of the equilibrium conditions for this ansatz one also introduces the components of the virtual displacement and rotation with respect to the new moving axes:

$$\begin{aligned} \delta\mathfrak{x} &= \alpha_1 \delta x + \beta_1 \delta y + \gamma_1 \delta z, \\ \delta\mathfrak{i} &= \alpha_1 \delta\pi + \beta_1 \delta\kappa + \gamma_1 \delta\varrho; \end{aligned} \qquad \begin{pmatrix} \mathfrak{x}, \mathfrak{y}, \mathfrak{z} \\ \mathfrak{i}, \mathfrak{j}, \mathfrak{k} \\ 1, 2, 3 \end{pmatrix}$$

then one has the "transition equations"

$$\begin{aligned} \delta\mathfrak{x}_a &= \frac{\partial \delta\mathfrak{x}}{\partial a} + \mathfrak{q}_a \delta\mathfrak{z} - \mathfrak{r}_a \delta\mathfrak{y}_a \delta\mathfrak{k} - \mathfrak{z}_a \delta\mathfrak{j}, \\ \delta\mathfrak{p}_a &= \frac{\partial \delta\mathfrak{i}}{\partial a} + \mathfrak{q}_a \delta\mathfrak{k} - \mathfrak{r}_a \delta\mathfrak{j} \end{aligned} \qquad \begin{pmatrix} \mathfrak{x}, \mathfrak{y}, \mathfrak{z} \\ \mathfrak{p}, \mathfrak{q}, \mathfrak{r};\ a, b, c \\ \mathfrak{i}, \mathfrak{j}, \mathfrak{k} \end{pmatrix}$$

and is therefore immediately able to compare the variation of the potential formulated with (17)

$$\delta\Phi = \iiint_{(V_0)} \sum_{\left(\substack{\mathfrak{x}, \mathfrak{y}, \mathfrak{z} \\ \mathfrak{p}, \mathfrak{q}, \mathfrak{r}} ; a, b, c\right)} \left( \frac{\partial \varphi}{\partial \mathfrak{x}_a} \delta\mathfrak{x}_a + \frac{\partial \varphi}{\partial \mathfrak{p}_a} \delta\mathfrak{p}_a \right) da\, db\, dc$$

---

[99] *E.* and *F. Cosserat*, "Corps déformables", p. 127.

unmittelbar mit der folgenden Form der virtuellen Arbeit vergleichen:

$$\delta A = \iiint\limits_{(V_0)} \sum_{\left(\substack{\mathfrak{X}\mathfrak{Y}\mathfrak{Z}\,;\,\mathfrak{L}\mathfrak{M}\mathfrak{N} \\ \mathfrak{x}\mathfrak{y}\mathfrak{z}\,;\,\mathfrak{i}\mathfrak{j}\mathfrak{k}}\right)} \left\{\varrho_0 \mathfrak{X}\delta\mathfrak{x} + \varrho_0 \mathfrak{L}\delta\mathfrak{i} - \sum_{(abc)} \left(\mathfrak{X}_a \frac{\partial \delta\mathfrak{x}}{\partial a} + \mathfrak{L}_a \frac{\partial \delta\mathfrak{i}}{\partial a}\right)\right\} da\, db\, dc,$$

in der die Komponenten der früher betrachteten Kräfte, Spannungen und Momente in bezug auf das bewegliche Koordinatenkreuz auftreten. Es ergeben sich danach Formeln[100].) vom Typus

$$(18) \qquad \begin{cases} \varrho_0 \mathfrak{X} = \sum\limits_{(abc)} \left(\mathfrak{r}_a \dfrac{\partial\varphi}{\partial\mathfrak{y}_a} - \mathfrak{q}_a \dfrac{\partial\varphi}{\partial\mathfrak{y}_a}\right), \quad \mathfrak{X}_a = \dfrac{\partial\varphi}{\partial\mathfrak{x}_a} \\ \varrho_0 \mathfrak{L} = \sum\limits_{(abc)} \left(\mathfrak{r}_a \dfrac{\partial\varphi}{\partial\mathfrak{q}_a} - \mathfrak{q}_a \dfrac{\partial\varphi}{\partial\mathfrak{r}_a} + \mathfrak{z}_a \dfrac{\partial\varphi}{\partial\mathfrak{y}_a} - \mathfrak{y}_a \dfrac{\partial\varphi}{\partial\mathfrak{z}_a}\right), \quad \mathfrak{L}_a = \dfrac{\partial\varphi}{\partial\mathfrak{p}_a}. \end{cases}$$

**7c. Der Potentialansatz für zwei- und eindimensionale Kontinua.** Für die zwei- und eindimensional ausgedehnten Kontinua im dreidimensionalen Raume kann man den Potentialansatz ohne Schwierigkeit durch ganz analoge Betrachtungen gewinnen.[101]) Die Energiedichte $\varphi$ — der als existierend vorausgesetzte Grenzwert des Quotienten aus dem Potential eines immer kleiner werdenden Teiles des Kontinuums und dessen Flächeninhalt bzw. Länge — wird eine gegebene Funktion der 6 Funktionen $x, y, z, \lambda, \mu, \nu$ von $a, b$ (bzw. von $a$) und ihrer Ableitungen, das Potential selbst also ein zwei- bzw. eindimensionales Integral:

$$\Phi = \iint\limits_{(S_0)} \varphi\, da\, db \quad \text{bzw.} \quad \Phi = \int\limits_0^l \varphi\, da.$$

Die Variation dieser Potentiale und daher die die auf die Anfangsparameter bezogenen Kraft-, Spannungs-, und Momentkomponenten ergeben sich unmittelbar aus den entsprechenden Formeln des dreidimensionalen Falles durch Fortlassen der auf $c$ bzw. $b$ und $c$ bezüglichen Glieder; der Übergang zu den auf den deformierten Zustand bezüglichen Grössen erfolgt dann nach Nr. **3**e, (16) und Nr. **4**b, (12).

Richtet man sein Augenmerk besonders auf orientierte Teilchen, so spielen die wie in Nr. **7**b definierten Winkelgeschwindigkeitskomponenten $p_a, \ldots$ wieder eine wichtige Rolle, und zwar hat man jetzt natürlich nur 2 bzw. 1 Tripel dieser Grössen. *E.* und *F. Cosserat* haben die Theorie solcher Medien unter Verwendung des jedem Teilchen

---

[100] *E.* u. *F. Cosserat*, l. c., p. 130 f.; vgl. auch IV **11** *K. Heun* Nr. **21**.

[101] Bereits *Lagrange* wendet ihn bei den von ihm behandelten Problemen aus diesem Gebiet[23]) an; er wurde dann in der Theorie der elastischen Fäden und Platten (vgl. IV 6 (*P. Stäckel*), Nr. **23**, **24** und IV 25, Kap. III, *Tedone-Timpe*), besonders aber auch in der Theorie der Kapillarität (vgl. V 9 (*Minkowski*), Nr. **2**) weiterentwickelt.

with the following form of the virtual work:

$$\delta A = \iiint\limits_{(V_0)} \sum_{\binom{\mathfrak{X}\mathfrak{Y}\mathfrak{Z}}{\mathfrak{x}\mathfrak{y}\mathfrak{z}};\binom{\mathfrak{L}\mathfrak{M}\mathfrak{N}}{\mathfrak{i}\mathfrak{j}\mathfrak{k}}} \left\{\varrho_0 \mathfrak{X}\delta\mathfrak{x} + \varrho_0 \mathfrak{L}\delta\mathfrak{i} - \sum_{(abc)} \left(\mathfrak{X}_a \frac{\partial \delta\mathfrak{x}}{\partial a} + \mathfrak{L}_a \frac{\partial \delta\mathfrak{i}}{\partial a}\right)\right\} da\, db\, dc,$$

in which the components of the earlier considered forces, stresses and torques appear with resepct to the moving coordinate triad. Accordingly, this results in formulas[100]) of the type

$$(18)\qquad \begin{cases} \varrho_0 \mathfrak{X} = \sum\limits_{(abc)} \left(\mathfrak{r}_a \dfrac{\partial \varphi}{\partial \mathfrak{y}_a} - \mathfrak{q}_a \dfrac{\partial \varphi}{\partial \mathfrak{y}_a}\right), \quad \mathfrak{X}_a = \dfrac{\partial \varphi}{\partial \mathfrak{x}_a} \\ \varrho_0 \mathfrak{L} = \sum\limits_{(abc)} \left(\mathfrak{r}_a \dfrac{\partial \varphi}{\partial \mathfrak{q}_a} - \mathfrak{q}_a \dfrac{\partial \varphi}{\partial \mathfrak{r}_a} + \mathfrak{z}_a \dfrac{\partial \varphi}{\partial \mathfrak{y}_a} - \mathfrak{y}_a \dfrac{\partial \varphi}{\partial \mathfrak{z}_a}\right), \quad \mathfrak{L}_a = \dfrac{\partial \varphi}{\partial \mathfrak{p}_a}. \end{cases}$$

**7c. The potential-based approach for two- and one-dimensional continua.** For the two- and one-dimensional extended continua in the tree-dimensional space, one can gain the potential-based approach without any difficulties using rather similar considerations.[101]) The energy density $\varphi$ — as the assumed existing limit of the quotient between the potential of a part of the continuum becoming continuously smaller and the area or length thereof — becomes a given function of the 6 functions $x, y, z, \lambda, \mu, \nu$ of $a, b$ (or of $a$) and the derivatives thereof, the potential itself [becomes] consequently a two- or one-dimensional integral:

$$\Phi = \iint\limits_{(S_0)} \varphi\, da\, db \quad \text{or} \quad \Phi = \int_0^l \varphi\, da.$$

The variation of these potentials and therefore the force, stress and torque components formulated with respect to the initial parameters are obtained immediately from the corresponding formulas of the three-dimensional case by omitting the terms concerning $c$ or $b$ and $c$; The transformation to the quantities formulated with respect to the deformed state follows then according to No. **3**e, (16) and No. **4**b, (12).

If one focuses on oriented particles, then the angular velocities $p_a, \ldots$ as defined in No. **7**b play again a crucial role, and indeed one has now naturally only 2 or 1 triple of these quantities. *E.* and *F. Cosserat* have widely developed the theory of such media using

[100] *E.* and *F. Cosserat*, l. c., p. 130 f.; cf. also IV **11** *K. Heun* No. **21**.

[101] Already *Lagrange* applied [the ansatz] to the problems in this area which he considered[23]); [the ansatz] was developed further in the theory of elastic wires and plates (cf. IV 6 (*P. Stäckel*), No. **23**, **24** and IV 25, Kap. III, *Tedone-Timpe*), but particularly also in the theory of capillarity (cf. V 9 (*Minkowski*), No. **2**).

zugeordneten Dreikantes als beweglichen Bezugssystemes weitgehend ausgebaut[102]) und haben auch hier die inneren Wirkungen betrachtet, die sich aus einem wie oben definierten *euklidischen Potential* herleiten; der Ausdruck dieses Potentials und die zugehörigen Kraft-, Spannungs-, und Momentformeln ergeben sich wiederum durch Spezialisierung der Gleichungen (16) ff. von Nr. 7b.

**7d. Die Bedeutung des wirklichen Minimums.** Ein wesentlicher Vorteil der Existenz eines Potentiales $\Phi$ der gesamten Kraftwirkungen ist die Möglichkeit, die Gleichgewichtsbedingungen ohne explizite Verwendung unendlichkleiner Verrückungen auszusprechen. Die Gleichgewichtsbedingung $\delta\Phi = 0$ ist nämlich die Bedingung dafür, dass $\Phi$ für die betrachtete Deformation einen Extremwert (Maximum oder Minimum, ev. aber auch einen sog. „Sattelwert") besitzt[103]): *Für eine Gleichgewichtslage des Quantums $V_0$ des Kontinuums wird also das Potential von $V_0$ ein Extremwert (im weitesten Sinne), verglichen mit den Werten für alle benachbarten nur den etwa stattfindenden Nebenbedingungen genügenden Deformationszuständen.* Damit ordnet sich die Gleichgewichtsbedingung genau dem normalen Problem der Variationsrechnung ein, innerhalb eines gegebenen Bereiches $V_0$ der Variablen $a, b, c$ die Funktionen $x, \ldots, \lambda, \ldots$ von ihnen so zu bestimmen, dass ein gewisses diese Funktionen und ihre Ableitungen enthaltendes Raum- und Oberflächenintegral ein Extremum wird — bei möglicherweise noch unbestimmten Randwerten; Differentialgleichungen und Randbedingungen, die hieraus nach den Regeln der Variationsrechnung entspringen, sind genau die früher aufgestellten Gleichgewichtsbedingungen.

Besonders hervorgehoben wird oft der Fall, dass $\Phi$ nur das Potential der im Inneren des Kontinuums angreifenden Wirkungen ist; dann tritt zu $-\delta\Phi$ in der Gleichgewichtsbedingung noch ein Oberflächenintegral, die virtuelle Arbeit der am Rande angreifenden äusseren Druckkräfte, hinzu, und $\delta\Phi$ selbst verschwindet notwendig nur für diejenigen virtuellen Verrückungen, die auf $S$ den Wert Null haben. *Für eine Gleichgewichtslage also wird das Potential $\Phi$ der im Innern von $V_0$ angreifenden Kräfte und Spannungen ein Extremum, verglichen mit allen benachbarten, den etwa bestehenden Nebenbedingungen*

[102] Man sehe die ausführliche Darstellung in chap. II, III der „corps déformables", wo die Gleichgewichtsbedingungen solcher Medien mit euklidischem Potential bei Verwendung der verschiedenen möglichen Koordinatensysteme und unter den mannigfachsten Spezialisierungen entwickelt sind.

[103] Die Bedeutung dieser Auffassung hat *Lagrange* auch für die Kontinua in der Méc. anal. nachdrücklich betont (s. 1. Part., sect. IV, § III).

the triad associated with every particle as moving frame of reference[102]) and have considered also here the internal actions, which are derived from a *euclidean potential* as defined above; the expression of this potential and the corresponding force, stress and torque formulas are again obtained by the specialization of the equations (16) ff. of No. 7b.

**7d. The relevance of the effective minimum.** An essential advantage of the existence of a potential $\Phi$ of the total force effects is the possibility to express the equilibrium conditions without explicitly using infinitesimal displacements. The equilibrium condition $\delta\Phi = 0$ is namely the condition that $\Phi$ has for the considered deformation an extremum (maximum or minimum, but possibly also a so called "saddle point")[103]): *For an equilibrium position of the portion $V_0$ of the continuum, the potential of $V_0$ therefore becomes an extremum (in the broadest sense), compared with the values for all neighboring states of deformations admissible with respect to possibly occurring constraints.* Hence, the equilibrium conditions can be classified just as the common problem of the calculus of variations, to determine inside a given domain $V_0$ of the variables $a, b, c$ the functions $x, \ldots, \lambda, \ldots$ thereof, such that a certain spatial and surface integral including these functions and their derivatives becomes an extremum — for possibly yet undetermined boundary values; differential equations and boundary conditions, which originate herefrom according to the rules of the calculus of variations, correspond exactly to the previously formulated equilibrium conditions.

Particularly emphasized is often the case, that $\Phi$ is only the potential of the effects applied within the continuum; then to $-\delta\Phi$ in the equilibrium condition a surface integral, the virtual work of the external compressive forces applied to the boundary, is added, and $\delta\Phi$ itself vanishes necessarily only for those virtual displacements, which have the value zero on $S$. *For an equilibrium position, thus the potential $\Phi$ of the forces and stresses applied within $V_0$ becomes an extremum, compared with all neighboring states of deformations, admissible with respect*

[102] One shall have a look at the extensive presentation in chap. II, III of "corps déformables", in which the equilibrium conditions of such media with euclidean potential are developed using the various possible coordinate systems and according to most manifold specializations.

[103] In the Méc. anal. (see 1. Part., sect. IV, § III) *Lagrange* has particularly emphasized the relevance of this perception also for continua.

*genügenden Deformationszuständen, für die jedes Grenzteilchen von $V_0$ denselben Ort innehat wie in der Gleichgewichtslage;* die Lösung dieses Variationsproblems ist natürlich nur dann bestimmt, wenn die Lage der Randteilchen, d. h. die Randwerte der Deformationsfunktionen, direkt gegeben sind.

Das Hauptinteresse, das sich mit diesen Formulierungen verknüpft, gehört der Frage, ob hier wie in der Mechanik diskreter Massen sich je nach der Art des Extremums von $\Phi$ auch die *Art des Gleichgewichts* bestimmt, insbesondere, ob das *Dirichlet*sche Stabilitätskriterium[104]) gilt, *dass für das Eintreten stabilen Gleichgewichts das Stattfinden eines wirklichen Minimums entscheidend ist.* Die allgemeine Beantwortung dieser Frage kann nur auf die Theorie der *Bewegung* des Kontinuums gegründet werden, und zwar kommt es darauf an, ob eine durch kleine Impulse aus einem Gleichgewichtszustand hervorgerufene Bewegung im Falle eines wirklichen Minimums von $\Phi$ stets in beliebiger Nähe eben dieses Deformationszustandes verläuft. Freilich kann man dabei den Begriff der „beliebigen Nähe" verschieden interpretieren, je nachdem man die Entfernung jedes einzelnen Teilchens von seiner Gleichgewichtslage beschränkt, oder diese Forderung nur im Mittel für das ganze Kontinuum oder für einzelne Teilbereiche stellt; man erhält danach verschiedene Arten von Stabilität.

Abgesehen von den Fällen der gewöhnlichen Elastizitätstheorie, wo die Verhältnisse sehr einfach liegen[105]), sind nur für wenige Probleme Stabilitätsuntersuchungen vollständig durchgeführt worden; und meist wurde ihnen überdies das Dirichletsche Kriterium oder ein äquivalenter Satz direkt zugrunde gelegt.[106]) Unter Hinweis auf diesen Sachverhalt und auf die Schwierigkeiten, die der direkten Übertragung des Dirichletschen Beweises auf Kontinua entgegenstehen, hat *A. Kneser*[107]) die Richtigkeit des Dirichletschen Kriteriums für die Kettenlinie gezeigt; für das Problem der elastischen Linie hat den Beweis *M. Born*[108]) unter ausdrücklicher Benutzung des *Osgood*schen Satzes[109]) der Variationsrechnung in einer auch auf andere eindimensionale Probleme übertragbaren Weise erbracht. Allgemein jedoch,

---

[104] *P. L. Dirichlet*, Journ. f. Math. 32 (1846), p. 85 = Werke II (Berlin 1897), p. 5.

[105] Vgl. die Übersicht in IV 25, Nr. **21**, *Tedone-Timpe*.

[106] S. IV 25, Nr. **21**, p. 211.

[107] *A. Kneser*, Journ. f. Math. 125 (1903), p. 189.

[108] *M. Born*, Untersuch. über die Stabilität der elastischen Linie. Preisschrift, Göttingen 1906, Anhang.

[109] *W. F. Osgood*, Amer. Trans. 2 (1901), p. 273; vgl. II A 8 a (*H. Hahn* u. *E. Zermelo*), Anm.[11]).

*to possibly occurring constraints, [and] for [those states of deformations in] which every boundary particle of $V_0$ is located at the same point as in the equilibrium position;* certainly, the solution of this variational problem is only determined when the position of the boundary particles, i. e. the boundary conditions of the deformation functions, are given directly.

The main interest, which is associated with this formulation, belongs to the question, if here, as in the mechanics of discrete masses, depending on the type of extremum of $\Phi$ also the *type of equilibrium* is determined, in particular, if *Dirichlet*'s stability criterion[104]) holds, *that for the appearance of a stable equilibrium the occurrence of an effective minimum is crucial.* The general answer of this question can only be founded on the theory of the *motion* of the continuum, and indeed it depends, if in the case of an effective minimum $\Phi$ a motion out of the equilibrium state caused by small impulses takes place always in the arbitrary neighborhood of exactly this state of deformation. However, in doing that one can interpret the notion of "arbitrary neighborhood" differently, depending on whether one bounds the distance of every individual particle from its equilibrium position, or [one] imposes this requirement only in average for the whole continuum or for individual subdomains; one obtains accordingly various types of stability.

Apart from the cases of the ordinary theory of elasticity, where the circumstances are very easy[105]), only for a few problems complete analyses of stability have been carried out; and moreover, mostly Dirichlet's criterion or an equivalent theorem is taken directly as a basis.[106]) With reference to this circumstance and to the difficulty, which the direct transition of Dirichlet's proof to continua is opposed to, *A. Kneser* [107]) has shown the validity of Dirichlet's criterion for the catenary; for the problem of the elastic line *M. Born*[108]) has elaborated the proof with explicit use of *Osgood*'s theorem[109]) of the calculus of variations in a way [which is] also applicable to other one-dimensional problems. Nevertheless in general,

---

104 *P. L. Dirichlet*, Journ. f. Math. 32 (1846), p. 85 = Werke II (Berlin 1897), p. 5.

105 Cf. the overview in IV 25, No. **21**, *Tedone-Timpe*.

106 See IV 25, No. **21**, p. 211.

107 *A. Kneser*, Journ. f. Math. 125 (1903), p. 189.

108 *M. Born*, Untersuch. über die Stabilität der elastischen Linie. Preisschrift, Göttingen 1906, Appendix.

109 *W. F. Osgood*, Amer. Trans. 2 (1901), p. 273; cf. II A 8 a (*H. Hahn* and *E. Zermelo*), Remark [11]).

für mehrdimensionale Integrale, dürfte der Osgoodsche Satz und daher auch das Dirichletsche Kriterium nicht ohne weiteres gelten.[110])

**7e. Direkte Bestimmung der Spannungskomponenten.** Für manche Zwecke wichtig ist eine Umformung des Prinzips vom Energieminimum, die der sog. *kanonischen Transformation* der Dynamik diskreter Medien analog ist.[111]) Sie besteht zunächst darin — wenn wir der Kürze halber uns nur auf den ersten Fall von Nr. **7**a beziehen — dass man an Stelle der 9 Ableitungen $x_a, \ldots, z_c$ als neue unbekannte Funktionen die 9 zugehörigen auf die Anfangsparameter bezogenen Spannungskomponenten

$$X_a = \frac{\partial \varphi}{\partial x_a}, \quad X_b = \frac{\partial \varphi}{\partial x_b}, \ldots, \; Z_c = \frac{\partial \varphi}{\partial z_c} \tag{19}$$

— unter Voraussetzung des Nichtverschwindens der entsprechenden Funktionaldeterminants — einführt. Bestimmt man sodann

$$H = \varphi - \sum_{(xyz\,;\,abc)} x_a X_a = H(x, \ldots; X_a, \ldots, Z_c) \tag{20}$$

als Funktion der $x, y, z$ und der neuen Grössen $X_a, \ldots, Z_c$ so zeigt man leicht mit Hilfe der bekannten Methoden der Variationsrechnung[112]), dass das Verschwinden von $\delta\Phi$ gleichbedeutend ist mit dem Verschwinden der ersten Variation des Integrales

$$\iiint\limits_{(V_0)} \Bigl(H(x, \ldots; X_a, \ldots, Z_c) + \sum_{(xyz\,;\,abc)} \frac{\partial x}{\partial a} X_a\Bigr) da\, db\, dc, \tag{21}$$

das als unbekannte Funktionen $x, y, z$ nebst ihren (linear auftretenden) Ableitungen und ausserdem $X_a, \ldots, Z_c$ ohne Ableitungen enthält. Daraus ergiebt sich dann die neue „kanonische" Form der im Innern geltenden Gleichgewichtsbedingungen:

$$\frac{\partial X_a}{\partial a} + \frac{\partial X_b}{\partial b} + \frac{\partial X_c}{\partial c} - \frac{\partial H}{\partial x} = 0 \quad \begin{pmatrix} X, Y, Z \\ x, y, z \end{pmatrix}, \tag{22a}$$

$$\frac{\partial H}{\partial X_a} + \frac{\partial x}{\partial a} = 0 \quad \begin{pmatrix} X, Y, Z \\ x, y, z \end{pmatrix}; a, b, c \Bigr). \tag{22b}$$

Die Gleichungen (22b) spielen dadurch, dass sie als Auflösung von (19) den expliziten Ausdruck der Deformation durch die Spannungskomponenten geben, in der Elastizitätstheorie eine wesentliche Rolle.

---

[110] Nach mündlicher Mitteilung von *A. Haar.* Haar hat jedoch bewiesen, dass ein analoger Satz wieder gilt sowie hinreichend hohe Ableitungen im Integranden des Variationsproblemes auftreten (vgl. den Bericht über einen Vortrag i. d. math. Ges. Göttingen, Jahresber. d. d. Math.-Ver. 19 (1910), p. 254.

[111] Vgl. IV 12, *P. Stäckel* sowie etwa die Darstellung der *Jacobi-Hamilion*schen Theorie in II A 5, Nr. **31**, *E. v. Weber.* Eine Ausdehnung auf mehrere unabhängige Veränderliche giebt *M. Born*[108]), Anhang.

[112] Vgl. *M. Born*, l. c. [108]) p. 91 ff.

for multidimensional integrals, Osgood's theorem and therefore also Dirichlet's criterion may not hold without further ado.[110])

**7e. Direct determination of the stress components.** For some purposes a transformation of the principle of minimum energy is important, which is analogue to the so called *canonical transformation* of the dynamics of discrete media.[111]) At first it involves — when for the sake of brevity, we refer only to the first case of No. **7**a — that one introduces in place of the 9 derivatives $x_a, \ldots, z_c$ as new unknowns, the 9 corresponding components of stress formulated with respect to the initial parameters

$$(19) \qquad X_a = \frac{\partial\varphi}{\partial x_a}, \quad X_b = \frac{\partial\varphi}{\partial x_b}, \ldots, \ Z_c = \frac{\partial\varphi}{\partial z_c}$$

— provided that the corresponding Jacobians do not vanish. If one determines then

$$(20) \qquad H = \varphi - \sum_{(xyz\,;\,abc)} x_a X_a = H(x, \ldots; X_a, \ldots, Z_c)$$

as a function of $x, y, z$ and the new quantities $X_a, \ldots, Z_c$, then one shows easily with the help of known methods form the calculus of variations[112]), that the vanishing $\delta\Phi$ is equivalent to the vanishing of the first variation of the integral

$$(21) \qquad \iiint\limits_{(V_0)} \Big(H(x, \ldots; X_a, \ldots, Z_c) + \sum_{(xyz\,;\,abc)} \frac{\partial x}{\partial a} X_a\Big)\, da\, db\, dc,$$

which contains as unknown functions $x, y, z$ together with their (linearly appearing) derivatives and moreover $X_a, \ldots, Z_c$ without derivatives. Thereof, the new "canonical" form of the equilibrium conditions follow, [which] hold in the interior:

$$(22a) \qquad \frac{\partial X_a}{\partial a} + \frac{\partial X_b}{\partial b} + \frac{\partial X_c}{\partial c} - \frac{\partial H}{\partial x} = 0 \quad \begin{pmatrix} X, Y, Z \\ x, y, z \end{pmatrix},$$

$$(22b) \qquad \frac{\partial H}{\partial X_a} + \frac{\partial x}{\partial a} = 0 \quad \begin{pmatrix} X, Y, Z \\ x, y, z \end{pmatrix}; a,\ b,\ c\Big).$$

In the theory of elasticity, the equations (22b) play a crucial role, since they give as solution of (19) an explicit expression of the deformation with respect to the stress components.

---

110 According to a private communication of *A. Haar*. However, Haar has proven, that a similar theorem holds again as soon as sufficiently high derivatives appear in the integrand of the variational problem (cf. the report on a presentation in the math. Ges. Göttingen, Jahresber. d. d. Math.-Ver. 19 (1910), p. 254.[)]

111 Cf. IV 12, *P. Stäckel* as well as for instance the presentation of the *Jacobi-Hamilion* theory in II A 5, No. **31**, *E. v. Weber*. An extension to several independent variables is given by *M. Born*[108]), Appendix.

112 Cf. *M. Born*, l. c. [108]) p. 91 ff.

Das Charakteristische dieses neuen *Variationsprinzips, dass in ihm nicht sowohl die Deformationsgrössen, als vielmehr die Spannungskomponenten hervortreten*, kommt noch deutlicher in dem speziellen Fall zum Ausdruck, *dass die Energiedichte $\varphi$ von den Werten der Deformationen $x, y, z$ selbst unabhängig ist*, also nur von der Formänderung (im weitesten Sinne) abhängt. Dann enthält also $H$ nur die Spannungskomponenten, und man kann (21) durch das folgende *Variationsprinzip mit Nebenbedingungen* ersetzen, das dem in der Theorie der Fachwerke als *Prinzip* von *L. F. Menabrea* und *A. Castigliano*[113]) bekannten analog ist: *Es soll die erste Variation des Integrales*

$$\iiint H(X_a, X_b, \ldots, Z_c) da\, db\, dc \tag{23}$$

*verschwinden, wobei zum Vergleich alle Systeme von Funktionen $X_a, \ldots, Z_c$ zugelassen werden, die den 3 Bedingungsgleichungen*

$$\frac{\partial X_a}{\partial a} + \frac{\partial X_b}{\partial b} + \frac{\partial X_c}{\partial c} = 0 \qquad (X, Y, Z) \tag{23a}$$

*genügen;* bezeichnet man mit $x, y, z$ drei diesen Nebenbedingungen zugeordnete Lagrangesche Faktoren, so ergeben sich hieraus in der Tat die Gleichungen (22b). Durch Elimination dieser Lagrangeschen Faktoren aus (22b) folgen für die 9 unbekannten Funktionen allein die 6 Gleichungen:

$$\frac{\partial}{\partial b}\left(\frac{\partial H}{\partial X_a}\right) = \frac{\partial}{\partial a}\left(\frac{\partial H}{\partial X_b}\right) \qquad (a, b, c; X, Y, Z); \tag{24}$$

das sind die sog. *Kompatibilitätsbedingungen*[114]) der Elastizitätstheorie, die ausdrücken, dass ein den Bedingungen (23a) genügendes Spannungssystem tatsächlich Gleichgewichtssystem in einem Kontinuum mit der Energiedichte $\varphi$ bzw. $H$ sein kann. — Dies *Castigliansche Prinzip* wird besonders in solchen Fällen bedeutsam, wo in einem Medium nur Spannungen gewisser Art stattfinden können; die diese Einschränkungen darstellenden Bedingungen können ihm ohne weiteres als Nebenbedingungen hinzugefügt werden.[115])

**7f. Die entsprechenden Ansätze für die Kinetik.** Auch bei bewegten Medien kommen in erster Linie die bisher betrachteten

[113] *L. P. Menabrea*, Torino Mem. (2) 25 (1871), p. 141 und *A. Castigliano*, Théorie d'équilibre des systèmes élastiques (Turin 1879); vgl. IV 29a, Nr. **7** ff., *M. Grüning*. Vgl. auch *E.* und *F. Cosserat*, Corps déform., p. 26 ff. für den Fall des eindimensionalen Kontinuums.

[114] S. IV 24, Nr. **7**a, *Tedone*; vgl. auch *A. Haar* u. *Th. v. Kármán*, Gött. Nachr., math.-phys. Kl. 1909, p. 204 ff.

[115] *Haar* u. *Kármán*, l. c. [114]), p. 212.

The characteristics of this new *variational principle, that not only the deformation quantities but rather the stress components do appear*, finds expression even more for the special case, *that the energy density $\varphi$ is independent of the values of the deformation functions $x, y, z$*, [that it] depends therefore only on the shape change (in the broadest sense). Then $H$ contains thus only the stress components, and one can substitute (21) with the following *variational problem with constraints*, which is analogous to the [principle] known in the theory of frameworks as the *principle of L. F. Menabrea* and *A. Castigliano*[113]): *The first variation of the integral*

$$\iiint H(X_a, X_b, \ldots, Z_c) da\, db\, dc \tag{23}$$

*shall vanish, where for comparison all systems of functions $X_a, \ldots, Z_c$ are allowed, which satisfy the 3 equations*

$$\frac{\partial X_a}{\partial a} + \frac{\partial X_b}{\partial b} + \frac{\partial X_c}{\partial c} = 0 \qquad (X, Y, Z); \tag{23a}$$

If one denotes the Lagrange multipliers associated with these three constraints by $x, y, z$, then herefrom the equations (22b) are obtained after all. By elimination of these Lagrange multipliers from (22b), for the 9 unknown functions just the 6 equations follow:

$$\frac{\partial}{\partial b}\left(\frac{\partial H}{\partial X_a}\right) = \frac{\partial}{\partial a}\left(\frac{\partial H}{\partial X_b}\right) \qquad (a, b, c; X, Y, Z); \tag{24}$$

these are the so called *compatibility conditions*[114]) of the theory of elasticity, which express, that a system of stresses being compatible with the conditions (23a) can in fact be an equilibrium system in a continuum with energy density $\varphi$ or $H$. — This *principle of Castigliano* becomes particularly important in such cases, where in the medium only stresses of a certain type can appear; the conditions representing these restrictions can easily be added [to the principle] as constraints.[115])

**7f. The appropriate approaches to kinetics.** In the first place, also for moving media the so far considered

[113] *L. P. Menabrea*, Torino Mem. (2) 25 (1871), p. 141 and *A. Castigliano*, Théorie d'équilibre des systèmes élastiques (Turin 1879); cf. IV 29a, No. **7** ff., *M. Grüning*. Cf. also *E.* and *F. Cosserat*, Corps déform., p. 26 ff. for the case of the one-dimensional continuum.

[114] See IV 24, No. 7a, *Tedone*; cf. also *A. Haar* and *Th. v. Kármán*, Gött. Nachr., math.-phys. Kl. 1909, p. 204 ff.

[115] *Haar* and *Kármán*, l. c. [114]), p. 212.

Wirkungen in Betracht, in die nur $t$ als Parameter eingeht. Fasst man zunächst den Ansatz von Nr. **7**a mit dem Ausdruck des *Hamiltonschen Prinzips* Nr. **5**, (5), (6) zusammen, so ergiebt sich der der Formulierung von **7**d analoge Satz: *Für die wirkliche Bewegung des Kontinuums* $V_0$ *im Zeitintervall* $t_0 \leqq t \leqq t_1$ *hat das vierfache Integral*

$$\int_{t_0}^{t_1} dt \iiint_{(V_0)} dV_0 \left\{\tfrac{1}{2}\varrho_0(x'^2 + y'^2 + z'^2) - \varphi\right\} \tag{25}$$

*einen Extremwert gegenüber seinen Werten für alle benachbarten, den etwa stattfindenden Nebenbedingungen genügenden Bewegungen, die zu den Zeiten* $t_0$ *und* $t_1$ *das Kontinuum in derselben Lage belassen.*

Diesen Ansatz kann man, wie man es im Falle endlich vieler Freiheitsgrade tut, sofort wesentlich ausdehnen, indem man von der speziellen Abhängigkeit des Integranden von den zeitlichen Ableitungen abgeht. Man braucht dazu, um sogleich auch den Fall orientierter Teilchen mit zu umfassen, nur an die Formeln (10), (12) von Nr. **5**d anzuschliessen und analog wie im Anfang von Nr. **7**a zu fordern: *Die virtuelle Arbeit des bewegten Kontinuums im Zeitintervall* $t_0, t_1$ *soll für jede virtuelle Verrückung gleich sein der Variation eines einzigen nur von dem jeweiligen Bewegungszustande abhängigen Ausdruckes*, der speziell ein vierfaches Integral über eine bekannte Funktion der Bewegungsfunktionen und ihrer zeitlichen und räumlichen Ableitungen sei:

$$\Phi = \int_{t_0}^{t_1} dt \iiint_{(V_0)} dV_0 \varphi(a, b, c, t; x, \ldots, \nu; x_a, \ldots, \nu_c; x', \ldots, \nu'; x'_a, \ldots, \nu'_c), \tag{26}$$

und *Wirkungsintegral (action)* heisse.[116]) Dann bleiben die Formeln für Kraft-, Spannungs- und Momentkomponenten im wesentlichen umgeändert, nur für die Impulskomponenten treten die Gleichungen hinzu

$$X_t = -\frac{\partial\varphi}{\partial x'}, \quad L_t = -\frac{\partial\varphi}{\partial\lambda'}l_1 - \frac{\partial\varphi}{\partial\mu'}l_2 - \frac{\partial\varphi}{\partial\nu'}l_3 \quad \left(x, y, z; \begin{matrix} L, M, N \\ l, m, n \end{matrix}\right). \tag{27}$$

*E.* und *F. Cosserat*[117]) haben auch hier die Annahme eines „*euklidischen Potentiales*" verfolgt, das sich bei einer jeden Bewegung des samt seinen Dreikanten erstarrt gedachten Kontinuums nicht ändert; es wird ausser den Grössen (16) noch die (nichtholonomen) Geschwindigkeitskoordinaten in bezug auf das bewegliche Koordinatensystem

$$\mathfrak{x} = \alpha_1 x' + \beta_1 y' + \gamma_1 z', \quad \mathfrak{p} = \alpha_3\alpha'_2 + \beta_3\beta'_2 + \gamma_3\gamma'_2 \quad \begin{pmatrix} \mathfrak{x}, \mathfrak{y}, \mathfrak{z} \\ \mathfrak{p}, \mathfrak{q}, \mathfrak{r} \\ 1, 2, 3 \end{pmatrix} \tag{28}$$

---

[116] Vgl. *E.* und *F. Cosserat*, Corps déform., p. 4.

[117] „Corps déformables", p. 156 ff.

effects, in which only $t$ enters as parameter, come into consideration. If one summarizes at first the ansatz of No. **7**a with the expression of *Hamilton's principle* No. **5**, (5), (6), then the theorem being analogous to the formulation of **7**d is obtained: *For the actual motion of the continuum $V_0$ within the time interval $t_0 \leqq t \leqq t_1$ the fourfold integral*

$$(25) \qquad \int_{t_0}^{t_1} dt \iiint_{(V_0)} dV_0 \left\{\tfrac{1}{2}\varrho_0(x'^2 + y'^2 + z'^2) - \varphi\right\}$$

*has an extremum with respect to its values for all neighboring motions being admissible with respect to possibly occurring constraints, which, at the time instants $t_0$ and $t_1$, leave the continuum in the same position.*

As one does it in the case of finitely many degrees of freedom, one can immediately extend this ansatz, by giving up the special dependency of the integrand on the [terms with] time derivatives. To readily include also the case of oriented particles, one only needs to make the connection to the formulas (10), (12) of No. **5**d and to demand analogously like in the beginning of No. **7**a: *For every virtual displacement, the virtual work of the moving continuum in the time interval $t_0, t_1$ shall be equal to the variation of a single expression depending only on the respective motion*, which shall specifically be a fourfold integral over a known function of the functions of motion and the temporal and spatial derivatives thereof:

$$(26) \qquad \Phi = \int_{t_0}^{t_1} dt \iiint_{(V_0)} dV_0 \varphi(a, b, c, t; x, \ldots, \nu; x_a, \ldots, \nu_c; x', \ldots, \nu'; x'_a, \ldots, \nu'_c),$$

and [shall] be called *action integral*.[116]) In that case, the formulas for the force, stress and torque components remain basically unchanged, only for the components of momentum additional equations do appear

$$(27) \qquad X_t = -\frac{\partial \varphi}{\partial x'}, \; L_t = -\frac{\partial \varphi}{\partial \lambda'} l_1 - \frac{\partial \varphi}{\partial \mu'} l_2 - \frac{\partial \varphi}{\partial \nu'} l_3 \quad \begin{pmatrix} x, \; y, \; z; & L, M, N \\ & l, \; m, \; n \end{pmatrix}.$$

Also here, *E.* and *F. Cosserat*[117]) have followed the assumption of a "*euclidean potential*", which does not change for an arbitrary motion of the continuum together with its triads being regarded as rigid; Besides the quantities (16) it will also include the (nonholonomic) velocity coordinates with respect to the movable coordinate system

$$(28) \qquad \mathfrak{x} = \alpha_1 x' + \beta_1 y' + \gamma_1 z', \; \mathfrak{p} = \alpha_3 \alpha'_2 + \beta_3 \beta'_2 + \gamma_3 \gamma'_2 \quad \begin{pmatrix} \mathfrak{x}, \mathfrak{y}, \mathfrak{z} \\ \mathfrak{p}, \mathfrak{q}, \mathfrak{r} \\ 1, 2, 3 \end{pmatrix},$$

116 Cf. *E.* and *F. Cosserat*, Corps déform., p. 4.

117 "Corps déformables", p. 156 ff.

enthalten, und hiernach lassen sich die in die Bewegungsgleichungen eingehenden Komponenten analog (18) unmittelbar bestimmen.

Auch hier lässt sich analog zu Nr. 7e die kanonische Transformation durchführen; transformiert man nur in Hinblick auf die zeitlichen Ableitungen, so entsteht, falls $\varphi$ von $t$ unabhängig ist, in

$$E = \varphi - \sum_{(\mathfrak{x}\mathfrak{y}\mathfrak{z})} \mathfrak{x}\frac{\partial\varphi}{\partial\mathfrak{x}} - \sum_{(\mathfrak{p}\mathfrak{q}\mathfrak{r})} \mathfrak{p}\frac{\partial\varphi}{\partial\mathfrak{p}}$$

die *Energiedichte des bewegten Systems.*[118])

Neben dieser weittragenden Verallgemeinerung ist noch eine speziellere Art des Eingehens der *zeitlichen* Ableitungen der Bewegungsfunktionen in die Wirkungskomponenten hervorzuheben, wie sie bei *Reibungswirkungen* u. dgl. auftritt und bei der ein dem Potentialansatz in gewisser Weise analoger Ansatz auftritt. Beschränken wir uns darauf, dass die Spannungsdyade einen von den zeitlichen Ableitungen der 9 Deformationsgrössen $x'_a, \ldots, z'_c$ abhängigen Teil enthält, so handelt es sich um die Besonderheit, dass die Spannungskomponenten gerade die Ableitungen einer bekannten Funktion $F(x'_a, x'_b, \ldots, z'_c)$ nach $x'_a, \ldots, z'_c$ sind:

$$(29) \qquad X_a = \frac{\partial F}{\partial x'_a}, \quad X_b = \frac{\partial F}{\partial x'_b}, \ldots, \quad Z_c = \frac{\partial F}{\partial z'_c}.$$

Zu der während der wirklichen Bewegung geleisteten Arbeit liefert diese Spannungsdyade den Beitrag (auf die Zeiteinheit berechnet):

$$(30) \qquad -\sum_{(xyz;abc)} X_a \frac{dx_a}{dt} = -\sum_{(xyz;abc)} \frac{\partial F}{\partial x'_a} \cdot x'_a = -D(x'_a, \ldots, z'_c).$$

Ist $D$ eine positiv definite Funktion seiner 9 Argumente, so wird während der Bewegung durch die Spannungen $X_a, \ldots$ stets Arbeit verzehrt, und zwar in einem durch die Funktion $D$ gemessenen Betrage; $D$ heisst die zu der Spannung gehörige *Dissipationsfunktion.*[119]) Tatsächlich wird übrigens lediglich der Fall benutzt, dass $F$ eine quadratische Funktion der $x'_a, \ldots$ ist; dann ist auch

$$(29') \qquad X_a = \frac{1}{2}\frac{\partial D}{\partial x'_a}, \quad X_b = \frac{1}{2}\frac{\partial D}{\partial x'_b} \quad \ldots, \quad Z_c = \frac{1}{2}\frac{\partial D}{\partial z'_c}.$$

In ganz ähnlicher Weise kann man auch die Abhängigkeit von höheren zeitlichen Ableitungen in Betracht ziehen und die zugehörigen Dissipationsfunktionen bestimmen; für lineare Abhängigkeit der $X_a, \ldots$ von den Ableitungen hat das *W. Voigt*[120]) durchgeführt.

---

[118] *M. Born*, l. c. p. 94 f.; *E.* und *F. Cosserat*, Corps déform., p. 171, 219.

[119] *Lord Rayleigh (J. W. Strutt)*, Lond. Math. Soc. Proc. 4 (1873), p. 357.

[120] *W. Voigt*, Kompendium I, p. 459 ff.; Lehrbuch der Krystallphysik, Leipzig 1910, p. 792 ff.

and according to this, the components entering the equations of motion can be determined immediately analogously to (18).

Also here a canonical transformation, analogous to No. 7e, can be carried out; if one transforms merely with respect to the time derivatives, then for $\varphi$ being independent of $t$ it appears

$$E = \varphi - \sum_{(\mathfrak{x}\mathfrak{y}\mathfrak{z})} \mathfrak{x}\frac{\partial\varphi}{\partial\mathfrak{x}} - \sum_{(\mathfrak{p}\mathfrak{q}\mathfrak{r})} \mathfrak{p}\frac{\partial\varphi}{\partial\mathfrak{p}}$$

the *energy density of the moving system.*[118])

Besides this far-reaching generalization there is to be pointed out additionally a special kind of emergence of the *temporal* derivatives of the functions of motion in the effects, which appears in *frictional effects* and similar ones and for which an ansatz appears being in a way analogous to the potential-based approach. If we restrict us, that the stress dyad contains a part depending on the time derivatives of the 9 deformation quantities $x'_a, \ldots, z'_c$, then it is about the specialty, that the stress components are just the derivatives of a known function $F(x'_a, x'_b, \ldots, z'_c)$ with respect to $x'_a, \ldots, z'_c$:

$$(29) \qquad X_a = \frac{\partial F}{\partial x'_a}, \quad X_b = \frac{\partial F}{\partial x'_b}, \ldots, \quad Z_c = \frac{\partial F}{\partial z'_c}.$$

Additional to the work done during the actual motion, the stress dyad contributes with (computed per unit of time):

$$(30) \qquad -\sum_{(xyz;abc)} X_a \frac{dx_a}{dt} = -\sum_{(xyz;abc)} \frac{\partial F}{\partial x'_a} \cdot x'_a = -D(x'_a, \ldots, z'_c).$$

If $D$ is a positive definite function of its 9 arguments, then the stresses $X_a, \ldots$ always use work, and indeed [they use work] in an amount measured by the function $D$; $D$ is called the *dissipation function* associated to the stresses.[119]) By the way, merely the case of $F$ being a quadratic function of $x'_a, \ldots$ is effectively used; then it follows

$$(29') \qquad X_a = \frac{1}{2}\frac{\partial D}{\partial x'_a}, \quad X_b = \frac{1}{2}\frac{\partial D}{\partial x'_b} \quad \ldots, \quad Z_c = \frac{1}{2}\frac{\partial D}{\partial z'_c}.$$

In a quite similar way one can also consider the dependency on higher time derivatives and determine the corresponding dissipation functions; this has been carried out by *W. Voigt*[120]) for linear dependency of $X_a, \ldots$ on the derivatives.

---

[118] *M. Born*, l. c. p. 94 f.; *E.* and *F. Cosserat*, Corps déform., p. 171, 219.

[119] *Lord Rayleigh (J. W. Strutt)*, Lond. Math. Soc. Proc. 4 (1873), p. 357.

[120] *W. Voigt*, Kompendium I, p. 459 ff.; Lehrbuch der Krystallphysik, Leipzig 1910, p. 792 ff.

**8. Grenzfälle des gewöhnlichen dreidimensionalen Kontinuums.** Endlich bleibt noch zu erörtern, wie man durch gewisse typische Grenzübergänge aus der Theorie des *freien dreidimensionalen* Kontinuums die bisher ohne direkten Zusammenhang mit ihr in rein formaler Analogie aufgestellten Ansätze für Kontinua anderer Art gewinnen kann; dabei genüge es, alles auf den Fall der Existenz eines Potentials der einfachsten Art (Nr. **7**a, Anfang) zu beziehen.

**8a. Unendlichdünne Platten und Drähte.** In erster Linie handelt es sich um die Theorie der Medien, deren Ausdehnung nach einer oder zwei Dimensionen hin als unendlichklein angesehen werden kann (Platten und Drähte). In Wahrheit liegt hier jedesmal ein *dreidimensional* ausgedehntes Gebiet $\mathfrak{V}$ vor, das von einem jene sehr kleinen Ausdehnungen messenden Parameter $\varepsilon$ abhängt; die abstrakten Grenzfälle *unendlichkleiner* Ausdehnung werden wir darstellen, wenn wir eine ganze Schar von Gebieten $\mathfrak{V}$ betrachten, die sich im Limes $\varepsilon = 0$ dem bei der direkten Behandlung (s. Nr. **2**c) zugrunde gelegten Flächen- oder Linienstück — wir dürfen noch annehmen: gleichmässig — annähern. Auf Grund dieser Vorstellung kann man die Theorie der Platten und Drähte an die Theorie der dreidimensionalen Kontinua anschliessen, und tatsächlich hat bereits *S. D. Poisson* in einem Falle[121]) konsequent diesen Weg eingeschlagen: Man wird die charakteristischen Grössen für das Gebiet $\mathfrak{V}$ als Funktion von $\varepsilon$ darstellen und dann durch ebenjenen Prozess $\lim \varepsilon = 0$ bzw. durch Beschränkung auf die ersten Glieder der Reihenentwicklung nach $\varepsilon$ zu den für den Grenzfall geltenden Gesetzen gelangen. Axiomatisch gesprochen würde dieses Verfahren die Konsequenz eines *allgemeinen Stetigkeitspostulates* sein, das man so formulieren kann: In einem Medium, dessen Gestalt oder physikalisches Verhalten von einem kontinuierlich variablen Parameter abhängt, ändern sich die Zustandsgleichungen ausnahmslos stetig mit diesem Parameter.

Die Ausführung dieses Verfahrens möge an das Variationsprinzip angeschlossen werden[122]). Als typisches Beispiel werde ein Medium

[121] Bei der Behandlung des Problems der elastischen Platte; Mém. de l'Acad., Paris 8 (1829), p. 523 ff.

[122] Solche Reihenentwicklungen und Grenzbetrachtungen liegen mehr oder weniger ausgesprochen allen Theorien der Platten und Drähte seit *Poisson* zugrunde (s. IV 25, Nr. **13** ff., *Tedone-Timpe*); nur wird die Übersicht dadurch erschwert, daß man sich von vornherein auf unendlichkleine Deformationen in kleinen Teilgebieten beschränkt und erst hinterher unter Heranziehung von Hilfehypothesen zu der endlichen Deformation des Ganzen übergeht. Der Text folgt der Darstellung, die *C. Carathéodory* in einer Göttinger Vorlesung im W.-S. 1906/7 für die elastischen Linie vorgetragen hat.

**8. Limit cases of the ordinary three-dimensional continuum.** Eventually it remains to be discussed how one can gain by certain typical limit processes from the theory of the *free three-dimensional* continuum the foundations of other classes of continua which [have been obtained] so far without direct connection by a purely formal analogy; thereby it is sufficient to relate everything to the case of an existing potential of the most simple form (No. **7**a, beginning).

**8a. Infinitely thin plates and wires.** Primarily, it is about the theory of media whose extension in one or two dimensions can be considered as infinitesimal (plates and wires). In reality there exists always a *three-dimensional* extended domain $\mathfrak{B}$, which depends on a parameter $\varepsilon$ measuring those very small extensions; we will express the abstract limit cases of *infinitesimal* extension, when we consider a whole family of domains $\mathfrak{B}$, which in the limit $\varepsilon = 0$ — furthermore we may assume: continuously — approach the surface or line element, which the direct approach (s. No. **2**c) is based on. Due to this perception, the theory of plates and wires can be connected to the theory of three-dimensional continua, and in fact already *S. D. Poisson* has chosen this way consistently for one case[121]): One is formulating the characteristic quantities for the domain $\mathfrak{B}$ as a function of $\varepsilon$ and arrives at the laws holding for the limit case by the just mentioned process $\lim \varepsilon = 0$ or else by the restriction to the first terms in the series expansion with respect to $\varepsilon$. From an axiomatic point of view, this approach would be the consequence of a *general continuity postulate*, which can be formulated as follows: In a medium, whose shape or physical property depends on a continuously variable parameter, the equations of state change without exception continuously with this parameter.

The presentation of this approach shall be based on the variational principle[122]). As a typical example a medium

[121] For the treatment of the problem of the elastic plate; Mém. de l'Acad., Paris 8 (1829), p. 523 ff.

[122] Such series expansions and limit processes are since *Poisson* the more or less declared basis for all theories of plates and wires (s. IV 25, No. **13** ff., *Tedone-Timpe*); but the overall view is complicated by restricting oneself from the beginning to infinitesimal deformations in small subdomains and only afterwards one changes over to the finite deformation of the whole under citation of auxiliary hypotheses. The text follows the exposition, which *C. Carathéodory* presented for the elastic line in a "Göttinger Vorlesung" in the winter term 1906/7.

betrachtet, das im Anfangszustande das über einem Flächenstück $S_0$ der $a$-$b$-Ebene gelegene Gebiet $-\varepsilon \leqq c \leqq +\varepsilon$ erfüllt; sein Potential sei:

$$\Phi = \iint\limits_{(S_0)} dadb \int\limits_{-\varepsilon}^{+\varepsilon} dc\varphi(a, b, c; x, \ldots; x_a, \ldots). \tag{1}$$

Die Gleichgewichtsfunktionen $x = x(a, b, c), \ldots$, die unter gewissen Randbedingungen $\Phi$ zum Minimum machen, werden nun von $\varepsilon$ abhängen; sie mögen in eine Potenzreihe in $\varepsilon$ und $c$ entwickelbar sein:

$$x = x^{(0)}(a, b) + cx_c^{(0)}(a, b) + \varepsilon x^{(1)}(a, b) + \varepsilon c x_c^{(1)}(a, b) + \cdots \qquad (x, y, z). \tag{2}$$

Führt man diese Ausdrücke in $\varphi$ ein und entwickelt danach $\varphi$ selbst nach Potenzen von $\varepsilon$ und $c$, so ergiebt sich für $\Phi$ eine Reihe

$$\Phi = \varepsilon \iint\limits_{(S_0)} \varphi_0 dadb + \varepsilon^2 \iint\limits_{(S_0)} \varphi_1 dadb + \cdots \tag{3}$$

wo

$$\varphi_0 = 2\varphi\Big(a, b, 0; x^{(0)}, \ldots; \frac{\partial x^{(0)}}{\partial a}, \frac{\partial x^{(0)}}{\partial b}, x_c^{(0)}, \ldots\Big)$$

lediglich von den Funktionen $x^{(0)}, \ldots$, ihren ersten partiellen Ableitungen nach $a, b$ und den Funktionen $x_c^{(0)}, \ldots$ abhängt, während in $\varphi_1, \ldots$ immer mehr der als Entwicklungskoeffizienten der Reihen (2) auftretenden Funktionen von $a, b$ eingehen können. Die eigentliche Aufgabe ist nun, die durch die Grenzfunktionen

$$\lim_{\varepsilon=0} x(a, b, 0) = x^{(0)}(a, b) \qquad (x, y, z)$$

bestimmte Gleichgewichtslage der „unendlichdünnen“ Platte (bzw. ihrer Mittelebene $c = 0$) zu ermitteln. Daneben kann aber auch die Bestimmung weiterer Glieder der Reihen (2) wichtig werden, z. B. der Funktionen $x_c^{(0)}(a, b)$, die die neue Lage der ursprünglichen Normalen der Platte d. h. die Verbiegung ihres Materials gegen die geometrische Gestalt der Mittelebene bestimmen. Diese Funktionen gehören tatsächlich zu den Bestimmungsstücken der Deformation, eben weil es sich in Wahrheit nicht um ein *streng* ein- bzw. zweidimensionales Medium handelt; bei der direkten Theorie werden sie durch das *Cosserat*sche Dreikant geliefert.

Da nun $\Phi$ für jedes $\varepsilon$ unter den angenommenen Randbedingungen ein Minimum werden soll, so muss nach (3) *in erster Linie* auch $\iint_{S_0} \varphi_0 dadb$ ein Minimum werden; dies ist aber gerade eine Bedingung für jene Funktionen $x^{(0)}(a, b), \ldots, x_c^{(0)}(a, b), \ldots$, wobei zum Vergleich alle die Funktionen zuzulassen sind, welche die aus den gegebenen Randbedingungen mittels (2) für $x^{(0)}, \ldots, x_c^{(0)}$ folgenden Randbedingungen erfüllen.

is considered, which in the initial state occupies the domain $-\varepsilon \leqq c \leqq +\varepsilon$ lying over the surface element $S_0$ of the $a$-$b$-plane; let its potential be:

$$(1) \qquad \Phi = \iint_{(S_0)} da db \int_{-\varepsilon}^{+\varepsilon} dc \varphi(a, b, c; x, \ldots; x_a, \ldots).$$

The functions of equilibrium $x = x(a, b, c), \ldots,$ which under certain boundary conditions make $\Phi$ to a minimum, will depend now on $\varepsilon$; let them be expandable in a series expansion of $\varepsilon$ and $c$:

$$(2) \qquad x = x^{(0)}(a, b) + cx_c^{(0)}(a, b) + \varepsilon x^{(1)}(a, b) + \varepsilon c x_c^{(1)}(a, b) + \cdots \quad (x, y, z).$$

If one introduces these expressions in $\varphi$ and subsequently expands $\varphi$ itself with respect to the powers of $\varepsilon$ and $c$, then the [following] series for $\Phi$ is obtained

$$(3) \qquad \Phi = \varepsilon \iint_{(S_0)} \varphi_0 da db + \varepsilon^2 \iint_{(S_0)} \varphi_1 da db + \cdots$$

where

$$\varphi_0 = 2\varphi\left(a, b, 0; x^{(0)}, \ldots; \frac{\partial x^{(0)}}{\partial a}, \frac{\partial x^{(0)}}{\partial b}, x_c^{(0)}, \ldots\right)$$

depends merely on the functions $x^{(0)}, \ldots,$ their first partial derivatives with respect to $a, b$ and the functions $x_c^{(0)}, \ldots,$ while in $\varphi_1, \ldots$ more and more coefficients can enter, [which] appear in the series expansion (2) as functions of $a, b$. The actual problem is now to calculate the equilibrium position of the "infinitely thin" plate (or rather its midsurface $c = 0$) determined by the limit function

$$\lim_{\varepsilon=0} x(a, b, 0) = x^{(0)}(a, b) \qquad (x, y, z).$$

Besides, also the determination of further terms in the series (2) can be important, for instance the functions $x_c^{(0)}(a, b)$, which determine the new position of the initial normals of the plate, i. e. the deflection of the material against the geometric shape of the midsurface. These functions belong in fact to the characteristic quantities of the deformation, just because in reality it is not about a *strict* one- or two-dimensional medium; in the direct theory they are given by the *Cosserat* triad.

Now, since $\Phi$ with respect to the considered boundary conditions shall become a minimum for every $\varepsilon$, according to (3) *primarily* $\iint_{S_0} \varphi_0 da db$ must become a minimum; but this is directly a condition for those functions $x^{(0)}(a, b), \ldots, x_c^{(0)}(a, b), \ldots,$ where for comparison all the functions are allowed, which satisfy the boundary conditions for $x^{(0)}, \ldots, x_c^{(0)}$ induced by the given boundary conditions together with (2).

Es ist nun möglich, dass hierdurch die Funktionen $x^{(0)}, \ldots$ noch nicht völlig bestimmt werden, sondern dass sich nur gewisse Relationen zwischen ihnen ergeben. Beschränkt man sich alsdann auf Funktionen, die diesen Relationen genügen, so folgt *zweitens*, dass $x^{(0)}(a, b), \ldots$ und die weiterhin noch in $\varphi_1$ eingehenden Funktionen auch das zweite Glied der Reihe (3), $\iint_{(S_0)} \varphi_1 da db$, zum Minimum machen, wobei sich die Randbedingungen analog wie vorhin ergeben; lassen jene Relationen etwa die Elimination von $x_c^{(0)}, \ldots$ zu, so kann dies neue Variationsprinzip höhere Ableitungen der Funktionen $x^{(0)}, \ldots$ enthalten. Fährt man ev. mit dieser Schlussweise fort, so bekommt man für die Funktionen $x^{(0)}, \ldots$ eine *Reihe zweidimensionaler Variationsprobleme*, die *höhere Ableitungen* enthalten und zu denen Nebenbedingungen hinzutreten können.

Bei der Durchführung dieses Ansatzes entsteht jedoch eine wesentliche Schwierigkeit: es wird hierbei für die Lösung des dreidimensionalen Problems Entwickelbarkeit in eine Reihe der Form (2) vorausgesetzt, d. h. es wird ein bestimmtes reguläres Verhalten dieser Lösungen als Funktionen eines in der Randgleichung des Kontinuums enthaltenen Parameters $\varepsilon$ gefordert. Nun braucht der Wert $\varepsilon = 0$ für Probleme dieser Art nicht nur keine reguläre Stelle zu sein, sondern er könnte sogar eine wesentlich singuläre Stelle sein[123]); die Möglichkeit einer Entwicklung (2) bleibt also zunächst durchaus problematisch. Solange daher nicht die Abhängigkeit der Lösungen von Parametern in den Randbedingungen eingehend erforscht ist, ist auf diesem Wege eine völlig befriedigende, über die Aufdeckung des formalen Zusammenhanges mit den Eigenschaften der dreidimensionalen Medien hinausgehende Theorie der Platten und Drähte nicht zu erzielen, und die direkten Ansätze, wie sie besonders *E.* und *F. Cosserat* ausgebildet haben (s. Nr. **3**e, **7**c) bleiben vorläufig das einzige Auskunftsmittel.

**8b. Medien mit einer kinematischen Nebenbedingung.** Prinzipiell gleichwertige Betrachtungen kann man anstellen, um aus der Theorie des frei deformierbaren Kontinuums die Gesetze solcher Medien abzuleiten, die Nebenbedingungen unterworfen sind, und für die der direkte Ansatz in Nr. **4c** gegeben wurde. Es handele sich, um wieder nur einen typischen einfachen Fall zu erörtern, um ein Medium $\mathfrak{M}$, zwischen dessen Deformationsgrössen die Nebenbedingung

$$\omega(x, \ldots; x_a, x_b, \ldots) = 0 \tag{4}$$

besteht, die übrigens ev. auch $a, \ldots$ explizit enthalten kann. In

[123] *E.* und *F. Cosserat*, Paris C. R. 145 (1907), p. 1139; 146 (1908), p. 169.

Now it is possible, that the functions $x^{(0)}, \ldots$ are hereby not yet completely determined, but that only certain relations between them emerge. If thereupon one restricts oneself to functions, which satisfy these relations, then it follows *secondly*, that $x^{(0)}(a, b), \ldots$ and the functions still entering $\varphi_1$ make the second term in the series $\iint_{(S_0)} \varphi_1 da db$ to a minimum, where the boundary conditions emerge analogously as before; if those relations allow for instance for the elimination of $x_c^{(0)}, \ldots$, then this new variational principle can contain higher derivatives of the functions $x^{(0)}, \ldots$. If one possibly continues with this procedure, then one obtains for the functions $x^{(0)}, \ldots$ a *series of two-dimensional variational problems*, which contain *higher derivatives* and to which constraints can be added.

Carrying out this ansatz, however, a crucial difficulty arises: for the solution of the three-dimensional problem, hereby the expansibility into a series of the form (2) is assumed, i. e. a certain regular behavior of these solutions as functions of a parameter $\varepsilon$ included in the boundary equation of the continuum is demanded. For problems of this kind, the value $\varepsilon = 0$ now does not only need to be not a regular point, but it could even be an essentially singular point[123]); the possibility of an expansion (2) remains therefore a priori quite questionable. Hence, as long as the dependency of the solutions on the parameters in the boundary conditions is not explored in detail, in this way, a completely satisfying theory of plates and wires, which goes beyond the disclosure of the formal connection with the properties of the three-dimensional media, is not obtained, and the direct approaches, which have especially been formulated by *E.* and *F. Cosserat* (see No. 3e, 7c) remain the only reference for now.

**8b. Media with one kinematic constraint.** In principle, one can make equivalent considerations to derive from the theory of the freely deformable continuum the laws of such media which are subjected to constraints, and for which the direct ansatz in No. 4c has been given. To discuss again only one typical easy case, it is about a medium $\mathfrak{M}$, for which there exists the constraint

$$\omega(x, \ldots; x_a, x_b, \ldots) = 0 \tag{4}$$

between its deformation quantities[;] by the way [the constraint] can also contain $a, \ldots$ explicitly. Now in

[123] *E.* and *F. Cosserat*, Paris C. R. 145 (1907), p. 1139; 146 (1908), p. 169.

Wahrheit wird nun ein solches Medium in der Natur niemals streng realisiert sein, vielmehr liegt hier wiederum nur eine Abstraktion aus der Betrachtung solcher Medien $\mathfrak{M}_\varepsilon$, vor, die die Relation (4) *nahezu* erfüllen. $\mathfrak{M}_\varepsilon$ mag durch ein Potential von der Gestalt Nr. **7**, (1) mit der Energiedichte $\varphi_\varepsilon$ charakterisiert sein, und es soll von einem Parameter $\varepsilon$ derart abhängen, dass für jede Gleichgewichtslage durchweg

$$|\omega(x, \ldots; x_a, \ldots, z_c)| < \varepsilon \tag{5}$$

bleibt. Solche Medien $\mathfrak{M}_\varepsilon$ betrachten wir nun für eine Schar gegen 0 konvergierender Werte des Parameters $\varepsilon$; nach dem oben ausgesprochenen allgemeinen Stetigkeitspostulat (S. 658) werden dann in der Grenze $\varepsilon = 0$ die Gesetze des Verhaltens von $\mathfrak{M}$ folgen.[124])

$\varphi_\varepsilon$ ist folgendermassen charakterisiert: es hänge ausser von den Deformationsfunktionen und ihren Ableitungen auch noch von dem Ausdruck $\omega$ explizit ab:

$$\varphi_\varepsilon = \varphi_\varepsilon(x, \ldots; x_a, \ldots; \omega(x, \ldots; x_a, \ldots)). \tag{6a}$$

Betrachtet man $\varphi_\varepsilon$ speziell als Funktion des letzten Argumentes $\omega$, so soll $\frac{\partial \varphi_\varepsilon}{\partial \omega}$ mit wachsendem $\omega$ stets wachsen, für $\omega = 0$ identisch in allen andern Argumenten verschwinden und in jedem 0 nicht enthaltenden Intervall für $\lim \varepsilon = 0$ gleichmässig den Grenzwert $\pm\infty$ (je nachdem $\omega \gtrless 0$) haben; ferner soll für den Wert $\omega = 0$ $\varphi_\varepsilon$ gleichmässig in dem in Betracht kommenden Variabilitätsbereich einen Limes haben:

$$\lim_{\varepsilon=0} \varphi_\varepsilon(x, \ldots; x_a, \ldots; 0) = \varphi_0(x, \ldots; x_a, \ldots). \tag{6b}$$

Ein Beispiel einer derartigen Funktion wäre $\varphi_\varepsilon = \varphi_0 + \frac{\omega^2}{2\varepsilon}$.

Die Gleichgewichtsdeformation von $\mathfrak{M}_\varepsilon$, wird nun, unter den betr. Randbedingungen, durch das Variationsprinzip

$$\delta \iiint\limits_{(V_0)} \varphi_\varepsilon(x, \ldots; x_a, \ldots; \omega(x, \ldots; x_a, \ldots)) da\, db\, dc = 0 \tag{7}$$

bestimmt. Zur Vorbereitung des Grenzüberganges dient eine der kano-

[124] Ein solcher Grenzübergang hat offenbar *Lagrange* vorgeschwebt, als er in seiner analytischen Mechanik den zu $\omega = 0$ gehörigen Multiplikator als „Kraft“ bezeichnet, die die Funktion $\omega$ zu ändern bestrebt ist; man vergleiche insbesondere die Sect. II, Nr. 9, Sect. IV, Nr. 6, 18, Sect. V, Nr. 53, Sect. VII, Nr. 21 des ersten Teiles, sowie die Noten von *J. Bertrand* hierzu — näher ausgeführt ist der Übergang indessen nicht. Die Darstellung des Textes ist nach Hinweisen ausgestaltet, die *D. Hilbert* in einer Göttinger Vorlesung im W.-S. 1906/7 für die Behandlung der inkompressiblen Flüssigkeiten gegeben hat.

reality such a medium will never be realized strictly in nature, moreover this here is again only an abstraction from considerations of such media $\mathfrak{M}_\varepsilon$, which *almost* satisfy the relation (4). $\mathfrak{M}_\varepsilon$ may be characterized by a potential of the form No. **7**, (1) with the energy density $\varphi_\varepsilon$, and it shall depend on a parameter $\varepsilon$, such that without exception for every equilibrium position

$$(5) \qquad |\omega(x, \ldots; x_a, \ldots, z_c)| < \varepsilon$$

holds. We consider now such media $\mathfrak{M}_\varepsilon$ for a family of values of the parameter $\varepsilon$ converging to 0; according to the above declared general continuity postulate (p. 658), in the limit $\varepsilon = 0$ the laws of the behavior of $\mathfrak{M}$ will follow.[124)]

$\varphi_\varepsilon$ is characterized as follows: besides the deformation functions and the derivatives thereof it depends also on the expression $\omega$ explicitly:

$$(6a) \qquad \varphi_\varepsilon = \varphi_\varepsilon(x, \ldots; x_a, \ldots; \omega(x, \ldots; x_a, \ldots)).$$

If one considers $\varphi_\varepsilon$ especially as a function of the last argument $\omega$, then with increasing $\omega$, $\frac{\partial \varphi_\varepsilon}{\partial \omega}$ shall increase continuously, for $\omega = 0$ [it shall] vanish identically for all other arguments and for every interval [which] does not contain 0, for $\lim \varepsilon = 0$ [it shall] have uniformly the limit $\pm\infty$ (depending on whether $\omega \gtrless 0$); furthermore, for the value $\omega = 0$, [the function] $\varphi_\varepsilon$ shall have uniformly a limit within the domain of variability coming into consideration.

$$(6b) \qquad \lim_{\varepsilon=0} \varphi_\varepsilon(x, \ldots; x_a, \ldots; 0) = \varphi_0(x, \ldots; x_a, \ldots).$$

An example of such a function would be $\varphi_\varepsilon = \varphi_0 + \frac{\omega^2}{2\varepsilon}$.

The equilibrium deformation of $\mathfrak{M}_\varepsilon$, considering the corresponding boundary conditions, is now determined by the variational principle

$$(7) \qquad \delta \iiint\limits_{(V_0)} \varphi_\varepsilon(x, \ldots; x_a, \ldots; \omega(x, \ldots; x_a, \ldots)) da\, db\, dc = 0$$

For the preparation of the limit process

124 Apparently, *Lagrange* had such a limit process in mind, as he denoted in his analytical mechanics the multiplier associated to $\omega = 0$ as "force", which tries to change the function $\omega$; one shall compare in particular Sect. II, No. 9, Sect. IV, No. 6, 18, Sect. V, No. 53, Sect. VII, No. 21 of the first part, as well as the notes of *J. Bertrand* hereto — meanwhile the transition is not carried out in more detail. The presentation of the text is formulated following suggestions, which *D. Hilbert* has given in a "Göttinger Vorlesung" in the winter term 1906/7 for the treatment of incompressible fluids.

nischen Transformation der Mechanik analoge Transformation[112]): aus

$$\text{(8a)} \qquad \frac{\partial \varphi_\varepsilon(x, \ldots; x_a, \ldots; \omega)}{\partial \omega} = \lambda$$

wird $\omega$ als Funktion von $\lambda$ sowie $x, \ldots; x_a, \ldots$ ausgedrückt:

$$\text{(8b)} \qquad \omega = \overline{\omega}_\varepsilon(x, \ldots; x_a, \ldots; \lambda)$$

und damit der Ausdruck

$$\text{(9)} \qquad \varphi_\varepsilon(x, \ldots; x_a, \ldots; \overline{\omega}) - \overline{\omega}_\varepsilon \cdot \lambda = H_\varepsilon(x, \ldots; x_a, \ldots; \lambda)$$

als Funktion von $\lambda, x, \ldots, x_a, \ldots$ gebildet. Dann folgt aus bekannten Methoden der Variationsrechnung[112]), dass (7) dem Variationsprinzip

$$\text{(10)} \qquad \delta \iiint\limits_{(V_0)} \{H_\varepsilon(x, \ldots; x_a, \ldots; \lambda) + \lambda \cdot \omega(x, \ldots; x_a, \ldots)\} da\, db\, dc = 0$$

für die *vier* unbekannten Funktion $x, y, z, \lambda$ äquivalent ist.

Hierin kann nun der Grenzübergang leicht vollzogen werden; nach den Annahmen über $\varphi_\varepsilon$ konvergiert $\overline{\omega}_\varepsilon(x, \ldots; x_a, \ldots; \lambda)$ mit abnehmendem $\varepsilon$ gleichmässig gegen 0, und da aus (9)

$$\frac{\partial H_\varepsilon(x, \ldots; x_a, \ldots; \lambda)}{\partial \lambda} = -\overline{\omega}$$

folgt, ergiebt sich unter Berücksichtigung von (6b) leicht die gleichmässige Existenz des Limes

$$\text{(11)} \qquad \lim_{\varepsilon=0} H_\varepsilon(x, \ldots; x_a, \ldots; \lambda) = \varphi_0(x, \ldots; x_a, \ldots),$$

der unabhängig von $\lambda$ ist. Also erhält man schliesslich als Grenzfall von (10) das Variationsprinzip

$$\text{(12)} \qquad \delta \iiint\limits_{(V_0)} \{\varphi_0(x, \ldots; x_a, \ldots) + \lambda \cdot \omega(x, \ldots; x_a, \ldots)\} da\, db\, dc = 0;$$

hierin aber kann man endlich $\lambda$ als Lagrangeschen Faktor ansehen und hat damit tatsächlich genau den Ansatz von Nr. **4**c für ein Medium mit der Energiedichte $\varphi_0$ und der Nebenbedingung (4) gewonnen. Obendrein kann man dieser Überlegung noch die Bedeutung des Lagrangeschen Faktors entnehmen: nach (8a) steht $\lambda$ zu der Verbindung $\omega$ der Deformationsgrössen in der gleichen Beziehung, wie die Spannungskomponente $X_a$ zu der Deformationsgrösse $x_a$ (s. Nr. **7**, (4)); es ist also gewissermassen die dieser Verbindung $\omega$ zugehörige Spannungskomponente, genauer: der Faktor von $\delta\omega$ im Ausdruck der virtuellen Arbeit bei einem „nahezu" der Bedingung $\omega = 0$ genügenden Medium.[124]) So sind die aus dem Stattfinden von Nebenbedingungen entspringenden *Reaktionswirkungen* als Grenzfälle den bisher durchgehends betrachteten *eingeprägten Wirkungen* eingeordnet.[125])

---

[125] Vgl. oben Nr. **7**e, S. 654 sowie Anm.[111]).

a transformation is used [which is] analogous to the canonical transformation of mechanics.[112]): Using

$$\text{(8a)} \qquad \frac{\partial \varphi_\varepsilon(x, \ldots; x_a, \ldots; \omega)}{\partial \omega} = \lambda$$

$\omega$ is expressed as a function of $\lambda$ as well as of $x, \ldots; x_a, \ldots$:

$$\text{(8b)} \qquad \omega = \overline{\omega}_\varepsilon(x, \ldots; x_a, \ldots; \lambda)$$

and thereby the expression

$$\text{(9)} \qquad \varphi_\varepsilon(x, \ldots; x_a, \ldots; \overline{\omega}) - \overline{\omega}_\varepsilon \cdot \lambda = H_\varepsilon(x, \ldots; x_a, \ldots; \lambda)$$

as a function of $\lambda, x, \ldots, x_a, \ldots$ is set up. Then from the well-known methods of the calculus of variations[112]) it follows that (7) is equivalent to the variational principle

$$\text{(10)} \qquad \delta \iiint_{(V_0)} \{H_\varepsilon(x, \ldots; x_a, \ldots; \lambda) + \lambda \cdot \omega(x, \ldots; x_a, \ldots)\} da\, db\, dc = 0$$

for the *four* unknown functions $x, y, z, \lambda$.

Herein the limit process can easily be carried out; according to the assumptions on $\varphi_\varepsilon, \overline{\omega}_\varepsilon(x, \ldots; x_a, \ldots; \lambda)$ converges with decreasing $\varepsilon$ uniformly to 0, and since from (9)

$$\frac{\partial H_\varepsilon(x, \ldots; x_a, \ldots; \lambda)}{\partial \lambda} = -\overline{\omega}$$

follows, under consideration of (6b), the uniform existence of the limit

$$\text{(11)} \qquad \lim_{\varepsilon=0} H_\varepsilon(x, \ldots; x_a, \ldots; \lambda) = \varphi_0(x, \ldots; x_a, \ldots),$$

is easily obtained, which is independent of $\lambda$. Hence, one obtains finally as limit case of (10) the variational principle

$$\text{(12)} \qquad \delta \iiint_{(V_0)} \{\varphi_0(x, \ldots; x_a, \ldots) + \lambda \cdot \omega(x, \ldots; x_a, \ldots)\} da\, db\, dc = 0;$$

herein one can consider $\lambda$ finally as Lagrange multiplier and has therewith in fact provided exactly the ansatz of No. **4**c for a medium with energy density $\varphi_0$ and constraint (4). Moreover one can gather from this consideration the relevance of the Lagrange multiplier: according to (8a), $\lambda$ is related to the connection of the deformation quantities $\omega$ in the same sense as the stress components $X_a$ [are related] to the deformation quantities $x_a$ (see No. **7**, (4)); it is in a way the stress component associated to this connection $\omega$, more precisely: the factor of $\delta\omega$ in the expression of the virtual work for a medium "almost" satisfying the constraint $\omega = 0$.[124]) Thus, the *reactive effects* originating from the occurrence of constraints are to be classified as limit cases of the *impressed effects* thoroughly considered so far.[125])

[125] Cf. above No. **7**e, p. 654 as well as remark[111]).

## B. Individualisierung für einzelne Gebiete.

**9. Eigentliche Elastizitätstheorie.** Es handelt sich nun darum aufzuweisen, an welchen Stellen der in Teil A entwickelten allgemeinen Schemata sich die für die Behandlung der einzelnen Disziplinen der Mechanik der Kontinua bisher hauptsächlich verwendeten Ansätze einordnen; beginnen wir mit der Elastizitätslehre im engeren Sinne, die ja dieser ganzen Entwicklung den Weg gewiesen hat.

Ein *vollkommen elastisches Medium* ist dadurch charakterisiert, dass der Spannungszustand in ihm jeweils lediglich abhängt von denjenigen Verbindungen der ersten Ableitungen der Deformationsfunktionen, die die *reine Formänderung* der kleinsten Teile gegenüber der Ausgangslage bestimmen:

$$(1) \qquad e_a = \tfrac{1}{2}(x_a^2 + y_a^2 + z_a^2 - 1), \qquad g_{bc} = x_b x_c + y_b y_c + z_b z_c \qquad (a, b, c);$$

diese Grössen bleiben bei rechtwinkligen Transformationen des $x$-$y$-$z$-Koordinatensystems, auf das die deformierte Lage bezogen ist, einzeln ungeändert, während sie sich bei Transformationen der Anfangskoordinaten $a$, $b$, $c$ wie Komponenten einer symmetrischen Dyade verhalten[126]). Bezieht man sich, wie man es in der Regel tut, auf den Fall der Existenz eines Potentiales $\Phi$ der einfachsten Form von Nr. **7**a, so leiten sich also die inneren Spannungen aus einer Energiedichtenfunktion $\varphi$ her, die lediglich von den 6 Deformationskomponenten (1) abhängt[127]):

$$(2) \qquad \varphi = \varphi(e_a, e_b, e_c, g_{bc}, g_{ca}, g_{ab});$$

dabei ist es irrelevant, ob man die Dichte pro Volumelement des deformierten oder undeformierten Zustandes rechnet, da die event. als Faktor hinzutretende Volumdilatation $\Delta$ selbst lediglich von den Grös-sen (1) abhängt. Aus den Formeln (4), (5) von Nr. **7** entnimmt man nun unmittelbar die verschiedenen Ausdrücke der Spannungs-

[126] Vgl. IV 14, Nr. **17**, **18**, *M. Abraham.*

[127] *G. Green* hat diesen Ansatz zuerst für unendlichkleine Deformationen entwickelt (Trans. Cambr. Phil. Soc. 1838 = Math. Pap., London 1871, p. 248 ff.); später (Trans. Cambr. Phil. Soc. 1839 = Math. Pap., p. 295 ff.) hat er ihn auch für endliche Deformationen ausgesprochen, ohne ihn indessen bis zur Aufstellung der Gleichgewichtsbedingungen durchzuführen. Das hat zuerst *G. Kirchhoff* (Sitzungsber. Wien, math.-phys. Kl. 9 (1852), p. 762) getan, allerdings nur im Hinblick auf isotrope Körper, und später allgemein *W. Thomson* (Phil. Trans. Royal Soc. 153 (1863) = Math. Phys. Pap., London 1910, vol. III, p. 386 = Appendix C. zu Vol. I, 2 des Treat. on natur. philos. von *Thomson* und *Tait*).

## B. Individualization for particular fields.

**9. Effective theory of elasticity.** Now, it is about to exhibit, at which places in the general schemes developed in part A, the fundamentals for the treatment of the particular fields of the mechanics of continua, mainly used so far, are integrated; let us start with the theory of elasticity in the narrower sense, which has pointed this whole development in the right direction.

A *purely elastic medium* is characterized in this way, that the stress state inside [the medium] depends in each case merely on those expressions of the first derivatives of the deformation functions, which determine the *pure shape change* of the smallest parts with respect to the initial position:

$$(1)\qquad e_a = \tfrac{1}{2}(x_a^2 + y_a^2 + z_a^2 - 1), \quad g_{bc} = x_b x_c + y_b y_c + z_b z_c \qquad (a, b, c);$$

each of these quantities remain unchanged for orthogonal transformations of the $x$-$y$-$z$-coordinate system being related to the deformed position, while they behave like the components of a symmetric dyad for transformations of the initial coordinates $a$, $b$, $c$[126]). If one refers, as one usually does, to the case of the existence of a potential $\Phi$ of the most simple form of No. **7**a, then the internal stresses are derived thus from the energy density function $\varphi$, which depends merely on the 6 deformation components (1)[127]):

$$(2)\qquad \varphi = \varphi(e_a, e_b, e_c, g_{bc}, g_{ca}, g_{ab});$$

thereby it is irrelevant, if one computes the density with respect to the volume element of the deformed or undeformed state, since the volume dilatation $\Delta$, appearing possibly as a factor, depends itself merely on the quantities (1). From the formulas (4), (5) of No. **7** one extracts immediately the various expressions of the stress

---

[126] Cf. IV 14, No. **17**, **18**, *M. Abraham*.

[127] *G. Green* has first developed this ansatz for infinitesimal deformations (Trans. Cambr. Phil. Soc. 1838 = Math. Pap., London 1871, p. 248 ff.); later (Trans. Cambr. Phil. Soc. 1839 = Math. Pap., p. 295 ff.) he also stated [this ansatz] for finite deformations, but without carrying out the derivation of the equilibrium conditions. This has been done first by *G. Kirchhoff* (Sitzungsber. Wien, math.-phys. Kl. 9 (1852), p. 762), however, only with respect to isotropic bodies, and later in general by *W. Thomson* (Phil. Trans. Royal Soc. 153 (1863) = Math. Phys. Pap., London 1910, vol. III, p. 386 = Appendix C. to Vol. I, 2 of the Treat. on natur. philos. of *Thomson* and *Tait*).

komponenten, insbesondere wird[128])

$$(3)\quad \begin{cases} X_x = \sum\limits_{(abc)} \frac{\partial\bar{\varphi}}{\partial e_a} x_a^2 + 2 \sum\limits_{(abc)} \frac{\partial\bar{\varphi}}{\partial g_{ab}} x_a x_b + \bar{\varphi} \\ X_y = \sum\limits_{(abc)} \frac{\partial\bar{\varphi}}{\partial e_a} x_a y_a + 2 \sum\limits_{(abc)} \frac{\partial\bar{\varphi}}{\partial g_{ab}} (x_a y_b + x_b y_a), \ldots.., \end{cases}$$

und hieraus oder aus der Bemerkung, dass bei jeder starren Rotation des Mediums sich die $e_a, \ldots$ nicht ändern, also auch $\delta\Phi$ und damit die virtuelle Arbeit der Spannungsdyade verschwindet, folgen die wichtigen Relationen

$$(3')\quad X_y = Y_x, \quad Y_z = Z_y, \quad Z_x = X_z.$$

Es ist nun die Aufgabe der speziellen Elastizitätstheorie zu untersuchen, welche Form die Funktion (2) von sechs Veränderlichen für die einzelnen Medien besitzt; indessen ist dieser allgemeine Fall endlicher Deformationen zugunsten der unendlichkleinen Deformationen in der Elastizitätslehre sehr zurückgetreten[129]). Speziell hervorgehoben sei hier nur der Fall des *isotropen* elastischen Mediums; wegen der Gleichwertigkeit der Richtungen im Medium können dann die 6 Formänderungskomponenten nur durch Vermittlung ihrer 3 Orthogonalinvarianten gegenüber Transformationen des Koordinatensystems $a$, $b$, $c$ in (2) eingehen, d. h. es wird

$$(4)\quad \varphi = \varphi(A, B, C)$$

wo $A$, $B$, $C$ die Koeffizienten der Fundamentalgleichung

$$(4a)\quad \begin{vmatrix} e_a - \Lambda, & \frac{1}{2}g_{ab}, & \frac{1}{2}g_{ac} \\ \frac{1}{2}g_{ab}, & e_b - \Lambda, & \frac{1}{2}g_{bc}, \\ \frac{1}{2}g_{ac}, & \frac{1}{2}g_{bc}, & e_c - \Lambda, \end{vmatrix} \equiv -\Lambda^3 + A\Lambda^2 - B\Lambda + C$$

sind, an deren Stellen natürlich auch die Wurzeln dieser Gleichung (Axenlängen des Deformationsellipsoides) treten können[130]). Diese Formeln umfassen ohne weiteres auch den Fall, dass das Medium im

---

[128] *J. Boussinesq*, Mém. prés. par div. sav., Paris 20 (1872), p. 594; die in 127) zitierten Autoren haben nur die Ausdrücke für die auf die Anfangsparameter bezogenen Spannungskomponenten $X_a, \ldots$. Vgl. auch Chap. III der zusammenfassenden Darstellung von *E.* u. *F. Cosserat*, Ann. de Toul. 10 (1896), p. J. 59.

[129] $\varphi$ als homogene quadratische Funktion der 6 Argumente hat *W. Thomson*, a. a. O.[127]), p. 390, andere für bestimmte Arten der Wellenfortpflanzung charakteristische Gestalten *J. Hamadard*, Leçons sur la propagation des ondes (Paris 1903), p. 257 ff. betrachtet.

[130] Konkrete Ansätze für den isotropen Körper finden sich bei *G. Kirchhoff*, a. a. O.[127]), p. 773 und *M. Brillouin*, C. R. Paris 112 (1891), p. 1500.

components, in particular [it] is[128])

$$(3)\quad \begin{cases} X_x = \sum\limits_{(abc)} \frac{\partial \overline{\varphi}}{\partial e_a} x_a^2 + 2 \sum\limits_{(abc)} \frac{\partial \overline{\varphi}}{\partial g_{ab}} x_a x_b + \overline{\varphi} \\ X_y = \sum\limits_{(abc)} \frac{\partial \overline{\varphi}}{\partial e_a} x_a y_a + 2 \sum\limits_{(abc)} \frac{\partial \overline{\varphi}}{\partial g_{ab}} (x_a y_b + x_b y_a), \ldots\ldots, \end{cases}$$

and herefrom or from the remark, that the $e_a, \ldots$ do not change for every rigid rotation of the medium [and that] therefore also $\delta\Phi$ and consequently the virtual work of the stress dyad vanishes, the important relations follows

$$(3')\qquad X_y = Y_x, \quad Y_z = Z_y, \quad Z_x = X_z.$$

It is now the task of the particular theory of elasticity to study the form of the function (2) of the six variables for the individual media; meanwhile this general case of finite deformations recedes much in the theory of elasticity in favor of the infinitesimal deformations[129]). Here only the case of the *isotropic* elastic medium shall be emphasized especially; due to the equivalence of the directions in the medium, the 6 components of the shape change can enter (2) just by using their 3 orthogonal invariants with respect to transformations of the coordinate system $a$, $b$, $c$, i. e. it becomes

$$(4)\qquad \varphi = \varphi(A, B, C)$$

where $A$, $B$, $C$ are the coefficients of the fundamental equation

$$(4a)\qquad \begin{vmatrix} e_a - \Lambda, & \frac{1}{2} g_{ab}, & \frac{1}{2} g_{ac} \\ \frac{1}{2} g_{ab}, & e_b - \Lambda, & \frac{1}{2} g_{bc}, \\ \frac{1}{2} g_{ac}, & \frac{1}{2} g_{bc}, & e_c - \Lambda, \end{vmatrix} \equiv -\Lambda^3 + A\Lambda^2 - B\Lambda + C,$$

which can be substituted certainly also by the square roots of this equation (lengths of axes of the deformation ellipsoid)[130]). These formulas include readily the case, that the medium

---

128 *J. Boussinesq*, Mém. prés. par div. sav., Paris 20 (1872), p. 594; the authors cited in 127) have only expressions for the stress components with respect to the initial parameters $X_a, \ldots$. Cf. also Chap. III of the summarizing presentation of *E.* and *F. Cosserat*, Ann. de Toul. 10 (1896), p. J. 59.

129 $\varphi$ as a homogeneous quadratic function of the 6 arguments has been considered by *W. Thomson*, op. cit.[127]), p. 390, other forms, being characteristic for certain types of wave propagation, [have been considered] by *J. Hamadard*, Leçons sur la propagation des ondes (Paris 1903), p. 257 ff.

130 Specific approaches for isotropic bodies can be found in *G. Kirchhoff*, op. cit.[127]), p. 773 and *M. Brillouin*, C. R. Paris 112 (1891), p. 1500.

undeformierten Anfangszustand „Selbstspannungen“ aufweist; andernfalls müssen die Spannungskomponenten (3) für verschwindende $e, g$ verschwinden, d. h. es muss die Potenzentwicklung von $\varphi$ nach seinen 6 Argumenten mit quadratischen Gliedern beginnen[131]).

Es sei noch erwähnt, dass *P. Duhem* seinen Potentialansatz (Nr. **7**, (7)), der eine direkte Einwirkung der Deformationszustände an je zwei verschiedenen Stellen aufeinander annimmt, speziell auf isotrope elastische Medien angewandt hat.[132]) Dabei sind die Variablen, die in $\varphi$ eingehen, neben der Entfernung der beiden betrachteten Stellen die $2 \cdot 3$ Invarianten der Formänderung an ihnen sowie die Bestimmungsstücke der Orientierung der Deformationsellipsoide an beiden Stellen gegeneinander und gegen die Verbindungsstrecke.

Die grösste Rolle in der Elastizitätstheorie spielt die Betrachtung *unendlichkleiner Deformationen*. Die ersten, in $\sigma$ linearen Glieder der Formänderungskomponenten (1) sind alsdann in den früheren Bezeichnungen (Nr. **6**, (6)) vom Faktor abgesehen:[126])

$$\text{(5)} \qquad \varepsilon_a = \frac{\partial u}{\partial a}, \quad \gamma_{bc} = \frac{\partial v}{\partial c} + \frac{\partial w}{\partial b} \qquad \begin{pmatrix} a, b, c \\ u, v, w \end{pmatrix};$$

die Funktion $\tilde{\varphi}$ aber, aus der sich gemäss Nr. **7**a, (9) die Spannungskomponenten

$$\text{(6)} \qquad X_a = \frac{1}{\sigma} \frac{\partial \tilde{\varphi}}{\partial u_a}, \dots$$

als lineare Funktion der Verrückungskomponenten ergeben, wird

$$\text{(6a)} \qquad \tilde{\varphi} = \sigma \varphi_1(\varepsilon_a, \gamma_{bc}) + \sigma^2 \varphi_1\left(\frac{u_a^2 + v_a^2 + w_a^2}{2}, u_b u_c + v_b v_c + w_b w_c\right) + \sigma^2 \varphi_2(\varepsilon_a, \gamma_{bc}),$$

wo $\varphi_1$ bzw. $\varphi_2$, die linearen bzw. quadratischen Terme der Potenzentwicklung von $\varphi$ nach seinen 6 Argumenten bedeutet, und der Kürze halber immer nur eines von je 3 Argumenten hingeschrieben ist[133]).

Treten keine Anfangsspannungen auf, so wird $\varphi$ eine quadratische Form der 6 Komponenten der unendlichkleinen Formänderung, und das ist der Fall, der den Ausgangspunkt der gewöhnlichen Elastizitätstheorie bildet (vgl. IV 24, Nr. **1**, (1), *O. Tedone*); dort wird dann insbesondere untersucht, welche Gestalten diese Funktion je nach den Symmetrieeigenschaften, die das Medium in bezug auf die Richtungen

---

[131] Auch für endliche Deformationen bereits angedeutet bei *G. Green*, a. a. O.[127]), p. 298. Vgl. auch *E.* und *F. Cosserat*, a. a. O.[128]), p. J. 70.

[132] *P. Duhem*, Ann. Éc. Norm., (3) 21 (1904), p. 117 ff.

[133] Eine solche Entwicklung benutzte schon *G. Green*, a. a. O.[127]), p. 299. Vgl. auch *H. Poincaré*, Leçons sur la théorie de l’élasticité, Paris 1892, p. 47 ff. sowie *E.* und *F. Cosserat*, a. a. O.[128]), p. J. 73 f.

has "residual stresses" in the undeformed initial state; otherwise for vanishing $e, g$, the stress components (3) must vanish, i. e. the series expansion of $\varphi$ with respect to its six arguments must begin with quadratic terms[131]).

Furthermore, it has to be mentioned, that *P. Duhem* has applied his potential based approach (No. **7**, (7)), which considers a direct effect of the states of deformation at two different points on each other, especially for isotropic elastic media.[132]) Thereby, the variables which enter $\varphi$ are besides the distance between the two considered points, the $2 \cdot 3$ invariants of the shape change [at those positions] as well as the characteristic quantities of the orientation of the deformation ellipsoids at both points with respect to each other and with respect to the connecting line segment.

The biggest issue in the theory of elasticity is the consideration of *infinitesimal deformations*. The first terms of the components of the shape change (1) being linear in $\sigma$ are then, apart from the factor, in the former notations (No. **6**, (6)):[126])

$$\text{(5)} \qquad \varepsilon_a = \frac{\partial u}{\partial a}, \quad \gamma_{bc} = \frac{\partial v}{\partial c} + \frac{\partial w}{\partial b} \qquad \begin{pmatrix} a, b, c \\ u, v, w \end{pmatrix};$$

the function $\tilde{\varphi}$ however, from which according to No. **7**a, (9) the stress components

$$\text{(6)} \qquad X_a = \frac{1}{\sigma} \frac{\partial \tilde{\varphi}}{\partial u_a}, \ldots$$

emerge as linear functions of the displacement components, becomes

$$\text{(6a)} \qquad \tilde{\varphi} = \sigma \varphi_1(\varepsilon_a, \gamma_{bc}) + \sigma^2 \varphi_1\left(\frac{u_a^2 + v_a^2 + w_a^2}{2}, u_b u_c + v_b v_c + w_b w_c\right) + \sigma^2 \varphi_2(\varepsilon_a, \gamma_{bc}),$$

where $\varphi_1$ and $\varphi_2$ denote the linear and quadratic terms in the series expansion of $\varphi$ with respect to its 6 arguments, respectively, and [where] due to the sake of brevity throughout only one of each 3 arguments is written[133]).

If there appear no residual stresses, then $\varphi$ becomes a quadratic form of the 6 components of the infinitesimal shape change, and this is the case which provides the starting point of the ordinary theory of elasticity (cf. IV 24, No. **1**, (1), *O. Tedone*); there it is then studied in particular of what forms this function will be, depending on the symmetry properties which the medium has with respect to the directions

---

131 Also already indicated for finite deformations by *G. Green*, op. cit.[127]), p. 298. Cf. also *E.* and *F. Cosserat*, op. cit.[128]), p. J. 70.

132 *P. Duhem*, Ann. Éc. Norm., (3) 21 (1904), p. 117 ff.

133 Such an expansion has already been used by *G. Green*, op. cit.[127]), p. 299. Cf. also *H. Poincaré*, Leçons sur la théorie de l'élasticité, Paris 1892, p. 47 ff. as well as *E.* and *F. Cosserat*,op. cit.[128]), p. J. 73 f.

durch einen Punkt etwa besitzt, von der allgemeinen Form mit 21 Konstanten (den *Elastizitätskoeffizienten*) bis hin zu der speziellsten mit 2 Konstanten (beim isotropen Medium) annehmen kann (vgl. IV 24, Nr. **2**b, **2**c). In diesem Falle nimmt das transformierte Variationsprinzip (23) von Nr. **7**e eine besonders einfache Form an, indem $H$ bis aufs Vorzeichen gleich der Energiedichte wird; seine Koeffizienten sind die *Elastizitätsmoduln* des Mediums.

Älter als dieser Gedankengang ist eine etwas andere Betrachtungsweise, die die Annahme, dass alle möglichen Deformationen des Mediums unendlichklein seien, mehr in den Vordergrund bringt. Das elastische Medium erscheint hier dadurch charakterisiert, dass sein Potential $\varphi$ lediglich von den Formänderungkomponenten $\varepsilon, \gamma$ der unendlichkleinen Deformation abhängt:[134])

$$\varphi = \varphi(\varepsilon_a, \varepsilon_b, \varepsilon_c, \gamma_{bc}, \gamma_{ca}, \gamma_{ab}), \tag{7}$$

während sich die Spannungskomponenten daraus als Ableitungen nach $u_a, \ldots$ ergeben. Im einfachsten Fall des Mediums ohne Selbstspannungen macht das freilich keinen Unterschied, sofern man sich wieder auf quadratische Glieder beschränkt. Man hat aber diesen Ansatz auch zur Behandlung von Selbstspannungen[135]) und auch zur Erzielung einer über das Hookesche Gesetz hinausgehenden Annäherung an die Naturvorgänge durch Berücksichtigung von Gliedern dritter und höherer Dimension[136]) verwendet; natürlich treffen diese Ansätze dann für andersartige Medien zu als die früheren.

Von einem ein wenig abweichenden Gesichtspunkte aus hat noch *J. Finger*[137]) die Grundformeln der Elastizitätstheorie endlicher Deformationen auszubauen versucht; er zieht nicht nur die Formänderungskomponenten (1) in Betracht, sondern lässt $\varphi$ von allen 9 Ableitungen $x_a, \ldots, z_c$ abhängen, wobei er — für ein isotropes Medium — lediglich symmetrisches Auftreten der drei Koordinatenrichtungen sowie Bestehen der Relationen (3′) voraussetzt und Glieder bis zur dritten Ordnung berücksichtigt.

Auch die Elastizitätslehre der *Körper mit einer oder zwei unendlichkleinen Dimensionen* ordnet sich dem Potentialsatz (für zwei bzw. eine

---

[134] Dies ist der ursprüngliche Ansatz von *G. Green*, a. a. O.[127]), p. 249; vgl. auch IV 23, Nr. **5**b, *Müller-Timpe*.

[135] Vgl. z. B. *H. von Helmholtz*, Dynamik kontinuierlich verbreiteter Massen (Leipzig 1902), p. 93.

[136] *W. Voigt* (Gött. Nachr., 1893, p. 534, math.-phys. Kl. 1894, p. 33; Ann. d. Phys, (3) 52 (1894), p. 536; Kompend. I, p. 339) zieht für isotrope Körper auch die Orthogonalinvariante dritter Ordnung der $\varepsilon, \gamma$ heran.

[137] *J. Finger*, Sitzungsber. Wien $103^{\mathrm{IIa}}$ (1894), p. 163, 231; s. speziell p. 175 ff.

through a point[; is it] of the most general form with 21 constants (the *elasticity constants*) [or is it] up to the most special [form] with 2 constants (for the isotropic medium) (cf. IV 24, No. **2**b, **2**c). In this case the transformed variational principle (23) of No. **7**e is of particular simple form, as $H$ corresponds up to the sign with the energy density; its coefficients are the *elasticity moduli* of the medium.

Older than this line of thought is another perspective, which gives priority to the assumption, that all possible deformations of the medium shall be infinitesimal. The elastic medium appears here to be characterized in this way, that its potential $\varphi$ depends merely on the components of the shape change $\varepsilon, \gamma$ of the infinitesimal deformation:[134])

$$\varphi = \varphi(\varepsilon_a, \varepsilon_b, \varepsilon_c, \gamma_{bc}, \gamma_{ca}, \gamma_{ab}), \tag{7}$$

while the stress components emerge thereout as derivatives with respect to $u_a, \ldots$ . In the most simple case of a medium without residual stresses, this makes certainly no difference, as far as one restricts oneself again to quadratic terms. However, one has also used this ansatz for the treatment of residual stresses[135]) and also for reaching an approximation of the natural processes going beyond Hooke's law by considering also terms of third and higher [polynomial] orders[136]); naturally, these approaches apply for different media than for the former ones.

Starting from a slightly different point of view, *J. Finger*[137]) has tried to extend the basic formulas of the theory of elasticity for finite deformations; He not only considers the components of the shape change (1), but lets also depend $\varphi$ on all 9 derivatives $x_a, \ldots, z_c$, whereby he assumes — for an isotropic medium — merely symmetric appearance of the three coordinate directions as well as the existence of the relation (3′) and [whereby he] considers terms up to third order.

Also the theory of elasticity of *bodies with one or two infinitesimal dimensions* are subordinate to the potential-based approach (for two or one

---

134 This is the original ansatz by *G. Green*, op. cit.[127]), p. 249; cf. also IV 23, No. **5**b, *Müller-Timpe*.

135 Cf. e. g. *H. von Helmholtz*, Dynamik kontinuierlich verbreiteter Massen (Leipzig 1902), p. 93.

136 *W. Voigt* (Gött. Nachr., 1893, p. 534, math.-phys. Kl. 1894, p. 33; Ann. d. Phys, (3) 52 (1894), p. 536; Kompend. I, p. 339) uses for isotropic bodies also the orthogonal invariants of third order of $\varepsilon, \gamma$.

137 *J. Finger*, Sitzungsber. Wien $103^{\mathrm{IIa}}$ (1894), p. 163, 231; see especially p. 175 ff.

Dimension; Nr. **7**c) unter; gegenüber den dreidimensionalen elastischen Medien ist dabei neu das Auftreten höherer Ableitungen der Deformationsfunktionen in der Energiedichte, wie es durch den Grenzübergang von Nr. **8**a erklärt ist. Dies Charakteristikum zeigt sich bereits bei dem Ausdruck des Potentiales

$$\Phi = \int_0^l \varphi\, da \tag{8}$$

der *ebenen Elastika*, d. h. eines an eine Ebene $z = 0$ gebunden gedachten elastischen Drahtes; es wird nämlich $\varphi$ eine Funktion des Krümmungsradius $\varrho$ der Kurve[138])

$$\varphi = \frac{E}{2} \cdot \frac{1}{\varrho^2} = \frac{E}{2} \left\{ \left( \frac{d^2x}{ds^2} \right)^2 + \left( \frac{d^2y}{ds^2} \right)^2 \right\} \tag{9}$$

wofern als Nebenbedingung noch Unausdehnbarkeit der Kurve $(s = a)$ zu der Bedingung des Potentialminimums hinzutritt — andernfalls erhält $\varphi$ noch ein von der Längsdilatation $\frac{ds}{da}$ abhängiges Glied. Durch den Grenzübergang von Nr. **8**b kann man die hier auftretenden Konstanten mit den Elastizitätskonstanten des dreidimensional ausgedehnten Mediums in Zusammenhang bringen.

Bei der *räumlichen Elastika* kommt die oben (S. 659) bereits angedeutete Tatsache hinzu, dass auch die Verschiebung des Materiales des Drahtes gegen die Lage seiner Zentralkurve die Energie beeinflusst. Die nähere Beschreibung geschieht am bequemsten mit Hilfe des Cosseratschen Dreikants. Man denke das jedem Teilchen der Kurve angeheftete rechtwinklige Dreikant in der Ruhelage so orientiert, dass die dritte Axe in die Kurventangente fällt, während die andern beiden die Grenzlagen der Hauptträgheitsaxen des Normalschnittes durch den betr. Punkt bei abnehmender Dicke des Drahtes markieren; fügt man dann noch die Nebenbedingung hinzu, dass die letzte Axe des Dreikants bei jeder Deformation die Kurve tangiert:

$$\alpha_3 : \beta_3 : \gamma_3 = \frac{dx}{ds} : \frac{dy}{ds} : \frac{dz}{ds} = x_a : y_a : z_a, \tag{10}$$

die von der in Nr. **4**c betrachteten Form ist, so hat man ein Cosseratsches Medium, das genau die Elastika darstellt. Der Eigenschaft elastischer Medien, ein nur von den Formänderungskomponenten abhängiges Potential zu besitzen, entspricht hier offenbar die Annahme eines *euklidischen Potentials* im Cosseratschen Sinne (Nr. **7**b, (17)), und

[138] *D. Bernoulli* in einem Brief an *Euler*; *P. H. Fuss*, Cerresp. mathém. et phys., T. II, St. Pétersbourg 1843, p. 507. Vgl. auch *L. Euler*, Methodus inveniendi lineas maximi minimive proprietate gaudentes, Lausannae 1744, im Anhang „de curvis elasticis“.

dimensions; No. **7**c); compared to the three-dimensional elastic media, thereby the appearance of higher derivatives of the deformation functions in the energy density is new, as it is explained by the limit process of No. **8**a. This characteristics arises already in the expression of the potential

$$\Phi = \int_0^l \varphi \, da \tag{8}$$

of the *planar elastica*, i. e. an elastic wire thought to be constraint to the plane $z = 0$; namely, $\varphi$ becomes a function of the curvature radius $\varrho$ of the curve[138])

$$\varphi = \frac{E}{2} \cdot \frac{1}{\varrho^2} = \frac{E}{2}\left\{\left(\frac{d^2x}{ds^2}\right)^2 + \left(\frac{d^2y}{ds^2}\right)^2\right\} \tag{9}$$

provided that the inextensibility of the curve ($s = a$) is added as a constraint to the requirement of the minimum of the potential — otherwise $\varphi$ is augmented by a term depending on the longitudinal dilatation $\frac{ds}{da}$. Due to the limit process of No. **8**b, one can relate the here appearing constants with the elasticity constants of the three-dimensional extended medium.

For the *spatial elastica* the above mentioned (p. 659) fact is added to, that the displacement of the material of the wire with respect to the position of the center curve influences the energy. The detailed description is most conveniently done with the help of the Cosserat triad. One thinks of the orthogonal triad attached to every particle of the curve being oriented in the position of rest such that the third axis coincides with the tangent of the curve, while the other two indicate the border locations of the principal axes of the normal cut through the considered point for decreasing thickness of the wire; if one adds then additionally the constraint, that for every deformation the last axis of the triad is tangent to the curve:

$$\alpha_3 : \beta_3 : \gamma_3 = \frac{dx}{ds} : \frac{dy}{ds} : \frac{dz}{ds} = x_a : y_a : z_a, \tag{10}$$

which is of the form considered in No. **4**c, then one has a Cosserat medium, which just describes the elastica. The property of elastic media, to have a potential depending only on the components of the shape change, corresponds here apparently with the assumption of a *euclidean potential* in the sense of the Cosserats (No. 7b, (17)), and

---

[138] *D. Bernoulli* in a letter to *Euler*; *P. H. Fuss*, Cerresp. mathém. et phys., T. II, St. Pétersbourg 1843, p. 507. Cf. also *L. Euler*, Methodus inveniendi lineas maximi minimive proprietate gaudentes, Lausannae 1744, in the appendix "de curvis elasticis".

da aus (10) und Nr. **7**, (16b) $\mathfrak{x}_a = \mathfrak{y}_a = 0$ folgt, kann die Energiedichte nur noch abhängen von $\mathfrak{z}_a$, das die Dehnung des Drahtes bestimmt, und den Winkelgeschwindigkeitskomponenten $\mathfrak{p}_a, \mathfrak{q}_a, \mathfrak{r}_a$, die die geometrische Krümmung des deformierten Drahtes und den *Drall* (*twist*) des Materiales messen (vgl. IV 25, Nr. **17**, *Tedone-Timpe*)[139]):

$$\varphi = \varphi(\mathfrak{z}_a, \mathfrak{p}_a, \mathfrak{q}_a, \mathfrak{r}_a) \tag{11}$$

Der spezielle Ansatz, der die Theorie der Elastika liefert, ist wiederum der einer quadratischen Form, und zwar — wenn wie oben noch die Nebenbedingung der Unausdehnbarkeit hinzugenommen wird — [140]):

$$\varphi = \frac{E}{2}(J_1 \mathfrak{p}_a^2 + J_2 \mathfrak{q}_a^2) + \frac{C}{2}\mathfrak{r}_a^2; \tag{12}$$

hierbei sind $E, J_1, J_2, C$ Materialkonstanten; ist speziell $J_1 = J_2$ (was einem kreisförmigen Querschnitt des Drahtes entspricht), so tritt wie in (9) die Krümmung $\frac{1}{\varrho^2} = \mathfrak{p}^2 + \mathfrak{q}^2$ der Kurve auf. Durch Annahmen ähnlicher Art über die Verknüpfung der Lage des Dreikants mit der Kurve kann man von dem gleichen Ansatz aus auch alle übrigen in der Elastizitätstheorie behandelten Typen von Stäben, Drähten, Fäden darstellen, wie das *E.* und *F. Cosserat* [141]) ausführlich entwickelt haben.

Ganz analoge Betrachtungen gelten für die Theorie der *Platten*; sie seien hier nur kurz angedeutet. Man kann die Platte ansehen als ein zweidimensionales Medium mit orientierten Teilchen, deren Dreikante mit der dritten Axe stets normal zu der vom Medium jeweils erfüllten Fläche stehen sollen[142]); dann ist $\mathfrak{z}_a = \mathfrak{z}_b = 0$ und

$$\varphi = \varphi(\mathfrak{x}_a, \mathfrak{y}_a, \mathfrak{x}_b, \mathfrak{y}_b; \mathfrak{p}_a, \ldots, \mathfrak{r}_c)$$

hängt von der Dehnung und Krümmung der deformierten Fläche und der inneren Verwindung der Materie auf ihr in orthogonal invarianter Weise ab. Auch hier haben *E.* und *F. Cosserat*[143]) im einzelnen ausgeführt, wie man daraus den Energieansatz für die übliche Näherungstheorie der elastischen Platte[144]) sowie überhaupt für alle behandelten Typen elastischer Platten, Membrane und Schalen[145]) herleiten kann.

**10. Dynamik idealer Flüssigkeiten.** Die idealen Flüssigkeiten ordnen sich ohne weiteres als Sonderfall denjenigen elastischen Medien

---

[139] *E.* und *F. Cosserat*, Corps déformables, p. 37 ff.

[140] *Thomson-Tait*, natural philos., new ed. I 2, p. 133 ff.; dort wird auch ein allgemeinerer quadratischer Ansatz in Betracht gezogen. Vgl. auch IV 25, Nr. **17**.

[141] *E.* und *F. Cosserat*, Corps déformables, Nr. 15—28.

[142] Vgl. *E.* und *F. Cosserat*, Corps déformables, p. 105 ff.

[143] *E.* und *F. Cosserat*, Corps déformables, Nr. 41—46.

[144] *Thomson-Tait*, a. a. O.[140]) I 2, p. 184 ff.

[145] Vgl. insbesondere die Angaben in IV 26, Nr. **5**, *H. Lamb*.

since from (10) and No. **7**, (16b) $\mathfrak{x}_a = \mathfrak{y}_a = 0$ follows, the energy density can only depend on $\mathfrak{z}_a$, which determines the elongation of the wire, and the components of the angular velocities $\mathfrak{p}_a, \mathfrak{q}_a, \mathfrak{r}_a$, which measure the geometric curvature of the deformed wire and the twist (*Drall*) of the material (cf. IV 25, No. **17**, *Tedone-Timpe*)[139]):

$$(11) \qquad \varphi = \varphi(\mathfrak{z}_a, \mathfrak{p}_a, \mathfrak{q}_a, \mathfrak{r}_a)$$

The special ansatz, which is provided by the theory of the elastica, is again the one of a quadratic form, and indeed — when as above additionally the inextensibility constraint is added — [140]):

$$(12) \qquad \varphi = \frac{E}{2}(J_1 \mathfrak{p}_a^2 + J_2 \mathfrak{q}_a^2) + \frac{C}{2}\mathfrak{r}_a^2;$$

hereby $E, J_1, J_2, C$ are material constants; If especially $J_1 = J_2$ (what corresponds to a circular cross section of the wire), then the curvature $\frac{1}{\varrho^2} = \mathfrak{p}^2 + \mathfrak{q}^2$ of the curve appears as in (9). By assumptions similar in kind about the relation between the position of the triad and the curve, one can represent all other types of bars, wires, strings treated in the theory of elasticity, how it has been developed extensively by *E.* and *F. Cosserat* [141]).

Very similar considerations apply to the theory of *plates*; which are treated here only briefly. One can consider a plate as a two-dimensional medium with oriented particles, whose triads shall stand [in such a way, that] the third axis stands always normal to the surface filled with the medium[142]); then $\mathfrak{z}_a = \mathfrak{z}_b = 0$ and

$$\varphi = \varphi(\mathfrak{x}_a, \mathfrak{y}_a, \mathfrak{x}_b, \mathfrak{y}_b; \mathfrak{p}_a, \ldots, \mathfrak{r}_c)$$

depends on the stretch and the curvature of the deformed surface and the internal twisting of the matter on [the surface] in an orthogonal invariant manner. Also here, *E.* and *F. Cosserat* [143]) have carried out in detail, how one can derive thereout the energy theorem for the common approximation theory of the elastic plate[144]) as well as anyway for all types of elastic plates, membranes and shells[145]).

**10. Dynamics of ideal fluids.** The ideal fluids are subordinate readily as a special case of those elastic media

[139] *E.* and *F. Cosserat*, Corps déformables, p. 37 ff.

[140] *Thomson-Tait*, natural philos., new ed. I 2, p. 133 ff.; there also a more general quadratic ansatz is considered. Cf. also IV 25, No. **17**.

[141] *E.* and *F. Cosserat*, Corps déformables, No. 15—28.

[142] Cf. *E.* and *F. Cosserat*, Corps déformables, p. 105 ff.

[143] *E.* and *F. Cosserat*, Corps déformables, No. 41—46.

[144] *Thomson-Tait*, op. cit.[140]) I 2, p. 184 ff.

[145] Cf. in particular the statement in IV 26, No. **5**, *H. Lamb*.

unter, die beliebige endliche Deformationen gestatten; sie sind dadurch charakterisiert, dass Arbeit nur für solche Deformationen aufgewendet werden muss, die mit einer Volumendilatation oder Kompression der kleinsten Teilchen verbunden sind[146]), und dass also die Energiedichte $\varphi$ allein von der durch die Funktionaldeterminante $\Delta$ gemessenen momentanen Volumdilatation an jeder Stelle abhängt[147]):

$$\varphi = \varphi(\Delta) \tag{1}$$

Da $\Delta$ als Orthogonalinvariante der Deformation eine Funktion der Grössen $A, B, C$ (Nr. **9**, (4a)) ist ($\Delta^2 = 1 + 2A + 4B + 8C$), so ist (1) tatsächlich nur ein spezieller Fall des Ansatzes (4) von Nr. **9**. Aus Nr. **7**, (5) ergeben sich leicht als Komponenten der zu diesem Potential gehörigen inneren Spannung:

$$\begin{cases} X_x = Y_y = Z_z = \frac{d\varphi}{d\Delta} = p, \\ X_y = Y_x = Y_z = Z_y = Z_x = X_z = 0, \end{cases} \tag{2}$$

d. h. die Spannungsdyade bestimmt einen in jeder Richtung gleichmässig wirkenden „*Flüssigkeitsdruck*“ $p$. Man erhält dasselbe Resultat auch direkter[148]), wenn man mit Hilfe der Relation (vgl. Nr. **2**, (8′))

$$\delta\Delta = \Delta \cdot \left(\frac{\partial \delta x}{\partial x} + \frac{\partial \delta y}{\partial y} + \frac{\partial \delta z}{\partial z}\right)$$

die Variation des Gesamtpotentiales $\iiint \varphi dV_0$ bestimmt und sie dem Ausdruck Nr. **3**, (1) der virtuellen Arbeit gleich setzt. — Diese Überlegungen gelten sowohl für die Hydrostatik als für die Hydrodynamik; durch Einsetzen von (2) in die Gleichungen von Nr. **3**c bzw. Nr. **5**a ergeben sich die bekannten Grundgleichungen.

Die Gleichung (1) erscheint in der Hydrodynamik gewöhnlich in einer etwas anderen Gestalt. Da nämlich $\Delta$ umgekehrt proportional der Dichte $\varrho$ des Mediums ist (Nr. **2**, (7)), so kann man sagen, dass sie $\varphi$ als Funktion von $\varrho$ giebt, und damit ist nach (2) auch der Druck als Funktion von $\varrho$ gegeben:

$$p = \frac{d\varphi}{d\Delta} = p(\varrho); \tag{3}$$

umgekehrt ist durch (3) auch die Relation (1) im wesentlichen bestimmt. In der Form (3) wird die „Zustandsgleichung“ der Hydrodynamik gewöhnlich gegeben[149]).

---

[146] *Lagrange*, Méc. anal., 1. part., sect. VIII, Nr. **1**

[147] *J. Hadamard*, Leçons sur la propagation des ondes (Paris 1903), p. 247 ff.

[148] Dies ist im wesentlichen das Verfahren von *Lagrange*[146]); vgl. auch sect. VII, Nr. 11.

[149] Vgl. die näheren Angaben in IV 15, Nr. **5**, *Love*.

which allow for arbitrary finite deformations; they are characterized in this way, that only work must be expended for such deformations which are related to the volume dilatation or compression of the smallest particles[146]), and that the energy density $\varphi$ depends consequently only on the current volume dilatation measured by the Jacobian $\Delta$ at every point[147]):

$$\varphi = \varphi(\Delta) \tag{1}$$

Since $\Delta$ is as an orthogonal invariant of the deformation, a function of the quantities $A, B, C$ (No. **9**, (4a)) ($\Delta^2 = 1 + 2A + 4B + 8C$), (1) is in fact only a special case of the ansatz (4) of No. **9**. From No. **7**, (5) the components of the internal stress related to this potential easily emerge as:

$$\begin{cases} X_x = Y_y = Z_z = \frac{d\varphi}{d\Delta} = p, \\ X_y = Y_x = Y_z = Z_y = Z_x = X_z = 0, \end{cases} \tag{2}$$

i.e. the stress dyad determines a *"fluid pressure"* $p$ acting uniformly in every direction. One obtains the same result also more directly[148]), when one determines with the help of the relation (cf. No. **2**, (8′))

$$\delta\Delta = \Delta \cdot \left( \frac{\partial \delta x}{\partial x} + \frac{\partial \delta y}{\partial y} + \frac{\partial \delta z}{\partial z} \right)$$

the variation of the total potential $\iiint \varphi dV_0$ and by equating [the variation] with the expression No. **3**, (1) of the virtual work. — These considerations are valid both for hydrostatics and for hydrodynamics; Using (2) in the equations of No. **3**c and No. **5**a, respectively, the well-known fundamental equations are obtained.

In hydrodynamics, equation (1) appears usually in a slightly different form. Since $\Delta$ is namely inversely proportional to the density $\varrho$ of the medium (No. **2**, (7)), one can say, that [equation (1)] gives $\varphi$ as a function of $\varrho$, so that according to (2) also the pressure is given as a function of $\varrho$:

$$p = \frac{d\varphi}{d\Delta} = p(\varrho); \tag{3}$$

the other way round, using (3), also the relation (1) is determined essentially. Usually, the "equation of state" of hydrodynamics is given in the form (3)[149]).

---

[146] *Lagrange*, Méc. anal., 1. part., sect. VIII, No. **1**

[147] *J. Hadamard*, Leçons sur la propagation des ondes (Paris 1903), p. 247 ff.

[148] This is basically the procedure of *Lagrange*[146]); cf. also sect. VII, No. 11.

[149] Cf. the details in IV 15, No. **5**, *Love*.

Eine grosse Rolle spielt bekanntlich der Fall der *inkompressiblen Flüssigkeit*, die durch die Nebenbedingung

$$\Delta = 1 \tag{4}$$

charakterisiert ist. Für ein solches Medium verliert die Zustandsgleichung (1) ihre Bedeutung; approximiert man es aber nach Nr. **8**b durch ein „nahezu inkompressibles“ Medium, so wird der Druck $p = \frac{d\varphi}{d\Delta}$ in der Grenze zum Lagrangeschen Faktor der Gleichung (4), wenn man sie direkt als Nebenbedingung dem Prinzip der virtuellen Arbeit oder dem d'Alembertschen Prinzip hinzufügt, in das dann freilich nur noch äussere bzw. Trägheitskräfte, keine inneren Spannungen mehr eingehen[150]).

Die üblichen Darstellungen der Hydrodynamik gehen meist nicht von dieser Auffassung der Flüssigkeitsbewegung als einer der Elastizitätslehre einzuordnenden endlichen Deformation aus, sondern stellen die sog. *Euler*sche Auffassung in den Vordergrund, d. h. die Betrachtung des Geschwindigkeitsvektors $x', y', z'$ an jeder Stelle. Der Flüssigkeitsdruck wird dann direkt gemäss den Gleichungen (2) zwischen den Spannungskomponenten definiert[151]) und die Bewegungsgleichungen aus dem d'Alembertschen oder aus dem Gaussschen Prinzip[152])

$$\delta \iiint\limits_{(V)} \tfrac{1}{2}\varrho \sum_{(x,y,z)} (x'' - X)^2 dV - \iiint\limits_{(V)} p \sum_{(x,y,z)} \frac{\partial \delta x''}{\partial x} dV = 0$$

gewonnen — bei Inkompressibilität wird $p$ Lagrangescher Faktor.

Auch speziell der Hydrodynamik hat *P. Duhem*[153]) seinen verallgemeinerten Potentialansatz Nr. **7**, (7) angepasst, indem er die Energiedichte $\varphi$ von den Dichtigkeiten an beiden betrachteten Stellen und deren Entfernung abhängen lässt; damit umfasst und verallgemeinert er Kräfte, die *H. A. E. Faye* [154]) zur Erklärung der Kometenschweife in Betracht gezogen hat, nämlich Attraktionskräfte, deren Intensität von der Dichte der wirkenden Teilchen abhängt.

**11. Innere Reibung und elastische Nachwirkung.** Bei *bewegten* elastischen Medien und Flüssigkeiten treten neben den bisher erörterten Spannungen und Drucken noch Zusatzspannungen auf, die

[150] In *Lagranges* Darstellung ist die inkompressible Flüssigkeit das primäre; man vgl. jedoch die Bemerkung in 124) (Nr. **8**).

[151] Das entspricht der Auffassung von *Euler*; vgl. IV 15, Nr. **2**, **8**, *Love*.

[152] Vgl. die ausführliche Darstellung von *A. Brill*, Mechanik raumerf. Massen[64]), p. 84 ff.

[153] *P. Duhem*, Ann. Éc. Norm. (3) 10 (1893), p. 183.

[154] *H. A. E. Faye*, Paris C. R. 47 (1858), p. 939. 1043.

As is generally known, the case of the *incompressible fluid* being characterized by the constraint

$$\Delta = 1 \tag{4}$$

looms large. For such a medium the equation of state (1) looses its meaning; however, if one approximates it according to No. **8**b as a "nearly incompressilbe" medium, the pressure $p = \frac{d\varphi}{d\Delta}$ becomes in the limit the Lagrangian multiplier of equation (4), if one adds [the equation] directly as a constraint to the principle of virtual work or to the principle of d'Alembert, in which then certainly only external and inertial forces and no more internal stresses enter[150]).

The common presentations of hydrodynamics do not start with this perception of the fluid motion as a finite deformation subordinated to the theory of elasticity, but give priority to the so called *Euler*ian perception, i. e. the consideration of the velocity vector $x', y', z'$ at every point. According to the equations (2), the fluid pressure is then directly defined between the stress components[151]) and the equations of motion are gained from the principle of d'Alembert or the principle of Gauss[152])

$$\delta \iiint\limits_{(V)} \tfrac{1}{2}\varrho \sum_{(x,y,z)} (x'' - X)^2 dV - \iiint\limits_{(V)} p \sum_{(x,y,z)} \frac{\partial \delta x''}{\partial x} dV = 0$$

— for incompressibility $p$ becomes a Lagrange multiplier.

Also especially for hydrodynamics *P. Duhem*[153]) has adapted his generalized potential-based approach No. **7**, (7), by letting the energy density $\varphi$ depend on the densities at both considered points and the distance between them; thereby he includes and generalizes forces, which *H. A. E. Faye* [154]) has taken into consideration for the explanation of the cometary train, in fact, attractive forces, whose intensity depend on the density of the acting particles.

**11. Internal friction and elastic hysteresis.** For *moving* elastic media and fluids there appear besides the so far discussed stresses and pressures also additional stresses, which

150 In *Lagrange*'s presentation the incompressible fluid is the primitive; however, one cf. the remark in 124) (No. **8**).

151 This corresponds to the perception of *Euler*; cf. IV 15, No. **2**, **8**, *Love*.

152 Cf. the extensive presentation of *A. Brill*, Mechanik raumerf. Massen[64]), p. 84 ff.

153 *P. Duhem*, Ann. Éc. Norm. (3) 10 (1893), p. 183.

154 *H. A. E. Faye*, Paris C. R. 47 (1858), p. 939. 1043.

durch innere Reibungen hervorgerufen werden, die also von den zeitlichen Ableitungen der Deformationsgrößen abhängen[155]). Verwendet man zur Darstellung der Bewegung nach der sog. Eulerschen Manier die Geschwindigkeitskomponenten als Funktionen des augenblicklichen Ortes jedes Teilchens

$$(1) \qquad \mathfrak{u} = x' = \mathfrak{u}(x, y, z; t) \qquad \begin{pmatrix} x, y, z \\ \mathfrak{u}, \mathfrak{v}, \mathfrak{w} \end{pmatrix},$$

so können die 9 Ableitungen $x'_a, \dots, z'_c$, die oben (Nr. **6**, S. 640 und Nr. **7**f, S. 657) verwendet wurden, auch ersetzt werden durch die 9 Ableitungen $\mathfrak{u}_x, \mathfrak{u}_y, \dots, \mathfrak{w}_z$ die lineare Funktionen von ihnen sind. Die Funktionen (1) bestimmen die unendlichkleine Deformation, die das Medium vermöge der Bewegung in einem Zeitelement erleidet; die Komponenten der zugehörigen reinen Formänderung (vgl. Nr. **9**, (5)) sind:

$$(2) \qquad \mathfrak{e}_x = \frac{\partial \mathfrak{u}}{\partial x}, \quad \mathfrak{g}_{yz} = \frac{\partial \mathfrak{w}}{\partial y} + \frac{\partial \mathfrak{v}}{\partial z} \qquad \begin{pmatrix} x, y, z \\ \mathfrak{u}, \mathfrak{v}, \mathfrak{w} \end{pmatrix},$$

und von diesen Grössen wird daher die innere Reibung allein abhängen, sofern man analog zu den Verhältnissen bei elastischen Medien annimmt, dass durch Drehgeschwindigkeiten der Teilchen keine Reibungswiderstände entstehen können[156]). Man erkennt übrigens leicht, dass die Komponenten der durch (2) bestimmten symmetrischen Dyade in Bezug auf das $a$-$b$-$c$-Koordinatensystem gerade die zeitlichen Ableitungen der Formänderungskomponenten Nr. **9**, (1), sind.

Die Theorie der Reibungskräfte bei endlichen Deformationen ist bisher nur in der Hydrodynamik vollständig ausgebildet; die Grundannahme dabei ist die der Existenz einer Dissipationsfunktion $D$, die eine homogene quadratische Funktion der Grössen (2) ist, und die obendrein — entsprechend der isotropen Konstitution der Flüssigkeit — nur von deren Orthogonalinvarianten abhängt[157]):

$$(3) \qquad D = a_1(\mathfrak{e}_x + \mathfrak{e}_y + \mathfrak{e}_z)^2 + a_2(\mathfrak{e}_x^2 + \mathfrak{e}_y^2 + \mathfrak{e}_z^2 + \tfrac{1}{2}(\mathfrak{g}_{xy}^2 + \mathfrak{g}_{yz}^2 + \mathfrak{g}_{zx}^2)).$$

Nach Nr. **7**f, (29′) und nach Nr. **3**c, (8) werden die zugehörigen, auf den deformierten Zustand bezogenen Spannungskomponenten

$$(4) \qquad \begin{cases} X_x^{(1)} = \frac{1}{2} \sum\limits_{(abc)} \frac{\partial D}{\partial x'_a} x_a = \frac{1}{2} \frac{\partial D}{\partial \mathfrak{u}_x} = a_1(\mathfrak{e}_x + \mathfrak{e}_y + \mathfrak{e}_z) + a_2 \mathfrak{e}_x \\ X_y^{(1)} = \frac{1}{2} \sum\limits_{(abc)} \frac{\partial D}{\partial x'_a} y_a = \frac{1}{2} \frac{\partial D}{\partial \mathfrak{u}_y} = \frac{1}{2} a_2 \mathfrak{g}_{xy}, \dots \end{cases}$$

---

[155] Vgl. die historischen Angaben zu Nr. **12** von IV 15, *Love*.

[156] *G. G. Stokes*, Cambr. Phil. Soc. Trans. 8 (1845) = Math. Phys. Papers **I**, p. 80.

[157] *W. Voigt*, Kompendium I, p. 462 ff.; einen allgemeineren Ansatz giebt *P. Duhem* Anm. Éc. Norm. (3) 21 (1904), p. 130 ff.

are caused by internal friction, and therefore depend on the time derivatives of the deformation quantities[155]). If one uses the velocity components as functions of the actual position of every particle for the representation of the motion in the sense of Euler

$$(1) \qquad \mathfrak{u} = x' = \mathfrak{u}(x, y, z; t) \qquad \begin{pmatrix} x, y, z \\ \mathfrak{u}, \mathfrak{v}, \mathfrak{w} \end{pmatrix},$$

then the 9 derivatives $x'_a, \ldots, z'_c$, which have been used above (No. **6**, p. 640 und No. **7**f, p. 657) can be substituted by the 9 derivatives $\mathfrak{u}_x, \mathfrak{u}_y, \ldots, \mathfrak{w}_z$ which are linear functions of the former [derivatives]. The functions (1) determine the infinitesimal deformation, which the medium undergoes due to the motion during one time element; the components of the associated pure shape change (cf. No. **9**, (5)) are:

$$(2) \qquad \mathfrak{e}_x = \frac{\partial \mathfrak{u}}{\partial x}, \quad \mathfrak{g}_{yz} = \frac{\partial \mathfrak{w}}{\partial y} + \frac{\partial \mathfrak{v}}{\partial z} \qquad \begin{pmatrix} x, y, z \\ \mathfrak{u}, \mathfrak{v}, \mathfrak{w} \end{pmatrix},$$

and from these quantities the internal friction will merely depend on, as long as one considers analogously to the conditions in elastic media, that no frictional resistances can occur due to angular velocities of the particles[156]). By the way, one recognizes easily, that the components of the symmetric dyad determined by (2) with respect to the *a*-*b*-*c*-coordinate system just correspond to the time derivatives of the shape change components of No. **9**, (1).

So far, the theory of friction forces for finite deformations is developed completely only in hydrodynamics; the basic assumption thereby is the one of the existence of a dissipation function $D$, which is a homogeneous quadratic function of the quantities (2), and which moreover — according to the isotropic constitution of the fluid — depends merely on the orthogonal invariants[157]):

$$(3) \qquad D = a_1(\mathfrak{e}_x + \mathfrak{e}_y + \mathfrak{e}_z)^2 + a_2(\mathfrak{e}_x^2 + \mathfrak{e}_y^2 + \mathfrak{e}_z^2 + \tfrac{1}{2}(\mathfrak{g}_{xy}^2 + \mathfrak{g}_{yz}^2 + \mathfrak{g}_{zx}^2)).$$

In accordance with No. **7**f, (29′) and No. **3**c, (8) the corresponding stress components with respect to the deformed state become

$$(4) \qquad \begin{cases} X_x^{(1)} = \frac{1}{2} \sum\limits_{(abc)} \frac{\partial D}{\partial x'_a} x_a = \frac{1}{2} \frac{\partial D}{\partial \mathfrak{u}_x} = a_1(\mathfrak{e}_x + \mathfrak{e}_y + \mathfrak{e}_z) + a_2 \mathfrak{e}_x \\ X_y^{(1)} = \frac{1}{2} \sum\limits_{(abc)} \frac{\partial D}{\partial x'_a} y_a = \frac{1}{2} \frac{\partial D}{\partial \mathfrak{u}_y} = \frac{1}{2} a_2 \mathfrak{g}_{xy}, \ldots \end{cases}$$

155 Cf. the historical details in No. **12** of IV 15, *Love*.

156 *G. G. Stokes*, Cambr. Phil. Soc. Trans. 8 (1845) = Math. Phys. Papers I, p. 80.

157 *W. Voigt*, Kompendium I, p. 462 ff.; a more general Ansatz is given by *P. Duhem* Anm. Éc. Norm. (3) 21 (1904), p. 130 ff.

Diese Spannungen treten also als Einfluss der inneren Reibung zu dem Flüssigkeitsdruck hinzu; häufig spezialisiert man die beiden Konstanten noch durch die Annahme, dass das arithmetische Mittel der drei resultierenden Normaldrucke $p + X_x^{(1)}, p + Y_y^{(1)}, p + Z_z^{(1)}$ gleich $p$ ist, was $a_1 = -\frac{a_2}{3}$ ergiebt[158]).

In der eigentlichen Elastizitätslehre hat man die innere Reibung nur erst für unendlichkleine Deformationen in Betracht gezogen. In diesem Falle unterscheiden sich die Grössen $\mathfrak{u}, \dots; \frac{\partial \mathfrak{u}}{\partial x}, \dots; \mathfrak{e}_x, \dots; \mathfrak{g}_{xy}, \dots$ nur durch den Faktor $\sigma$ von $u', \dots;$ $u'_a, \dots; \varepsilon'_a, \dots; \gamma'_{bc}, \dots$ (in der Bezeichnung von Nr. **6**, (6) und Nr. **9**, (5)), und demgemäss wird die Dissipationsfunktion eine quadratische, die Spannungskomponenten also lineare Formen der zeitlichen Ableitungen der Formänderungskomponenten der unendlichkleinen Deformation. *W. Voigt*[159]) hat die Abhängigkeiten, die hier auftreten können, eingehend untersucht.

In naher Beziehung zu diesen Ansätzen stehen die Versuche, die Erscheinungen der *elastischen Nachwirkung* im Rahmen der Mechanik der Kontinua theoretisch zu fassen, die freilich bisher an den grossen Komplex der hier zu umspannenden Tatsachen noch nicht vollständig herangekommen sind[160]). Typisch ist hier in erster Linie der Ansatz *L. Boltzmanns*[161]), der den elastischen Spannungskomponenten ein Zeitintegral von der in Nr **6**, (5) erörterten Form hinzufügt; er nimmt dabei — was natürlich nur für unendlichkleine Deformationen gilt — den Integranden als lineare Funktion der Formänderungskomponenten Nr. **9**, (5) an von analoger Form, wie sie die Spannungskomponenten im isotropen Medium haben:

$$(5) \quad \begin{cases} X_x = \int\limits_{-\infty}^{t} [a_1(t-\tau)\{\varepsilon_a(\tau) + \varepsilon_b(\tau) + \varepsilon_c(\tau)\} + 2a_2(t-\tau)\cdot\varepsilon_a(\tau)]d\tau, \\ X_y = \int\limits_{-\infty}^{t} a_1(t-\tau)\gamma_{ab}(\tau)d\tau. \end{cases}$$

*E. Wiechert*[162]) hat diese Formeln durch spezielle Annahmen über die

[158] *Stokes*, a. a. O.[156]). Vgl. auch IV 15, *Love*, Nr. **12—14** und für jene Relationen *H. Lamb*, Hydrodynamik (deutsche Ausg. Leipzig 1907), § 314.

[159] Abhandl. Ges. d. Wiss. Göttingen 36 (1889); Kompendium I, p. 456 ff., 467 ff.; Lehrbuch der Krystallphysik, Leipzig 1910, p. 792 ff.

[160] Vgl. IV 31, Nr. **13** u. **19** (*v. Kármán*).

[161] Ann. d. Phys. u. Chem., Ergänzungsb. 7 (1876), p. 630.

[162] Ann. d. Phys. (3) 50 (1893), p. 335.

Anyway, these stresses are added to the fluid pressure as influence of the internal friction; often one specifies the two constants in addition with the assumption that the arithmetic average of the three resulting normal pressures $p+X_x^{(1)}, p+Y_y^{(1)}, p+Z_z^{(1)}$ is equal to $p$, what results in $a_1 = -\frac{a_2}{3}$ [158]).

In the effective theory of elasticity, so far, one has taken into consideration the internal friction only for infinitesimal deformations. In this case the quantities $\mathfrak{u}, \ldots; \frac{\partial \mathfrak{u}}{\partial x}, \ldots; \mathfrak{e}_x, \ldots; \mathfrak{g}_{xy}, \ldots$ and $u', \ldots; u'_a, \ldots; \varepsilon'_a, \ldots; \gamma'_{bc}, \ldots$ (in the notation of No. **6**, (6) and No. **9**, (5)) differ only by the factor $\sigma$, and accordingly the dissipation function becomes a quadratic [form and] the stress components [become] consequently linear forms of the time derivatives of the shape change components of the infinitesimal deformation. *W. Voigt*[159]) has thoroughly studied the dependences which can occur here.

In close relation to these approaches are the efforts to theoretically conceive the appearance of the *elastic hysteresis* in the context of the mechanics of continua, which however have not reached yet completely the large set of issues being treated here[160]). Typically is here in the first place the ansatz of *L. Boltzmann*[161]), which adds to the elastic stress components a time integral of the form as discussed in No. **6**, (5); thereby he assumes — what is certainly only valid for infinitesimal deformations — the integrand as linear function of the shape change components No. **9**, (5) to be of similar form as the stress components are in the isotropic medium:

$$(5)\quad \begin{cases} X_x = \int\limits_{-\infty}^{t} [a_1(t-\tau)\{\varepsilon_a(\tau)+\varepsilon_b(\tau)+\varepsilon_c(\tau)\} + 2a_2(t-\tau)\cdot\varepsilon_a(\tau)]d\tau, \\ X_y = \int\limits_{-\infty}^{t} a_1(t-\tau)\gamma_{ab}(\tau)d\tau. \end{cases}$$

*E. Wiechert*[162]) has developed these formulas by special assumptions on the

158 *Stokes*, op. cit.[156]). Cf. also IV 15, *Love*, No. **12—14** and for the latter relation *H. Lamb*, Hydrodynamik (German edition Leipzig 1907), § 314.

159 Abhandl. Ges. d. Wiss. Göttingen 36 (1889); Kompendium I, p. 456 ff., 467 ff.; Lehrbuch der Krystallphysik, Leipzig 1910, p. 792 ff.

160 Cf. IV 31, No. **13** and **19** (*v. Kármán*).

161 Ann. d. Phys. u. Chem., Ergänzungsb. 7 (1876), p. 630.

162 Ann. d. Phys. (3) 50 (1893), p. 335.

Funktionen $a_1, a_2$ von $t - \tau$ ausgestaltet. Eine Reihe hierin gehöriger Probleme hat neuerdings *V. Volterra* behandelt[163]) (vgl. S. 641).

Für den Fall bleibender Formänderungen, für plastische Medien also, haben *A. Haar* und *Th. v. Kármán*[164]) aus ganz andern Gesichtspunkten Ansätze abgeleitet. Sie gehen aus von dem Variationsprinzip Nr. **7**, (23), in dem (vgl. S. 655) für isotrope Medien $H$ die Energiedichte und gleich einer homogenen quadratischen Funktion der ersten beiden Orthogonalinvarianten der (symmetrischen) Spannungsdyade wird:

$$(6) \qquad 2H = a_1(X_x + Y_y + Z_z)^2 + a_2(X_y^2 + Y_z^2 + Z_x^2 - X_xY_y - Y_yZ_z - Z_zX_x)$$

Zu diesem Variationsproblem mit seinen drei Nebenbedingungen (23a), Nr. **7** tritt nun als die für plastische Medien charakteristische Eigenschaft die Bedingung hinzu, dass die grösste irgendwo auftretende Schubspannung einen festen Wert $K$ nicht überschreitet, d. h. dass die Differenzen je zweier Wurzeln der Gleichung

$$\begin{vmatrix} X_x - \Lambda, & X_y, & X_z \\ Y_x, & Y_y - \Lambda & Y_z \\ Z_x, & Z_y & Z_z - \Lambda \end{vmatrix} = 0$$

absolut genommen unterhalb $K$ bleiben:

$$(7) \qquad |\Lambda_1 - \Lambda_2| \leqq K, \quad |\Lambda_2 - \Lambda_3| \leqq K, \quad |\Lambda_1 - \Lambda_3| \leqq K.$$

Eine Lösung dieses Variationsproblemes mit drei Gleichungs- und drei Ungleichungsnebenbedingungen wird in verschiedenen Teilgebieten verschiedene Eigenschaften haben, je nachdem für sie in den Bedingungen (7) das Gleichheits- oder Ungleichheitszeichen gilt. Gelten alle drei Ungleichheitszeichen, so kommt man auf die Gleichgewichtsbedingungen der gewöhnlichen Elastizitätstheorie zurück, andernfalls kommt man auf neue charakteristische „halbplastische" oder „vollplastische" Zustände.

Prinzipiell wäre es ein leichtes, diesen Ansatz auf *sandartige Massen (Erddrucktheorie)* zu übertragen; an Stelle von (7) treten als Nebenbedingungen andere Ungleichungen, die ausdrücken, dass die Richtung der Spannung auf jedes Flächenelement nicht ausserhalb eines gewissen „Reibungskegels" fällt. Indessen fehlt es hier an sicherer Kenntnis eines Ausdruckes (6) der Verzerrungsenergie, so dass dieser Ansatz zunächst nur in dem extremen Fall brauchbar ist,

---

[163] Rom. Acc. Linc. Rend. (5) 18, 2 (1909), p. 295, 577; (19) 1 (1910), p. 107, 239; (22) 1 (1913), p. 529. Acta math. 35 (1912), p. 295.

[164] Gött. Nachr., math.-phys. Kl., 1909, p. 212.

functions $a_1, a_2$ of $t - \tau$. A series of problems, belonging here, has been treated recently by *V. Volterra*[163]) (cf. p. 641).

For the case of remaining shape changes, i. e. for plastic media, *A. Haar* and *Th. v. Kármán*[164]) have formulated the foundations concerning completely different aspects. They start with the variational principle No. **7**, (23), in which (cf. p. 655) $H$ corresponds for isotropic media to the energy density and becomes a homogeneous quadratic function of the first two orthogonal invariants of the (symmetric) stress dyad:

$$2H = a_1(X_x + Y_y + Z_z)^2 + a_2(X_y^2 + Y_z^2 + Z_x^2 - X_xY_y - Y_yZ_z - Z_zX_x) \tag{6}$$

To this variational problem with its three constraints (23a), No. **7**, as characteristic property for plastic media, now the condition is added, that the largest shear stress appearing somewhere does not exceed a constant value $K$, i. e. that the differences between each of two roots of the equation

$$\begin{vmatrix} X_x - \Lambda, & X_y, & X_z \\ Y_x, & Y_y - \Lambda & Y_z \\ Z_x, & Z_y & Z_z - \Lambda \end{vmatrix} = 0$$

remain in absolute value below $K$:

$$|\Lambda_1 - \Lambda_2| \leqq K, \quad |\Lambda_2 - \Lambda_3| \leqq K, \quad |\Lambda_1 - \Lambda_3| \leqq K. \tag{7}$$

A solution of this variational problem with three equality and three inequality constraints will have various properties in various cases, depending on whether in the conditions (7) the equalities or inequalities hold. If all three inequalities hold, then one comes back to the common theory of elasticity, otherwise one arrives at the newly characteristic "semi-plastic" or "fully-plastic" states.

Basically, it would be a simple task to transfer this ansatz to *sandy matter (theory of lateral earth pressure)*; in place of (7) other inequalities appear as constraints[. Inequalities] which express that the direction of the stress at every surface element lies not outside a certain "cone of friction". Meanwhile, there is missing here reliable information about the expression (6) of the deformation energy, such that this ansatz is so far only useful in the extreme case,

[163] Rom. Acc. Linc. Rend. (5) 18, 2 (1909), p. 295, 577; (19) 1 (1910), p. 107, 239; (22) 1 (1913), p. 529. Acta math. 35 (1912), p. 295.

[164] Gött. Nachr., math.-phys. Kl., 1909, p. 212.

wo zwei der Ungleichheitsnebenbedingungen als Gleichungen erfüllt sind; dann resultieren nämlich Differentialgleichungen, die von der speziellen Form des Energieausdruckes unabhängig sind[165]).

**12. Kapillarität.** Die Phänomene der Kapillarität enthalten den zuletzt betrachteten Erscheinungen gegenüber insofern ein wesentlich neues Moment, als sie an das Auftreten von *Grenzflächen* verschiedenartiger Medien gegeneinander geknüpft sind. Demgemäss wird man, sofern man an der Existenz eines Potentiales festhält, die Kapillaritätswirkungen aus einem Potentialbestandteil der Gestalt (6) von Nr. **7**a, nämlich einem Integral über jene Grenzflächen herleiten:

(1) $$\Phi = \iint\limits_{(S)} \bar{\psi}\, dS = \iint\limits_{(S_0)} \psi\, dS_0.$$

Der Ansatz für $\bar{\psi}$, den *Gauss*[166]) durch den oben (S. 647)[93]) angedeuteten Grenzübergang hergeleitet hat, ist, dass $\psi$ nur von der Beschaffenheit der aneinandergrenzenden Medien, nicht von den Deformationsfunktionen abhängt; dann wird, falls nur homogene Medien auftreten, $\Phi$ gleich einem linearen Aggregat der Inhalte $S_1, S_2, \ldots$ der verschiedenen Grenzflächen (im deformierten Zustande)[167]):

(2) $$\Phi = C_1 S_1 + C_2 S_2 + \cdots .$$

Die Umformung von $\delta\Phi$ auf die Gestalt Nr. **3**e, (15) ergibt die folgenden Wirkungen: eine innerhalb der Fläche $S_i$ senkrecht zu jedem Linienelement $ds$ wirkende Spannung $C_i ds$, die nur an den Grenzkurven der Flächenteile $S_i$, zur Geltung kommt, und eine normal zu jedem inneren Flächenelement gerichtete und bis auf den Faktor $2C_i$ seiner mittleren Krümmung gleiche Druckkraft.[168])

Will man den Ansatz (1) enger mit der sonst im Vordergrunde stehenden Vorstellung räumlicher Verteilung der Energie verknüpfen, als es durch die S. 646 erwähnte rechnerische Transformation des Flächenintegrales in ein Raumintegral geschehen kann, so gelingt das

---

[165] *Haar* u. v. Kármán a. a. O.[164]), S. 217. Über die Erddrucktheorie vgl. IV 27 (*Reissner*), im übrigen ausser der dort gegebenen Litteratur auch *J. Sylvester* Phil. Mag. (4) 20 (1860), p. 489 = Collected Papers, vol. 2, Cambridge 1908, p. 215 und *J. Massau*, Mémoire sur l'intégration graphique des équations aux dérivées partielles, fasc. 2 et 3. Mons 1902 und 1904. (Extrait des Annales des Ingénieurs sortis des Écoles spéciales de Gand.)

[166] *C. F. Gauss*, Princ. generalia theoriae figurae fluiderum Comment. soc. reg. scient. Gotting. recent. 7 (1830) = Werke V, p. 29.

[167] A. a. O. Nr. **18**.

[168] Vgl. die ausführliche Darstellung dieser Entwicklung in V 9, Nr. **2** ff. (*Minkowski*).

where two of the inequality constraints hold as equality; then [this] results namely in differential equations which are independent of the special form of the energy expression[165]).

**12. Capillarity.** The phenomena of capillarity include in contrast to the lastly considered phenomena insofar an essential new aspect, as they are related to the occurrence of *interfaces* of various media against each other. Hence, one will, as long as one holds on to the existence of a potential, derive the effects of capillarity from a potential constituent of the form (6) of No. **7**a, namely from an integral over those interfaces:

$$\Phi = \iint_{(S)} \bar{\psi}\, dS = \iint_{(S_0)} \psi\, dS_0. \tag{1}$$

The ansatz for $\bar{\psi}$, which *Gauss*[166]) has derived by the above (p. 647)[93]) mentioned limit process, is, that $\psi$ depends not on the deformation functions but only on the constitution of the media adjacent to one another; then, when only homogeneous media occur, $\Phi$ is equal to a linear aggregate of the areas $S_1, S_2, \ldots$ of the various interfaces (in the deformed state)[167]):

$$\Phi = C_1 S_1 + C_2 S_2 + \cdots . \tag{2}$$

The transformation of $\delta\Phi$ into the form No. **3**e, (15) leads to the following efforts: a stress $C_i ds$ which appears only on the boundary curves of the surface patches $S_i$ [and which] acts within the surface $S_i$ orthogonally to every line element $ds$, and a pressure force, oriented normally to each internal surface element, [which is] up to the factor $2C_i$ equal to its mean curvature.[168])

If one likes to relate the ansatz (1) closer to the usually prior perception of a spatially distributed energy, as it can be achieved by the computational transformation of the surface integral into a volume integral mentioned on p. 646, then one succeeds

[165] *Haar* and v. Kármán op. cit.[164]), p. 217. On the theory of lateral earth pressure cf. IV 27 (*Reissner*), besides the literature given there [cf.] also *J. Sylvester* Phil. Mag. (4) 20 (1860), p. 489 = Collected Papers, vol. 2, Cambridge 1908, p. 215 and *J. Massau*, Mémoire sur l'intégration graphique des équations aux dérivées partielles, fasc. 2 et 3. Mons 1902 and 1904. (Extrait des Annales des Ingénieurs sortis des Écoles spéciales de Gand.)

[166] *C. F. Gauss*, Princ. generalia theoriae figurae fluiderum Comment. soc. reg. scient. Gotting. recent. 7 (1830) = Werke V, p. 29.

[167] Op. cit. No. **18**.

[168] Cf. the extensive presentation of this derivation in V 9, No. **2** ff. (*Minkowski*).

mit Hilfe eines Grenzüberganges, der dem in Nr. **8** zu verwandten Zwecken benutzten analog ist.[169]) Beschränkt man sich der Einfachheit halber auf ein System von zwei durch die Fläche S getrennten Medien, die die Raumteile $V_1$, $V_2$ erfüllen, so kann man an seine Stelle setzen den den tatsächlichen Verhältnissen näher kommenden Fall *eines* Kontinuums, dessen Zustand sich stetig, aber in der Nähe von S ausserordentlich rasch ändert und das als abstrakten Grenzfall jenes System aus zwei Medien einschließt. Die Energiedichte $\varphi$ eines solchen Mediums wird (vgl. Nr. **7**a, S. 645) auch von den lokalen Ableitungen der Deformationsgrössen, d. h. von den zweiten Ableitungen der Funktionen $x(a, b, c), \ldots$ abhängen; man wird diese Abhängigkeit nur in einem kleinen $S$ umschliessenden Gebiete $V^{(\varepsilon)}$ zu berücksichtigen brauchen, während in den Restgebieten $V_1^{(\varepsilon)}$ und $V_2^{(\varepsilon)}$ die Betrachtung der Abhängigkeit von den Deformationsgrössen erster Ordnung genügt. Approximiert man nun mit dem so beschriebenen Kontinuum das ursprüngliche System, indem man $V^{(\varepsilon)}$ sich unbegrenzt um $S$ zusammenziehen und gleichzeitig $V_1^{(\varepsilon)}$, $V_2^{(\varepsilon)}$ gegen $V_1$, $V_2$ konvergieren lässt, so wird in der Grenze bei passender Verfügung über $\varphi$ im Gesamtpotential neben dem räumlichen Potential von $V_1$ und $V_2$ gerade ein Flächenintegral vom Typus (1) auftreten. Lässt man speziell, was von dem Ansatz Nr. **10**, (1) der Hydrodynamik aus naheliegt, $\varphi$ innerhalb $V^{(\varepsilon)}$ von der Ableitung $\frac{\partial \varrho}{\partial n}$ der Dichte normal zu einem $V^{(\varepsilon)}$ erfüllenden System von Parallelflächen zu $S$ abhängen, setzt also etwa $\varphi = C \cdot \frac{\partial \varrho}{\partial n}$, so tritt im Limes

$$C \cdot \iint\limits_{(S)} (\varrho_1 - \varrho_2) dS$$

zum Potential hinzu, wobei $\varrho_1, \varrho_2$ die Randwerte der Dichte in $V_1$, $V_2$ sind — d. i. bei konstanten Dichten gerade die Form (2). Für die genaue Durchführung dieses Ansatzes ist natürlich wieder (vgl. S. 660) Vorbedingung, dass die Gleichgewichtslage des approximierenden Systems in ihrer Abhängigkeit von dem Parameter $\varepsilon$ untersucht ist.

**13. Optik.** Um die optischen Erscheinungen dem Schema der allgemeinen Mechanik der Kontinua einzufügen, sieht man bekanntlich die Komponenten $u$, $v$, $w$ des *Lichtvektors* als Verschiebungskomponenten der Teilchen eines deformierbaren raumerfüllenden Me-

[169] Für die folgende Darstellung vgl. eine Bemerkung am Anfang der Nr. **5** in *H. Minkowskis* Referat V 9; den gleichen Weg hat D. Hilbert in einer Göttinger Vorlesung im W.-S. 1906/07 eingeschlagen.

with the help of a limit process, which is similar to the one being used in No. **8** for related purposes.[169]) By restricting oneself for the sake of simplicity to a system with two media divided by a surface $S$, [media] which occupy the spatial parts $V_1$, $V_2$, one can substitute [the system] with *one* continuum representing the actual circumstances better[. One continuum] whose state changes continuously, but in the neighborhood of $S$ extraordinary fast and which includes as an abstract limit the system of the two media. The energy density $\varphi$ of such a medium (cf. No. **7**a, p. 645) will depend also on the local derivatives of the deformation quantities, i. e. on the second derivatives of the functions $x(a, b, c), \ldots$; one will need to consider this dependence only in a small region $V^{(\varepsilon)}$ which surrounds $S$, while in the remaining domains $V_1^{(\varepsilon)}$ and $V_2^{(\varepsilon)}$ the consideration of the dependence on the deformation quantities of first order is enough. If one approximates now with such a described continuum the original system, by contracting $V^{(\varepsilon)}$ around $S$ indefinitely and simultaneously by letting $V_1^{(\varepsilon)}$, $V_2^{(\varepsilon)}$ converge to $V_1$, $V_2$, then by appropriately controlling $\varphi$ there will appear in the total potential besides the spatial potential $V_1$ and $V_2$ just a surface integral of the kind (1). If within $V^{(\varepsilon)}$, what from the ansatz No. **10**, (1) of hydrodynamics immediately suggests itself, one specially lets $\varphi$ depend on the derivatives $\frac{\partial \varrho}{\partial n}$ of the density normal to a system of parallel surfaces to $S$ occupying $V^{(\varepsilon)}$, by setting something like $\varphi = C \cdot \frac{\partial \varrho}{\partial n}$, then in the limit

$$C \cdot \iint\limits_{(S)} (\varrho_1 - \varrho_2) dS$$

is added to the potential, whereby $\varrho_1$, $\varrho_2$ are the boundary conditions of the densities in $V_1$, $V_2$ — this corresponds for constant densities just to the form (2). For the specific computation of this Ansatz, certainly there is again (cf. p. 660) the assumption, that the dependence of the equilibrium position of the approximated system on the parameter $\varepsilon$ is studied.

**13. Optics.** In order to introduce the optical phenomena within the scheme of the general mechanics of continua, one considers as is well known the components $u$, $v$, $w$ of the *light vector* as displacement components of particles of a deformable space-occupying me-

[169] For the following presentation cf. a note at the beginning of No. **5** in *H. Minkowskis* paper V 9; the same procedure has been followed by D. Hilbert in a Göttinger Vorlesung in the winter term 1906/07.

diums (*Lichtäther*) an; es genügt für die Zwecke der Optik, wenn man sich dabei auf unendlichkleine Deformation beschränkt.[170]) Bei dieser Auffassung ist es aber keinesfalls erforderlich, dem Lichtäther — wie in der eigentlichen elastischen Lichttheorie — die Eigenschaften eines elastischen Mediums im engeren Sinne zuzuschreiben; vielmehr erhält man die richtigen Formeln der Optik gerade dann in einfachster Weise, wenn man nicht den Komponenten der reinen Formänderung (Nr. **9**, (5)), sondern denen der Rotation der Volumelemente

$$(1) \qquad \tfrac{1}{2}\xi = \tfrac{1}{2}\left(\frac{\partial w}{\partial b} - \frac{\partial v}{\partial c}\right), \quad \tfrac{1}{2}\eta = \tfrac{1}{2}\left(\frac{\partial u}{\partial c} - \frac{\partial w}{\partial a}\right), \quad \tfrac{1}{2}\zeta = \tfrac{1}{2}\left(\frac{\partial v}{\partial a} - \frac{\partial u}{\partial b}\right)$$

die bestimmende Rolle für den Wert der Deformationsenergie zuschreibt. Diesen Gedanken hat zuerst *J. Mac Cullagh*[171]) durchgeführt, und es gelang ihm auf diese Weise nicht nur, die Differentialgleichungen, sondern auch — über die elastische Lichttheorie hinaus — die richtigen Grenzbedingungen der Optik zu gewinnen.

Für *isotrope durchsichtige Medien* besteht *Mac Cullaghs* Ansatz darin, im Lichtäther eine Energiedichte proportional dem Quadrate des Betrages des Rotationsvektors (1) anzunehmen[172]):

$$(2) \qquad \varphi = \frac{1}{2}A(\xi^2 + \eta^2 + \zeta^2) = \frac{1}{2}A \sum_{\binom{abc}{uvw}} (w_b - v_c)^2.$$

Dann folgt aus Nr. **9**, (6) für die Spannungskomponenten

$$X_a = Y_b = Z_c = 0$$
$$Z_b = -Y_c = A\xi, \; X_c = -Z_a = A\eta, \; Y_a = -X_b = A\zeta,$$

und die Gleichungen Nr. **5**, (2) für den Bewegungszustand lauten daher

$$(3\text{a}) \qquad \varrho u'' = A\left(\frac{\partial\eta}{\partial c} - \frac{\partial\zeta}{\partial b}\right) = A\left(\Delta u - \frac{\partial(u_a + v_b + w_c)}{\partial a}\right) \quad \begin{pmatrix} u,v,w \\ a,b,c \\ \xi,\eta,\zeta \end{pmatrix};$$

das sind, wenn man noch die Bedingung $u_a + v_b + w_c = 0$ der Inkompressibilität hinzufügt, genau die Schwingungsgleichungen der Optik. Ebenso aber sind die Randbedingungen der Optik in den Randbedingungen enthalten, die sich analog Nr. **3**c, (5b) ergeben und die z. B. für die Grenzfläche zweier Medien mit verschiedenen Konstanten $A$

---

[170] Von der Hinzufügung des unendlichkleinen Faktors $\sigma$ wird im folgenden der Kürze halber abgesehen.

[171] *Mac Cullagh*, An essay towards a dynam. theory of cryst. reflexion and refraction, Trans. Roy. Irish Acad., 21 (1839) = Coll. Works (Dublin 1880), p. 145. — Vgl. auch V 21, Nr. **24** (*Wangerin*) und V 22, Nr. **1** (*W. Wien*).

[172] Vgl. auch die Darstellung von *W. Voigt*, Kompendium II, p. 563.

dium (*light ether*); it is enough for the purposes of optics to restrict oneself thereby to infinitesimal deformations.[170]) For this perception it is not at all necessary, to attribute to the light ether — as in the effective elastic theory of light — the property of an elastic medium in the narrower sense; on the contrary one obtains the correct formulas of optics just then in the most simple way, when one does not attribute to the components of the pure shape change (No. **9**, (5)) but to the one of the rotation of the volume elements

$$(1)\qquad \tfrac{1}{2}\xi = \tfrac{1}{2}\left(\frac{\partial w}{\partial b} - \frac{\partial v}{\partial c}\right), \quad \tfrac{1}{2}\eta = \tfrac{1}{2}\left(\frac{\partial u}{\partial c} - \frac{\partial w}{\partial a}\right), \quad \tfrac{1}{2}\zeta = \tfrac{1}{2}\left(\frac{\partial v}{\partial a} - \frac{\partial u}{\partial b}\right)$$

the determining role for the value of the deformation energy. This idea originates from *J. Mac Cullagh*[171]), and he succeeded in this manner not only to achieve the differential equations, but also — beyond the elastic theory of light — [to achieve] the correct boundary conditions of optics.

For *isotropic transparent media Mac Cullagh*'s Ansatz lies therein to assume an energy density proportional to the squares of the absolute value of the rotation vector (1) within the light ether[172]):

$$(2)\qquad \varphi = \frac{1}{2}A(\xi^2 + \eta^2 + \zeta^2) = \frac{1}{2}A \sum_{\binom{abc}{uvw}} (w_b - v_c)^2.$$

Then it follows from No. **9**, (6) for the stress components

$$X_a = Y_b = Z_c = 0$$
$$Z_b = -Y_c = A\xi,\ X_c = -Z_a = A\eta,\ Y_a = -X_b = A\zeta,$$

and the equations No. **5**, (2) for the motion reads therefore as

$$(3a)\qquad \varrho u'' = A\left(\frac{\partial \eta}{\partial c} - \frac{\partial \zeta}{\partial b}\right) = A\left(\Delta u - \frac{\partial(u_a + v_b + w_c)}{\partial a}\right) \quad \begin{pmatrix} u,v,w \\ a,b,c \\ \xi,\eta,\zeta \end{pmatrix};$$

which are, when one adds the condition $u_a + v_b + w_c = 0$ of the incompressibility, precisely the oscillation equations of optics. Likewise, the boundary conditions of optics are included in the boundary conditions, which are obtained similarly to No. **3**c, (5b) and which express for instance for the interface between two media with different constants $A$,

---

[170] In what follows, we refrain for the sake of brevity from adding the infinitesimal factor $\sigma$.

[171] *Mac Cullagh*, An essay towards a dynam. theory of cryst. reflexion and refraction, Trans. Roy. Irish Acad., 21 (1839) = Coll. Works (Dublin 1880), p. 145. — Cf. also V 21, No. **24** (*Wangerin*) and V 22, No. **1** (*W. Wien*).

[172] Cf. also the presentation of *W. Voigt*, Kompendium II, p. 563.

aussagen, dass die für beide gebildeten Ausdrücke

$$A(\eta \cos nc - \zeta \cos nb) \qquad \begin{pmatrix} \xi, \eta, \zeta \\ a, b, c \end{pmatrix} \tag{3b}$$

übereinstimmen.

Das Integral der Energiedichte (2) gestattet eine Transformation, der zufolge es bis auf Randintegrale mit dem Raumintegral von

$$\frac{1}{2} A \sum_{\binom{uvw}{abc}} \left\{ (w_b + v_c)^2 - 4 v_b w_c \right\}$$

übereinstimmt, d. i. aber die Energiedichte eines rein elastischen isotropen Mediums, dessen Lameésche Konstanten $\lambda, \mu$ in der Beziehung $\lambda = -2\mu = -2A$ stehen. Ein Medium dieser Konstitution gerade hat *W. Thomson* (Lord *Kelvin*) zur Erklärung der optischen Phänomene herangezogen[173])

*Mac Cullagh* hat seinen Ansatz insbesondere für die Optik *kristallinischer Medien* durchgeführt, indem er $\varphi$ gleich einer quadratischen Form von $\xi, \eta, \zeta$ (mit konstanten Koeffizienten)[174]) setzt:

$$2\varphi = A_{11}\xi^2 + 2A_{12}\xi\eta + \cdots + 2A_{23}\eta\zeta + A_{33}\zeta^2. \tag{4}$$

Ganz analog wie oben folgen dann als Differentialgleichungen

$$\varrho u'' = \frac{\partial \mathrm{H}}{\partial c} - \frac{\partial \mathrm{Z}}{\partial b}, \quad \text{wo} \quad \Xi = \frac{\partial \varphi}{\partial \xi} \qquad \begin{pmatrix} \Xi, \mathrm{H}, \mathrm{Z} \, ; \, \xi, \eta, \zeta \\ u, v, w \, ; \, a, b, c \end{pmatrix} \tag{4a}$$

während in den Randbedingungen die Ausdrücke auftreten:

$$\mathrm{H} \cos nc - \mathrm{Z} \cos nb. \qquad \begin{pmatrix} \Xi, \mathrm{H}, \mathrm{Z} \\ a, \ b, \ c \end{pmatrix} \tag{4b}$$

*E.* und *F. Cosserat* haben darauf hingewiesen[175]), daß ihr „Euklidisches Potential" auch diese *Mac Cullagh*schen Ansätze umfaßt.

Man kann auf dieser Grundlage versuchen, durch Erweiterung des Potentialansatzes nach einer der in Nr. **7** erörterten Richtungen die sämtlichen für die verschiedenen optischen Probleme notwendigen Gleichungen zu umfassen; in dieser Weise ist *W. Voigt* in seinem Kompendium[176]) systematisch vorgegangen.

In erster Linie gewinnt er den Übergang zu der *Abhängigkeit der optischen Erscheinungen von der Farbe* (Schwingungsdauer $\tau$), indem

---

[173] *W. Thomson*, Phil. Mag. (5) 26 (1888), p. 414 ff. Vgl. auch V 21, Nr. **31** (*Wangerin*).

[174] Vgl. *Mac Cullagh*, works[171]), p. 156, wo (4) sogleich auf eine Summe von Quadraten transformiert erscheint. Siehe auch die Darstellung in *P. Volkmann*, Vorles. über die Theorie des Lichtes, Leipzig 1891, p. 260, 294.

[175] *E.* und *F. Cosserat*, Corps déform., p. 151.

[176] S. namentlich V. Teil (Optik), § 7 (Bd. II, p. 563 ff.) sowie Kap. II, III dieses Teiles und vgl. auch II. Teil, § 34 (Band I, p. 486 ff.), wo die Kraftwirkungen direkt ohne Vermittlung eines Potentiales angesetzt werden.

that both generated expressions

(3b) $$A(\eta \cos nc - \zeta \cos nb) \qquad \begin{pmatrix} \xi, \eta, \zeta \\ a, b, c \end{pmatrix}$$

coincide.

The integral of the energy density (2) allows for a transformation, by virtue of which [the result] coincides up to a boundary integral with the volume integral of

$$\frac{1}{2} A \sum_{\binom{u\,v\,w}{a\,b\,c}} \left\{ (w_b + v_c)^2 - 4 v_b w_c \right\},$$

but this is the energy density of a purely elastic isotropic medium, whose Lamé parameters $\lambda, \mu$ are related according to $\lambda = -2\mu = -2A$. *W. Thomson* (Lord *Kelvin*) has used a medium of just this constitution for the explanation of optical phenomena[173])

*Mac Cullagh* has carried out his ansatz in particular for the optics of *crystalline media*, by setting $\varphi$ equal to a quadratic form of $\xi, \eta, \zeta$ (with constant coefficients)[174]):

(4) $$2\varphi = A_{11}\xi^2 + 2A_{12}\xi\eta + \cdots + 2A_{23}\eta\zeta + A_{33}\zeta^2.$$

Completely analogous as above the differential equations follow as

(4a) $$\varrho u'' = \frac{\partial \mathrm{H}}{\partial c} - \frac{\partial \mathrm{Z}}{\partial b}, \text{ where } \Xi = \frac{\partial \varphi}{\partial \xi} \qquad \begin{pmatrix} \Xi, \mathrm{H}, \mathrm{Z} \\ u, v, w \end{pmatrix}; \begin{pmatrix} \xi, \eta, \zeta \\ a, b, c \end{pmatrix}$$

while in the boundary conditions the [subsequent] expressions appear:

(4b) $$\mathrm{H} \cos nc - \mathrm{Z} \cos nb. \qquad \begin{pmatrix} \Xi, \mathrm{H}, \mathrm{Z} \\ a, b, c \end{pmatrix}$$

*E.* und *F. Cosserat* have indicated[175]), that their "euclidean potential" includes also these fundamental approaches of *Mac Cullagh*.

Based on this foundation, one can try, with an enhancement of the potential-based approach according to the direction discussed in No. **7**, to include all equations required for the various optical problems; in this way *W. Voigt* proceeded systematically in his Kompendium.[176])

In the first place, he gains the transition to the *dependence of the optical appearance of the color* (oscillation duration $\tau$), by

---

[173] *W. Thomson*, Phil. Mag. (5) 26 (1888), p. 414 ff. Cf. also V 21, No. **31** (*Wangerin*).

[174] Cf. *Mac Cullagh*, works[171]), p. 156, where (4) readily emerges as a sum of squares. See also the presentation in *P. Volkmann*, Vorles. über die Theorie des Lichtes, Leipzig 1891, p. 260, 294.

[175] *E.* und *F. Cosserat*, Corps déform., p. 151.

[176] See particularly V. Teil (Optik), § 7 (Bd. II, p. 563 ff.) as well as Kap. II, III of this part and cf. also II. Teil, § 34 (Band I, p. 486 ff.), where the force effects are formulated directly without the communication of a potential.

er den Gliedern von (4) ebenso gebildete quadratische Formen der zeitlichen Ableitungen $\xi', \eta', \zeta'$ oder $\xi'', \eta'', \zeta''$ usw. hinzufügt, freilich unter gleichzeitiger Beschränkung darauf, dass der Lichtvektor durchweg Sinusschwingungen mit der Periode $\tau$ ausführt. Er verwendet nun das Hamiltonsche Prinzip in der Form Nr. **7**, (25), und kann durch partielle Integration nach der Zeit diese Zusatzglieder derart umformen[177]), dass schliesslich wiederum eine quadratische Form von $\xi, \eta, \zeta$ genau wie (4) die Stelle der Energiedichte einnimmt, nur daß ihre Koeffizienten $A$ nun *Funktionen von* $\tau$ sind; die Art dieser Funktionen hängt von dem Medium ab und bestimmt sein Verhalten gegenüber den verschiedenen Farben.

In ähnlicher Weise zieht *Voigt* auch quadratische mit zeitlichen Ableitungen verschiedener Ordnung der $\xi, \eta, \zeta$ gebildete Terme heran und zeigt, dass man sie auf *einen* wesentlich neuen charakteristischen Bestandteil der Energiedichte zurückführen kann:

$$B_1(\zeta'\eta - \eta'\zeta) + B_2(\xi'\zeta - \zeta'\xi) + B_3(\eta'\xi - \xi'\eta); \tag{5}$$

dabei sind $B_1, B_2, B_3$ gegebene Konstanten oder Funktionen von $\tau$.[178]) Die Zusatzglieder, die hiernach zu den Differentialgleichungen und Randbedingungen zu treten haben, sind den allgemeinen Formeln leicht zu entnehmen; sie beschreiben die Veränderung, die die Lichtbewegung durch ein magnetisches Feld erleidet (*magnetische Aktivität*), und zwar hängen die Grössen $B$, die sich wie Komponenten eines axialen Vektors transformieren, von der Lage der magnetischen Axe an der betrachteten Stelle und der magnetischen Feldstärke ab.[179])

An dritter Stelle zieht *Voigt* endlich noch Aggregate von Produkten aus je einer zeitlichen Ableitung von $u, v, w$ selbst und einer von $\xi, \eta, \zeta$ in Betracht. Auch sie haben, wie durch ähnliche Umformungen gezeigt wird[180]), das Auftreten einfacherer Glieder im Ausdruck der virtuellen Arbeit zur Folge, für die typisch ist

$$C(u\delta\xi + v\delta\eta + w\delta\zeta). \tag{6}$$

Die Differentialgleichungen hierzu sind leicht herzustellen; sie liefern die Phänomene in den *natürlich aktiven Medien.*[181])

*Mac Cullagh* selbst hatte diese Medien gleichfalls in seine Betrachtungen einbezogen, indem er der Energiedichte ein Ableitungen

[177] A. a. O. [176]), p. 569.

[178] A. a. O. p. 568 ff.

[179] A. a. O. p. 572, 679 ff.

[180] A. a. O. p. 572 ff.

[181] A. a. O. p. 574, 687 ff.

adding to the terms of (4) equally generated quadratic forms of the time derivatives $\xi', \eta', \zeta'$ or $\xi'', \eta'', \zeta''$ and so on, certainly with the simultaneous restriction that the light vector realizes throughout sine oscillations with period $\tau$. Now he uses Hamilton's principle in the form No. **7**, (25), and by integration by parts with respect to time he can transform this additional terms such[177]), that eventually again a quadratic form of $\xi, \eta, \zeta$ just like (4) acquires the position of the energy density, save that their coefficients $A$ are now *functions of* $\tau$; The type of these functions depend on the medium and determine its behavior with respect to different colors.

In a similar way *Voigt* uses also quadratic terms formed with time derivatives of different orders of $\xi, \eta, \zeta$ and shows, that one can reduce them to *one* essentially new characteristic constituent of the energy density:

$$B_1(\zeta'\eta - \eta'\zeta) + B_2(\xi'\zeta - \zeta'\xi) + B_3(\eta'\xi - \xi'\eta); \tag{5}$$

thereby $B_1, B_2, B_3$ are given constants or functions of $\tau$.[178]) The additional terms, which appear consequently in the differential equations and the boundary conditions, are extracted easily from the general formulas; they describe the change, which the motion of light undergoes under [the influence] of a magnetic field (*magnetic activity*), and indeed, the quantities $B$, which transform like the components of an axial vector, depend on the position of the magnetic axis at the considered position and the magnetic field strength.[179])

In the third place, *Voigt* considers finally also aggregates of products between a first time derivative of $u, v, w$ and one of $\xi, \eta, \zeta$. Also they have, which is shown by similar transformations[180]), as a result the appearance of more simple terms in the expression of the virtual work, for which it is typical [that]

$$C(u\delta\xi + v\delta\eta + w\delta\zeta). \tag{6}$$

The differential equations hereto are easily obtained; they provide the phenomena of the *naturally active media.*[181])

*Mac Cullagh* himself has considered these media in the same way, by giving the energy density an additional term

177 Op. cit.[176]), p. 569.

178 Op. cit. p. 568 ff.

179 Op. cit. p. 572, 679 ff.

180 Op. cit. 572 ff.

181 Op. cit. p. 574, 687 ff.

zweiter Ordnung enthaltendes Zusatzglied gab

(6′) $$\frac{1}{2}C \sum_{\binom{u\,v\,w}{\xi\,\eta\,\zeta}} \xi\left(\frac{\partial^2 u}{\partial a^2} + \frac{\partial^2 u}{\partial b^2} + \frac{\partial^2 u}{\partial c^2}\right).$$

Dann erhalten die Differentialgleichungen Ableitungen dritter statt wie bei Voigt erster Ordnung als Zusatzglieder.[182])

Den Übergang zu *absorbierenden Medien* gewinnt*Voigt*, indem er unter Heranziehung einer als quadratische Form der Ableitungen $\xi', \eta', \zeta'$ gegebenen Dissipationsfunktion (Nr. **7**f, S. 657, Nr. **11**, S. 671)

(7) $$2D = m_{11}\xi'^2 + 2m_{12}\xi'\eta' + \cdots + m_{33}\xi'^2 (sollte \zeta sein)$$

der virtuellen Arbeit

(7a) $$-\left(\frac{\partial D}{\partial \xi'}\delta\xi + \frac{\partial D}{\partial \eta'}\delta\eta + \frac{\partial D}{\partial \zeta'}\delta\zeta\right)$$

hinzufügt; das bewirkt einfach ein Hinzutreten der Komplexe $\frac{\partial D}{\partial \xi'}, \ldots$ zu den $\Xi, \ldots$ in den Formeln (4a), (4b).[183])

Während alle diese Betrachtungen den Ausdruck der potentiellen Energie betreffen, kann man ebenso auch versuchen, Verallgemeinerungen des einfachsten Ausdruckes $\frac{1}{2}\varrho(u'^2 + v'^2 + w'^2)$ der kinetischen Energie, wie sie in Nr. **5**d diskutiert sind, in der Optik zu benutzen. In dieser Richtung liegt der bereits erwähnte Ansatz von *J. W. Strutt* (*Lord Rayleigh*)[184]), die kinetische Energie pro Volumelement des Lichtäthers als allgemeine quadratische Form der Geschwindigkeiten $u', v', w'$ anzunehmen; dabei treten dann auf den linken Seiten der optischen Gleichungen lineare Kombinationen der Beschleunigungen $u'', v'', w''$ auf.

**14. Beziehungen zur Elektrodynamik.** Die Grundgleichungen der Elektrodynamik sind ihrer Form nach bekanntlich im wesentlichen in den optischen Grundgleichungen und damit in dem allgemeinen Schema der Mechanik der Kontinua enthalten. Deutet man nämlich, um nur vom isotropen Medium zu reden, die zeitlichen Ableitungen der Komponenten $u, v, w$ des soeben betrachteten Lichtvektors bis auf einen konstanten Faktor als Vektor der elektrischen Feldstärke $\mathfrak{E}$:

$$u' = \gamma_1 \mathfrak{E}_a, \quad v' = \gamma_1 \mathfrak{E}_b, \quad w' = \gamma_1 \mathfrak{E}_c,$$

und ebenso die Komponenten der Rotation eines Volumelements als

---

[182] *Mac Cullagh*, Proc. R. Irish Ac. II (1841), p. 96 = Works, p. 187. Vgl. auch *P. Volkmann*, Theorie des Lichtes, p. 414 ff.

[183] A. a. O. p. 575 f., 708 ff.

[184] *J. W. Strutt*, Phil Mag. (4) 41—43 (1871, 1872). Vgl. auch V 21, Nr. **29** (*Wangerin*).

containing derivatives of second order

$$(6') \qquad \frac{1}{2}C \sum_{\binom{uvw}{\xi\eta\zeta}} \xi\left(\frac{\partial^2 u}{\partial a^2} + \frac{\partial^2 u}{\partial b^2} + \frac{\partial^2 u}{\partial c^2}\right).$$

Then the differential equations obtain as additional terms derivatives of third instead of first order as in Voigt.[182])

*Voigt* gains the transition to *absorbing media*, by using a dissipation function (No. **7**f, p. 657, No. **11**, p. 671) given as a quadratic form of derivatives $\xi', \eta', \zeta'$

$$(7) \qquad 2D = m_{11}\xi'^2 + 2m_{12}\xi'\eta' + \cdots + m_{33}\xi'^2 (sollte \zeta sein)$$

[and] adding to the virtual work

$$(7a) \qquad -\left(\frac{\partial D}{\partial \xi'}\delta\xi + \frac{\partial D}{\partial \eta'}\delta\eta + \frac{\partial D}{\partial \zeta'}\delta\zeta\right);$$

this leads simply to the addition of the complexes $\frac{\partial D}{\partial \xi'}, \ldots$ to the $\Xi, \ldots$ in the formulas (4a), (4b).[183])

While all this considerations concern the expression of the potential energy, one can likewise try to use in optics generalizations of the most simple expression $\frac{1}{2}\varrho(u'^2 + v'^2 + w'^2)$ of the kinetic energy, as they are discussed in No. **5**d. In this direction lies the already mentioned ansatz of *J. W. Strutt* (*Lord Rayleigh*)[184]), to assume the kinetic energy per unit volume of the light ether as a general quadratic form of the velocities $u', v', w'$; thereby linear combinations of accelerations $u'', v'', w''$ do appear on the left hand side of the optical equations.

**14. Relations to electrodynamics.** The fundamental equations of electrodynamics are in their form, as it is well known, essentially included in the optical fundamental equations and thereby [included] in the general scheme of the mechanics of continua. To speak only of the isotropic medium, if one interprets namely the time derivatives of the components $u, v, w$ of the just considered light vector up to a constant factor as vector of the electric field strength $\mathfrak{E}$:

$$u' = \gamma_1 \mathfrak{E}_a, \quad v' = \gamma_1 \mathfrak{E}_b, \quad w' = \gamma_1 \mathfrak{E}_c,$$

and likewise the components of the rotation of a volume element as

[182] *Mac Cullagh*, Proc. R. Irish Ac. II (1841), p. 96 = Works, p. 187. Cf. also *P. Volkmann*, Theorie des Lichtes, p. 414 ff.

[183] Op. cit. p. 575 f., 708 ff.

[184] *J. W. Strutt*, Phil Mag. (4) 41—43 (1871, 1872). Cf. also V 21, No. **29** (*Wangerin*).

Komponenten der magnetischen Feldstärke $\mathfrak{H}$:

$$\xi = \gamma_2 \mathfrak{H}_a, \quad \eta = \gamma_2 \mathfrak{H}_b, \quad \zeta = \gamma_2 \mathfrak{H}_c,$$

so gehen die Gleichungen (3a) und (1) von Nr. **13** bei passender Wahl der Konstanten $\gamma_1, \gamma_2, A, \varrho$ über in

$$(1) \qquad \begin{cases} \frac{\partial \mathfrak{H}_c}{\partial b} - \frac{\partial \mathfrak{H}_b}{\partial c} = \frac{\varepsilon}{c} \mathfrak{E}'_a & (a, b, c), \\ \frac{\partial \mathfrak{E}_c}{\partial b} - \frac{\partial \mathfrak{E}_b}{\partial c} = -\frac{\mu}{c} \mathfrak{H}'_a & (a, b, c), \end{cases}$$

und das sind gerade die *Maxwellschen Grundgleichungen* im freien Äther.[185]) Ein weiteres, äusseren Kräften entsprechendes Glied, das die Gleichungen (3a) noch enthalten können, findet im ersten Tripel (1) seine Deutung als elektrischer Strom. Ähnlich kann man auch die elektromagnetischen Gleichungen für nichtisotrope Medien gewinnen.

Bei den Darstellungen, die die allgemeinen Ansätze der Elektrodynamik jetzt meist finden, geht man indessen in der Regel nicht von dieser Auffassung aus, die die elektrischen und magnetischen Grössen mit den Verschiebungen eines Mediums in so direkte Verbindung bringt; man sieht vielmehr diese Grössen als „physikalische Parameter" im Sinne von Nr. **2b** an, die den Stellen des Kontinuums als Ortsfunktionen zugeordnet sind, und von denen man allenfalls einige als abhängig von den Bewegungsfunktionen eines immateriellen Mediums — der Elektrizität — deutet. Daneben können dann noch die Bewegungsfunktionen des materiellen Mediums, in dem der Vorgang sich abspielt, in Betracht kommen. Die Gleichungen der Elektrodynamik verknüpfen nun alle diese Größen direkt mit den Kräften, Spannungen, Energiedichten. Die Variationsprinzipe, in die man sie nach dem Vorgange von *H. A. Lorentz*[186]) und *H. v. Helmholtz*[187]) vielfach zusammengefaßt hat, sind dann in gewisser Weise den mechanischen analog, nur daß sie durch die größere Anzahl der in sie eingehenden Größen sehr viel komplizierter sind. Über diese Probleme der speziellen Elektrodynamik vergleiche man die Referate von *H. A. Lorentz*, insbesondere V 13, Nr. **35—39** und V 14, Nr. **8, 9**. Nur ein besonderer Fall sei noch hervorgehoben, als typisch dafür,

---

[185] Vgl. *W. Thomson*, Math. phys. pap. 3 (London 1890), p. 436 ff. Man kann auch die Rolle von elektrischer und magnetischer Feldstärke gerade vertauschen; vgl. über die verschiedenen möglichen Deutungen V 13, Nr. **42**, *H. A. Lorentz*.

[186] *H. A. Lorentz*, La théorie électromagn. de Maxwell (Leiden 1892), § 55 ff.)

[187] *H. v. Helmholtz*, Das Prinzip der kleinsten Wirkung in der Elektrodynamik. Ann. d. Phys. 47 (1892) p. 1 = Wissensch. Abh. III (Leipzig 1895), p. 476.

components of the magnetic field strength $\mathfrak{H}$:

$$\xi = \gamma_2 \mathfrak{H}_a, \quad \eta = \gamma_2 \mathfrak{H}_b, \quad \zeta = \gamma_2 \mathfrak{H}_c,$$

then for a suitable choice of the constants $\gamma_1, \gamma_2, A, \varrho$ the equations (3a) and (1) of No. **13** transform to

$$(1) \qquad \begin{cases} \frac{\partial \mathfrak{H}_c}{\partial b} - \frac{\partial \mathfrak{H}_b}{\partial c} = \frac{\varepsilon}{c} \mathfrak{E}'_a & (a, b, c), \\ \frac{\partial \mathfrak{E}_c}{\partial b} - \frac{\partial \mathfrak{E}_b}{\partial c} = -\frac{\mu}{c} \mathfrak{H}'_a & (a, b, c), \end{cases}$$

and these are just *Maxwell's Fundamental Equations* in the free ether.[185]) A further term corresponding to external forces that can be additionally included in equations (3a) finds its interpretation as electric current in the first three equations of (1). Similarly one can also obtain the electromagnetic equations for anisotropic media.

In the presentations, in which the general fundamentals of electrodynamics are found, however, usually one does not start with the perception to relate the electric and magnetic quantities with the displacements of a medium in such a direct way; one considers these quantities rather as "physical parameters" in the sense of No. **2**b, from which, if necessary, one interprets some as dependent on the motion of an immaterial medium — of the electricity. In addition also the motion of the material medium, in which the process takes place, can be taken into consideration. The equations of electrodynamics now relate all these quantities directly with forces, stresses, energy densities. The variational principles, into which one has often condensed them according to the approach of *H. A. Lorentz*[186]) and *H. v. Helmholtz*[187]), are then in a certain sense analogous to the mechanical one, save that they are very much more complicated due to the bigger amount of involved quantities. One shall compare the papers of *H. A. Lorentz* on these problems of special electrodynamics, in particular V 13, No. **35—39** and V 14, No. **8, 9**. Only a special case shall be highlighted in addition, in being typical

[185] Cf. *W. Thomson*, Math. phys. pap. 3 (London 1890), p. 436 ff. One can also interchange the role of the electric and magnetic field strength; cf. about the various possible interpretations V 13, No. **42**, *H. A. Lorentz*.

[186] *H. A. Lorentz*, La théorie électromagn. de Maxwell (Leiden 1892), § 55 ff.)

[187] *H. v. Helmholtz*, Das Prinzip der kleinsten Wirkung in der Elektrodynamik. Ann. d. Phys. 47 (1892) p. 1 = Wissensch. Abh. III (Leipzig 1895), p. 476.

wie solche physikalischen Parameter in die Stoffgleichungen eines materiellen Mediums im früheren Sinne eingehen können.

In einem elastischen Medium sei ein elektrisches Feld $\mathfrak{E}$ erregt; die Verallgemeinerung des früheren Ansatzes ist dann die, dass die Energiedichte $\varphi$ ausser von den Deformationsgrössen $e_a, \ldots, g_{ab}, \ldots$ noch von den Komponenten der Feldstärke abhängt[188]):

$$\varphi = \varphi(e_a, \ldots, g_{ab}, \ldots; \mathfrak{E}_a, \mathfrak{E}_b, \mathfrak{E}_c). \tag{2}$$

Aus den früheren Formeln Nr. **7**, (4) ergeben sich unverändert die Spannungskomponenten, die also von den elektrischen Feldstärken abhängig werden; andererseits aber ist das Potential auch gegenüber Variationen der Feldstärke $\mathfrak{E}$ zum Minimum zu machen, und daraus ergeben sich Gleichungen für die „elektrischen Momente“

$$\mathfrak{P}_a = \frac{\partial\varphi}{\partial\mathfrak{E}_a}, \quad \mathfrak{P}_b = \frac{\partial\varphi}{\partial\mathfrak{E}_b}, \quad \mathfrak{P}_c = \frac{\partial\varphi}{\partial\mathfrak{E}_c}, \tag{3}$$

die eine Abhängigkeit des elektrischen Zustandes von der Deformation zeigen. In beiden Formelsystemen bzw. in den aus ihnen folgenden Relationen vom Typus

$$\frac{\partial X_a}{\partial\mathfrak{E}_a} = \frac{\partial\mathfrak{P}_a}{\partial x_a} \tag{4}$$

sind die sog. *Reziprozitätssätze*[189]) enthalten, die in allen diesen verschiedenartige Gebiete verknüpfenden Erscheinungen eine wesentliche Rolle spielen; hat eine Änderung des einen physikalischen Parameters eine Änderung der einem anderen zugeordneten Spannungskomponente zur Folge, so bewirkt auch eine Variation dieses Parameters eine bestimmte Änderung der jenem ersten zugehörigen Spannungskomponente. In diesen Formeln sind die Erscheinungen der *Piezoelektrizität* enthalten, die mit Hilfe einfacher Ansätze für $\varphi$ entsprechend den Symmetrieverhältnissen der kristallinischen Medien genau untersucht worden sind.[190])

---

188 Nachdem *W. Voigt* zuerst die Theorie auf Grund direkt angesetzter Abhängigkeit der Spannungs- und Momentkomponenten von Deformation und Feldstärke behandelt hatte (Abhandl. Ges. d. Wiss. Göttingen, 36, 1890), haben *P. Duhem*, Leçons sur l'électricité 2 (1892), p. 467, *E. Riecke* (Nachr. Ges. d. Wiss. Göttingen 1893, p. 19) und *W. Voigt* (ebenda, math.-phys. Kl. 1894, p. 343) den Potentialansatz verwendet; näheres siehe in V 16, Nr. **8**, *F. Pockels*.

189 Vgl. hierzu *Voigts* Kompendium II, p. 106. — Man kann diese Reziprozitätssätze, die meist nur für den Fall endlich vieler Freiheitsgrade behandelt werden (vgl. *J. J. Thomson*, Anwendungen der Dynamik auf Physik und Chemie [Leipzig 1890] und *H. von Helmholtz*, Journ. f. Math. 100 (1887), p. 137 = Wiss. Abh. III, p. 203 ff.) in weitem Umfang auf Kontinua übertragen.

190 Vgl. die ausführliche Darstellung in V 16, Nr. **8—10**, *F. Pockels*.

how such physical parameters can be treated in the former sense within the constitutive laws of a material medium.

Let an electric field $\mathfrak{E}$ be excited in an elastic medium; the generalization of the former ansatz is then the one, that the energy density $\varphi$ depends apart from the deformation quantities $e_a, \ldots, g_{ab}, \ldots$ also on the components of the field strength[188]):

$$\varphi = \varphi(e_a, \ldots, g_{ab}, \ldots; \mathfrak{E}_a, \mathfrak{E}_b, \mathfrak{E}_c). \tag{2}$$

Using the former formulas No. **7**, (4), the stress components appear unchanged, which depend consequently also on the electric field strength; Apart from that, the potential has to be minimized also with respect to the variation of the field strength $\mathfrak{E}$, and thereof the equations for the "electric torques" are obtained

$$\mathfrak{P}_a = \frac{\partial \varphi}{\partial \mathfrak{E}_a}, \quad \mathfrak{P}_b = \frac{\partial \varphi}{\partial \mathfrak{E}_b}, \quad \mathfrak{P}_c = \frac{\partial \varphi}{\partial \mathfrak{E}_c}, \tag{3}$$

which show a dependence of the electric state on the deformation. In both systems of formulas or in the consequently following relation of the kind

$$\frac{\partial X_a}{\partial \mathfrak{E}_a} = \frac{\partial \mathfrak{P}_a}{\partial x_a} \tag{4}$$

the so-called *reciprocity theorems*[189]) are included, which play a crucial role in all these phenomena which relate various fields; if a change of one physical parameter causes the change of another related stress component, then also a variation of this parameter causes a certain change of the stress component related to the first. In these formulas the phenomena of *piezoelectricity* are included, which have been studied thoroughly with the help of simple forms of $\varphi$ according to the symmetry conditions of crystalline media.[190])

---

[188] After *W. Voigt* had treated first the theory based on the directly formulated dependence of the stress and torque components on the deformation and field strength (Abhandl. Ges. d. Wiss. Göttingen, 36, 1890), *P. Duhem*, Leçons sur l'électricité 2 (1892), p. 467, *E. Riecke* (Nachr. Ges. d. Wiss. Göttingen 1893, p. 19) and *W. Voigt* (ibid., math.-phys. Kl. 1894, p. 343) have used the potential-based approach; For more details see V 16, No. **8**, *F. Pockels*.

[189] Cf. hereto *Voigts* Kompendium II, p. 106. — One can transmit these reciprocity theorems, which are mostly treated for the case of finitely many degrees of freedom (cf. *J. J. Thomson*, Anwendungen der Dynamik auf Physik und Chemie [Leipzig 1890] and *H. von Helmholtz*, Journ. f. Math. 100 (1887), p. 137 = Wiss. Abh. III, p. 203 ff.) in the wider extent to continua.

[190] Cf. the extensive presentation in V 16, No. **8**—**10**, *F. Pockels*.

**15. Einfügung der thermodynamischen Ansätze.** Es giebt zwei Wege, von den bisher entwickelten Grundformeln der Mechanik der Kontinua zu den umfassenderen Ansätzen der Thermodynamik aufzusteigen, die im Rahmen dieses Artikels nur in aller Kürze zu skizzieren sind. Der eine schliesst an die Gleichungen der Kinetik, etwa an das Hamiltonsche Prinzip in der verallgemeinerten Gestalt Nr. **7**, (26) an und geht von der Annahme aus, dass irgendeine Verbindung $\omega$ der Bewegungsfunktionen und ihrer räumlichen Ableitungen selbst nicht explizit im Integranden $\varphi$ auftritt, vielmehr lediglich ihre zeitliche Ableitung. Eine solche „verborgene Koordinate“, die man an Stelle einer der Bewegungsfunktionen als bewegungsbestimmend ansehen kann, kann man dann gerade so behandeln, wie man es in der *Helmholtz*schen Theorie[191]) der zyklischen Systeme in der Mechanik der Systeme mit endlichvielen Freiheitsgraden tut: Mit Hilfe der Eliminationsmethoden der Variationsrechnung, wie sie in der Theorie der kanonischen Transformation der Dynamik gehandhabt werden[192]), führt man im Variationsprinzip statt $\omega'$ die Ableitung $\pi = \frac{\partial \varphi}{\partial \omega'}$ ein und erhält dann für den Grenzfall solcher Zustandsänderungen, bei denen Geschwindigkeiten und Beschleunigungen der übrigen Koordinaten (ausser $\omega$) unendlichklein sind, ein Variationsprinzip, das sich von dem Prinzip der virtuellen Verrückungen nur durch das Hinzutreten eines Termes $\omega' \cdot \delta\pi$ unterscheidet. Das Raumintegral dieses Termes findet nun seine Deutung als die bei der virtuellen Verrückung zugeführte Wärmemenge, während $\omega'$ und $\pi$ Temperatur und Entropie des Systems darstellen. Die analogen Betrachtungen finden in der Thermodynamik der Systeme mit endlich vielen Freiheitsgraden stets ausführlich Platz[193]); übrigens scheint aber eine explizite Anwendung innerhalb der Mechanik der Kontinua nicht vorzuliegen.

Der zweite Weg ist wesentlich mehr formaler Natur und schliesst sich den bisher zum Ausdruck gebrachten formalen Auffassungen aufs nächste an. Den Deformationsfunktionen wird — wir beschränken uns der Einfachheit halber auf die Statik — ein „physikalischer“ Parameter im Sinne von Nr. **2b**

$$s = s(a, b, c)$$

hinzugefügt, dessen Wert an jeder Stelle den „thermischen Zustand“

[191] *H. v. Helmholtz*, J. f. Math. 97 (1884), p. 111 = Wiss. Abhandl. III, p. 119 ff. Vgl. IV 11, Nr. **23**, *Heun*.

[192] Vgl. die Anwendungen derselben Methoden oben in Nr. **7**e, S. 654 und Nr. **8**b, S. 662 sowie Anm.[111]).

[193] Siehe die Referate V 3, Nr. **28** ff. (*Bryan*) und IV 1, Nr. **48** (*Voss*).

**15. Introduction of the thermodynamical foundations.** From the so far developed basic formulas of the mechanics of continua, there are two ways to climb up to the more comprehensive foundations of thermodynamics, which, within the scope of this article, are outlined only in a nutshell. One [way] builds on the equations of kinetics, e. g. Hamilton's principle of the generalized form No. **7**, (26), and relies on the assumption that some relation $\omega$ of the motion and the spatial derivatives thereof does not appear explicitly in the integrand $\varphi$, but rather its time derivative. One can treat such a "hidden coordinate", which one can see instead of the motion as motion determining, just as one does it in *Helmholtz*'s theory[191]) of cyclic systems in the mechanics with finitely many degrees of freedom: With the help of elimination methods of the calculus of variations which are used in the theory of canonical transformations of dynamics[192]), one introduces in the variational principle instead of $\omega'$ the derivative $\pi = \frac{\partial\varphi}{\partial\omega'}$ and obtains consequently for the limit case of a state change, for which the velocities and accelerations of the remaining coordinates (except $\omega$) are indefinitely small, a variational principle which differ from the principle of virtual displacements only by the additional term $\omega' \cdot \delta\pi$. The volume integral of this term has then the interpretation of the added heat quantity for a virtual displacement, while $\omega'$ and $\pi$ represent temperature and entropy[, respectively]. Similar considerations are always treated extensively within the thermodynamics of systems with finitely many degrees of freedom[193]); but after all, an explicit application within the mechanics of continua seems not to be available.

The second way is of much more formal nature and builds directly on the formal understanding being expressed so far. To the deformation functions — we restrict us for the sake of simplicity to statics — a "physical parameter" in the sense of No. **2b**

$$s = s(a, b, c)$$

is added, whose value at every point describes the "thermal state"

[191] *H. v. Helmholtz*, J. f. Math. 97 (1884), p. 111 = Wiss. Abhandl. III, p. 119 ff. Cf. IV 11, No. **23**, *Heun*.

[192] Cf. the applications of the very same method above in No. **7**e, p. 654 and No. **8**b, p. 662 as well as remark[111]).

[193] See the articles V 3, No. **28** ff. (*Bryan*) and IV 1, No. **48** (*Voss*).

des Mediums beschreibt; man bezeichnet ihn als *Entropie* des Mediums an dieser Stelle, berechnet auf die Masseneinheit, indem man Entropie eines Quantums $V_0$ des Mediums das Integral:

$$\iiint_{(V_0)} s(a,b,c)\varrho_0 dV_0 = \iiint_{(V)} s(x,y,z)\varrho dV$$

nennt. Bei einer virtuellen Verrückung des Kontinuums wird auch $s$ eine unendlichkleine virtuelle Änderung $\delta s$ zu erfahren haben. Man kann dann, entsprechend dem *zweiten Hauptsatz* der Thermodynamik, das Prinzip der virtuellen Verrückungen in folgender Weise erweitern[194]):

Zu der virtuellen Arbeit $\delta A$ der gesamten Kraftrichtungen tritt gleichberechtigt die „*Wärmezufuhr*" bei einer virtuellen Verrückung:

(1) $$\delta Q = \iiint_{(V)} \Theta\, \delta s\, \varrho\, dV;$$

dabei bedeutet die „*Temperatur*" $\Theta$ einen den Spannungskomponenten gleichberechtigten Faktor, der für jedes Medium in charakteristischer Weise in seiner Abhängigkeit von den Deformationsfunktionen und der Entropie sowie von deren Ableitungen gegeben ist. *Thermodynamisches Gleichgewicht wird bedingt durch die Variationsgleichung*

(2) $$\delta Q + \delta A = 0,$$

die genau im alten Sinne zu verstehen ist; dabei können Nebenbedingungen auch das thermische Verhalten des Mediums, d. h. die Funktion $s$ betreffen.

Die volle Bedeutung der Thermodynamik kommt indessen erst zum Vorschein, wenn man in diesem Ansatz den sog. *ersten Hauptsatz* zur Geltung bringt, der einen *allgemeingültigen* Zusammenhang der sämtlichen Wirkungskomponenten einschliesslich der Temperatur mit einer *einzigen* Funktion der Zustandsgrössen statuiert, von der Art wie er in Nr. **7** für einzelne Fälle diskutiert wurde.[195]) Zieht man nämlich in $\delta A$ nur die inneren Wirkungen innerhalb des Mediums in Betracht, *so soll $\delta Q + \delta A$ für jede virtuelle Verrückung bis aufs Vorzeichen gleich der Variation eines in bestimmter, für jedes Medium charakteristischer Weise nur von den jeweiligen Deformationsfunktionen und der Entropie abhängigen Ausdruckes, der potentiellen Energie $\Phi$, sein.* Was die Gestalt von $\Phi$ angeht, so ist der einfachste Fall der, dass $\Phi$ ein

[194] Für die Ansätze der gewöhnlichen Thermodynamik, die sich im folgenden genau wiederholen, vgl. das Referat V 3 (*Bryan*).

[195] Für kontinuierliche Medien hat namentlich *P. Duhem* diese Ansätze nach den verschiedensten Richtungen hin angewendet; man vergleiche die zusammenfassende Darstellung in seinem Traité d'Énergétique, T. II (Paris 1911), Chap. XIV.

of the medium; one denotes it as *entropy* of the medium at this point, evaluated per unit mass, by calling the integral

$$\iiint\limits_{(V_0)} s(a,b,c)\varrho_0 dV_0 = \iiint\limits_{(V)} s(x,y,z)\varrho dV$$

entropy of a portion $V_0$ of the medium. For a virtual displacement of the continuum, also $s$ will undergo an infinitesimal virtual change $\delta s$. According to the *second law* of thermodynamics, one can then enhance the principle of virtual displacements in the following way[194]):

Additionally to the virtual work $\delta A$ of all force contributions, equitably the "heat supply" appears for a virtual displacement:

(1) $$\delta Q = \iiint\limits_{(V)} \Theta\, \delta s\, \varrho\, dV;$$

thereby "*temperature*" $\Theta$ denotes a factor being similar to the stress components[. A factor], which is given for every medium in its characteristic way in its dependence on the deformation functions and on the entropy as well as on the derivatives thereof. *Thermodynamic equilibrium is determined by the variational equation*

(2) $$\delta Q + \delta A = 0,$$

which needs to be understood exactly in the old way; thereby also constraints can affect the thermal behavior of the medium, i. e. the function $s$.

However, the full significance of thermodynamics appears only, when one asserts to this ansatz the so-called *first law* [of thermodynamics], which states a *generally valid* connection of all effects including the temperature with a *single* function of the state variables of the kind as discussed in No. **7** for individual cases.[195]) Namely, if one considers in $\delta A$ only the internal effects within the medium, *then* $\delta Q + \delta A$ *shall be up to the sign equal to the variation of a certain expression, the potential energy* $\Phi$, *depending for every medium in a characteristic way only on the respective deformation functions and the entropy.* Concerning the form of $\Phi$, then the most simple case is, that $\Phi$

---

[194] For the foundations of common thermodynamics, which are just repeated in the following, cf. the article V 3 (*Bryan*).

[195] For continuous media nominally *P. Duhem* has applied these fundamental approaches in various directions; one shall confer the summarizing presentation in his Traité d'Énergétique, T. II (Paris 1911), Chap. XIV.

Raumintegral über eine Funktion $\varphi$ von $x, y, z$, ihren ersten Ableitungen und $s$ ist[196]):

$$\delta Q + \delta A = -\delta \iiint\limits_{(V_0)} \varphi(x, \ldots; x_a, \ldots; s)\, da\, db\, dc; \tag{3}$$

dann folgt speziell für die Temperatur

$$\Theta = -\frac{1}{\varrho_0}\frac{\partial\varphi}{\partial s}, \tag{4}$$

was den im wesentlichen unverändert bleibenden Gleichungen (4) von Nr. **7a** zur Seite tritt. Aus diesen Gleichungen folgen wieder reziproke Relationen der Art

$$-\varrho_0 \frac{\partial\Theta}{\partial x_a} = \frac{\partial X_a}{\partial s} \tag{5}$$

zwischen je einem Paare thermischer und elastischer Parameter — in analoger Bedeutung, wie oben in einem anderen Fall erörtert (Nr. **14**, (4)) wurde.

Es ist häufig zweckmässig, an Stelle von $s$ die Temperatur $\Theta$ als bestimmenden Parameter einzuführen; das ist wiederum der Form nach die in Nr. **7e** (S. 654) angewandte kanonische Transformation: Berechnet man aus (4) $s$ als Funktion von $\Theta$ und bestimmt damit

$$\psi = \varphi + \varrho_0\Theta s = \psi(x, \ldots; x_a, \ldots; \Theta),$$

so erhält man statt (3) für alle willkürlichen Variationen $\delta x, \delta y, \delta z$ und $\delta\Theta$ die Identität:

$$-\iiint\limits_{(V)} s\,\delta\,\Theta\,\varrho dV + \delta A = -\delta \iiint\limits_{(V_0)} \psi da\, db\, dc. \tag{3'}$$

Man nennt $\psi$ das „thermodynamische Potential bei gegebenem Deformationszustand“; zieht man gleichzeitig noch die Transformation von Nr. **7e** heran, die die Deformationsgrössen durch die Spannungskomponenten ersetzt, so erhält man die anderen Arten thermodynamischer Potentiale in völliger Analogie zu den üblichen Betrachtungen der Thermodynamik der Systeme mit endlichvielen Freiheitsgraden.[197])

Indem man rechter Hand in (3) das Auftreten der Deformationsgrössen in geeigneter Weise spezialisiert, erhält man die thermodynamischen Ansätze für die einzelnen im vorigen behandelten Gebiete; dabei bedingt die Art, wie $s$ in $\varphi$ (oder $\Theta$ in $\psi$) mit den einzelnen Deformationsgrössen verknüpft ist, natürlich den thermischen Effekt

---

[196] Dieser Ansatz hat für den speziellen Fall der reinen Elastizitätstheorie zuerst *W. Thomson*, Quart. Journ of Math. 1 (1857) ausgebildet; vgl. V 3, Nr. **21**, *Bryan*.

[197] S. V 3, Nr. **16** (*Bryan*).

is a volume integral of a function $\varphi$ depending on $x, y, z$, the first derivatives thereof and $s$[196]):

$$\delta Q + \delta A = -\delta \iiint\limits_{(V_0)} \varphi(x, \ldots; x_a, \ldots; s)\, da\, db\, dc; \tag{3}$$

then especially for the temperature it follows

$$\Theta = -\frac{1}{\varrho_0}\frac{\partial \varphi}{\partial s}, \tag{4}$$

which stands aside to the equations (4) of No. **7a** [which] remain basically unchanged. From these equations again reciprocal relations of the kind

$$-\varrho_0 \frac{\partial \Theta}{\partial x_a} = \frac{\partial X_a}{\partial s} \tag{5}$$

follow between a pair of thermal and elastic parameter — with a similar meaning, as discussed above for another case (No. **14**, (4)).

Often it is useful to introduce the temperature $\Theta$ as determining parameter instead of $s$; according to the form, this corresponds again with the canonical transformation applied in No. 7e (p. 654): If one computes $s$ as a function of $\Theta$ using (4) and if one determines therewith

$$\psi = \varphi + \varrho_0 \Theta s = \psi(x, \ldots; x_a, \ldots; \Theta),$$

then one obtains instead of (3) for all arbitrary variations $\delta x, \delta y, \delta z$ and $\delta \Theta$ the identity:

$$-\iiint\limits_{(V)} s\, \delta\, \Theta\, \varrho dV + \delta A = -\delta \iiint\limits_{(V_0)} \psi\, da\, db\, dc. \tag{3'}$$

One denotes $\psi$ the "thermodynamic potential for a given state of deformation"; If one considers at the same time also the transformation of No. **7e**, which substitutes the deformation quantities with the stress components, then one obtains the other types of thermodynamic potential in complete analogy to the common considerations of thermodynamics of systems with finitely many degrees of freedom.[197])

By specializing on the right hand side of (3) the appearance of the deformation quantities in a convenient way, one obtains the thermodynamic fundamentals for the before treated individual fields; Thereby the way how $s$ in $\varphi$ (or $\Theta$ in $\psi$) is related with the particular deformation quantities determines certainly the thermal effect

[196] For the special case of the pure theory of elasticity, this ansatz has been formulated originally by *W. Thomson*, Quart. Journ of Math. 1 (1857); cf. V 3, No. **21**, *Bryan*.

[197] See V 3, No. **16** (*Bryan*).

der einzelnen Arten der Deformationen bzw. die Art der Deformationen, die durch thermische Wirkungen hervorgerufen werden. Für die *Elastizitätstheorie* und die *Hydrodynamik* sind diese Zusammenhänge vielfach untersucht worden.[198])

Man hat in (3) für die potentielle Energie $\Phi$ auch andere der in Nr. **7** untersuchten Ansätze verwendet, wobei zu den früheren Formeln nur die Berücksichtigung der Abhängigkeit von $s$ neu hinzukommt. Neben den Integralen vom Typus Nr. **7**, (7), die *P. Duhem*[199]) in dieser Richtung vielfach verwendet hat, sei hier nur der Fall hervorgehoben, dass $\Phi$ als Summanden ein *Flächenintegral* etwa über die Trennungsfläche verschiedener in $V$ enthaltenen Medien besitzt; entsprechend wird man dann auf dieser Fläche auch eine *Flächendichte der Entropie* und demgemäss ein Flächenintegral als Beitrag zur Wärmezufuhr anzunehmen haben. Diese Ansätze stellen die *thermischen Wirkungen der Kapillarität*[200]) dar.

Die weitere Ausbildung dieser thermodynamischen Ansätze erfolgt dann so, dass man in der Fundamentalgleichung neue die Konstitution des betrachteten Mediums beschreibende physikalische Parameter auftreten lässt, also etwa $\varphi$ von ihnen abhängen lässt. Es genüge, als Beispiel hier zu erwähnen, dass man so durch Aufnahme der elektrischen Feldstärke wie in Nr. **14**, (2) zu den Erscheinungen der *Pyroelektrizität*, der Wechselwirkung von Wärme, Druck und elektrischer Erregung, in Kristallen geführt wird.[201])

Endlich sind hier noch die an *J. W. Gibbs*[202]) anknüpfenden *thermochemischen* Untersuchungen zu erwähnen, die auf der Vorstellung mehrerer denselben Raum simultan ausfüllender Medien beruhen, deren Zustandsparameter gleichzeitig in $\varphi$ eingehen; man hat hier freilich sich bisher durchweg auf den Fall von endlich vielen Freiheitsgraden beschränkt: man nimmt die einzelnen Medien (Phasen) homogen an, so dass ihr Zustand durch eine Reihe nicht mehr vom Ort abhängiger Variabler charakterisiert wird.[203])

**16. Beziehungen zur Relativitätstheorie.** Es soll zum Schluss noch die Frage aufgenommen werden, die schon wiederholt gelegent-

[198] Vgl. z. B. *Voigt*, Kompendium I, p. 523 ff.; *Voigt*, Lehrbuch der Kristallphysik, Leipzig 1910, p. 276 ff., p. 763 ff.; *Duhem*, Traité d'énergétique II, Paris 1911, p. 115 ff.; *G. Hamel*, Elementare Mechanik, Leipzig 1912, p. 571 ff.

[199] S. insbes. Ann. Éc. Norm. (3) 10 (1893), p. 183 ff. und 21 (1904), p. 99 ff. und Traité, a. a. O.[195])

[200] Vgl. V 9, Nr. **18** (*Minkowski*).

[201] Siehe V 16, Nr. **11**, *F. Pockels*.

[202] Trans. Connect. Acad. III (1876—1878) = Scient. Papers I (1906), p. 55.

[203] Vgl. V 3, Nr. **26**, (*Bryan*) and IV 11, Nr. **22—24** (*K. Heun*).

of the individual kinds of deformation or the kind of deformation, which is caused by thermal effects. These relations have been studied in many ways for the *theory of elasticity* and *hydrodynamics*.[198])

In (3) one has used for the potential energy $\Phi$ also other approaches studied in No. **7**, where to the former formulas now the consideration of the dependence on $s$ is added anew. Besides the integrals of the kind No. **7**, (7), which *P. Duhem*[199]) has used frequently for this kind of applications, let us highlight here the case, that $\Phi$ has as a summand a *surface integral* for instance over the interface between different media contained in $V$; accordingly, one needs to assume on this surface also a *surface density of the entropy* and consequently a surface integral as a contribution to the heat supply. This fundamental approaches represent the *thermal effects of capillarity*[200]).

The further development of these thermodynamic approaches follows in the way that one let appear in the fundamental equation new physical parameters describing the constitution of the considered medium. It is enough to mention here as an example that by adding the electric field strength as in No. **14**, (2) one is led to the occurrence of *pyroelectricity* in crystals, [i. e.] the interaction between heat, pressure and electric excitation.[201])

Finally, the *thermochemical* studies following *J. W. Gibbs*[202]) have to be mentioned which are based on the perception of various media occupying the same space simultaneously, whose state variables appear in $\varphi$ at once; here, certainly one has restricted oneself throughout to the case of finitely many degrees of freedom: one assumes the individual media (phases) to be homogeneous, such that their state is characterized by a series of variables not depending on the position anymore.[203])

**16. Relations to the theory of relativity.** At the end, the question shall be incorporated, which already has been touched repeatedly once in a while,

[198] Cf. e. g. *Voigt*, Kompendium I, p. 523 ff.; *Voigt*, Lehrbuch der Kristallphysik, Leipzig 1910, p. 276 ff., p. 763 ff.; *Duhem*, Traité d'énergétique II, Paris 1911, p. 115 ff.; *G. Hamel*, Elementare Mechanik, Leipzig 1912, p. 571 ff.

[199] See in particular Ann. Éc. Norm. (3) 10 (1893), p. 183 ff. and 21 (1904), p. 99 ff. and Traité, op. cit.[195])

[200] Cf. V 9, No. **18** (*Minkowski*).

[201] See V 16, No. **11**, *F. Pockels*.

[202] Trans. Connect. Acad. III (1876—1878) = Scient. Papers I (1906), p. 55.

[203] Cf. V 3, No. **26**, (*Bryan*) and IV 11, No. **22**—**24** (*K. Heun*).

lich gestreift wurde, *wie sich die verschiedenen Ansätze der Mechanik der Kontinua bei Transformationen des verwendeten Koordinatensystems verhalten*; von hier aus wird auch die Verbindung zu den Ansätzen der modernen Relativitätstheorie hergestellt werden.

Unsere Vorstellung von der Homogenität und Isotropie des gewöhnlichen Raumes verlangt zunächst, dass die Gesetze jedes physikalischen Vorganges ungeändert bleiben, wenn man sie auf irgend ein anderes rechtwinkliges Koordinatensystem bezieht und gleichzeitig *alle* in den Vorgang eingreifenden Grössen der entsprechenden Transformation unterwirft; man sagt kurz, *dass die gesamte Physik invariant ist gegenüber der Gruppe der sämtlichen rechtwinkligen Koordinatentransformationen der gewöhnlichen Geometrie, der sog. „Hauptgruppe“ oder „Euklidischen Gruppe“*. Hieraus folgt speziell, dass die virtuelle Arbeit der sämtlichen inneren Wirkungen in einem kontinuierlichen System bei der einer unendlichkleinen Abänderung des Koordinatensystems entsprechenden virtuellen Verrückung notwendig verschwindet, oder dass das gesamte Potential dieser Wirkungen bei jeder solchen Verrückung des Kontinuums ungeändert bleibt, d. h. im Sinne von *E.* und *F. Cosserat* ein *Euklidisches Potential* ist (vgl. Nr. **7**b, S. 650).

In analoger Weise kann man nun in der Kinetik fragen, ob es auch Transformationen der Zeitvariablen $t$ oder gar simultane Transformationen der Zeit- und Raumvariablen gibt, die die physikalischen Gesetze ungeändert lassen. Legt man die kinetischen Grundansätze in ihrer ursprünglichen Gestalt (Nr. **5**, (1), (5), (4), (6)) zugrunde, so ergiebt sich, dass eine Verschiebung des Nullpunktes der Zeitrechnung

$$(1) \qquad \bar{t} = t + \beta$$

sowie eine gleichförmige Bewegung des rechtwinkligen Koordinatensystemes parallel mit sich

$$(2) \qquad \bar{x} = x + \alpha_1 t, \quad \bar{y} = y + \alpha_2 t, \quad \bar{z} = z + \alpha_3 t$$

die kinetischen Glieder im wesentlichen nicht ändert; nur die zeitlichen Ableitungen 1. Ordnung, beispielsweise die kinetische Energie $T$, werden bei der Substitution (2) zunächst modifiziert, aber man sieht leicht, dass die Zusatzglieder bei der Variation fortfallen und daher die Bewegungsgesetze ungeändert bleiben. Also sind *die Theoreme der Mechanik der Kontinua in gewissem Umfange invariant gegenüber einer zehnparametrigen Gruppe linearer Transformationen der Raum- und Zeitkoordinaten*[204]), die sich aus den rechtwinkligen Koordinatentransforma-

[204] Vgl. hierzu die Darlegungen in IV 1, Nr. **13**—**17**, *Voss*.

*how the various fundamental approaches of the mechanics of continua behave under the transformation of the used coordinate system*; from here, also the relation to the foundations of the modern theory of relativity are established.

Our perception of the homogeneity and isotropy of the ordinary space demands at first that the laws of every physical process remain unchanged if one relates them to another orthogonal coordinate system and [if], at the same time, [one] subjects *all* quantities involved in the process to the corresponding transformation; it is said briefly, *that the entire physics is invariant with respect to the group of all orthogonal coordinate transformations of the ordinary geometry, the so-called "basic group" or "Euclidean group"*. Herefrom it follows in particular, that the virtual work of all internal effects within a continuous system necessarily vanishes for a virtual displacement corresponding to an infinitesimal change of the coordinate system, or that the total potential of these effects remain unchanged for any such displacement of the continuum, i. e. [that the potential] is a *euclidean potential* in the sense of *E.* and *F. Cosserat* (cf. No. **7**b, p. 650).

In a similar way, one can ask in kinetics if there are also transformations of the time variable $t$ or even simultaneous transformations of time and space variables which leave the physical laws unchanged. If one bases on the fundamental laws of kinetics in the original form (No. **5**, (1), (5), (4), (6)), then it follows, that a displacement of the origin in the computation of time

(1) $$\bar{t} = t + \beta$$

as well as a uniform motion of the orthogonal coordinate system parallel to itself

(2) $$\bar{x} = x + \alpha_1 t, \quad \bar{y} = y + \alpha_2 t, \quad \bar{z} = z + \alpha_3 t$$

do not change the kinetic terms essentially; only the time derivatives of 1. order, for instance the kinetic energy $T$, are modified at first by the substitution (2), but one sees easily that the additional terms vanish for the variation and thus the laws of motion remain unchanged. Hence, *the theorems of the mechanics of continua are in a certain range invariant with respect to a ten-parameter group of linear transformations in space and time coordinates*[204]*)*, which are composed of an orthogonal coordinate transforma-

[204] Cf. hereto the explanation in IV 1, No. **13—17**, *Voss*.

tionen, aus den Parallelverschiebungen des Axensystems mit konstanter Geschwindigkeit sowie den Änderungen des Nullpunktes der Zeitrechnung zusammensetzt; die allgemeine Substitution dieser sog. *Galilei*schen oder *Newton*schen *Gruppe* lautet:

$$(3) \quad \begin{aligned} \bar{x} &= \alpha_{11}x + \alpha_{12}y + \alpha_{13}z + \alpha_1 t + \beta_1 \\ \bar{y} &= \alpha_{21}x + \alpha_{22}y + \alpha_{23}z + \alpha_2 t + \beta_2 \\ \bar{z} &= \alpha_{31}x + \alpha_{32}y + \alpha_{33}z + \alpha_3 t + \beta_3 \\ \bar{t} &= \phantom{\alpha_{31}x + \alpha_{32}y + \alpha_{33}z + \alpha_3} t + \beta \end{aligned}$$

wo die 9 Grössen $\alpha_{ik}$ ein orthogonales Koeffizientensystem bilden. Die Substitutionen (3) haben die charakteristische Eigenschaft, dass sie das Differential $dt$ ungeändert lassen, wenn aber $dt = 0$ ist, auch das Quadrat der Linienelemente $dx^2+dy^2+dz^2$. Eine besondere Bedeutung in der Mechanik hat auch die durch Hinzunahme aller *Ähnlichkeitstransformationen* des Raumes einerseits und der Zeitaxe andererseits erweiterte *zwölfgliedrige Gruppe*; ihre Anwendung lässt die physikalischen Grössen nicht mehr absolut invariant, sondern setzt ihre *Dimensionen* in bezug auf Längen- und Zeiteinheit in Evidenz.[205])

In der modernen Entwicklung der Optik und Elektrodynamik hat die Tatsache besondere Wichtigkeit gewonnen, dass durchaus nicht alle Gesetze der Physik diese Invarianz gegenüber der *Galilei*schen Gruppe aufweisen. Das kann einmal so zustande kommen, dass gemäss den allgemeinen Ansätzen von Nr. **5**d und **7**f ganz andersartige kinetische Glieder, als die der klassischen Mechanik den Vorgang bestimmen, andererseits aber auch dadurch, dass bei sonst unverändertem Ansatz durch den physikalischen Sachverhalt für gewisse Grössen eine andere Deutung und damit auch eine andere Behandlung bei der Transformation nahegelegt wird. So müssen z. B. die optischen Grundgleichungen (3a) von Nr. **18**, da sie genau aus dem normalen Ansatz des *d'Alembert*schen Prinzips entstehen, gegenüber der *Galilei*transformation (2) invariant seien, wenn man nur den $x$-$y$-$z$-Raum transformiert und $a, b, c$ als die jedes substantielle Teilchen charakterisierenden Parameter ungeändert lässt. Dem entgegen giebt die Optik Anlass, den Lichtvektor $u, v, w$ als Funktion der Stelle $a, b, c$ zu betrachten und demgemäss diese Variablen $a, b, c$ der *Galilei*transformation (2) zu unterwerfen; alsdann ist nach der Transformation die zeitliche Differentiation bei konstantem $\bar{a} = a+\alpha_1 t, \ldots$ zu vollziehen, und in diesem Sinne sind die Gleichungen der Optik nicht mehr invariant.

---

[205] Vgl. hierüber IV 1, Nr. **10**, *Voss*.

tion, a parallel displacement of the system of axes with constant velocity, as well as the changes of the origin in the computation of time; the general substitution of this so-called *Galilean* or *Newtonian group* formulates as:

$$(3)\quad \begin{aligned} \bar{x} &= \alpha_{11}x + \alpha_{12}y + \alpha_{13}z + \alpha_1 t + \beta_1 \\ \bar{y} &= \alpha_{21}x + \alpha_{22}y + \alpha_{23}z + \alpha_2 t + \beta_2 \\ \bar{z} &= \alpha_{31}x + \alpha_{32}y + \alpha_{33}z + \alpha_3 t + \beta_3 \\ \bar{t} &= \qquad\qquad\qquad\qquad\qquad t + \beta \end{aligned}$$

where the 9 quantities $\alpha_{ik}$ form an orthogonal system of coefficients. The substitutions (3) have the characteristic property to leave the differential $dt$ unchanged, but when $dt = 0$, then also the square of the line element $dx^2 + dy^2 + dz^2$ [remains unchanged]. In mechanics also the extended *twelve-parameter group* [obtained by] adding all *similarity transformations* of the space on the one hand and of the time axis on the other hand is of particular importance; its application does not leave the physical quantities to be invariant, but relates their *dimensions* with respect to the unit length and time.[205])

In the modern development of optics and electrodynamics the fact that definitely not all laws of physics show this invariance with respect to the *Galilean* group has gained particular importance. This can be achieved on the one hand that according to the general foundations of No. **5**d and **7**f completely different kinetic terms as in classical mechanics determine the process, or on the other hand also thereby that for an unchanged ansatz a physical circumstance suggests for certain quantities a different interpretation and consequently a different treatment under transformations. Thus, for instance the fundamental equations of optics (3a) of No. **18**, as they emerge exactly from the usual ansatz of *d'Alembert*'s principle, must be invariant with respect to *Galilean* transformations, if one transforms only the $x$-$y$-$z$-space and [if one lets] unchanged $a, b, c$ as parameters characterizing every substantial particle. Contrary to this, optics gives rise to consider the light vector $u, v, w$ as a function of the point $a, b, c$ and to subject these variables $a, b, c$ accordingly to the *Galilean* transformation (2); therupon after the transformation, the time derivative for constant $\bar{a} = a + \alpha_1 t, \ldots$ has to be carried out, and in this sense the equations of optics are not anymore invariant.

205 Cf. about this IV 1, No. **10**, *Voss*.

Bestimmt man nun aber diejenigen ganzen linearen Transformationen der Variablen $a, b, c, t$, bei denen die Grundgleichungen der Optik in diesem Sinne invariant bleiben[206]), so ergeben sich die nach *H. Poincarés*[207]) Vorschlag als *Lorentz-Transformationen* bezeichneten Transformationen, deren fundamentale Bedeutung für die Elektrodynamik und die Physik die Untersuchungen von *H. A. Lorentz*[208]), *A. Einstein*[209]), *H. Poincaré*[207]), *H. Minkowski*[210]) erwiesen haben. Es sind diejenigen „Affinitäten" des vierdimensionalen $x$-$y$-$z$-$t$-Raumes, der *Minkowski*schen „Welt":

$$\text{(4)} \qquad \begin{aligned} \bar{x} &= \alpha_{11}x + \alpha_{12}y + \alpha_{13}z + \alpha_{14}t + \alpha_{15} \\ \bar{y} &= \alpha_{21}x + \alpha_{22}y + \alpha_{23}z + \alpha_{24}t + \alpha_{25} \\ \bar{z} &= \alpha_{31}x + \alpha_{32}y + \alpha_{33}z + \alpha_{34}t + \alpha_{35} \\ \bar{t} &= \alpha_{41}x + \alpha_{42}y + \alpha_{43}z + \alpha_{44}t + \alpha_{45}, \end{aligned}$$

welche die Differentialform $dx^2+dy^2+dz^2-c^2dt^2$ (in der $c$ die *Lichtgeschwindigkeit* bedeutet) in sich selbst transformieren:

$$\text{(5)} \qquad d\bar{x}^2 + d\bar{y}^2 + d\bar{z}^2 - c^2 d\bar{t}^2 = dx^2 + dy^2 + dz^2 - c^2dt^2$$

und für die obendrein gilt:

$$\text{(6)} \qquad \frac{d\bar{t}}{dt} = \alpha_{44} > 0;$$

sie bilden wiederum eine *zehngliedrige Gruppe*, die *Lorentzgruppe*. Bemerkt man, dass wegen (5) die Transformation (4) eine orthogonale Substitution im Raume der Koordinaten $x, y, z, ct\sqrt{-1}$ darstellt, so kann man die Relationen für die Koeffizienten von (4) und die Invarianten der Gruppe leicht angeben.[211]) Übrigens kann eine etwas umfassendere „erweiterte *Lorentz*gruppe" geometrisch dadurch charakterisieren, dass sie die quadratische Fläche $x^2 + y^2 + z^2 - c^2t^2 = 0$ in

---

[206] *W. Voigt*, Nachr. Ges. d. W. Göttingen 1887, p. 41.

[207] *H. Poincaré*, Rendic. Circ. mat. Palermo 21 (1906), p. 129.

[208] Versuch einer Theorie der elektrischen und optischen Erscheinungen in bewegten Körpern, § 89—92 (Leiden 1895). Amsterdam Acad. Sc. Proc. 6 (1904), p. 809. Abgedr. im Heft 2 der „Fortsch. d. math. Wissenschaften" (Leipzig 1913; hrsg. *v. O. Blumenthal*).

[209] *A. Einstein*, Ann. d. Phys. (4) 17 (1905), p. 891. Abgedruckt am selben Orte.

[210] *H. Minkowski*, a) Die Grundgleichungen für die elektrodynamischen Vorgänge in bewegten Körpern, Nachr. Ges. d. W. Göttingen, math.-phys. Kl., 1908, p. 53 = Math. Ann. 68 (1910), p. 472; auch abgedr. in Fortschr. d. math. Wiss. (Leipzig 1910), Heft 1. b) Raum und Zeit, Jahresber. d. D. M. V. 18 (1909), p. 75 = Phys. Z. 10 (1909), p. 104; auch separat Leipzig 1909 und in dem in 208) genannten Heft.

[211] *H. Minkowski*,[210]) a) § 5; vgl. auch *A. Sommerfeld*, Ann. d. Phys. (4) 32 (1910), p. 749; 33 (1910), p. 649.

If one determines however those linear transformations of the variables $a, b, c, t$ for which the fundamental laws of optics remain invariant in this sense[206]), then transformations emerge being denoted as *Lorentz-transformations* according to *H. Poincaré*'s[207]) suggestion, whose fundamental relevance for electrodynamics and physics has been approved by the studies of *H. A. Lorentz*[208]), *A. Einstein*[209]), *H. Poincaré*[207]), *H. Minkowski*[210]). There are these "affinities" of the four-dimensional $x$-$y$-$z$-$t$-space, the *Minkowski*an "world":

$$\begin{aligned} \bar{x} &= \alpha_{11}x + \alpha_{12}y + \alpha_{13}z + \alpha_{14}t + \alpha_{15} \\ \bar{y} &= \alpha_{21}x + \alpha_{22}y + \alpha_{23}z + \alpha_{24}t + \alpha_{25} \\ \bar{z} &= \alpha_{31}x + \alpha_{32}y + \alpha_{33}z + \alpha_{34}t + \alpha_{35} \\ \bar{t} &= \alpha_{41}x + \alpha_{42}y + \alpha_{43}z + \alpha_{44}t + \alpha_{45}, \end{aligned} \tag{4}$$

which transform the differential form $dx^2 + dy^2 + dz^2 - c^2dt^2$ (in which $c$ denotes the *speed of light*) into itself:

$$d\bar{x}^2 + d\bar{y}^2 + d\bar{z}^2 - c^2d\bar{t}^2 = dx^2 + dy^2 + dz^2 - c^2dt^2 \tag{5}$$

and for which moreover it holds:

$$\frac{d\bar{t}}{dt} = \alpha_{44} > 0; \tag{6}$$

they form again a *ten-parameter group*, the *Lorentz group*. If one notes, that due to (5) the transformation (4) represents an orthogonal substitution in the space of coordinates $x, y, z, ct\sqrt{-1}$, then one can easily give the relation for the coefficients of (4) and the invariants of this group.[211]) After all one can characterized a somehow more encompassing "extended *Lorentz* group" geometrically thereby, that it transforms into itself the quadratic area $x^2 + y^2 + z^2 - c^2t^2 = 0$ of

206 *W. Voigt*, Nachr. Ges. d. W. Göttingen 1887, p. 41.

207 *H. Poincaré*, Rendic. Circ. mat. Palermo 21 (1906), p. 129.

208 Versuch einer Theorie der elektrischen und optischen Erscheinungen in bewegten Körpern, § 89—92 (Leiden 1895). Amsterdam Acad. Sc. Proc. 6 (1904), p. 809. Published in Heft 2 of "Fortsch. d. math. Wissenschaften" (Leipzig 1913; hrsg. *v. O. Blumenthal*).

209 *A. Einstein*, Ann. d. Phys. (4) 17 (1905), p. 891. Published at the same place.

210 *H. Minkowski*, a) Die Grundgleichungen für die elektrodynamischen Vorgänge in bewegten Körpern, Nachr. Ges. d. W. Göttingen, math.-phys. Kl., 1908, p. 53 = Math. Ann. 68 (1910), p. 472; also published in Fortschr. d. math. Wiss. (Leipzig 1910), Heft 1. b) Raum und Zeit, Jahresber. d. D. M. V. 18 (1909), p. 75 = Phys. Z. 10 (1909), p. 104; also separately Leipzig 1909 and in the number referred to in 208).

211 *H. Minkowski*,[210]) a) § 5; cf. also *A. Sommerfeld*, Ann. d. Phys. (4) 32 (1910), p. 749; 33 (1910), p. 649.

dem dreidimensionalen unendlichfernen Gebilde des vierdimensionalen $x$-$y$-$z$-$t$-Raumes in sich transformiert, und man kann daher ihre Theorie aus bekannten Untersuchungen der projektiven bzw. affinen Geometrie entnehmen.[212]) Diese erweiterte Gruppe enthält elf Parameter statt zehn, und ihre Transformationen erfüllen die Identität (5) nur bis auf einen konstanten Faktor; bestimmt man diesen etwa durch die Determinantenbedingung

$$|\alpha_{ik}| = 1 \qquad (i, k = 1, 2, 3, 4),$$

so zerfällt sie noch in zwei getrennte Kontinua, von denen das eine mit der durch (6) charakterisierten Lorentzgruppe identisch ist. Der elfte Parameter der erweiterten Gruppe entspricht einer Änderung der Masseinheit im $x$-$y$-$z$-$t$-Raum; in ihm ist tatsächlich nur *eine* Masseinheit verfügbar, da Raum- und Zeitkoordinaten durch die Forderung der Invarianz der Form (5), d. h. durch die Festlegung der Lichtgeschwindigkeit verknüpft sind, während bei der Galileigruppe durch Erweiterung um zwei Parameter über Zeit- und Raumeinheit getrennt verfügt werden konnte.[213])

Lässt man nun $c$ gegen $\infty$ konvergieren, so geht die erweiterte *Lorentz*gruppe über in die Gesamtheit der linearen Transformationen, welche die durch die beiden Gleichungen $x^2 + y^2 + z^2 = 0$, $t = 0$ im Unendlichfernen des $x$-$y$-$z$-$t$-Raumes bestimmte quadratische Kurve (d. i. der imaginäre Kugelkreis der Räume $t$ = konst.) in sich überführen; das ist aber gerade die erweiterte *Galilei*gruppe[214]), und es ist sonach die *Galileigruppe der Grenzfall der Lorentzgruppe bei unendlich wachsender Konstante $c$*. Dies hat *Minkowski* veranlasst, dem sog. *Relativitätsprinzip*, das als Forderung der *Invarianz gegenüber der Lorentzgruppe* zunächst für die Gesetze der Elektrodynamik ausgesprochen wurde, als „*Postulat der absoluten Welt*" einen weiteren Gültigkeitsbereich zu geben[215]): Was zunächst als Invarianz gegenüber der *Galilei*gruppe erscheint, ist in Wahrheit nur eine empirische Approximation an die exakte Invarianz gegenüber der *Lorentz*gruppe mit einem im Vergleich zu den gewöhnlich auftretenden Geschwindigkeiten sehr grossen $c$.

Die Ansätze für die Dynamik eines Kontinuums, das diesem Relativitätspostulat unterliegt, sind in den früher aufgestellten allgemeinen

---

[212] *F. Klein*, Die geometr. Grundlagen der Lorentzgruppe, Jahresber. d. D.M.V. 19 (1910), p. 281.

[213] Vgl. *F. Klein*, a. a. O., p. 295 f.

[214] *F. Klein*, a. a. O., p. 291 f.

[215] *H. Minkowski*, [210]) a) Anhang; b) Cap. I, II.

the indefinitely far-away three-dimensional shape of the four-dimensional $x$-$y$-$z$-$t$-space, and therefore one can take the theory thereof from known studies of projective or affine geometry.[212]) This extended group contains eleven instead of ten parameters, and their transformations fulfill the identity (5) only up to a constant factor; If one determines this [factor] for instance by the determinant condition

$$|\alpha_{ik}| = 1 \qquad (i, k = 1, 2, 3, 4),$$

then [the group] decomposes into two separate continua [i. e. connected components], from which one is identical with the Lorentz group characterized by (6). The eleventh parameter of the extended group corresponds to a change of the measuring unit in the $x$-$y$-$z$-$t$-space; in [this space] there is only *one* measuring unit available, since space and time coordinates are related by the requirement of the invariance of the form (5), i. e. by fixing the speed of light, while for an extension of the Galilean group by two parameters, we could separately decide about time and spatial unit.[213])

If one lets $c$ converge towards $\infty$, then the extended *Lorentz* group changes into the totality of linear transformations, which transform into itself the quadratic curve in the infinity of the $x$-$y$-$z$-$t$-space determined by the two equations $x^2 + y^2 + z^2 = 0$, $t = 0$ (this is the imaginary spherical circle of the spaces $t =$ const.); this is just the extended *Galil*ean group[214]), and consequently the *Galilean group is the limit case of the Lorentz group for indefinitely growing constant c*. This has motivated *Minkowski* to give to the so-called *relativity principle*, which has been stated at first as a requirement of the *invariance with respect to the Lorentz group* for the laws of electrodynamics, a further range of validity as "postulate of the absolute world"[215]): What seems at first as invariance with respect to the *Galil*ean group, is in fact only an empirical approximation of the exact invariance with respect to the *Lorentz* group with a very large $c$ compared to the usually appearing velocities.

The foundations of the dynamics of a continuum which respects this relativity postulate are included in the previously formulated general

[212] *F. Klein*, Die geometr. Grundlagen der Lorentzgruppe, Jahresber. d. D.M.V. 19 (1910), p. 281.

[213] Cf. *F. Klein*, op. cit., p. 295 f.

[214] *F. Klein*, op. cit., p. 291 f.

[215] *H. Minkowski*, [210]) a) Anhang; b) Cap. I, II.

Formen enthalten; es sind nur alle eingehenden Zustandsfunktionen als Invarianten bzw. Kovarianten der *Lorentz*gruppe zu wählen. Ist die Bewegung des Kontinuums wieder wie in Nr. **2**, (5) gegeben, so gestalten sich die Formeln homogener, wenn man für jedes Teilchen $a, b, c$ als Funktion von $t$ eine „Ortszeit"

$$\tau = \tau(a, b, c, t)$$

einführt; setzt man noch der Symmetrie halber

$$\begin{aligned} x &= x_1, \quad & y &= x_2, \quad & z &= x_3, \quad & t &= x_4, \\ a &= \xi_1, \quad & b &= \xi_2, \quad & c &= \xi_3, \quad & \tau &= \xi_4, \end{aligned}$$

so schreiben sich die Bewegungsgleichungen

(7) $$x_i = x_i(\xi_1, \xi_2, \xi_3, \xi_4) \qquad (i = 1, 2, 3, 4);$$

sie stellen bei variablem $\xi_4$ ein den vierdimensionalen Raum einfach überdeckendes System von Kurven (*Weltlinien*) dar, deren Gesamtverlauf ein vollständiges Bild der Bewegung giebt.[216])

Eine wesentliche Ergänzung des Relativitätspostulates bildet die Forderung, *dass alle überhaupt möglichen Geschwindigkeiten unterhalb der Lichtgeschwindigkeit c liegen*, d. h. wenn wir allgemein:

(7a) $$\frac{\partial x_i}{\partial \xi_k} = x_{ik} \qquad (i, k = 1, 2, 3, 4)$$

setzen, dass

$$x_{14}^2 + x_{24}^2 + x_{34}^2 < c^2 x_{44}^2$$

oder — geometrisch gesprochen — dass jede Tangente einer Weltlinie *innerhalb* des Kegels der Richtungen $dx^2 + dy^2 + dz^2 = c^2 dt^2$ liegt. Äquivalent damit ist die Tatsache, dass man jedes Teilchen zu jeder Zeit durch eine passende *Lorentz*transformation (4) „auf Ruhe transformieren" kann, d. h. dass man zu einem solchen neuen Koordinatensystem $\bar{x}_i$ übergehen kann, in dem bei analoger Bezeichnung wie in (7a) für den betrachteten Wertekomplex $\xi_1, \dots \xi_4$

$$\bar{x}_{14} = \bar{x}_{24} = \bar{x}_{34} = 0$$

wird. Alle möglichen „Ruhtransformationen" unterscheiden sich voneinander, abgesehen von den willkürlich bleibenden Größen $\alpha_{15}, \dots \alpha_{45}$, nur durch eine gewöhnliche orthogonale Transformation der drei Koordinaten $\bar{x}_1, \bar{x}_2, \bar{x}_3$; in den verschiedenen „Ruhkoordinatensystemen" werden also die Deformationsgrössen erster Ordnung $\bar{x}_{ik}$ wohl von-

216 *H. Minkowski*, 210) b).

forms; we only have to choose all relevant state functions as invariants or covariants of the *Lorentz* group. If the motion of the continuum is given again as in No. **2**, (5), then the formulas arrange more homogeneously when one introduces for every particle $a, b, c$ a "local time"

$$\tau = \tau(a, b, c, t)$$

as function of time $t$; by setting for the sake of symmetry

$$\begin{array}{llll} x = x_1, & y = x_2, & z = x_3, & t = x_4, \\ a = \xi_1, & b = \xi_2, & c = \xi_3, & \tau = \xi_4, \end{array}$$

then the equations of motion write as

$$x_i = x_i(\xi_1, \xi_2, \xi_3, \xi_4) \qquad (i = 1, 2, 3, 4); \tag{7}$$

they represent for varying $\xi_4$ a system of curves (*world lines*) simply covering the four-dimensional space [. Curves] whose whole courses give a complete picture of the motion.[216])

An essential supplement of the relativity postulate forms the requirement, *that all generally possible velocities are below the speed of light*, i. e. when we generally set

$$\frac{\partial x_i}{\partial \xi_k} = x_{ik} \qquad (i, k = 1, 2, 3, 4), \tag{7a}$$

that

$$x_{14}^2 + x_{24}^2 + x_{34}^2 < c^2 x_{44}^2$$

or — geometrically spoken — that every tangent to a world line lies *within* the cone of the directions $dx^2 + dy^2 + dz^2 = c^2 dt^2$. Equivalently to this is the fact that one can "transform to rest" every particle for every time by a suitable *Lorentz* transformation (4), i. e. that one can change over to a new coordinate system $\bar{x}_i$, in which for similar labeling as in (7a) for the considered tuple $\xi_1, \ldots \xi_4$

$$\bar{x}_{14} = \bar{x}_{24} = \bar{x}_{34} = 0.$$

All possible "transformations of rest" differ, apart from the arbitrarily remaining quantities $\alpha_{15}, \ldots \alpha_{45}$, only by an ordinary orthogonal transformation of the three coordinates $\bar{x}_1, \bar{x}_2, \bar{x}_3$; in the different "coordinate systems of rest" consequently all deformation quantities of first order $\bar{x}_{ik}$ possibly will be

---

[216] *H. Minkowski*, [210]) b).

einander verschieden sein können, hingegen werden die den orthogonalen Koordinatentransformationen gegenüber invarianten eigentlichen „Formänderungskomponenten" (Nr. **9**, (1))

$$\text{(8)} \qquad \begin{aligned} e_i &= \tfrac{1}{2}(\bar{x}_{1i}^2 + \bar{x}_{2i}^2 + \bar{x}_{3i}^2 - 1), \\ g_{ik} &= \bar{x}_{1i}\bar{x}_{1k} + \bar{x}_{2i}\bar{x}_{2k} + \bar{x}_{3i}\bar{x}_{3k} \end{aligned} \qquad (i, k = 1, 2, 3)$$

unabhängig von der speziell gewählten Ruhtransformation sein. Denkt man die $\bar{x}_{ik}$ durch ihre Ausdrücke in den ursprünglichen Bewegungsfunktionen ersetzt, so sind diese *Ruhdeformationen* die *einzigen Invarianten erster Ordnung*, die das System der Weltlinien (7) gegenüber der *Lorentz*gruppe aufweist; sie sind zugleich auch von der willkürlichen Wahl des Parameters $\xi_4 = \tau$ unabhängig.[217])

Eine virtuelle Variation der Bewegung des Kontinuums stellt sich nun durch vier Funktionen $\delta x_i$ $(i = 1, \ldots 4)$ dar; da die Variable $\xi_4 = \tau$ willkürlich ist, bedeutet das für die Bewegung des Kontinuums selbst, d. h. für das System der Weltlinien nur wieder drei willkürliche Funktionen. Die virtuelle Arbeit irgendwelcher am Kontinuum angreifender Volumkräfte im Intervall $\tau_1 \leqq \tau \leqq \tau_2$ erhält dann den Ausdruck

$$\text{(9)} \qquad \delta A = \int_{\tau_1}^{\tau_2} d\tau \iiint_{(V_0)} \sum_{i=1}^{4} X_i \delta x_i \varrho_0 d\xi_1 d\xi_2 d\xi_3 .$$

Dabei bedeuten $X_1, X_2, X_3$ analog zu Nr. **3a**, S. 613, wenn man ihnen noch den Faktor $\frac{d\tau}{dt} = \frac{1}{x_{44}}$ hinzufügt, die auf die Masseneinheit des undeformierten Mediums im $\xi_1$-$\xi_2$-$\xi_3$-Raume berechneten Kraftkomponenten; bemerkt man ferner, dass eine Variation, für die an jeder Stelle

$$\text{(10)} \qquad \delta x_1 : \delta x_2 : \delta x_3 : \delta x_4 = x_{14} : x_{24} : x_{34} : x_{44}$$

ist, nur eine Verschiebung der Weltlinien in sich, also eine Änderung des Parameters $\tau$ bedeutet, und dass für sie also $\delta A$ identisch verschwinden muss, so folgt, dass

$$\text{(9a)} \qquad -X_4 = \frac{1}{x_{44}} \sum_{i=1}^{3} x_{i4} X_i = X_1 \frac{dx_1}{dt} + X_2 \frac{dx_2}{dt} + X_3 \frac{dx_3}{dt}$$

— wiederum bis auf den Faktor $\frac{1}{x_{44}}$ — die in der Zeiteinheit an der Masseneinheit des undeformierten Mediums geleistete Arbeit bedeutet.[218])

---

[217] *M. Born*, Ann. d. Phys. (4) 30 (1909), p. 1; speziell § 2. — *G. Herglotz*, Ann. d. Phys. (4) 36 (1911), p. 493; speziell § 1, 2.

[218] *H. Minkowski*,[210]) a) Anhang; *G. Herglotz*, a. a. O., p. 506.

different from each other, whereas the effective "shape change components" (No. **9**, (1))

$$e_i = \tfrac{1}{2}(\bar{x}_{1i}^2 + \bar{x}_{2i}^2 + \bar{x}_{3i}^2 - 1), \qquad (i,k = 1,2,3) \tag{8}$$
$$g_{ik} = \bar{x}_{1i}\bar{x}_{1k} + \bar{x}_{2i}\bar{x}_{2k} + \bar{x}_{3i}\bar{x}_{3k}$$

being invariant with respect to the orthogonal coordinate transformations will be independent of the specially chosen transformations of rest. If one thinks of $\bar{x}_{ik}$ being substituted by their expressions of the original motion, then these *deformations of rest* are the *only invariants of first order*, which the system of world lines (7) shows with respect to the *Lorentz* group; they are likewise also independent of the arbitrary choice of the parameter $\xi_4 = \tau$.[217])

A virtual variation of the motion of the continuum is represented now by the four functions $\delta x_i$ ($i = 1, \ldots 4$); since the variable $\xi_4 = \tau$ is arbitrary, this implies that the motion of the continuum itself, i. e. the the system of world lines, [is described] only [by] three arbitrary functions. The virtual work of any volume forces applied in the interval $\tau_1 \leqq \tau \leqq \tau_2$ is then assumed to be given by the expression

$$\delta A = \int_{\tau_1}^{\tau_2} d\tau \iiint_{(V_0)} \sum_{i=1}^{4} X_i \delta x_i \varrho_0 d\xi_1 d\xi_2 d\xi_3. \tag{9}$$

Thereby by introducing the factor $\frac{d\tau}{dt} = \frac{1}{x_{44}}$ to $X_1, X_2, X_3$, they denote similarly to No. **3a**, p. 613, the force components computed with respect to unit mass of the undeformed medium in the $\xi_1$-$\xi_2$-$\xi_3$-space; If one notices furthermore, that a variation for which at every point

$$\delta x_1 : \delta x_2 : \delta x_3 : \delta x_4 = x_{14} : x_{24} : x_{34} : x_{44} \tag{10}$$

implies only a displacement of the world line in itself, thus [implies] a change of the parameter $\tau$, and that for [this variation] $\delta A$ consequently has to vanish identically, then it follows that

$$-X_4 = \frac{1}{x_{44}} \sum_{i=1}^{3} x_{i4} X_i = X_1 \frac{dx_1}{dt} + X_2 \frac{dx_2}{dt} + X_3 \frac{dx_3}{dt} \tag{9a}$$

— again up to the factor $\frac{1}{x_{44}}$ — which denotes the work done at the mass unit of the undeformed medium in the unit of time.[218])

---

[217] *M. Born*, Ann. d. Phys. (4) 30 (1909), p. 1; especially § 2. — *G. Herglotz*, Ann. d. Phys. (4) 36 (1911), p. 493; especially § 1, 2.

[218] *H. Minkowski*,[210]) a) Anhang; *G. Herglotz*, op. cit., p. 506.

Analog wird (vgl. Nr. **5**, (10)) als Arbeit irgendwelcher am Kontinuum angreifenden Spannungen bei einer virtuellen Verrückung das Integral

$$\delta A_1 = -\int_{\tau_1}^{\tau_2} d\tau \iiint_{(V_0)} \sum_{i,k=1}^{4} X_{ik} \frac{\partial \delta x_i}{\delta \xi_k} d\xi_1 d\xi_2 d\xi_3, \tag{11}$$

die „*Spannungswirkung*" von *H. Minkowski* [219]), angesetzt; da für die virtuelle Verrückung (10) auch $\delta A_1$ identisch verschwinden muss, ergeben sich die Identitäten

$$\sum_{i=1}^{4} x_{i4} \sum_{k=1}^{4} \frac{\partial X_{ik}}{\partial \xi_k} = 0 \tag{11 a}$$

im Inneren des Bereiches $V_0$ und

$$\sum_{i=1}^{4} x_{i4} \sum_{k=1}^{3} \frac{\partial \omega}{\partial \xi_k} X_{ik} = 0 \tag{11 b}$$

am Rande, wofern dessen Gleichung

$$\omega(\xi_1, \xi_2, \xi_3) = 0$$

ist. [220]) Die schon mehrfach angewandten Umformungen gestatten aus dem verallgemeinerten *Hamilton*schen Prinzip, das das Verschwinden von

$$\delta A + \delta A_1 = 0 \tag{12}$$

für alle willkürlichen nur für $\tau = \tau_1$ und $\tau = \tau_2$ identisch verschwindenden virtuellen Verrückungen $\delta x_i$ fordert, die Bewegungsgleichungen zu entnehmen:

$$\varrho_0 X_i + \sum_{k=1}^{4} \frac{\partial X_{ik}}{\partial \xi_k} = 0 \quad \text{innerhalb } V_0, \tag{12 a}$$

$$(i, 1, \ldots 4)$$

$$\sum_{k=1}^{3} \frac{\partial \omega}{\partial \xi_k} X_{ik} = 0 \quad \text{auf dem Rande von } V_0. \tag{12 b}$$

Vermöge der Identitäten (11a), (11b) ist je eine dieser vier Gleichungen von den drei anderen abhängig. Analog wie früher (Nr. **3**c, S. 617 f.) kann man in (11) statt der $\xi_i$ die $x_i$ als unabhängige Variable einführen, und man erhält dann eine den Gleichungen (5) von Nr. **3** entsprechende Form der Bewegungsgleichungen, wie sie von *Minkowski* angegeben wurde. [221])

---

219 *H. Minkowski*, [210]) a) Anhang, Formel (17)

220 *G. Herglotz*, a. a. O., p. 506 f.

221 *H. Minkowski*, [210]) a) Anhang, Formel (20). Die Gleichungen erscheinen

Analogously (cf. No. **5**, (10)), the integral

$$(11) \qquad \delta A_1 = -\int_{\tau_1}^{\tau_2} d\tau \iiint_{(V_0)} \sum_{i,k=1}^{4} X_{ik} \frac{\partial \delta x_i}{\partial \xi_k} d\xi_1 d\xi_2 d\xi_3,^{\dagger}$$

is assumed as work [expression] for any stresses applied at the continuum for a virtual displacement [i. e.] the "*stress action*" of *H. Minkowski* [219]; since for the virtual displacement (10) also $\delta A_1$ has to vanish identically, the [following] identities are obtained

$$(11\text{ a}) \qquad \sum_{i=1}^{4} x_{i4} \sum_{k=1}^{4} \frac{\partial X_{ik}}{\partial \xi_k} = 0$$

in the interior of $V_0$ and

$$(11\text{ b}) \qquad \sum_{i=1}^{4} x_{i4} \sum_{k=1}^{3} \frac{\partial \omega}{\partial \xi_k} X_{ik} = 0$$

at the boundary, provided that the equation thereof is[220])

$$\omega(\xi_1, \xi_2, \xi_3) = 0.$$

The transformations applied already several times allow to extract the equations of motion from the generalized *Hamilton*'s principle, which demands the vanishing of

$$(12) \qquad \delta A + \delta A_1 = 0$$

for all arbitrary virtual displacements [which] vanish identically only for $\tau = \tau_1$ and $\tau = \tau_2$:

$$(12\text{ a}) \qquad \varrho_0 X_i + \sum_{k=1}^{4} \frac{\partial X_{ik}}{\partial \xi_k} = 0 \quad \text{in } V_0, \qquad (i, 1, \ldots 4)$$

$$(12\text{ b}) \qquad \sum_{k=1}^{3} \frac{\partial \omega}{\partial \xi_k} X_{ik} = 0 \quad \text{on the boundary of } V_0.$$

Due to the identities (11a), (11b) each one of these four equations is dependent on the other three. Analogously to before (No. **3**c, p. 617 f.) in (11) one can introduce instead of the $\xi_i$ the $x_i$ as independent variables, and one obtains then a form of the equations of motion corresponding to the equations (5) of No. **3**, as they have been given by *Minkowski*.[221])

† The $\delta$ in the denominator of the original is exchanged by a $\partial$ – (TN)

[219] *H. Minkowski*, [210]) *a) Anhang, formula (17)*

[220] *G. Herglotz*, op. cit., p. 506 f.

[221] *H. Minkowski*, [210]) a) appendix, formula (20). The equations appear

Über die Art der Abhängigkeit der Spannungskomponenten von den Bewegungsfunktionen (7) kann zunächst ganz frei verfügt werden, wenn nur die Relationen (11a), (11b) erfüllt sind; es sei hier nur auf den Potentialansatz Nr. **7**, (26) eingegangen, der auf die von *G. Herglotz* [222]) angegebene Übertragung der Formeln der gewöhnlichen Elastizitätslehre (Nr. **9**, (2), (3)) in die Relativitätstheorie der Lorentzgruppe führt. Es sei also

(13) $$-\delta A_1 = +\delta\Phi = \delta \int_{\tau_1}^{\tau_2} dt \iiint_{(V_0)} \varphi(x_{ik}) d\xi_1 d\xi_2 d\xi_3,$$

wo $\varphi$ nur von den ersten Ableitungen der Bewegungsfunktionen abhängen möge; dann bleibt $\Phi$ bei allen *Lorentz*transformationen nur dann ungeändert, wenn $\varphi$ bis auf den Faktor $x_{44}$ lediglich von den sechs Ruhdeformationen (8) abhängt:

(13 a) $$\varphi = \varphi(e_1, e_2, e_3, g_{12}, g_{23}, g_{31}) \cdot x_{44}.$$

Durch Vergleich von (13) und (11) folgt nun

(13 b) $$X_{ik} = \frac{\partial\varphi}{\partial x_{ik}},$$

und die Substitution dieser Werte in (12a), (12b) liefert die von *Herglotz* angegebenen Grundgleichungen.[223])

*Minkowski* hat die Analogie mit dem klassischen *Hamilton*schen Prinzip noch weiter getrieben, indem er von den Arbeitsausdrücken, in die hier von vornherein die kinetischen Glieder mit eingehen, allgemein einen rein kinetischen Teil abtrennt.[224]) Betrachtet man die Umgebung einer bestimmten Stelle $x_i$ in einem dieser Stelle zugehörigen Ruhkoordinatensystem $\bar{x}_i$ und misst Volumen und Masse des Mediums in dem $\bar{x}_1$-$\bar{x}_2$-$\bar{x}_3$-Raume, so heißt die für den Punkt $\bar{x}_i$ sich ergebende Massendichte $\bar{\varrho}$ die Ruhdichte dieser Stelle. *Minkowski* lässt alsdann die Variation des als *Massenwirkung* bezeichneten, über den betrachteten vierdimensionalen Raum erstreckten Integrales

(14) $$P = \iiiint \bar{\varrho}\, dx\, dy\, dz\, dt$$

in ein wenig modifizierter Form, da bei ihm der Parameter $\tau$ stets die nur durch (15) definierte „Eigenzeit" ist und also bei der Variation stets eine Nebenbedingung und ein Lagrangescher Faktor auftritt.

[222] *G. Herglotz*, a. a. O. p. 503 ff. — Vgl. auch für den speziellen Fall der Hydrodynamik — analog Nr. **10** — die Ansätze von *E. Lamla* Ann. d. Phys. (4) 37 (1912), p. 772.

[223] a. a. O., p. 505 f.

[224] *H. Minkowski*,[210]) a) Anhang, Formel (7) bis (14).

At first, the class with dependence of the stress components on the motion (7) can be chosen completely freely, as long as the relations (11a), (11b) are fulfilled; here only the potential-based approach No. **7**, (26) is considered, which leads to the transmission of the formulas from the ordinary theory of elasticity (No. **9**, (2), (3)) to the theory of relativity of the Lorentz group given by *G. Herglotz* [222]). Let therefore

$$(13) \qquad -\delta A_1 = +\delta\Phi = \delta \int_{\tau_1}^{\tau_2} dt \iiint_{(V_0)} \varphi(x_{ik}) d\xi_1 d\xi_2 d\xi_3,$$

where $\varphi$ may depend only on the first derivatives of the motion; then $\Phi$ remains unchanged for all *Lorentz* transformations only if $\varphi$, up to the factor $x_{44}$, depends merely on the six rest deformations (8):

$$(13\text{ a}) \qquad \varphi = \varphi(e_1, e_2, e_3, g_{12}, g_{23}, g_{31}) \cdot x_{44}.$$

Comparing (13) and (11) it follows now that

$$(13\text{ b}) \qquad X_{ik} = \frac{\partial \varphi}{\partial x_{ik}},$$

and the substitution of these values into (12a), (12b) leads to the fundamental equations given by *Herglotz*.[223])

*Minkowski* pushed the analogy with the classical *Hamilton*'s principle further by separating from the work expressions, in which here from the beginning the kinetic terms are contained, generally a purely kinetic part.[224]) If one considers in the neighborhood of a certain point $x_i$ in a corresponding rest coordinate system $\bar{x}_i$ and [if one] measures volume and mass of the medium in this $\bar{x}_1$-$\bar{x}_2$-$\bar{x}_3$-space, then the mass density $\bar{\varrho}$ which is given for the point $\bar{x}_i$ is called rest density at this point. *Minkowski* lets then add the variation of the so-called *mass action*, [i. e.] the integral extended over the considered four-dimensional space

$$(14) \qquad P = \iiiint \bar{\varrho}\, dx\, dy\, dz\, dt$$

in slightly modified form, since for him the parameter $\tau$ is always the "proper time" defined by (15) and thus for a variation it does appear always a constraint and a Lagrange multiplier.

222 *G. Herglotz*, op. cit. p. 503 ff. — Cf. also for the special case of hydrodynamics — analogous to No. **10** — the foundations of *E. Lamla* Ann. d. Phys. (4) 37 (1912), p. 772.

223 op. cit., p. 505 f.

224 *H. Minkowski*,[210]) a) appendix, formula (7) to (14).

additiv zu (12) hinzutreten, wobei während der Variation die Masse konstant zu halten ist. Verwendet man nun eine spezielle Ortszeit $\tau$, nämlich eine solche, die der Relation

$$c^2\left(\frac{\partial x_4}{\partial \tau}\right)^2-\left(\frac{\partial x_1}{\partial \tau}\right)^2-\left(\frac{\partial x_2}{\partial \tau}\right)^2-\left(\frac{\partial x_3}{\partial \tau}\right)^2=c^2 \tag{15}$$

genügt, so wird $\delta P$ bis auf Randglieder gleich

$$\iiiint \bar{\varrho}\left(\frac{\partial^2 x_1}{\partial \tau^2}\delta x_1+\frac{\partial^2 x_2}{\partial \tau^2}\delta x_2+\frac{\partial^2 x_3}{\partial \tau^2}\delta x_3-\frac{\partial^2 x_4}{\partial \tau^2}\delta x_4\right)dx\,dy\,dz\,dt,$$

und die Faktoren der $\delta x_i$ treten — in völliger Analogie zu den Grundgleichungen der *Newton*schen Mechanik — zu den Gleichungen (12) hinzu. *M. Born* [225]) hat gezeigt, wie man den Massenfaktor auch als *Lagrange*schen Faktor der Nebenbedingung (15) einführen kann.

---

(Abgeschlossen im August 1913.)

[225] Ann. d. Phys. (4) 28 (1909), p. 571.

to (12), whereas during the variation the mass has to be kept constant. If one uses now a special proper time $\tau$, namely the one, which satisfies the relation

$$c^2\left(\frac{\partial x_4}{\partial \tau}\right)^2 - \left(\frac{\partial x_1}{\partial \tau}\right)^2 - \left(\frac{\partial x_2}{\partial \tau}\right)^2 - \left(\frac{\partial x_3}{\partial \tau}\right)^2 = c^2, \tag{15}$$

then $\delta P$ becomes up to boundary terms equal to

$$\iiiint \bar{\varrho}\left(\frac{\partial^2 x_1}{\partial \tau^2}\delta x_1 + \frac{\partial^2 x_2}{\partial \tau^2}\delta x_2 + \frac{\partial^2 x_3}{\partial \tau^2}\delta x_3 - \frac{\partial^2 x_4}{\partial \tau^2}\delta x_4\right) dx dy dz dt,$$

and the factors of $\delta x_i$ are added — in complete analogy to the fundamental equations of the *Newton*ian Mechanics — to the equations (12). *M. Born* [225]) has shown, how one can introduce the mass factor also as *Lagrange* multiplier of the constraint (15).

---

(Completed in August 1913.)

[225] Ann. d. Phys. (4) 28 (1909), p. 571.

## 3.3 Translator's commentaries

The contribution of Hellinger is extraordinary and strongly contrasts with others because of its sharp, precise and meaningful development of a physical theory of continuous media. The side-by-side translation allows Hellinger's concise argumentations to speak for themselves. However, some statements shall not remain uncommented, either to highlight or to relate them to more current developments in the field. Only the topics to which the author is most familiar with are commented. Scientists from complementary fields should be encouraged to do the same for the remaining parts. Besides the mere rediscovery and diffusion of the work, a thorough discussion among scholars is exactly the motivation of this translation.

### *The notion of a continuum*

In order to formulate a general field theory in such a compact form as done by Hellinger, one must accept the concept of a continuum. While many predecessors of Hellinger such as Cauchy, Navier or Piola had been involved in discussions about the atomistic structure of the matter, Hellinger abstained from making any relation to the microscopic composition of the material. In No. 2, a three-dimensional extended continuous medium is introduced as a subset of the three-dimensional space representing the subset of spatial points occupied by the material points in the current configuration. Hellinger introduced in a very modern way the nonlinear kinematics of the three-dimensional continuous body in terms of placement functions relating the subset of material points with the current configuration and even formulated the restriction that the determinant of the corresponding gradient should never be zero. All further introduced field objects are then functions which have as domain either the set of material points $(a, b, c)$ or the set of the spatial points occupied by the material points in the current configuration $(x, y, z)$.

Even though the atomistic structure is not of interest in this article, Hellinger discussed the relation between discrete and continuous systems right after the postulation of the principle of virtual displacements for the three-dimensional continuum on p. 615. There, it is written how discrete systems can approximate the behavior of continuous ones. Moreover, it is pointed out that there is still no systematic and rigorous theory that can deal with such an idea. In fact, Hellinger anticipated the problem of homogenization of discrete systems, which consists to prove that suitable $\varepsilon$-families of solutions of discrete problems converge, when suitable continuation processes are introduced, to the solution of a continuous problem. In this kind of rigorous problems, the concept of Gamma-convergence is now playing a crucial role. The literature on this subject is becoming immense: we quote here [125, 23] among the most interesting papers obtaining first gradient continua as continuous limit while, for what concerns the papers where a higher gradient continuum limit is obtained, we cite [119] and [9, 10, 138, 27].

A less technical approach was already formulated by Hellinger's predecessor Gabrio Piola, who considered a continuum as an approximation needed to deduce results with tractable mathematics. Piola's idea was simple. The "true", or say, the most accurate mathematical model for matter is given by a discrete molecular theory. However, the problems to be solved in using this theory directly are too difficult. Therefore, Piola suggested to homogenize the discrete micro-theory and to deduce the most suitable macro-theory, see Capitolo VI of [120] or pp. 146–164 of [42] for the English translation and [38] for further comments. The therein proposed heuristic variational asymptotic procedure is thus called Piola's micro-macro identification procedure and can be summarized by the following three steps:

1. specify the "most likely" macro-motion once a micro-motion with a scaling parameter $\varepsilon$ is chosen;
2. formulate the principle of virtual displacements at micro- and macro-level;
3. identify the virtual work contributions of the micro- and macro-level when the scaling parameter $\varepsilon$ tends to zero;

It is then the continuous macro-theory which Piola hoped to use for formulating and solving deformation problems of interest in applications. For some examples of different variants of Piola's micro-macro identification procedure, we refer to [46, 66, 17, 16].

## *Variational principles*

In the introduction, No. 1, one finds maybe one of the most precise discussions on variational principles ever written. In an extraordinary concise way Hellinger explained what he means by variational principles; and what should be understood under variational principles. His statement is reinforced by the following lines. The calculus of variations is a mathematical theory whose aim is to find extrema for functionals, usually expressed by means of integrals. To find these extrema one can calculate the first variation of the integral operators involved, by obtaining some linear functionals of the variations of the unknown fields. To base continuum mechanics on an extremum principle may be regarded as a too hazardous choice. Therefore, following Lagrange, we prefer to base the postulation of mechanics by formulating a principle *having the form* of the necessary criterion for being an extremum. This point is rather abstract, but its implications have a marvelous impact, allowing for a very general postulation of physical theories (see for instance [122, 124, 123, 33, 118, 56, 108, 19, 22, 26, 24, 130, 58]). This postulation is based on the principle of virtual displacements also known as principle of virtual work, principle of virtual velocities, principle of virtual power, howsoever you want to call this principle. Note that the first form of this principle has been attributed to the Pythagorean philosopher Archytas of Tarentum, see [165].

Hellinger clearly stated that the unifying mathematical form for all individual physical theories is given by variational principles. He gave plenty of examples and formulated the principle of virtual displacements for the three-dimensional continuum for statics and dynamics, see Eq. (4) in No. 3b and Eq. (1) in No. 5a, for media with oriented particles, see Eq. (2) in No. 4b or No. 7b for the hyperelastic case, for thermodynamics Eq. (2) in No. 15, and many more. Another advantage of the variational formulation is that the principle is formulated using a single formula. This had already been recognized by Gabrio Piola who talked about "quel principio uno, di dove emanano tutte le equazioni che comprendono innumerabili verità", i.e., that fundamental principle from which are emanating all those equations which include innumerable truths, see [42, Chapter 1, p. 110]. Remark that Hellinger accepts in a paper dated 1913 a statement as obvious which is still nowadays denied by some authors. This statement is the following: when formulating a variational principle (in the wider sense given to this expression by Lagrange and Piola, but also when accepting the more restrictive sense considered by Hamilton) one gets "for free" the required boundary conditions. Note that those who refuse variational principles sometimes claim that boundary conditions need to be determined on "physical grounds".

## *The principle of virtual displacements*

Even though "Principle of Virtual Work" is the more contemporary name for the fundamental variational principle in mechanics, we will stick in this commentaries with Hellinger's terminology. Hellinger was extremely precise in the use of terminology, which can be underlined by the following word choices. For the English "displacement", in German there are the two synonyms "Verschiebung" and "Verrückung". While "Verrückung" is a rather old-fashioned word, nowadays it is more common to use "Verschiebung". Nevertheless, throughout his paper, Hellinger distinguished between actual and virtual displacements by attributing to them the words "Verschiebung" and "Verrückung", respectively. In modern literature mainly the word "Verschiebung" is in use. And it is this virtual displacement, which Hellinger introduced very rigorously in No. 2a in terms of Gâteaux derivatives, i.e., variations of the current placement field. So, there is absolutely nothing obscure about virtual displacements as sometimes claimed by opposers of variational principles. Virtual displacements can be as rigorously defined as velocities or infinitesimal displacements in the linearized theory. All of which can be expressed as functions with the set of material points or the current configuration as domain.

In No. 3a, Hellinger wrote unambiguously that the notion of work is the elemental quantity to base on a mechanical theory. With this conception of mechanics, he is a follower of the ideas advocated by d'Alembert and Lagrange. These ideas are explained in a nutshell as follows. First one discusses the kinematics and only then one introduces with the notion of work the laws of dynamics governing the motion of the considered system. In fact, Hellinger started formulating the fundamental problem of mechanics guided by the conceptual frame set up by d'Alembert, see the first edition of the "Traité de dynamique" [34], page viij,ix (end,beginning):

«Mais comment arrive-t'il que le Mouvement d'un corps suive telle ou telle loi particuliére? C'est sur quoi la Geométrie seule ne peut rien nous apprendre, & c'est aussi ce qu'on peut regarder comme le premier Problême qui appartienne immédiatement à la Méchanique.

On voit d'abort fort clairement, qu'un Corps ne peut se donner le Mouvement lui-même. Il ne peut donc être tiré du repos, que par l'action de quelque cause étrangére.»[1]

The external cause ("cause étrangére") evoked by d'Alembert was called by Hellinger force or stress. He used the word force exactly in the same spirit and with the same intentions as d'Alembert. The forces and stresses "applied on" or "applied in" a continuous body have in common, or, are characterized by the fact that they expend a virtual work on virtual displacements. Indeed on page xxv (loc. cit.) d'Alembert warns the reader:

«Au reste, comme cette seconde Partie est destinée principalement à ceux, qui déja instruits du calcul différentiel & intégral, se seront rendus familiers les principes établis dans la premiére, ou seront déja exercés à la solution des Problêmes connus & ordinaires de la Méchanique ; je dois avertir que pour éviter les circonlocutions, je me suis souvent servi du terme obscur de force, & de quelques autres qu'on employe communément quand on traite du Mouvement des Corps ; mais je n'ai jamais prétendu attacher à ces termes d'autres idées que celles qui résultent des principes que j'ai établis, soit dans cette Préface, soit dans la premiére Partie de ce Traité.»[2]

On p. 611, Hellinger gave maybe one of the mathematical most sophisticated and rigorous definition of the virtual work at that time and defined the virtual work as a coordinate independent linear homogeneous function on the space of all possible virtual displacements. This very modern statement in its brief clarity should not need any comment if it were accepted without controversies. Unfortunately, many debates were started about its content even in relatively more modern works and conference discussions. Hence, we remark here that:

- The forces and stresses appear naturally in variational postulations as the dual quantities with respect to virtual displacements and virtual deformations, respectively. They are univocally characterized by the work functionals, so that they do not need to be introduced as independent concepts. For more details see for instance [47, 73, 75, 74, 76, 44, 77, 55].
- Hellinger's mathematical knowledge becomes apparent in the elegant way in which he treats this point. He is a contemporary of Fréchet and Gâteaux and therefore it is most likely that he knew and mastered their ideas and methods.

---

1 «But how it happens that the motion of a body follows this or this other particular law? This is where the Geometry, alone, cannot teach us anything and this is what one can regard as the first Problem which belongs immediately to Mechanics.

One can see immediately [and] really clearly that a body cannot give to him-self a motion. It therefore can be subtracted from a state of rest only by the action of some external cause.»

2 «On the other hand, as this second part is addressed mainly to those who being already learned in differential and integral calculus managed to become familiar with the principles established in the first one, I must warn [these readers] that for avoiding the circumlocutions I have often used the obscure term "force", and some other terms which one commonly employs when he treats the motion of bodies; but I never wanted to attribute to these terms any other ideas [different] from those which result from the principles which I have established, either in this Preface, or in the first Part of this Treatise.»

Moreover, he shows a vision on the concept of distributions which is anticipating the revolutionary results by Laurent Schwartz and their applications to continuum mechanics. Concerning this point the reader is referred to [47, 74].

In Eq. (1) of No. 2, Hellinger introduced a virtual work that consists of three parts and restricted himself to a special case sufficient for the treatment of classical continua. However, he was absolutely aware of possible generalizations by adding more expressions depending on the virtual displacements and their derivatives at certain locations of the continuum as well as line, surface and volume integrals of such expressions. Thus, Hellinger had already anticipated the structure theorems that were proven by Laurent Schwartz, [137], for distributions and that were explicitly considered in [47] and [46, 43, 41] to develop an *N*th-gradient theory of continuum mechanics.

Hellinger formulated the virtual work of the continuum in the actual configuration. On pp. 614, considering an arbitrary subdomain of the continuum, he showed, by integrating by parts, how the stress distribution induces surface contact forces, i.e. the stress vectors, on the boundary of the subdomain. In fact, this leads to the linear relation between the stress vector and the outward pointing unit normal of the subdomain's boundary surface also known as Cauchy's tetrahedron theorem. The integration by parts had already been performed by Piola (see [42, 38]) because of exactly the same reason: to transform the volume expression of internal work into the expression of work expended by surface forces. The delicate question concerning the priority between Piola and Cauchy in the introduction of surface contact forces (we mean the concept generalizing the concept of pressure to solids) needs a very detailed scrutiny, if ever one will be able to solve it. However, the priority of the introduction of the aforementioned integration by parts process for "deducing" the existence of contact forces inside a deformable continuous body has to be attributed to Piola, in both the reference and the current configuration.

In No. 3b., the principle of virtual displacement for the three-dimensional continuum is given. And as promised by Hellinger in the introduction, the local equilibrium equations, Eq. (5a), together with the force boundary conditions, Eq. (5b), are a direct consequence of the principle of virtual displacements. Using the coordinate independence of the work expressions, the equilibrium equations are presented also in terms of the coordinates of the set of material points . In particular, the relation between the Cauchy stress and the 1$^{st}$ Piola-Kirchhoff stress is given in Eq. (8).

In No. 3d, the balance of forces and moments for the body and all its subbodies in integral form are obtained by applying smoothened discontinuous virtual displacements. The section gives thus a connection to the so called "rigidification principle", which states that every part cut out of the deformable continuum exposed to the volume forces applied within the part and the forces applied on the surface must be in equilibrium like a rigid body, see also the discussion in [55]. Using the presented limit argument in the construction of smoothened discontinuous virtual displacements, it seems possible to give an answer to the following question: In formulating the principle of virtual work, do we need to assume that the virtual work vanishes for *all* (regular) virtual displacements of *all* (suitably regular) subbodies of the considered body? Or is it sufficient to assume that it vanishes for *all* regular displacements of the whole body only? Indeed, Hellinger had mastered the concept of mollifiers in three-dimensional Euclidean space whose existence Urysohn proved in a more general setting few years later.[3]

In the last part of No. 3d, Hellinger mentioned also his understanding of Piola's approach to continuum mechanics or rather to the mechanics of rigid bodies. It is not clear if he was aware of the true content of Piola's works. What Hellinger referred to are statements which can be found in Piola's works. However, Piola developed the Lagrangian theory of deformable bodies and, by considering the subset of rigid virtual displacements, he proved that balance of forces and moments are necessary conditions for the equilibrium. Moreover, Piola proved that introducing the constraint of rigidity makes the stress undetermined and therefore he assessed the logical necessity of the introduction of the theory of deformable bodies. It is not clear how the linguistic barrier prevented Hellinger to appreciate completely the value of Piola's works (see [42, 38]).

We close this section by a further comment to clarify that there is not a "petitio principii"[4] hidden in Hellinger's statement of the principle of virtual displacements, how unfortunately too often sustained by the opposers of d'Alembertian-Lagrangian postulation of mechanics and in particular in [155] p. 595 where one can read in the first footnote:

> «The derivation given by HELLINGER [. . . ] fails through *petitio principi*, since the stress components appear in the original variational principle. [. . . ] Existence of the stress tensor can be proved from variational principles which assume the existence of an internal energy having a special functional form.»

The footnote is a comment on the following passage:

> «[...] no variational principle has ever been shown to yield Cauchy's fundamental theorem in its basic sense as asserting that existence of the stress vector implies the existence of the stress tensor.»

Simply the authors of the aforementioned statements do not want to follow the reasonings presented in the works by d'Alembert, Lagrange, Piola and finally Hellinger: the fundamental, primitive concept in mechanics is *(virtual) work* while contact force is a derived concept. One postulates that work is a linear and continuous functional on a set of test functions, i.e., virtual displacements, and then, via the celebrated theory of distributions by L. Schwartz or via a suitable series of regularity ansatz, one gets a representation of work in terms of $N$th order stresses which are defined as

---

[3] Urysohn lemma: For any two disjoint closed sets $A$ and $B$ of a normal space $X$ there exists a real-valued function $f$, continuous at all points, taking the value 0 at all points of $A$, the value 1 at all points of $B$ and for all $x \in X$ satisfying the inequality $0 \le f(x) \le 1$. See for instance [13, 101].

[4] We resist to use in this context the most common English expression "begging the question", as it is usually phrased, as unfortunately it originated in the 16th century as a wrong translation of the Latin correct expression "petitio principii". A correct English translation could be: "assuming the initial point" or even better "a fallacy in which a conclusion is taken for granted in the premises". Remark that obviously very often the conclusion may be accepted in an indirect way such that its presence within the premise is hidden or at least not easily apparent.

the dual in work of $N$th gradient of virtual displacements. There is no logical reason for which contact actions (in the case of first gradient continua they reduce to contact surface forces) must be the most fundamental concept. Actually Piola, Hellinger and many others (see [75, 94, 47, 38, 66, 48, 133] and references cited therein) prefer to consider the stress as the dual to the gradient of the virtual displacement field and to deduce the contact actions as concepts derived in terms of stresses.

To be more precise, d'Alembertian postulation of Mechanics is based on the principle of virtual displacements which is formulated following the subsequent steps:

1. to introduce an admissible set of configurations and an admissible kinematics, specifying the set of all possible motions,
2. to introduce the required work functionals in order to model ALL interactions of the system, including the inertial work, which was considered explicitly by d'Alembert and is given in terms of kinematic quantities including accelerations.[5]
3. to postulate that the sum of internal work plus external work plus inertial work, hence the total virtual work, is vanishing.

In this postulation scheme, the word force and stress is simply used to describe the structure of the work functionals and should not be considered as primitive concept. In particular, there is no need to postulate any balance of forces. Certainly, these balances can be derived as it was done by Hellinger in No. 3d or in [66]. It seems that Cauchy's tetrahedron theorem, which was considered as the only possible way for founding continuum mechanics in [155] (see there p. 595), although very interesting and meaningful, cannot be regarded as the "unavoidable" basis of continuum mechanics (see e.g. [46]). In the works of Cauchy one cannot find such a strong statement:[6] Cauchy followers seem much more extreme than Cauchy himself.

## *Axiom of power of internal forces*

For one point, Hellinger can be criticized. Interestingly, he left the variational postulation scheme, when arguing below Eq. (11) of No. 3d about the symmetry of the stress components. There, he wrote that one has to postulate the law of equal areas, i.e., the balance of moments, to obtain the symmetry of the stress components.

[5] This treatment of the inertia forces is a little unsatisfactory as it already relates force quantities with kinematic quantities, i.e., it includes the relation that the inertia forces are proportional to the accelerations. In No. 5d of Hellinger's work, a general principle for dynamics is introduced in which momentum is considered as dual quantity to the time derivative of the virtual displacement field. Thus, in such an ansatz the relation between momentum and kinematical quantities remains unspecified.

[6] While Cauchy's lemma and the symmetry of the stress tensor are formulated in [28] as "Théorème I" and "Théorème II", respectively. The celebrated stress theorem of Cauchy has to be extracted out of the text and the formulas on pp. 68-69.

So it seems that he was not aware of the "Axiom of power of internal forces", which later was postulated by Germain, see [73, 74, 75][7]. This axiom is in fact the variational postulate that generalizes the law of action-reaction stated by Newton for systems of point masses. Once more, the power of variational postulates becomes striking here, since the same postulate can be used for systems of point masses, systems of rigid bodies, or, as used by Germain, for continua. All these systems differ in the admissible kinematics, the corresponding work functionals and consequently the appearing force quantities. The axiom of power of internal forces leads then to restrictions of the internal force effects. These are for systems of point masses that forces between two point masses are equal in length and opposite in direction and that the direction is given by the connection line between the two points. The interaction between rigid bodies is given by forces and moments, where the forces are equal in length and opposite in directions, and the moments together with the induced moments of the interaction forces are equal in length and opposite in direction. For continua, in which only stress models the internal force effects, the symmetry of the Cauchy stress follows.

In No. 6, starting on p. 637, Hellinger included a lucid distinction between internal and external force effects, which is precisely in the sense of Germain's definition [73, 74] (see also the very useful textbooks of J. Salençon [132, 131, 133]). And it is in the final section of the article, on p. 686, where Hellinger redeemed himself and surprises with the formulation of the "axiom of power of internal forces":

> «It is said briefly, that the entire physics is invariant with respect to the group of all orthogonal coordinate transformations of the ordinary geometry, the so-called "basic group" or "Euclidean group". Herefrom it follows in particular, that *the virtual work of all internal effects within a continuous system necessarily vanishes for a virtual displacement corresponding to an infinitesimal change of the coordinate system*, or that the total potential of these effects remain unchanged for any such displacement of the continuum, i.e. [that the potential] is a euclidean potential in the sense of E. and F. Cosserat (cf. No. 7b, p. 650).»

## *Second-gradient materials*

The massive potential of variational principles becomes apparent in No. 4, where extensions of the virtual work contributions are discussed. With an intriguing naturalness, Hellinger introduced second-gradient continua by augmenting the virtual work by a linear form on the 18 second derivatives of the virtual displacements. In these lines one can read one of the first traces of the second-gradient continua theory which after many decades were developed in detail by Toupin and Germain, see [149, 74]. Even though the details are not carried out, Hellinger had already recognized the appearing surface tension contributions as well as contributions dual to the derivatives normal to the tangent. These described contact actions were intensively studied by Germain [74] and rediscovered in [71]. Interestingly, Hellinger wrote that

[7] The English translation of [74] can be found in [77].

there is not yet an application of the force effects dual to the derivatives normal to the tangent.

On pp. 639–640 one can read the following:

> «The state of deformation at a position is described more precisely, if one uses besides the first also *higher spatial derivatives* of the functions (1), i. e. the deformation in the neighborhood is approximated by a transformation of higher order instead of a linear one; the dependence of the stresses on the deformation will be represented more completely, if one includes also these higher derivatives in the material laws. In fact, one has considered so far [derivatives which are] not higher than second derivatives, this is namely required not until then, when the state of the medium varies *very quickly* in space; the stresses at a position then depends also on the spatial slope of the common deformation quantities of 1. order.»

The senior author of the exegetic series [59, 61, 62], who cannot easily understand German, was rather astonished by the just quoted paragraph. Indeed, because of the paper of Gurtin [95] and all papers influenced by it, for a long time it was believed, in a certain group of scientists and in a certain cultural milieu, that second- and a fortiori higher-gradient materials were logically NOT possible. In the aforementioned paper, one finds on p. 341 the following very clear statement: «One might ask the question: is it possible to have a material which obeys (1.6) but is not a simple elastic material? Here we prove that it is not.» What is astonishing is that the footnote 8 in the same paper WAS APPARENTLY not read by many followers of Gurtin (and sometimes one has the impression that Gurtin himself for a long period forgot his own footnote). This footnote reads:

> «Thus the stress cannot depend upon the gradients of F and $\eta$[8]. Of course this does not mean that higher order elasticity theories which include multipolar stresses are incorrect because of the dependence of these stresses on the higher order gradients. It simply means that one should not include such higher gradients if multipolar stresses are not included.»

After a long neglect of his own footnote, the late works of Gurtin came back to higher order gradient theories (see e.g. [71] and all related and subsequent papers), where the contribution by Toupin were reevaluated. There is also a large amount of papers available which prove the applicability of higher-gradient theories, cf. for instance [63, 12, 8, 7, 2, 1, 113, 102, 3, 81, 80]. A very interesting field of applications for higher gradient theories lies in the description of pantographic structures, see [40, 17, 16, 18, 45, 49, 86, 141, 168, 25] to mention just a few.

On p. 645, Hellinger came back to second-gradient materials for the case of hyperelasticity. In fact, he considered volume energy densities that depend also on the second derivatives of the placement functions. As a particular subset of such continua, he discussed continua whose potential is augmented by a potential with a surface energy density depending only on first derivatives of the placement functions. As mentioned by Hellinger, this surface energy density can be transformed to a volume energy density depending on the second derivatives of the placement functions. He focused on this particular form in order to formulate capillarity in No. 12. With this comments, Hellinger sketched a series of results which, in a completely analytical form, were later developed in [98].

[8] With $F$ and $\eta$ the deformation gradient and the entropy are meant.

## *Media with oriented particles*

On p. 610, the kinematics of a generalized continuum is introduced augmenting the placement functions by angle functions that describe for each material point an orthonormal director triad. Hellinger had perceived the importance of the pioneering works by the Cosserat brothers, [32]. It has to be remarked that for many years their results have been nearly ignored: A conjecture is (see [107]) that this circumstance is related to the fact that their presentation is systematically based on Hamilton's principle.

In No. 4b and at the end of No. 5d, the static and dynamic theory of an oriented three-dimensional continuum is discussed. The hyperelastic case is treated in No. 7b. Again, the ideas presented by Hellinger are bound to be developed later for instance by Toupin and Germain (see [149, 74]) and have a huge impact in contemporary research. Without trying to write a complete list of the papers developing more recently the subject, we refer here to the following papers and textbooks: [144, 11, 57, 82, 157, 110, 88, 160, 148] and the references cited therein. All of the cited papers accept the point of view of Hellinger and base their treatment on the solid ground of suitable variational principles. The reader should remark that the extended kinematics considered by Hellinger includes micro-rotations, but does not consider micro-deformations. A clear variational treatment of continua with micro-stretch is presented in [75], where the ideas of Hellinger were fully developed. How much of the so called "modern" theory of continua with directors was already available to Hellinger is again surprising. Except for, maybe, the notation which often became more compact using tensorial algebra, the equations of the present subsection have been rediscovered several times, since 1913.

## *Lower-dimensional continua*

The theory of media with oriented particles is closely related to lower-dimensional continua. The theory of two- and one-dimensional continua presented in No. 3e can describe only membranes and strings. The application range of this kind of continua is rather limited and also quite delicate due to the loss of stiffness for certain configurations. For instance, a horizontally, straight string cannot resist to an applied vertical force. However, some interesting applications of the continuum models introduced here have been found (see e.g. [65, 31, 72]).

Lower-dimensional continua can be indeed used as "reduced-order" approximate models for three-dimensional bodies having one or two dimensions preponderant with respect to the other two or one, even if some relevant deformation energy is stored in the changes of shape "along" neglected dimensions. In a direct theory, some extra kinematical descriptors can be added to the two- or one-dimensional continuum to include the effect of the neglected dimensions. In the last part of No. 4b, Hellinger augmented in the style of the Cosserat brothers each material point with a director triad and stated the corresponding virtual work contribution. The presented continua have orientations that remain unaffected by the change of the base curve or surface. To obtain a virtual work expression that couples these effects, the easiest is to be guided by a theory of hyperelastic one- or two-dimensional continua as sketched by Hellinger on pp. 666–668. We refer to [147, 69, 67] for a variational formulation in the case of one-dimensional continua generalized by an orthonormal director triad. For the numerical treatment and some interesting applications see among others [97, 96, 68, 90, 91, 156, 79, 21, 84, 83, 140, 159, 29, 30, 92, 53, 64]. The extra kinematics can also be due to higher-gradient effects such as in the pantographic beam, [17, 158, 89], which is a generalization of the Euler-Bernoulli beam similar to [150]. Higher-gradient elasticity effects can also be introduced for two-dimensional continua [50, 146, 87, 134, 136, 143, 52, 5, 162, 161, 36].

In contrast to the direct theory, one can try to deduce the governing equations via a reduction process. This is what Hellinger suggested in No. 8 giving a short but elegant resumé of the most important results in asymptotic analysis as applied to reduced order and reduced dimension mechanical models. The ontological intrinsic three-dimensional nature of deformable bodies is here, p. 658, synthetically described «In reality there exists always a three-dimensional extended domain» and the mathematical nature of the abstraction which leads to lower dimensional continua is clearly stated by means of the introduction of the concept of "families" of models depending on a small parameter $\varepsilon$ and in the calculation of suitable limits when this parameter is vanishing. Remark the very elegant observation about the interpretation of the gradients in the orthogonal direction as Cosserat triad. Hellinger gave also a careful list of difficulties which may arise in the asymptotic expansions and that some research is required in to understand this approach better.

## Constitutive laws

The structure of the encyclopedia article is pretty clear. After the introduction of the kinematics and the corresponding virtual work contributions of the considered systems, the principle of virtual displacements is postulated. With the virtual work contributions, it becomes apparent what force effects model the interaction mechanisms of the continuous system. However, there is yet no need for specifying the relation between these force effects and the placement functions of the continuum. These relations are then discussed in Part III of the article and define eventually the individual fields of continuous media. Hellinger had thus already recognized that the theory of continuous media includes and generalizes the theory of elasticity, No. 9, and hydrodynamics, No. 10.

Hellinger accepted the constitutive laws for the stress, i.e., the material laws, to be of a very general form, but gives one important restriction, which can be read in the last paragraph of the following quote from p. 638.

> «The values of the stress components $X_x, \ldots, Z_z$ corresponding to the particle $a, b, c$ located at time $t$ at the position

(1) $$x = x(a, b, c; t), \quad y = y(a, b, c; t), \quad z = z(a, b, c; t),$$

must be given by the material laws for every possible motion of the continuum; hence [the values] are represented explicitly as expressions of any kind depending on $a$, $b$, $c$, $t$ and the functions (1). [These expressions] also include besides the values of the functions [(1)] and their spatial and time derivatives at the positions $a, b, c, t$ possibly values at other positions $\bar{a}, \bar{b}, \bar{c}, \bar{t}$ and in general the complete history in the domain of variability of the four variables (integrals and similar ones) — Hence, symbolically written in the form:

(2) $$F(a, b, c, t; x(\bar{a}, \bar{b}, \bar{c}, \bar{t}), \ldots\ldots).$$

Changing over to another orthogonal coordinate system $x$, $y$, $z$, then these nine expressions of the stress components have to be transformed like the components of a dyad (and similarly the expressions for $X, Y, Z$ like vector components and so on); if it concerns internal force effects, then there must exist equations between the transformed components and the new coordinates [which are] exactly of the old form.»

This last sentence is a concise statement of the *principle of objectivity* (or *principle of material frame-indifference*) for constitutive equations for stress. The reader is invited to compare the present section by Hellinger to the statements found in Truesdell's First Course in Rational Mechanics [153]. Once the difference of notation is taken into account, the reader will remark the substantial coincidence of the content presented in both works – obviously the change of notations may make a content clearer but for sure it is not changing the attribution of scientific priority. In [153] (Chap. IV Constitutive Equations, Sect. 2 Constitutive Equations. Noll's Axioms) on p. 200, one reads: «The further development of continuum mechanics in this book will fall within the axioms laid down by Noll in 1958.»[9] On p. 202, one finds the "Axiom N3. Principle of Material Frame-Indifference.", where capital N stands for Noll. The attentive reader will immediately remark that Truesdell claims that Noll has written in formulas exactly what Hellinger said in words. This transcription into formulas does not seem enough to attribute the axiom to Noll. Consider that Hellinger finished his work in 1913 and did not attribute this axiom to himself. If one gives a glance to the historical overview about the principle of material-frame indifference in [154] and by considering Truesdell's awareness of Hellinger's article (see again the first footnote on p. 595 of [155]), then Truesdell can be criticized in his own words found in [151] on p. 152: «Lagranges historische Angaben beziehen sich gewöhnlich auf die richtigen Quellen, verdrehen oder verringern jedoch ihren Inhalt.»[10]

On p. 640, Hellinger presented also a precise statement of material symmetry.

«For all laws of the class (3) the question, how these equations behave under a transformation of the directions of the $a$-$b$-$c$ parameter lines through these points $a, b, c$, while the $x$-$y$-$z$-coordinates remain unchanged, is of fundamental evidence. Thereby it is determined namely, if and which different directions through a point of the medium are tantamount for its constitution, provided that it is expressed in the considered material laws, i. e. it is decided on *isotropy or aeolotropy of the medium*;»

[9] Noll 1958 corresponds to reference [114] in the chapter at hand.

[10] Most of [151], even in English, can be found in [152], including the hypothesis about Lagrange on p. 247: «Lagrange's histories usually give the right references but misrepresent or slight the contents.»

So, Hellinger clearly distinguished between objectivity and material symmetry. The theoretical basis established by Hellinger contributed to the establishment of the modern classification of constitutive equations based on their symmetry group: we find remarkable the recent contributions by [117, 116, 6, 126, 57].

The variational formulation of continuum mechanics allowed Hellinger to introduce in No. 7 hyperelastic material laws without a direct consideration of any thermodynamical theory. Even though he is absolutely aware of the connection to thermodynamics. On p. 643, one reads:

> «[...], *that the virtual work coming into question is, up to sign, for every virtual displacement equivalent to the variation of a single scalar expression depending only on the corresponding state of deformation, [which is] the "potential" or the "potential energy" of the acting forces and stresses*; this assumption can be traced back to general theorems of thermodynamics.»

More important is the relation to the calculus of variations and to extremum principles which follow naturally in the variational postulation scheme and which were discussed by Hellinger in No. 7d. In fact, he formulated the principle of stationary potential energy and discussed the possibility for the existence of a minimum principle. Moreover, he described the methods to be used in order to characterize the stability of configurations for infinite dimensional systems. He is aware of the fact that not all norms are equivalent in infinite dimensional systems: indeed, he warns the reader about the fact that different concepts of "neighborhood" are possible in the considered context. Hellinger proved himself once more as a first class mathematician.

In No. 9, Hellinger introduced the foundations of finite elasticity theory, which is based on the choice of a strain energy density that depends on the Green-Lagrange strain measure, see Eq. (1) and (2) on p. 663. Hellinger recognized the objectivity of this strain measure, which leads directly to symmetric stress components. Moreover, for isotropic materials, he showed the important result that the strain energy density can only depend on the three invariants of the Green-Lagrange strain. For a contemporary introduction to finite elasticity theory, we refer to [145]. The variational approach enlightens the essential role of deformation energy in constitutive theory [50, 85] and gives a guidance to the developments of identification methods. Therefore the recents works [104, 164, 70, 15, 4, 35, 121, 105, 163] seem to follow a research program envisioned by Hellinger.

## *Hellinger–Reissner principle*

Without telling it explicitly, with No. 7e, Hellinger has left an original contribution to mechanical science, which has especially impacted computational mechanics [135, 127]. Note that the formulated variational principle holds in the general nonlinear regime. Due to the presentations of the same variational principle, including also boundary terms, in Reissner [128, 129], this principle is generally referred to as the Hellinger–Reissner principle. On pp. 2.14–2.15 of [100], Reissner translated the present No. 7e into English and gives an astonishing commentary in which he questioned the historical relevance of Hellinger's contribution due to technical details:

> «While the absence of any consideration of boundary integrals in the above is generally known, other difficulties appear not to have been noted previously. These include the entirely casual reference to the matter of the invertibility of the relations $s_{ij} = \partial\phi/\partial z_{j,i}$[11] (which is, of course, a much more significant restriction than the corresponding condition for $\sigma_{ij} = \partial\Sigma/\partial\varepsilon_{ij}$), the absence of a concern with conditions on $\phi$ or $H$ so as to ensure moment equilibrium, and, most importantly, the unqualified conclusion concerning the statement of a general variational theorem for *stresses* alone, as an obvious consequence of (1.38)[12], with this clearly being the purpose of this section, given the wording of the heading of the section. Altogether, these difficulties make it questionable whether it is in fact historically meaningful to consider Hellinger's considerations as a stepping-stone to the variational theorem for displacements and stresses in Ref. [15][13].»

Concretely, Reissner demanded priority about this variational principle for himself. We dare to construct a provocative hypothesis concerning mechanical science after World War II. Scientists in mechanics from the United States, thus scientists from the victorious power, tried to demolish the scientific heritage of Europe by slighting the contents of the earlier contributions or by not even citing the correct references. The rise of the English language as the new lingua franca, as discussed in the first part of the exegetic series of Hellinger's article, [59], played amongst others into the hands of Truesdell and Reissner[14] to rewrite the recent history of mechanics. Thus, one could claim that also the "history of mechanical science is written by the victors". Certainly, this provocative statement should be investigated further and more scientifically.

Hellinger's article supplies, more generally, the conceptual framework for a wide class of numerical integrations schemes to be used in generalized continuum mechanics. Following Hellinger's spirit, it is clear that there is not a preferred way for discretizing the evolutionary equations of the continua. While in first gradient continua, finite element method with piece-wise polynomials can be considered a universally efficient tool, in the case of generalized continua more sophisticated discretization techniques must be developed. Therefore, we can consider that the papers [169, 115, 93] are continuing a research stream started by Hellinger.

## *Peridynamics*

Another interesting example of the question of priority concerns the field of peridynamics, which according to Silling [139] gives «a new framework for the basic

---

[11] With $\phi$ and $s_{ij}$ the potential $\varphi$ and the stress components $X_a, X_b, \ldots, Z_c$ of Hellinger are meant.

[12] This corresponds to Eq. (20) on p. 654.

[13] This corresponds to reference [129] in current chapter.

[14] Even though Reissner is of German origin, he got his scientific education in the United States and received his United States citizenship in 1945 at the age of 32.

equations of continuum mechanics». Peridynamics is a non-local theory, where each material point of the body can interact with all other points of the body [99, 166, 167, 112]. For the case of elasticity, i.e., when the interaction between two material points is described by a potential, one can find such a formulation on pp. 646–647 in Hellinger's article. In previous papers, see [38], the credit as first founder of peridynamics is given to Piola. Hellinger credits Duhem [54] and also claims that Duhem has found results to assure when peridynamics reduces to classical elasticity. Therefore not only in the Italian mechanical literature, but also in the German and French literature, peridynamics was known long before its modern formulation by Silling [139].

## *Closing remarks*

Hellinger's article is, as one could expect from an encyclopedia article, definitively a treasure of ideas. He not only collected and ordered the concepts available at his time, but he contributed immensely to the variational formulation of continuum mechanics. He can definitively be set in one line with d'Alembert, Lagrange, Piola and the Cosserat brothers.

The author must admit that he was only able to comment a small part of Hellinger's work. Some underestimated sections are definitively the treatment of dynamics, where Gauss' principle of least constraint is discussed and where the connection to the principle of Hamilton is made. Also the treatment of constraints, which in the variational formulation of mechanics is a natural concept, would have deserved more attention. Extremely interesting would be an evaluation of the Nos. 13–16 by contemporary experts from the fields of optics, electrodynamics, thermodynamics as well as from relativity theory.

It is hoped that the reader could recognize the sharpness and precision of Hellinger's contribution, which reflects also the spirit of the scientific environment at his time. The aspiration to "concise brevity" is completely coherent with the concept of "economy of science" by Ernst Mach.[15] Mach's positivistic views were greatly influenced by the Vienna Circle and by Ludwig Wittgenstein, whose rigorous style was adopted also by Mach, when dealing with the history of mechanics. On page 481 of [103], having as a synoptic side note "The basis of science, economy of thought." one can read:

> «It is the object of science to replace, or *save*, experiences, by the reproduction and anticipation of facts in thought. Memory is handier than experience, and often answers the same purpose. This economical office of science, which fills its whole life, is apparent at first glance; and with its full recognition all mysticism in science disappears. Science is communicated by instruction, in order that one man may profit by the experience of another and be spared the trouble of accumulating it for himself; and thus, to spare posterity, the experiences of whole generations are stored up in libraries.»

[15] See e.g. Section 4 of Chapter IV of the book "The science of mechanics; a critical and historical account of its development", [103].

Maybe Mach's most important statement starts at the bottom of page 489 of [103]:

> «But, as a matter of fact, within the short span of a human life and with man's limited powers of memory, any stock of knowledge worthy of the name is unattainable except by the greatest mental economy.»

It seems that this greatest mental economy has been reached in the presentation by Hellinger and can be achieved in mechanics only when working with variational formulations.

## References


1. Abali, B.E., Barchiesi, E.: Additive manufacturing introduced substructure and computational determination of metamaterials parameters by means of the asymptotic homogenization. Continuum Mechanics and Thermodynamics pp. 1–17 (2020)
2. Abali, B.E., Müller, W.H., dell'Isola, F.: Theory and computation of higher gradient elasticity theories based on action principles. Archive of Applied Mechanics **87**(9), 1495–1510 (2017)
3. Abali, B.E., Müller, W.H., Eremeyev, V.A.: Strain gradient elasticity with geometric nonlinearities and its computational evaluation. Mechanics of Advanced Materials and Modern Processes **1**(1), 1–11 (2015)
4. Abali, B.E., Wu, C.C., Müller, W.H.: An energy-based method to determine material constants in nonlinear rheology with applications. Continuum Mechanics and Thermodynamics **28**(5), 1221–1246 (2016)
5. Abbas, I.A., Abdalla, A.E.N.N., Alzahrani, F.S., Spagnuolo, M.: Wave propagation in a generalized thermoelastic plate using eigenvalue approach. Journal of Thermal Stresses **39**(11), 1367–1377 (2016)
6. Abdoul-Anziz, H., Auffray, N., Desmorat, B.: Symmetry classes and matrix representations of the 2D flexoelectric law. Symmetry **12**(4), 1–29 (2020)
7. Abdoul-Anziz, H., Seppecher, P.: Strain gradient and generalized continua obtained by homogenizing frame lattices. Mathematics and Mechanics of Complex Systems **6**(3), 213–250 (2018)
8. Abdoul-Anziz, H., Seppecher, P., Bellis, C.: Homogenization of frame lattices leading to second gradient models coupling classical strain and strain-gradient terms. Mathematics and Mechanics of Solids **24**(12), 3976–3999 (2019)
9. Alibert, J.J., Della Corte, A.: Second-gradient continua as homogenized limit of pantographic microstructured plates: a rigorous proof. Z. Angew. Math. Phys. **66**(5), 2855–2870 (2015)
10. Alibert, J.J., Seppecher, P., dell'Isola, F.: Truss modular beams with deformation energy depending on higher displacement gradients. Mathematics and Mechanics of Solids **8**(1), 51–73 (2003)
11. Altenbach, J., Altenbach, H., Eremeyev, V.A.: On generalized Cosserat-type theories of plates and shells: a short review and bibliography. Archive of Applied Mechanics **80**(1), 73–92 (2010)
12. Andreaus, U., dell'Isola, F., Giorgio, I., Placidi, L., Lekszycki, T., Rizzi, N.L.: Numerical simulations of classical problems in two-dimensional (non) linear second gradient elasticity. International Journal of Engineering Science **108**, 34–50 (2016)
13. Arkhangel'Skii, A., Ponomarev, V.I.: Fundamentals of general topology: problems and exercises, vol. 13. Springer Science & Business Media (2001)
14. Auffray, N., dell'Isola, F., Eremeyev, V.A., Rosi, G.: Analytical continuum mechanics à la Hamilton–Piola least action principle for second gradient continua and capillary fluids. Mathematics and Mechanics of Solids **20**(4), 375–417 (2013)

15. Auger, P., Lavigne, T., Smaniotto, B., Spagnuolo, M., dell'Isola, F., Hild, F.: Poynting effects in pantographic metamaterial captured via multiscale dvc. The Journal of Strain Analysis for Engineering Design pp. 1–16 (2020)
16. Barchiesi, E., Eugster, S.R., dell'Isola, F., Hild, F.: Large in-plane elastic deformations of bi-pantographic fabrics: asymptotic homogenization and experimental validation. Mathematics and Mechanics of Solids **25**(3), 739–767 (2020)
17. Barchiesi, E., Eugster, S.R., Placidi, L., dell'Isola, F.: Pantographic beam: a complete second gradient 1D-continuum in plane. Zeitschrift für angewandte Mathematik und Physik **70**(5), 135–1–24 (2019)
18. Barchiesi, E., Harsch, J., Ganzosch, G., Eugster, S.R.: Discrete versus homogenized continuum modeling in finite deformation bias extension test of bi-pantographic fabrics. Continuum Mechanics and Thermodynamics pp. 1–14 (2020)
19. Barchiesi, E., Khakalo, S.: Variational asymptotic homogenization of beam-like square lattice structures. Mathematics and Mechanics of Solids **24**(10), 3295–3318 (2019)
20. Barchiesi, E., Spagnuolo, M., Placidi, L.: Mechanical metamaterials: a state of the art. Mathematics and Mechanics of Solids **24**(1), 212–234 (2019)
21. Baroudi, D., Giorgio, I., Battista, A., Turco, E., Igumnov, L.A.: Nonlinear dynamics of uniformly loaded elastica: Experimental and numerical evidence of motion around curled stable equilibrium configurations. Zeitschrift für Angewandte Mathematik und Mechanik **99**(7), 1–20 (2019)
22. Bersani, A.M., Caressa, P.: Lagrangian descriptions of dissipative systems: a review. Mathematics and Mechanics of Solids pp. 1–19 (2020)
23. Braides, A., Gelli, M.S.: Continuum limits of discrete systems without convexity hypotheses. Mathematics and Mechanics of Solids **7**(1), 41–66 (2002)
24. Capobianco, G., Eugster, S.R.: Time finite element based Moreau-type integrators. International Journal for Numerical Methods in Engineering **114**(3), 215–231 (2018)
25. Capobianco, G., Eugster, S.R., Winandy, T.: Modeling planar pantographic sheets using a nonlinear Euler–Bernoulli beam element based on B-spline functions. Proceedings in Applied Mathematics and Mechanics **18**(1), e201800220:1–2 (2018)
26. Capobianco, G., Winandy, T., Eugster, S.: The principle of virtual work and Hamilton's principle on Galilean manifolds. Journal of Geometric Mechanics pp. 1–27 (2021)
27. Carcaterra, A., dell'Isola, F., Esposito, R., Pulvirenti, M.: Macroscopic description of microscopically strongly inhomogenous systems: A mathematical basis for the synthesis of higher gradients metamaterials. Archive of Rational Mechanics and Analysis **218**(3), 1239–1262 (2015)
28. Cauchy, A.: De la pression ou tension dans un corps solide. In: Œuvres Complètes, 2, vol. 7, pp. 60–78. Gauthier-Villars et Fils (1827)
29. Cazzani, A., Malagù, M., Turco, E.: Isogeometric analysis of plane-curved beams. Mathematics and Mechanics of Solids **21**(5), 562–577 (2016)
30. Cazzani, A., Malagù, M., Turco, E., Stochino, F.: Constitutive models for strongly curved beams in the frame of isogeometric analysis. Mathematics and Mechanics of Solids **21**(2), 182–209 (2016)
31. Cazzani, A., Wagner, N., Ruge, P., Stochino, F.: Continuous transition between traveling mass and traveling oscillator using mixed variables. International Journal of Non-Linear Mechanics **80**, 82–95 (2016)
32. Cosserat, E., Cosserat, F.: Théorie des Corps déformables. Librairie Scientifique A. Hermann et Fils (1909)
33. Cuomo, M., Contrafatto, L., Greco, L.: A variational model based on isogeometric interpolation for the analysis of cracked bodies. International Journal of Engineering Science **80**, 173–188 (2014)
34. d'Alembert, J.B.l.R.: Traité de Dynamique (1743)
35. De Angelo, M., Barchiesi, E., Giorgio, I., Abali, B.E.: Numerical identification of constitutive parameters in reduced-order bi-dimensional models for pantographic structures: application to out-of-plane buckling. Archive of Applied Mechanics **89**(7), 1333–1358 (2019)

36. De Angelo, M., Spagnuolo, M., D'Annibale, F., et al.: The macroscopic behavior of pantographic sheets depends mainly on their microstructure: experimental evidence and qualitative analysis of damage in metallic specimens. Continuum Mechanics and Thermodynamics **31**(4), 1181–1203 (2019)
37. Del Vescovo, D., Giorgio, I.: Dynamic problems for metamaterials: review of existing models and ideas for further research. International Journal of Engineering Science **80**, 153–172 (2014)
38. dell'Isola, F., Andreaus, U., Placidi, L.: At the origins and in the vanguard of peridynamics, non-local and higher-gradient continuum mechanics: An underestimated and still topical contribution of Gabrio Piola. Mathematics and Mechanics of Solids **20**(8), 887–928 (2015)
39. dell'Isola, F., Della Corte, A., Giorgio, I.: Higher-gradient continua: The legacy of Piola, Mindlin, Sedov and Toupin and some future research perspectives. Mathematics and Mechanics of Solids **22**(4), 852–872 (2017)
40. dell'Isola, F., Giorgio, I., Pawlikowski, M., Rizzi, N.L.: Large deformations of planar extensible beams and pantographic lattices: heuristic homogenization, experimental and numerical examples of equilibrium. Proceedings of the Royal Society A: Mathematical, Physical and Engineering Sciences **472**(2185), 20150790:1–23 (2016)
41. dell'Isola, F., Madeo, A., Seppecher, P.: Cauchy tetrahedron argument applied to higher contact interactions. Archive of Rational Mechanics and Analysis **219**(3), 1305–1341 (2015)
42. dell'Isola, F., Maier, G., Perego, U., Andreaus, U., Esposito, R., Forest, S.: The complete works of Gabrio Piola: Volume I: Commented English Translation. No. 38 in Advanced Structured Materials. Springer (2014)
43. dell'Isola, F., Seppecher, P.: Edge contact forces and quasi-balanced power. Meccanica **32**(1), 33–52 (1997)
44. dell'Isola, F., Seppecher, P.: "Hypertractions and hyperstresses convey the same mechanical information Continuum Mech. Thermodyn. (2010) 22:163–176" by Prof. Podio Guidugli and Prof. Vianello and some related papers on higher gradient theories. Continuum Mechanics and Thermodynamics **23**(5), 473–478 (2010)
45. dell'Isola, F., Seppecher, P., Alibert, J.J., et al.: Pantographic metamaterials: an example of mathematically driven design and of its technological challenges. Continuum Mechanics and Thermodynamics **31**(4), 851–884 (2019)
46. dell'Isola, F., Seppecher, P., Della Corte, A.: The postulations à la D'Alembert and à la Cauchy for higher gradient continuum theories are equivalent: a review of existing results. P. Roy. Soc. Lond. A Mat. **471**(2183) (2015)
47. dell'Isola, F., Seppecher, P., Madeo, A.: How contact interactions may depend on the shape of Cauchy cuts in nth gradient continua: approach "à la D'Alembert". Zeitschrift für Angewandte Mathematik und Physik **63**(6), 1119–1141 (2012)
48. dell'Isola, F., Seppecher, P., Placidi, L., Barchiesi, E., Misra, A.: Least action and virtual work principles for the formulation of generalized continuum models. In: F. dell'Isola, D. Steigmann (eds.) Discrete and Continuum Models for Complex Metamaterials, pp. 327–394. Cambridge University Press (2020)
49. dell'Isola, F., Seppecher, P., Spagnuolo, M., et al.: Advances in pantographic structures: design, manufacturing, models, experiments and image analyses. Continuum Mechanics and Thermodynamics **31**(4), 1231–1282 (2019)
50. dell'Isola, F., Steigmann, D.: A two-dimensional gradient-elasticity theory for woven fabrics. Journal of Elasticity **118**(1), 113–125 (2015)
51. dell'Isola, F., Steigmann, D.J. (eds.): Discrete and Continuum Models for Complex Metamaterials. Cambridge University Press (2020)
52. Desmorat, B., Spagnuolo, M., Turco, E.: Stiffness optimization in nonlinear pantographic structures. Mathematics and Mechanics of Solids **25**(12), 2252–2262 (2020)
53. Deutschmann, B., Eugster, S.R., Ott, C.: Reduced models for the static simulation of an elastic continuum mechanism. IFAC-PapersOnLine **51**(2), 403 – 408 (2018). 9th Vienna International Conference on Mathematical Modelling
54. Duhem, P.: Le potentiel thermodynamique et la pression hydrostatique. Annales scientifiques de l'École Normale Supérieure **3e série, 10**, 183–230 (1893)

55. Epstein, M., Smelser, R.E.: An appreciation and discussion of Paul Germain's "The method of virtual power in the mechanics of continuous media, I: Second-gradient theory". Mathematics and Mechanics of Complex Systems **8**(2), 191–199 (2020)
56. Eremeyev, V.A., Lebedev, L.P., Cloud, M.J.: The Rayleigh and Courant variational principles in the six-parameter shell theory. Mathematics and Mechanics of Solids **20**(7), 806–822 (2015)
57. Eremeyev, V.A., Pietraszkiewicz, W.: Material symmetry group and constitutive equations of micropolar anisotropic elastic solids. Mathematics and Mechanics of Solids **21**(2), 210–221 (2016)
58. Eugster, S.R.: Geometric Continuum Mechanics and Induced Beam Theories, *Lecture Notes in Applied and Computational Mechanics*, vol. 75. Springer (2015)
59. Eugster, S.R., dell'Isola, F.: Exegesis of the introduction and sect. I from "Fundamentals of the mechanics of continua" by E. Hellinger. Zeitschrift für angewandte Mathematik und Mechanik **97**(4), 477–506 (2017)
60. Eugster, S.R., dell'Isola, F.: An ignored source in the foundations of continuum physics "Die Allgemeinen Ansätze der Mechanik der Kontinua" by E. Hellinger. Proceedings in Applied Mathematics and Mechanics **17**(1), 413–414 (2017)
61. Eugster, S.R., dell'Isola, F.: Exegesis of sect. II and III.A from "Fundamentals of the mechanics of continua" by E. Hellinger. Zeitschrift für angewandte Mathematik und Mechanik **98**(1), 31–68 (2018)
62. Eugster, S.R., dell'Isola, F.: Exegesis of sect. III.B from "Fundamentals of the mechanics of continua" by E. Hellinger. Zeitschrift für angewandte Mathematik und Mechanik **98**(1), 69–105 (2018)
63. Eugster, S.R., dell'Isola, F., Steigmann, D.: Continuum theory for mechanical metamaterials with a cubic lattice substructure. Mathematics and Mechanics of Complex Systems **7**(1), 75–98 (2019)
64. Eugster, S.R., Deutschmann, B.: A nonlinear Timoshenko beam formulation for modeling a tendon-driven compliant neck mechanism. Proceedings in Applied Mathematics and Mechanics **18**(1), e201800208:1–2 (2018)
65. Eugster, S.R., Glocker, Ch.: Constraints in structural and rigid body mechanics: a frictional contact problem. Annals of Solid and Structural Mechanics **5**(1-2), 1–13 (2013)
66. Eugster, S.R., Glocker, Ch.: On the notion of stress in classical continuum mechanics. Mathematics and Mechanics of Complex Systems **5**(3-4), 299–338 (2017)
67. Eugster, S.R., Harsch, J.: A variational formulation of classical nonlinear beam theories. In: B.E. Abali, I. Giorgio (eds.) Developments and Novel Approaches in Nonlinear Solid Body Mechanics, pp. 95–121. Springer International Publishing (2020)
68. Eugster, S.R., Hesch, C., Betsch, P., Glocker, Ch.: Director-based beam finite elements relying on the geometrically exact beam theory formulated in skew coordinates. International Journal for Numerical Methods in Engineering **97**(2), 111–129 (2014)
69. Eugster, S.R., Steigmann, D.J.: Variational methods in the theory of beams and lattices. In: H. Altenbach, A. Öchsner (eds.) Encyclopedia of Continuum Mechanics, pp. 1–9. Springer (2018)
70. Fedele, R., Raka, B., Hild, F., Roux, S.: Identification of adhesive properties in GLARE assemblies using digital image correlation. Journal of the Mechanics and Physics of Solids **57**(7), 1003–1016 (2009)
71. Fried, E., Gurtin, M.E.: Tractions, balances, and boundary conditions for nonsimple materials with application to liquid flow at small-length scales. Archive of Rational Mechanics and Analysis **182**(3), 513–554 (2006)
72. Gavrilov, S., Eremeyev, V., Piccardo, G., Luongo, A.: A revisitation of the paradox of discontinuous trajectory for a mass particle moving on a taut string. Nonlinear Dynamics **86**(4), 2245–2260 (2016)
73. Germain, P.: Sur l'application de la méthode des puissances virtuelles en mécanique des milieux. Comptes Rendus de l'Académie des Sciences Paris **A**, 1051–1055 (1972)
74. Germain, P.: La méthodes des puissances virtuelles en mécanique des milieux continus - 1ère partie, théorie du second gradient. Journal de Mécanique **12**, 235–274 (1973)

75. Germain, P.: The method of virtual power in continuum mechanics. part 2: Microstructure. SIAM Journal on Applied Mathematics **25**, 556–575 (1973)
76. Germain, P.: Functional concepts in continuum mechanics. Meccanica **33**(5), 433–444 (1998)
77. Germain, P.: The method of virtual power in the mechanics of continuous media, I: Second-gradient theory. Mathematics and Mechanics of Complex Systems **8**(2), 153–190 (2020)
78. Germain, P., Nguyen, Q.S., Suquet, P.: Continuum thermodynamics. Journal of Applied Mechanics **105**(50), 1010–1020 (1983)
79. Giorgio, I.: A discrete formulation of Kirchhoff rods in large-motion dynamics. Mathematics and Mechanics of Solids **25**(5), 1081–1100 (2020)
80. Giorgio, I.: Lattice shells composed of two families of curved Kirchhoff rods: an archetypal example, topology optimization of a cycloidal metamaterial. Continuum Mechanics and Thermodynamics pp. 1–20 (2020)
81. Giorgio, I., Ciallella, A., Scerrato, D.: A study about the impact of the topological arrangement of fibers on fiber-reinforced composites: some guidelines aiming at the development of new ultra-stiff and ultra-soft metamaterials. International Journal of Solids and Structures **203**, 73–83 (2020)
82. Giorgio, I., De Angelo, M., Turco, E., Misra, A.: A Biot–Cosserat two-dimensional elastic nonlinear model for a micromorphic medium. Continuum Mechanics and Thermodynamics **32**, 1357–1369 (2019)
83. Giorgio, I., Del Vescovo, D.: Non-linear lumped-parameter modeling of planar multi-link manipulators with highly flexible arms. Robotics **7**(4), 60:1–13 (2018)
84. Giorgio, I., Del Vescovo, D.: Energy-based trajectory tracking and vibration control for multilink highly flexible manipulators. Mathematics and Mechanics of Complex Systems **7**(2), 159–174 (2019)
85. Giorgio, I., dell'Isola, F., Steigmann, D.J.: Axisymmetric deformations of a 2nd grade elastic cylinder. Mechanics Research Communications **94**, 45–48 (2018)
86. Giorgio, I., Harrison, P., Dell'Isola, F., Alsayednoor, J., Turco, E.: Wrinkling in engineering fabrics: a comparison between two different comprehensive modelling approaches. Proceedings of the Royal Society A: Mathematical, Physical and Engineering Sciences **474**(2216), 1–20 (2018)
87. Giorgio, I., Rizzi, N., Turco, E.: Continuum modelling of pantographic sheets for out-of-plane bifurcation and vibrational analysis. Proceedings of the Royal Society A: Mathematical, Physical and Engineering Sciences **473**(2207), 20170636:1–21 (2017)
88. Giorgio, I., Spagnuolo, M., Andreaus, U., Scerrato, D., Bersani, A.M.: In-depth gaze at the astonishing mechanical behavior of bone: A review for designing bio-inspired hierarchical metamaterials. Mathematics and Mechanics of Solids pp. 1–30 (2020)
89. Greco, L.: An iso-parametric $G^1$-conforming finite element for the nonlinear analysis of Kirchhoff rod. Part I: the 2D case. Continuum Mechanics and Thermodynamics pp. 1–24 (2020)
90. Greco, L., Cuomo, M.: B-spline interpolation of Kirchhoff-Love space rods. Computer Methods in Applied Mechanics and Engineering **256**(0), 251–269 (2013)
91. Greco, L., Cuomo, M.: Consistent tangent operator for an exact Kirchhoff rod model. Continuum Mechanics and Thermodynamics **27**(4-5), 861–877 (2015)
92. Greco, L., Cuomo, M.: An isogeometric implicit G1 mixed finite element for Kirchhoff space rods. Computer Methods in Applied Mechanics and Engineering **298**, 325–349 (2016)
93. Greco, L., Cuomo, M., Contrafatto, L.: A reconstructed local B formulation for isogeometric Kirchhoff–Love shells. Computer Methods in Applied Mechanics and Engineering **332**, 462–487 (2018)
94. Green, A.E., Rivlin, R.S.: Multipolar continuum mechanics. Archive of Rational Mechanics and Analysis **17**(2), 113–147 (1964)
95. Gurtin, M.E.: Thermodynamics and the possibility of spatial interaction in elastic materials. Archive for Rational Mechanics and Analysis **19**(5), 339–352 (1965)
96. Harsch, J., Capobianco, G., Eugster, S.R.: Finite element formulations for constrainted spatial nonlinear beam theories. Mathematics and Mechanics of Solids pp. 1–26 (2021)

97. Harsch, J., Eugster, S.R.: Finite element analysis of planar nonlinear classical beam theories. In: B.E. Abali, I. Giorgio (eds.) Developments and Novel Approaches in Nonlinear Solid Body Mechanics, pp. 123–157. Springer International Publishing (2020)
98. Javili, A., Dell'Isola, F., Steinmann, P.: Geometrically nonlinear higher-gradient elasticity with energetic boundaries. Journal of the Mechanics and Physics of Solids **61**(12), 2381–2401 (2013)
99. Javili, A., Morasata, R., Oterkus, E., Oterkus, S.: Peridynamics review. Mathematics and Mechanics of Solids **24**(11), 3714–3739 (2019)
100. Kardestuncer, H., Norrie, D., Brezzi, F.: Finite element handbook. McGraw-Hill reference books of interest: Handbooks. McGraw-Hill (1987)
101. Kelley, J.: General Topology. Graduate Texts in Mathematics. Springer (1975)
102. Laudato, M., Ciallella, A.: Perspectives in generalized continua. In: B.E. Abali, I. Giorgio (eds.) Developments and Novel Approaches in Biomechanics and Metamaterials, *Advanced Structured Materials*, vol. 132, pp. 1–13. Springer (2020)
103. Mach, E.: The Science of Mechanics: A Critical and Historical Account of Its Development. Open court publishing Company, Chicago (1919). Translated from the German by Thomas J. McCromack.
104. Maier, G., Bocciarelli, M., Bolzon, G., Fedele, R.: Inverse analyses in fracture mechanics. International Journal of Fracture **138**(1-4), 47–73 (2006)
105. Marmo, F., Sessa, S., Vaiana, N., De Gregorio, D., Rosati, L.: Complete solutions of three-dimensional problems in transversely isotropic media. Continuum Mechanics and Thermodynamics **32**(3), 775–802 (2020)
106. Maugin, G.: Continuum Mechanics Through the Eighteenth and Nineteenth Centuries: Historical Perspectives from John Bernoulli (1727) to Ernst Hellinger (1914). Solid Mechanics and Its Applications. Springer International Publishing (2014)
107. Maugin, G.A., Metrikine, A.V.: Mechanics of generalized continua. Advances in mechanics and mathematics **21** (2010)
108. Mawassy, N., Reda, H., Ganghoffer, J.F., Eremeyev, V.A., Lakiss, H.: A variational approach of homogenization of piezoelectric composites towards piezoelectric and flexoelectric effective media. International Journal of Engineering Science **158**, 103410 (2020)
109. Meissner, H.: Ernst Hellinger. In: Topics in operator theory: Ernst D. Hellinger memorial volume, *Operator Theory: Advances and Applications*, vol. 48. Birkhäuser Verlag, Basel (1990)
110. Misra, A., Poorsolhjouy, P.: Identification of higher-order elastic constants for grain assemblies based upon granular micromechanics. Mathematics and Mechanics of Complex Systems **3**(3), 285–308 (2015)
111. Mühlich, U., Abali, B.E., dell'Isola, F.: Commented translation of Erwin Schrödinger's paper 'On the dynamics of elastically coupled point systems' (Zur Dynamik elastisch gekoppelter Punktsysteme). Mathematics and Mechanics of Solids **26**(1), 133–147 (2020)
112. Nguyen, C.T., Oterkus, S., Oterkus, E.: An energy-based peridynamic model for fatigue cracking. Engineering Fracture Mechanics **241**, 107373 (2021)
113. Nguyen, T.H., Niiranen, J.: A second strain gradient damage model with a numerical implementation for quasi-brittle materials with micro-architectures. Mathematics and Mechanics of Solids **25**(3), 515–546 (2020)
114. Noll, W.: A mathematical theory of the mechanical behavior of continuous media. Archive for Rational Mechanics and Analysis **2**(1), 197–226 (1958)
115. Obohat, M.A., Tahvilian, E., Yildizdag, M.E., Ergin, A.: Three-dimensional multi-patch isogeometric analysis of composite laminates with a discontinuous galerkin approach. Proceedings of the Institution of Mechanical Engineers, Part M: Journal of Engineering for the Maritime Environment pp. 1–14 (2020)
116. Olive, M.: Effective computation of SO(3) and O(3) linear representation symmetry classes. Mathematics and Mechanics of Complex Systems **7**(3), 203–237 (2019)
117. Olive, M., Auffray, N.: Symmetry classes for even-order tensors. Mathematics and Mechanics of Complex Systems **1**(2), 177–210 (2013)

118. Pepe, G., Carcaterra, A., Giorgio, I., Del Vescovo, D.: Variational feedback control for a nonlinear beam under an earthquake excitation. Mathematics and Mechanics of Solids **21**(10), 1234–1246 (2016)
119. Pideri, C., Seppecher, P.: A second gradient material resulting from the homogenization of an heterogeneous linear elastic medium. Continuum Mechanics and Thermodynamics **9**(5), 241–257 (1997)
120. Piola, G.: Intorno alle equazioni fondamentali del movimento di corpi qualsivogliono considerati secondo la naturale loro forma e costituzione. Memorie di matematica e di fisica della Società italiana delle scienze **24**, 1–186 (1848)
121. Placidi, L., Andreaus, U., Della Corte, A., Lekszycki, T.: Gedanken experiments for the determination of two-dimensional linear second gradient elasticity coefficients. Zeitschrift für angewandte Mathematik und Physik **66**(6), 3699–3725 (2015)
122. Placidi, L., Barchiesi, E., Misra, A.: A strain gradient variational approach to damage: a comparison with damage gradient models and numerical results. Mathematics and Mechanics of Complex Systems **6**(2), 77–100 (2018)
123. Placidi, L., Barchiesi, E., Misra, A., Andreaus, U.: Variational methods in continuum damage and fracture mechanics. Encyclopedia of continuum mechanics pp. 2634–2643 (2020)
124. Placidi, L., Misra, A., Barchiesi, E.: Two-dimensional strain gradient damage modeling: a variational approach. Zeitschrift für angewandte Mathematik und Physik **69**(3), 56:1–19 (2018)
125. Pradel, F., Sab, K.: Homogenization of discrete media. Le Journal de Physique IV **8**(PR8), 317–324 (1998)
126. Ranaivomiarana, N., Irisarri, F.X., Bettebghor, D., Desmorat, B.: Optimal orthotropy and density distribution of two-dimensional structures. Mathematics and Mechanics of Complex Systems **6**(4), 293–305 (2018)
127. Reiher, J.C., Giorgio, I., Bertram, A.: Finite-element analysis of polyhedra under point and line forces in second-strain gradient elasticity. Journal of Engineering Mechanics **143**(2), 1–13 (2017)
128. Reissner, E.: On a variational theorem in elasticity. Journal of Mathematics and Physics **29**(1), 90–95 (1950)
129. Reissner, E.: On a variational theorem for finite elastic deformations. Journal of Mathematics and Physics **32**, 129–135 (1953)
130. Sailer, S., Eugster, S.R., Leine, R.I.: The tippedisk: a tippetop without rotational symmetry. Regular and Chaotic Dynamics **25**(6), 553–580 (2020)
131. Salençon, J.: Handbook of Continuum Mechanics: General Concepts, Thermoelasticity. Physics and Astronomy Online Library. Springer (2001)
132. Salençon, J.: Mécanique des milieux continus: Concepts généraux. Mécanique des milieux continus. École Polytechnique (2005)
133. Salençon, J.: Virtual Work Approach to Mechanical Modeling. Wiley (2018)
134. Scerrato, D., Giorgio, I.: Equilibrium of two-dimensional cycloidal pantographic metamaterials in three-dimensional deformations. Symmetry **11**(12), 1523:1–20 (2019)
135. Scerrato, D., Zhurba Eremeeva, I.A., Lekszycki, T., Rizzi, N.L.: On the effect of shear stiffness on the plane deformation of linear second gradient pantographic sheets. Zeitschrift für Angewandte Mathematik und Mechanik **96**(11), 1268–1279 (2016)
136. Schulte, J., Dittmann, M., Eugster, S.R., Hesch, S., Reinicke, T., dell'Isola, F., Hesch, C.: Isogeometric analysis of fiber reinforced composites using Kirchhoff–Love shell elements. Computer Methods in Applied Mechanics and Engineering **362**, 112845:1–34 (2020)
137. Schwartz, L.: Théorie des Distributions. Hermann (1973)
138. Seppecher, P., Alibert, J.J., dell'Isola, F.: Linear elastic trusses leading to continua with exotic mechanical interactions. Journal of Physics: Conference Series **319**(1), 1–13 (2011)
139. Silling, S.: Reformulation of elasticity theory for discontinuities and long-range forces. Journal of the Mechanics and Physics of Solids **48**(1), 175 – 209 (2000)
140. Spagnuolo, M., Andreaus, U.: A targeted review on large deformations of planar elastic beams: extensibility, distributed loads, buckling and post-buckling. Mathematics and Mechanics of Solids **24**(1), 258–280 (2019)

141. Spagnuolo, M., Peyre, P., Dupuy, C.: Phenomenological aspects of quasi-perfect pivots in metallic pantographic structures. Mechanics Research Communications **101**, 103415:1–6 (2019)
142. Spagnuolo, M., Scerrato, D.: The mechanical diode: On the tracks of James Maxwell employing mechanical–electrical analogies in the design of metamaterials. In: Developments and Novel Approaches in Biomechanics and Metamaterials, pp. 459–469. Springer (2020)
143. Spagnuolo, M., Yildizdag, M.E., Andreaus, U., Cazzani, A.M.: Are higher-gradient models also capable of predicting mechanical behavior in the case of wide-knit pantographic structures? Mathematics and Mechanics of Solids **26**(1), 18–29 (2020)
144. Steigmann, D.J.: Theory of elastic solids reinforced with fibers resistant to extension, flexure and twist. International Journal of Non-Linear Mechanics **47**(7), 734 – 742 (2012)
145. Steigmann, D.J.: Finite Elasticity Theory. Oxford University Press (2017)
146. Steigmann, D.J., dell'Isola, F.: Mechanical response of fabric sheets to three-dimensional bending, twisting, and stretching. Acta Mechanica Sinica **31**(3), 373–382 (2015)
147. Steigmann, D.J., Faulkner, M.G.: Variational theory for spatial rods. Journal of Elasticity **33**(1), 1–26 (1993)
148. Timofeev, D., Barchiesi, E., Misra, A., Placidi, L.: Hemivariational continuum approach for granular solids with damage-induced anisotropy evolution. Mathematics and Mechanics of Solids pp. 1–33 (2020)
149. Toupin, R.A.: Elastic materials with couple-stresses. Archive of Rational Mechanics and Analysis **11**, 385–414 (1962)
150. Tran, L.V., Niiranen, J.: A geometrically nonlinear Euler–Bernoulli beam model within strain gradient elasticity with isogeometric analysis and lattice structure applications. Mathematics and Mechanics of Complex Systems **8**(4), 345–371 (2020)
151. Truesdell, C.: Die Entwicklung des Drallsatzes. Zeitschrift für Angewandte Mathematik und Mechanik **44**(4-5), 149–158 (1964)
152. Truesdell, C.: Essays in the History of Mechanics. Springer, New York (1968)
153. Truesdell, C.: A First Course in Rational Continuum Mechanics. Academic Press (1977)
154. Truesdell, C., Noll, W.: The non-linear field theories of mechanics. In: S. Flügge (ed.) The non-linear field theories of mechanics, *Encyclopedia of Physics*, vol. III/3. Springer (1965)
155. Truesdell, C., Toupin, R.: The classical field theories. In: S. Flügge (ed.) Principles of Classical Mechanics and Field Theory, *Encyclopedia of Physics*, vol. III/1. Springer (1960)
156. Turco, E.: Discrete is it enough? The revival of Piola–Hencky keynotes to analyze three-dimensional Elastica. Continuum Mechanics and Thermodynamics **30**(5), 1039–1057 (2018)
157. Turco, E.: In-plane shear loading of granular membranes modeled as a Lagrangian assembly of rotating elastic particles. Mechanics Research Communications **92**, 61–66 (2018)
158. Turco, E., Barchiesi, E.: Equilibrium paths of Hencky pantographic beams in a three-point bending problem. Mathematics and Mechanics of Complex Systems **7**(4), 287–310 (2019)
159. Turco, E., Barchiesi, E., Giorgio, I., dell'Isola, F.: A Lagrangian Hencky-type non-linear model suitable for metamaterials design of shearable and extensible slender deformable bodies alternative to Timoshenko theory. International Journal of Non-Linear Mechanics pp. 1–19 (2020)
160. Turco, E., dell'Isola, F., Misra, A.: A nonlinear lagrangian particle model for grains assemblies including grain relative rotations. International Journal for Numerical and Analytical Methods in Geomechanics **43**(5), 1051–1079 (2019)
161. Turco, E., Giorgio, I., Misra, A., dell'Isola, F.: King post truss as a motif for internal structure of (meta) material with controlled elastic properties. Royal Society open science **4**(10), 1–20 (2017)
162. Turco, E., Misra, A., Pawlikowski, M., dell'Isola, F., Hild, F.: Enhanced Piola–Hencky discrete models for pantographic sheets with pivots without deformation energy: numerics and experiments. International Journal of Solids and Structures **147**, 94–109 (2018)
163. Vaiana, N., Sessa, S., Marmo, F., Rosati, L.: A class of uniaxial phenomenological models for simulating hysteretic phenomena in rate-independent mechanical systems and materials. Nonlinear Dynamics **93**(3), 1647–1669 (2018)

164. Valoroso, N., Fedele, R.: Characterization of a cohesive-zone model describing damage and de-cohesion at bonded interfaces. Sensitivity analysis and mode-i parameter identification. International Journal of Solids and Structures **47**(13), 1666–1677 (2010)
165. Winter, T.N.: The mechanical problems in the corpus of Aristotle. Faculty Publications, Classics and Religious Studies Department (2007)
166. Yang, Z., Oterkus, E., Oterkus, S.: Peridynamic formulation for higher-order plate theory. Journal of Peridynamics and Nonlocal Modeling pp. 1–26 (2020)
167. Yang, Z., Oterkus, E., Oterkus, S.: A state-based peridynamic formulation for functionally graded Kirchhoff plates. Mathematics and Mechanics of Solids pp. 1–22 (2020)
168. Yildizdag, M.E., Barchiesi, E., dell'Isola, F.: Three-point bending test of pantographic blocks: numerical and experimental investigation. Mathematics and Mechanics of Solids **25**(10), 1965–1978 (2020)
169. Yildizdag, M.E., Demirtas, M., Ergin, A.: Multipatch discontinuous Galerkin isogeometric analysis of composite laminates. Continuum Mechanics and Thermodynamics **32**(3), 607–620 (2020)

# Chapter 4
# The loss and recovery of the works by Piola and the Italian tradition of Mechanics

Mario Spagnuolo, Alessandro Ciallella, and Daria Scerrato

## 4.1 Orthodoxy in Continuum Mechanics: social phenomenology and its naive explanations

Continuum Mechanics, or rather Generalized Continuum Mechanics, is not so recent as it is usually believed: already in the last pages of the celebrated textbook by Lagrange [92, 93], one can find the first variational version of this theory, including some basic formulations of the concept of deformation and stress. We postpone the detailed discussion of this source to further investigations and remark here that already Piola discussed the importance of the contribution by Lagrange, also in this context (see [34, 36, 42, 43, 57, 59, 60]).

Exactly when, in the first half of XIX Century, the fundamental assumptions for the Theory of Elasticity had been chosen, immediately a rather lively debate was started. The debate was mainly focused, since the very beginning, on the Postulation Scheme to be used as a guidance for the novel theories to be formulated.

Mario Spagnuolo, e-mail: mario.spagnuolo@unica.it
DICAAR, Università degli Studi di Cagliari, Cagliari, Italy.
International Research Center on Mathematics and Mechanics of Complex Systems, University of L'Aquila, L'Aquila, Italy.
Alessandro Ciallella e-mail: alessandro.ciallella@univaq.it
International Research Center on Mathematics and Mechanics of Complex Systems, University of L'Aquila, L'Aquila, Italy.
Dipartimento di Ingegneria Civile, Edile-Architettura e Ambientale (DICEAA), University of L'Aquila, Italy.
Daria Scerrato, e-mail: daria.scerrato@gmail.com
International Research Center on Mathematics and Mechanics of Complex Systems, University of L'Aquila, L'Aquila, Italy.
Dipartimento di Scienze di Base ed Applicate per l'Ingegneria (SBAI), University of Rome La Sapienza, Italy.



F. dell'Isola et al. (eds.), *Evaluation of Scientific Sources in Mechanics*,
Advanced Structured Materials 152, https://doi.org/10.1007/978-3-030-80550-0_4

## *Two competing schools debate about the Foundation of Continuum Mechanics*

One school, mainly headed by Gabrio Piola in Italy, strongly supported the point of view of Archytas of Tarentum (428-360 BC) as rediscovered by D'Alembert and Lagrange. In the context of this book, Archytas' storyline is doubly relevant (see [139, 88]). Then, as reported by Diogenes Laertius and currently recognized by modern scholars, Archytas can be considered the founder of Mechanics. And not only that. Diogenes points out that Archytas was «the first to systematize Mechanics using mathematical principles». In his treatise *Mechanical Problems*, one can recognize the first insights for the future development of the Principle of Virtual Work. As in Lagrange's *Analytical Mechanics*, Archytas' interest was more for the mathematical generalization of Mechanics than for the individual problems related to the construction of machines (which was underlined by Plutarco and Pappus [31]). A second important aspect of Archytas' fame vicissitude consists in the fact that, like Piola, his scientific results have been erased and have come down to us only because for a long time they were wrongly attributed to Aristotle. As we will see below, parts of Piola's results have been transmitted, obviously not attributing authorship to the Italian scholar, because they were partially contained and reorganized (often in a very unclear way) in Truesdell's compendium [135].

Piola believed that every physical theory had to be formulated in terms of a Virtual Work Principle, and when possible, in terms of a Least Action Principle. In fact, this second principle reduces to a particular case of the first one, once calculating the action's first variation and imposing this variation to vanish (see [47, 19]). Nowadays, variational principles have proven successful as a very powerful tool in formulating theories. Especially, their computational capabilities can be experienced in the numerical descriptions of Continuum Mechanics [21, 25, 23, 22, 24, 72], of generalized continua [137, 138, 79, 80], in structural mechanics [66, 83, 82, 48, 62, 18, 63, 126], for the development of time-integration schemes [17].

The other school, mainly headed by Navier, Cauchy and Poisson in France, strongly supported a point of view in which the concept of force had to be regarded as the most fundamental one, and the balance of force and the balance of moment of force had to be postulated at the beginning of the development of Mechanics. In [118, p. 400], Poisson even criticized the methods of Lagrange for being not suitable for continuum mechanics.

In Continuum Mechanics, this second approach became dominant, especially in XX Century Mechanics and in Applied Mechanics. At the center of this version of Continuum Mechanics stands Cauchy's tetrahedron argument that allows for the deduction of the concept of stress starting from the concept of force.

## *Balance laws-postulation tried to impose an orthodoxy in mechanical sciences*

The wide range of applicability of the version of Continuum Mechanics as formulated by Cauchy and his followers, and the impulse to engineering applications given by the brilliant developments of the Theory of Elasticity, managed to "eclipse" the ideas and the results by Gabrio Piola.

A careful reading of the works by Piola shows that he used the Principle of Virtual Work to formulate a much more general version of Continuum Mechanics, as he knew that its applicability range is much wider than the version developed by Cauchy, although it needs a more complicated mathematical formulation.

Sometimes it has been argued that, as the computational tools of XX Century were not sophisticated enough to use effectively the more general Piola's Continuum Mechanics, his ideas, and also those subsequently used by the Cosserat Brothers for the formulation of their generalized version of Cauchy continua (see [27, 99, 100, 71, 4, 51, 6, 108, 2, 5, 52, 3, 50]) were bound to remain unheard and unexploited.

Although, for sure, the lack of suitable computational tools did represent a blockage in the development of applications for generalized Continuum Mechanics, one cannot ignore the effects and the influence of the dominant "ideology" of dominant schools. Exactly as the ideas of Archytas of Tarentum did manage to reach us only by being transmitted as part of the Corpus of the works of Aristotle (and the fake attribution to Aristotle is rejected nowadays nearly unanimously, see e.g. [139]), it has to be underlined that Piola's vision about the foundations of Mechanics did manage to reach a modern audience mainly because of the references to his works due to Hellinger (see [86, 57, 59, 60, 58, 56]) or because of the fundamentally negative remarks by Truesdell (see [135]).

We cannot, however, exclude that the transmission of the ideas and postulation choices developed by Piola did reach the modern scholars also by means of oral transmission (for a detailed discussion of the role of oral transmission also in modern Science see e.g. [32]). Probably in many European schools (Italian, French, German, but also Russian) the variational principles championed by Piola were instilled in the mind of the most eminent mechanicians by the inspired lectures of their "Maestri", albeit these Maestri did refrain from publishing textbooks.

As it has been shown in the other Chapters of the present book, the difficulties to get closed form solution for Cosserat or Piola continua was not the only reason for the partial erasure and removal of these models in the literature of mechanical sciences. In fact, the social processes, to which we referred when Vitruvius' and Tartaglia's ways of transmitting previously acquired knowledge was discussed, did activate also in this case.

A Publish or Perish Culture did actively participate to the partial erasure of Lagrange's and Piola's results and viewpoints, insofar their epistemological ideas were not accepted by a dominant cultural and scientific paradigm in the sense of Kuhn [91]. More generally, Lagrange's and Piola's "Weltanschauung" were rebutted by what can be called the "Truesdellian school". One can say that, as a group of influential mechanicians did decide to choose exclusively the less efficient Cauchy's paradigm for developing new models and theories, the research in mechanical sciences did loose a lot of its potential innovative push. Their stubborn orthodoxy control of the postulation scheme to be used for formulating new models did block creativity in Continuum Mechanics. This orthodoxy did not allow for the systematic use of the Principle of Virtual Work, of the Principle of Least Action, for obtaining the governing equations of a studied system, or of the Hamilton-Rayleigh Principle, for introducing in said system dissipation.

This statement may seem too extreme to many. However, it is firmly grounded, as indicated, for instance, by a very strange circumstance, which can be explained only in terms of social science phenomena and not in terms of mechanical science logics. At least two very original mechanicians, that is Richard Toupin and Allen C. Pipkin, were obliged to hide their truly variational inspiration in their papers [111, 134].

In fact, when reading Pipkin's paper, it is clear that the author could find his novel mathematical model for Tchebychev nets only by using variational principles. However, in order to publish more easily the paper in Truesdell's journal, he changed the order of presentation and wrote (apparently out of the blue) balance equations before discussing the variational principle from which these equations can be deduced.

Toupin was obliged to behave similarly when he treated in [134] the most general second-gradient continuum. While the title of his paper refers explicitly only to the particular case of second-gradient continua whose evolution equation can be deduced by using the balance of force and moment of forces, inside the paper the attentive reader finds a more complete theory, based on the Principle of Virtual Work. The reason of this misleading choice is simple: the more general equations for second-gradient continua (when stress and couple stress are not enough to describe contact interactions) cannot be deduced from the balance of forces and moment of forces (see [46, 20, 41, 45]). The title of Toupin's paper is surely misleading and willingly reductive. Because of this title, too often the theory of second-gradient continua is attributed, in its modern form, exclusively to Paul Germain [75]. In fact, Germain did greatly contribute to its development and its mathematical framing by the systematic use of Tensor Calculus and Functional Analysis [76, 74, 73, 98]. However, the astonished reader will see that the complete second-gradient theory of continua is fully developed in Toupin's paper, notwithstanding its apparently limiting title.

## 4.2 In other scientific groups Variational Principles remained the mainly used conceptual tool

One positive aspects of the fragmentation of knowledge (for a discussion of this phenomenon see for instance [35]) observed in modern Science consists in the possibility, for ideas and scientists, to get around orthodoxies by simply changing research group.

In fact, the negative attitude against variational principles was circumscribed to some specific (and luckily not so influential) groups of scholars. Instead, it was never even considered in other, more qualified, scientific milieux. In Theoretical Physics, different scientists used alternative textbooks and developed other viewpoints: physicists continued to teach to their students the basic variational principles and maintained the persuasion that they are the most fundamental tool for founding physical theories.

For instance, in the textbooks [94] and [70], it is plainly explained why Variational Principles are to be considered the most powerful tool for conjecturing novel theories and models for getting predictive insight in not-yet-described physical phenomena. Landau and Feynman (to name some of the most famous physicists) were persuaded that the Lagrangian postulation scheme supplies the most powerful tool of innovative research in Mechanical Sciences and more generally in Physics. We recall here that:

i) Quantum Mechanics [125, 26, 121, 69, 68, 13, 84],
ii) General Relativity [49, 28, 132, 133]

but also

iii) Generalized Continuum Mechanics [97, 64, 87, 102, 120, 115, 40, 38, 44, 116, 39, 37, 112, 7, 78, 90, 136, 10, 1, 61, 11, 9, 55, 65, 67]

are all parts of physical sciences whose impetuous development has been pushed by the systematic application of variational principles, regarded as the most basic heuristic tool to be used when formulating new theories.

Not only theoretical physicists share explicitly Piola's vision of science: one has to mention that Hellinger (see [86] and for the just recent translation into English [57, 59, 60] or Chapter 3 of this volume, [56]), which is a famous applied mathematician, was one among the greatest and more effective supporters of Variational Principles and his celebrated encyclopedia article undoubtedly is an effective witness of the results and effectiveness of the Lagrangian School of Mechanics.

It is somehow surprising that Hellinger, who probably could not read Italian, cites in one footnote the works by Piola. Notwithstanding our efforts in finding the list of sources connecting Piola to Hellinger, we could not, yet, establish if Hellinger's citation is due to a direct knowledge of Piola's works, which had been continuously available in all main mathematics and physics libraries in the world, or if the information reached Hellinger indirectly: there are hints that an intermediate connection could have been the encyclopedia article by Müller and Timpe [105], also cited by Hellinger. Hellinger did not come to universal fame in the scientific milieu of mathematical-physics because of two sad circumstances, [103]:

i) his career was disrupted by the Nazi prosecution. As he was Jewish, he was imprisoned in Dachau and, only because of a quick job offer from the Northwestern University at Evanston, Illinois, and the required U.S.-Visa, he was released after six weeks, as he was in condition to emigrate immediately;
ii) his position as lecturer in Mathematics at Northwestern University did not give him the possibility to be as authoritative as if he had been a German full professor. He got the U.S. citizenship only in 1944 and promoted to professorship in 1945.

Notwithstanding many negative circumstances, one can say that Piola's vision of science did manage to influence the great majority of open minded scholars in mechanical sciences. In a similar way as the works by Archytas did reach us via a fake attribution of his works to Aristotle, it is clear that the very critical remarks about Piola's appreciation of variational principles (as reported in [135] and due to Toupin's stubbornness in having them discussed) allowed the survival of Piola's ideas also in the Continuum Mechanics community.

A careful study of sources is necessary to establish which were the cultural processes that made Paul Germain to became the modern French champion of Variational Principles. We observe that, in Germain's sources, one finds explicitly the names: Mindlin, Toupin, Casal, Green, Rivlin and, above all, the Cosserat Brothers.

A very interesting problem in sociology of Science could be formulated as follows: *Did the tradition of Variational Principles arrived to influence Paul Germain directly from the French scientific descendance stemming from Lagrange (transmission which was made possible notwithstanding the influence of the Balance-Laws School started by Cauchy, Navier and Poisson) or Paul Germain was somehow influenced by the Italian tradition as started by Piola, as a direct "pupil" of Lagrangia (the true Italian family name of Lagrange, always used by Piola, in his citations)?*

A naive explanation of the social phenomena linked to the "castling" inside the orthodoxy of balance-laws postulation of a particular subgroup of scientists can be resumed as follows: the concepts of Functional Analysis needed to formulate variational formulations are not fully mastered, in general, and therefore balance-laws postulation is preferred for the apparent simplification of its formal apparatus. However, this simplification limits the capability of formulating models more general than those stemming from the conceptual frame as established by Cauchy. The innate tendency of human beings to follow the trends accepted by the "majority" caused, consequently, the partial removal of the tradition started by Lagrange to base the postulation of mechanical theories on variational principles.

## 4.3 How Piola's works were transmitted and how they were – locally in space and time – lost in the Mechanics literature

It is now clear that Gabrio Piola's contribution to mechanical sciences has been greatly underestimated in both the more theoretical mathematical-physics literature and in the more applied and engineering oriented one. However, and undoubtedly, an attentive reader (before the translations presented in [42, 43] such a reader had not only to be attentive, but also he had to be able to understand the elegant Italian language used by Piola) immediately discovers that Piola's works are original, mathematically deep and far-reaching in their physical applications. In some aspect they are, even nowadays, at the edge of the most advanced research in generalized Continuum Mechanics.

A further and closer analysis shows that Piola's contributions to mechanical sciences, in fact, were not completely removed and erased in the mechanical literature.

Moreover one can say that the greatest part of them did, albeit indirectly, provided a relevant impetus and influenced substantially the ideas and the work of many among the most preeminent Piola's successors in Continuum Mechanics.

As usually happens in the History of Science, only the memory of the existence of Piola was almost completely removed: Piola's name and the relation between his name and his ideas were generally ignored. Probably also Hellinger, in his encyclopedia article [86] referred to Piola only incidentally, most probably because of a secondary source. Hellinger did not seem to dedicate many efforts to unveil the quality and quantity of the contributions by Gabrio Piola to the fundations of Generalized Continuum Mechanics. In our opinion it is not possible, nowadays, to doubt that the non-local and higher-gradient continuum mechanics had been conceived by Piola in his works, cf. [36, 101].

However, more than 150 years after the death of Piola and the publication of his last work, many of Piola's results are being rediscovered and reformulated, sometimes in a partial and even incomplete form, notwithstanding the fact that the written copies of his contributions were available in many libraries without interruptions during all these years.

We must try here to give an explanation, albeit naive, for this social phenomenon: naturally we are aware of the fact that deeper studies in social Science may be necessary to fully explain it (some hints on the nature of these studies can be found in [33]). While we cannot hope to explain easily the reasons of the erasure of the memory of Piola's contribution, we do believe that this explanation must be searched, as the erasure phenomena occurred to Piola did occur very often in the literature.

## *The linguistic barrier: As Archimedes used Doric Greek, Gabrio Piola used Italian*

As we have remarked in Chapter 2 about Heiberg's Prolegomena translation and as it is very clear in the Prolegomena text itself, an astounding fact about the transmission of Archimedes' ideas after the end of the Hellenistic Age is that his main ideas were lost immediately. Differently from Piola, Archimedes' name was never forgotten around the centuries, but already in Vitruvius (which was the nearest thing to a scholar in Roman Age) Archimedes' name was mostly related to folkloristic stories without great interest to the scientific aspects of his works.

During the Middle Ages the name of Archimedes was mainly known as that of the man who had succeeded in providing Syracuse with unimaginable defenses and the figure of the scientist was completely confused with that of a magician. We probably have to owe to Leo the Mathematician for the production of the codices A, B and C, copies that were sent from Constantinople to Europe and then to William of Moerbeke, who translated them into Latin, (we have extensively discussed this translation and its vicissitudes in Chapter 2) the transmission of Archimedes' works to the present day.

One of the possible causes of the removal of Archimedes' scientific results, in addition to the cultural decay that occurred at the end of the Hellenistic age, is to be found in the difficulties related to the use of the Doric dialect. The language barrier certainly constituted a disincentive to reading and the consequent transmission of Archimedes' works.

Piola decided to use the Italian language for writing his works. This can be regarded as the most important cause to the lack of full recognition of Piola's contribution to Continuum Mechanics. Piola paid for his "ideological" choice: he wanted to show that Italian could be regarded as a fully mature language, the language of a great Nation (he hoped for the Unification of the Italian Nation but died 10 years before it was realized), a language in which deep mathematical theories could be easily expressed. To this aim, and probably also for his personal inclination, he composed his texts in an extremely elegant and erudite style. This made difficult also to Italian scientists to understand his work. Only few specialists in Mechanics, having a solid background in Classical Studies, were capable to fully appreciate its true content.

Also Piola did not care to have his works translated into other more popular languages. We can formulated a reasonable conjecture about Piola's linguistic choice: albeit he was fluent in French, he wanted "per la gloria dell'Italia", i.e., for the glory of Italy, to use Italian. This patriotic choice was not repaid with a big consideration of his compatriots. They completely neglected, in general and with the exception of few among his pupils, to recognize his contribution to Mechanical Science. Activated by said ideological motivations Piola established a linguistic barrier between his ideas and his successors. This barrier plays a very strange role in the phenomenon of spreading of concepts, models and theories.

It is very well-known that the growth of Hellenistic science (see e.g. [124, 89, 123, 15]) was favored by the use of a unique *lingua franca* but also that its propulsive force was stopped by the change in the dominant language in the Mediterranean sea: Latin replaced Greek when Rome became the dominant political power. The scientific knowledge did manage to flow from East to West but, because of the change of language, the translation process of the fundamental textbooks was too slow. It needed some centuries, and, in order to obtain the translated texts in Latin, it was needed to incorporate the Greek nomenclature and terminology. The process of information transmission was so long that Hellenistic did manage to pollinate Europe only via the Italian Renaissance. While the basics of Hellenistic Science was, notwithstanding the delays and the misunderstandings, preserved (the reader will remind what we have written about the role of Tartaglia in the transmission of Archimedes' works in Europe) to humanity, one has to admit that, in general, the process of book translations induced a nearly complete loss of the true names of the scientists who had first formulated some of the most fundamental ideas of the modern scientific knowledge.

Therefore one should not be surprised that:

i) the contribution of Piola to Generalized Continuum Mechanics still is permeating modern literature [34];

ii) Piola's results have been repeatedly rediscovered by more and more modern scholars and
iii) the true content of his results is generally misunderstood and underestimated, also by the few scholars who knew about his existence.

Indeed linguistic barriers, in this case represented by the differences between the Italian style written by Piola and modern English, become too often insurmountable.

## *The continuity of personal relationships between Maestro and Pupil is necessary to maintain a body of knowledge: the genealogy of scholars started by Piola*

We can try, here, to develop some more reasonings concerning the processes of erasure and removal in the body of knowledge available for a certain social group during intergenerational transmission. The question is: *How is it possible that well-established scientific knowledge can be lost?*

We conjecture that this loss can be associated to the simultaneous occurrence of many social phenomena including specifically the associated:

i) loss of continuity in the succession sequence of Maestro and Pupil in the academic institutions and
ii) loss of the financial support of academic institutions because of the loss of confidence in political institutions on science. This loss of confidence occurs when politicians doubt about the indissoluble connection existing between theoretical science (including the most abstract one, that is Mathematics) and technology.

In those societies where these two losses (that obviously are also correlated) occur at the same time, the political power decide that investing resources in the preservation and transmission of theoretical knowledge is useless. One of the consequences of this lack of resources is that the personal contact between Maestro and Pupil is interrupted. Even when good quality textbooks remain available, the younger generations do not manage to reach easily the highest levels possible of scientific knowledge: nothing can replace the effectiveness of a living scientist who teaches to his pupils the most difficult and important theories that he mastered.

In said societies a sequence of phenomena can be generally observed: at first the theoretical knowledge is lost, generally being considered useless. The focus of the intellectuals of these societies, that are at the beginning of their decadence, is on technological applications. In a second moment, that occurs after a more or less long time interval, such societies become explicitly decadent as, unavoidably, also their technological capabilities are lost.

Differently from what happened to the works of those Hellenistic scientists, whose fame was not preserved also during decadent times as luckily happened to Archimedes, the works by Piola were not materially lost: In fact, they were only forgotten, and possibly removed from the scientific "memory".

One of the reason for which they could be recovered is related to the existence of "a direct genealogy" stemming from Gabrio Piola and arriving to the founders of absolute tensor calculus. Here we want to shortly reconstruct it. We warn the reader that we do not believe that all the scientists in this line of descendence were aware of the fact that they were completing the scientific program of Piola. Only his direct Pupil, Francesco Brioschi, was aware of the importance of his work, as a continuation of that started by Piola.

Gabrio Piola never accepted a university chair, even if it is documented that some chairs were offered to him. Piola supported scientifically and politically his Pupil Francesco Brioschi, who was the founder of the Politecnico di Milano. Brioschi edited the last work by Piola, which appeared posthumously in 1856.

Francesco Brioschi was the Maestro of Enrico Betti and Eugenio Beltrami. Ulisse Dini was Pupil of Enrico Betti, while Gregorio Ricci Curbastro was Pupil of Ulisse Dini, Eugenio Beltrami and Enrico Betti. Finally, Tullio Levi-Civita was pupil of Gregorio Ricci Curbastro.

Everybody who has studied (Continuum) Mechanics or Applied Mathematics has met many among the most important results originated by this handful of scientists. However, differently from what happens in the French School, the Italian School of Mathematical Physics, Mathematical Analysis and Differential Geometry reflects the National spirit of political and cultural fragmentation, possibly inherited from similar habits widely spread in Greek societies.

Albeit Napoleon, behaving as King of Italy, did try to favor the establishment of a genuine Italian Scientific School, he clearly could not manage in this seemingly impossible task. In fact, he could not change the Italian spirit. The Italian scientists never managed to follow the habit, so deeply rooted in the French School, that leads French Scientists to support their compatriots, and to recognize and to develop their contributions. On the contrary, the Italian scientists are inclined to follow the behavior of those scientists of Greek language who did produce Hellenistic science. In the Hellenistic tradition one observes the intentional and systematic removal and contempt of the contribution due to the compatriots. Italians added to this tradition also the systematic preference for the work of foreign scientists, who are considered, just because of their ethnic origin, better than the Italian colleagues. Italian scientists seem always to consider the other national scientific groups more original, more qualified and more productive: at the same time they actively operate to sabotage any activity and to limit the growth possibilities of the compatriot scientists.

Napoleon managed to impress to Italian Science a remarkable momentum and his choices are to be, probably, regarded as the first cause of the birth of Tensor Calculus. However, this momentum was wasted quickly because of the very peculiar habit that Italians have, consisting in creating obstacles to their compatriots. A sadly exemplary action carried against a great Italian scientist by his compatriots is represented by the removal of Levi-Civita form his university chair, because of Mussolini's racial laws. Big maneuvers were started by small minds: they had no problems in declaring unanimously that the Italian Science could, easily, do without Levi-Civita. His chair in Rome was immediately occupied by Antonio Signorini.

Whichever may be the social phenomena involved in the growth, in an intergenerational effort, of the mathematical theories needed to develop Continuum Mechanics and in their establishment as parts of Mathematics, independently of their applicability in one specific field of Physics, one must recognize that:

i) since the very first beginning (see [34]) one can observe a very strict relationship between Differential Geometry and Continuum Mechanics;
ii) the Italian School in (Continuum) Mechanics founded by Piola and culminating in Levi-Civita did perfection the Tensor Calculus and important parts of Differential Geometry;
iii) Riemannian geometry experienced its biggest advancement when it was possible to prove (a theorem due to Levi-Civita) that a unique parallel transport is compatible with a Riemannian metric: this results has its deepest roots in the results presented already by Piola (we refer here to the concept of Piola's Transformation from reference to current configuration);
iv) Ulisse Dini proved formally the famous implicit function theorem, which gives a solid mathematical ground to the concept of constraint, that was intensively exploited by Piola. In particular, the fundamental definition of independent constraints (defined as those constraints represented by set of functions whose Jacobian is not singular) was fully understood thanks to the studies of Dini, many decades after that their importance had been recognized by Piola. Moreover, the concept of independent constraints has been fundamental in the study of embedded manifolds, another concept whose growth was pushed by Continuum Mechanics and which, also, greatly contributed to Continuum Mechanics growth (see e.g. [46, 45, 41]).

## *Rediscoveries of the same body of knowledge in different places and times*

It can be observed very often, in the History of Science, that some theories needed to be rediscovered (possibly being reformulated more or less drastically) in different times and in different places. The typical examples are often considered when comparing the few sources of Hellenistic Science with the results obtained later, during Renaissance.

Aristarchus of Samos conceived a heliocentric model for the solar system probably in Alexandria already in the third century B.C. . He also conjectured, as Anaxagoras had done before him, that the stars were other suns, located at larger distances away from Earth. Plutarch refers that:

i) Aristarchus «postulated heliocentrism only as a hypothesis», and that
ii) the astronomer Seleucus of Seleucia[1], who lived a century after Aristarchus, «gave a demonstration of it» and considered heliocentrism as an undisputed fact.

[1] Concerning the place where Seleucus lived, we refer to [106]: «Among several cities named Seleukia, the best known is Seleukia on the Tigris, the capital of the Seleucid kingdom. It is

One can see that already in Hellenistic Science the same theory was reformulated in different places and different times, by members of different social groups.

In the secondary source given to us by Pliny and Seneca, one finds an explanation of the observed retrograde motion of some planets as an *apparent* phenomenon: these sources claim that *reality* is different, in this way implying that their primary sources were based on heliocentrism and not on geocentrism and that a relationship between these primary sources and Seleucus is likely. The heliocentric theory was rediscovered by Copernicus and his theoretical efforts made possible also the more careful description of the planetary motions by Kepler (see [77, 131, 85]).

The previous and maybe the most famous example shows clearly that theories are not *set in stone* once forever and that the scientific knowledge, like every human institution, is subject to different transformations, including regression, removal and erasures, and that a theory, in fact, is bound not to become a universally recognized part of knowledge (the reader is invited to think to the modern revival of «Flat Earth Societies», see e.g. [122, 16, 95, 96]).

Another interesting phenomenon occurs with a similar frequency. Some theories are accepted, in a given time and in a given society, by some groups of scholars and are erased, removed and completely ignored by other groups. We are not talking about the fundamentally healthy debates and comparison of different competing theories. We refer to the fact that some theories are (unwillingly or sometimes willingly) ignored by some subgroups of certain scientific societies while are considered well-established by other subgroups.

An example may be given by Functional Analysis and its most fundamental concept of functional, usually defined using the sentence "a function whose argument is a function". Erik Ivar Fredholm's and Jacques Hadamard's works fully exploited this concept that, however, had been introduced already by Vito Volterra. The theory of nonlinear functionals was developed by Maruice René Fréchet and Paul Lévy. While linear Functional Analysis was continued by Frigyes Riesz and the school headed by Stefan Banach. Rather surprisingly, it is to be observed that Werner Karl Heisenberg and Paul Dirac had to rediscover and redevelop this already known theory, until John von Neumann remarked that Quantum Mechanics was based on the well-known concept of Hilbert Spaces.

The causes of such kind of *co-existence* of different *advancement stages* of one and the same theory can be various:

i) the standard tendency towards laziness of human beings;
ii) the objective difficulty in understanding the mathematical concepts as developed by other scientists;
iii) lack of sufficient economical means for accessing to the works of other scientists;
iv) difficulties in reading texts written in another language;
v) need of establishing an effective control of the careers of the members of the same power group: in order to be able to claim that this group is producing high

possible that the astronomer Seleukos lived or was born in this city, but it is also possible that his native town was Seleukia on the Erythrean Sea.».

quality science, its members willingly ignore the contributions to the discipline produced by other groups;

vi) the need of hiding one own results to other scientist for reasons linked to political or economical competition.

All these causes, and together with some others which may deserve to be discussed, may have as an effect that otherwise erudite scientists do actually ignore results, relevant to their own researches, already known to other scientists. In fact one can observe that even the mathematicians listed in Piola's descendence did rediscover several times the results that some of their predecessors had already obtained.

An exemplary case, which we want to cite here, of misattribution and repeated re-discovery of a result is given by a formula obtained by Piola (see for more details [34, 42, 43]) when he solved the problem of transporting the stationarity condition characterizing the equilibrium of continua from the reference configuration to the current configuration. We refer here to the case of so-called "Nanson's Formula". In the studies by Piola the demonstration of this formula is a necessary step in solving a more general scientific problem, whose motivation is clear. Starting from a variational principle that characterizes the (possibly) stable equilibrium configurations, Piola calculates from the Principle of Virtual Work formulated for continuous systems (that will be later called the weak formulation of equilibrium conditions) the set of PDEs plus boundary conditions that give the corresponding Euler-Lagrange conditions, so obtaining what will be later called the strong formulation of equilibrium conditions. The transport of the obtained PDEs requires the transformation of the divergence operator in the reference configuration into the divergence operator in the current configuration, while the transport of the boundary conditions require the transformation of the normal to the boundary of the body in the reference configuration into its normal in the current configuration. This transformation formula for mentioned normals is already published in Piola's works in 1848 (see [42, 43, 34]) and it is not clear how it happened that in some Anglo-Saxon literature this formula was attributed to Edward John Nanson who was born 1850 exactly in the same year when Piola died.

## 4.4 Mathematical difficulties inherent to Variational Formulation

While the formulation of the Principle of Minimum Energy is very clear and simply formulated, the mathematical procedure needed to deduce from this Principle the consequent local PDEs and Boundary Conditions is, instead, rather intricate. While Lagrange and Piola, for obtaining this deduction, could find a line of reasoning that is fundamentally correct, a rigorous mathematical setting allowing the formalization of precise mathematical results has been established only more recently, fundamentally by Sergej Lvovich Sobolev.

## *Functional Analysis in Lagrange's and Piola's works*

To calculate the first variation of the total energy functional characterizing the physically admissible equilibrium configurations of a continuous system requires:

i) the determination of the functional space in which the total energy functional is defined and, after having proven the existence of minima for the specific functional (eventually by means of a more general result generalizing to infinite dimensional spaces the well-know existence theorem usually called Weierstrass extreme value theorem);
ii) the computation of its Fréchet derivative. Once the mentioned first variation is computed, one has to find some necessary conditions assuring that, when estimated in the candidate equilibrium configuration, it vanishes for every "small" variation of the equilibrium configuration.

In the works by Piola and Lagrange a particular attention is spent to the concept of "commutativity" of the delta symbol with the derivation operator. The delta symbol, in the nomenclature by Lagrange, denotes "small variations of configurations", i.e. variation of configurations which are close, in a sense, to the zero variation.

Albeit both Lagrange and Piola could not use the concepts of Banach (or Fréchet) spaces, concepts which allowed Sobolev to develop his successful formalization of the subject, it has to be remarked that their analysis was careful and somehow "prudent". They had a remarkable capacity to avoid the many difficult points that one finds when generalizing the results valid in finite dimensional spaces to infinite dimensional spaces. Roughly speaking Lagrange was aware of the fact that, when dealing with a function whose derivative plays a role in the calculation of the energy (or action) functionals, it is necessary to bound not only the variation of the values of the function itself, but also the variation of the values of its derivative. This is exactly the starting point of the analysis developed by Sobolev.

In fact, the concept of smallness is made rigorous by introducing a norm in the considered functional space. The mathematical frame makes precise the concept of "virtual displacement", cleverly introduced by Lagrange and Piola. Many scholars, not aware of the rigorous mathematical setting established by Sobolev and Fréchet, have criticized Lagrangian Postulation of Mechanics by claiming that it is based on the obscure concept of "virtual displacement". This concept is rather standard in mathematical analysis when minimization problems are considered: one can think of the standard argument used for proving that in a value of the independent variable that is a minimum for a function the function's first derivative must vanish. Roughly speaking one considers the Taylor expansion of the function truncated to the first order and estimates the function values close to the candidate minimum value, by adding to this value a small (or if you like: a virtual) increment. The logical process which we have shortly described here is rather old: the analysis of the ancient book Mechanical Problems (the text is called in Greek $M\eta\chi\alpha\nu\iota\kappa\alpha$ and in Latin Mechanica, has been traditionally attributed to Aristotle but now it is recognized as a work by Archytas of Tarentum) by Thomas Winter [139] shows that, most likely, already in the 4th century B.C. the Principle of Virtual Work was applied to study the problems of equilibrium of bodies.

The concept of "small" variation of the independent variable in the neighborhood of the candidate value for the independent variable to supply a minimum is needed to find the stationarity condition by using the methods of modern Calculus. Originally, in the context of the application of minima principles to Mechanics, the small variations of configurations needed to get the stationarity conditions were called "virtual displacements". This nomenclature was suggestive: when one varies the stable equilibrium configuration with a "virtual small displacement" the total energy increases. As the concept of virtual displacement was somehow unsatisfactory to more refined scholars in mathematical analysis, both Lagrange and Piola preferred to call the Stationarity Condition Principle as «The Principle of Virtual Velocities». Later the name of this principle was changed into the Principle of Virtual Work.

## *The mathematical tools used by Piola*

The careful study of the works by Piola shows that the mathematical methods that he uses are very often "modern". This statement is particularly true for what concerns mathematical analysis: Piola, being a careful reader of Lagrange's works, uses skillfully the concept of functional, albeit he could not use consciously the methods of modern Functional Analysis that had not been formalized then. It is, however, rather surprising that Piola's works could be forgotten by the majority of the scholars in Mechanics: we discussed the causes of this phenomenon in the previous sections. Here we want to underline that the removal process we are referring to cannot be simply justified by the fact that Piola wrote in Italian. The mathematical difficulties intrinsic in his approach to the postulation of the basic principles of Mechanics also play a major role.

In a mathematical aspect, however, Piola's work seems rather primitive, as he wrote all his equations component-wise, without any explicit notation allowing him to easily handle objectivity. Even though he managed to include in his expressions for deformation measures the correct objective quantities and he was aware of the need of writing equations which are independent of the observer (see [34]), he had not the time (he died 56 years old, in 1850, being arrived, in our opinion, very close to the invention of a very compact matrix notation) to conceive a *tensorial* formulation of Mechanics. In fact, Levi-Civita's absolute calculus needed to wait many decades before being formalized.

Notwithstanding the dramatic lack of mathematical tools, Piola's works show a rigorous sequence of results, albeit made difficult to follow because of his heavy component-wise notation. To some modern mechanician (not all, as many are still refusing Levi-Civita notation) the calculation notation used by Piola may convey the wrong impression of primitiveness. It is really surprising to see how many advancements of Continuum Mechanics he did manage to obtain without using tensor calculus.

Piola tried, during all his scientific life, to find the correct notation for presenting his results in the best possible way. In many of his works, he proved to know the importance, in mathematics, of the choice of the right notation and, in mechanics, of the right conceptual tools. He was open minded: in fact, he fully accepted the nominalistic and conceptual novelties that Lagrange introduced in Mathematical Analysis: novelties that were fiercely opposed by many of his contemporaries.

## *Piola has been «forgotten» for a time period: Peridynamics in Piola's work*

In his fundamental work [109], Gabrio Piola formulates a homogenized continuum theory by means of a micro-macro identification process. He tries to solve the problem of finding the equations governing the phenomena of deformation by starting, at a micro-level, from a discrete micro-model. He assumes to have many basic components of matter, that he calls molecules, which constitute the bodies, and that these systems of molecules are governed by Lagrange's equations of motion. Then he assumes that, once *macro*-descriptors of the kinematics of the considered bodies are chosen, a reasonably detailed *micro*-motion can be determined, representing the effective motion occurring at micro-level. Subsequently, he identifies macro-motions by formulating two times the Principle of Virtual Work, bot at macro and micro level and by identifying micro and macro powers, by using the introduced discrete micro-model and the continuous macro-model.

As Piola's assumptions are general enough, we can say that his continuum theory is a non-local theory (in the nomenclature used by Eringen [53, 54]). For sure it is more general than the one that was introduced later by Cauchy and Navier (see [14]).

Piola's starting point is what, in modern nomenclature, is called a system of material particles with long range interactions. Of course his model is the classical model developed by Lagrange for a finite set of particles: he introduces a kinetic energy and an interaction energy between every pair of particles. Therefore, albeit with an energy which quickly decreases with distance, at the micro-level there is the possibility of having clusters of particles which are interacting. As a consequence, in the macro theory developed by Piola the so-called principle of locality is valid only when certain specifications are made clear. We believe that a clear study of this point can be found in [127].

A similar idea has later been presented by Germain, by considering a microstructured continuum as composed by many material particles which can be modeled as continua of small extension themselves.

It is rather surprising, also, that many Italian authors (see for instance [119]), also those who did produce very interesting novel results in the field, seem not to be aware of the fact that they are continuing Piola's work. The reader who is interested in getting evidence about this last statement can consult [34, 42, 43] where many excerption of Piola's work are translated that prove it.

In [34], Piola's formulas are translated into a more modern notation: we believe that the greatest part of Piola's argument are, even nowadays, at the edge of the research in Continuum Mechanics.

We will stress here how the work of Piola have surely influenced the following works, by attempting the use of some techniques taken from philology, in the spirit of the present volume, by carefully analyzing the written sources. In fact, in Piola [110] (Capo I, pag. 8) the attentive reader will verify that the reference configuration of the studied deformable body is described, with a very careful series of statements. The labelling symbols introduced by Piola for characterizing each material particle with the three Cartesian coordinates is: «(a, b, c)». It is really surprising that exactly the same notation is used in Hellinger [86], see e.g., pag. 605. The reader will remember that similar observations did persuade Heiberg (and all experts about Archimedes) about the hierarchy of the sources concerning Archimedes works.

We intend, in future investigations, to continue this kind of observations to fully track how Piola did really influence the development of modern Generalized Continuum Mechanics. In fact, we are rather lucky, as all sources are available and can be easily consulted. We do not believe that such a challenge is useless. Already in the fundamental textbook by Lagrange [92], differently from many other authors, the presented results are framed in a correct historical perspective. Lagrange did try to give due credit to all his predecessors, including Greek mechanicians and his friend D'Alembert. At the beginning of the *Mécanique Analytique* one finds a very interesting historical introduction: we believe that it is the first source of modern History of Mechanics. We regret that also this aspect of the Lagrangian lesson has been too often forgotten by too many scholars in Continuum Mechanics.

As a fundamental example, we recall that in some modern papers one finds announced a «very new and powerful» theory: Peri-dynamics. We believe that many among the ideas presented there really have big merits and deserve to be developed. In particular, they can allow great advancements in fracture and damage Mechanics. In Peri-dynamics papers, Cauchy continua are generalized in a clever form and we believe that they could give an interesting framework to develop the theory for studying e.g. in crack formation and growth (for some interesting relevant literature see also [112, 113, 114, 116, 115, 117, 104, 29, 12, 81, 30]).

It is, however, a pity that even all the many scientists whose mother language is Italian and who work in this novel part of Continuum Mechanics seem to be unaware of the fundamental works by Gabrio Piola exactly in this field. We believe that Piola has to be credited to be the father of Peri-dynamics and that the loss of memory observed also in this case may be very detrimental to the advancement of science. The lack of the credit due to the major scientific sources is very dangerous: one looses, because of said lack, the important information about the motivations which contributed to the genesis of considered new ideas. To believe that everything which is modern is better than more ancient sources is wrong also for many other reasons: for instance one looses the needed informations concerning the influences that a model had on the development of other models. The tendency towards a "modernism" unfortunately is becoming more and more trendy. There are many papers in Continuum Mechanics written during the XX century that are less advanced than Piola's ones.

In fact, after many years in which Piola's ideas were not developed, in [128] the study by Piola has been somehow completed and the bases for more developments established. One could believe that Silling was one of Piola's direct pupils: to this aim the reader is invited to compare Silling arguments with those presented by Piola (see again [34]). A comparison between the works by Piola and those of his modern pupils can be very fruitful: it is interesting to see that some concepts and modeling ideas formulated more than 150 years ago by Piola may be topical even nowadays: the reader is invited to consult, for instance, the abstract of the papers [8, 107].

Also by reading the literature that was originated from the papers by Silling one can remark that there is a strict relationship between non-local continuum theories and the theories of discrete systems of interacting particles. This connection was clear to Piola: how deep was his understanding can be verified by consulting [42, 43] and the more modern literature (see e.g. [128, 129, 130]).

## 4.5 Conclusions

The process of systematically removing references to the name of Gabrio Piola in Continuum Mechanics (and part of his results) is just one of many examples of how some social groups have, over the centuries and in different cultural fields, rewritten more or less relevant parts of the cultural knowledge of a society. At the basis of this phenomenon we can always find common features:

i) a sectarian vision of cultural progress;
ii) the conviction that the point of view of the own social group is clearly superior to that of all the others;
iii) the occupation with own members of positions not always of cultural relevance but certainly in key positions for the management of power;
iv) the inability to face a fruitful discussion with representatives of groups of different cultural orientation.

The fundamental difference with groups or individuals that produce effective cultural advancement for mankind, such as Piola, Archytas or Archimedes (to cite the few examples we refer to in this book), lies in the fact that those who are actually engaged in the advancement of scientific knowledge do not have the time to occupy positions of power. So, the competition between the two groups of scholars (those who study and those who occupy power) is lost, without any hope, by those who, like Piola, are essentially dedicated to research. Those who, instead, less competent have had the possibility to spend their time to maneuver in order to occupy key positions in power management do effectively have the power to decide the destiny of the scholars who are more competent. In this sense, the scientific advancement of a society is largely distorted by the presence of unscrupulous and incompetent people in key positions.

When this social process becomes prevalent in the subgroup of scholars, the decline of the whole society is already in an advanced state.

The sad story of how the enormous scientific advances made by Gabrio Piola in the mid-XIX century (and which are gradually being rediscovered only now) is paradigmatic of these social mechanisms of elimination of original ideas but, above all, of the name of their inventors. This is the manifestation of a form of real thought dictatorship. At the end of Hellenism, this kind of mechanisms allowed the cancellation of the enormous scientific progress achieved by Greek society in the third century BC. While the new strong vision of the expansion of the Roman Empire became dominant, the cultural quality of dominant classes became dramatically lower. The germs of the Middle Ages were already present in that process of cancellation. A thousand years of ignorance and obscurantism derived from the loss of the scientific results of Hellenistic Science.

But when those who have achieved results in any field of knowledge have also had the wisdom to leave more copies (and in different places) of these results, then there is hope that sooner or later someone will be inspired by the novelty and originality of these results and scientific progress can finally return. We must thank Leo the Mathematician for having ordered the copy of Archimedes' works and for having caused their fortuitous consequent diffusion in Europe. Sometimes we also have to thank those who (like Tartaglia, although not understanding much in Archimedes' writings and crediting himself with a translation not his own, as demonstrated by Heiberg and as it has been widely discussed in Chap. 2) contributed, unconsciously, to spread at least the knowledge of works that perhaps would have been less known otherwise. In this regard, even Piola has benefited from a similar treatment, having been, although not understood and openly criticized, referred to in Truesdell's compendium.

## Acknowledgments

Mario Spagnuolo was supported by P.O.R. SARDEGNA F.S.E. 2014-2020 - Asse III "Istruzione e Formazione, Obiettivo Tematico: 10, Obiettivo Specifico: 10.5, Azione dell'accordo di Partenariato:10.5.12" Avviso di chiamata per il finanziamento di Progetti di ricerca - Anno 2017.

## References

1. Abdoul-Anziz, H., Auffray, N., Desmorat, B.: Symmetry classes and matrix representations of the 2d flexoelectric law. Symmetry **12**(4), 674:1–29 (2020)
2. Altenbach, H., Bîrsan, M., Eremeyev, V.A.: Cosserat-type rods. In: Generalized Continua from the Theory to Engineering Applications, pp. 179–248. Springer (2013)
3. Altenbach, H., Eremeyev, V.A.: On the theories of plates based on the cosserat approach. In: Mechanics of generalized continua, pp. 27–35. Springer (2010)

4. Altenbach, H., Eremeyev, V.A.: Cosserat media. In: Generalized Continua from the Theory to Engineering Applications, pp. 65–130. Springer (2013)
5. Altenbach, H., Eremeyev, V.A.: Cosserat-type shells. In: Generalized continua from the theory to engineering applications, pp. 131–178. Springer (2013)
6. Altenbach, J., Altenbach, H., Eremeyev, V.A.: On generalized Cosserat-type theories of plates and shells: a short review and bibliography. Archive of Applied Mechanics **80**(1), 73–92 (2010)
7. Andreaus, U., Dell'Isola, F., Giorgio, I., Placidi, L., Lekszycki, T., Rizzi, N.L.: Numerical simulations of classical problems in two-dimensional (non) linear second gradient elasticity. International Journal of Engineering Science **108**, 34–50 (2016)
8. Askari, E., Bobaru, F., Lehoucq, R., Parks, M., Silling, S., Weckner, O., et al.: Peridynamics for multiscale materials modeling. Journal of Physics: Conference Series **125**(1), 012078:1–11 (2008)
9. Barchiesi, E., Eugster, S.R., dell'Isola, F., Hild, F.: Large in-plane elastic deformations of bi-pantographic fabrics: asymptotic homogenization and experimental validation. Mathematics and Mechanics of Solids **25**(3), 739–767 (2020)
10. Barchiesi, E., Eugster, S.R., Placidi, L., dell'Isola, F.: Pantographic beam: a complete second gradient 1D-continuum in plane. Zeitschrift für angewandte Mathematik und Physik **70**(5), 135:1–24 (2019)
11. Barchiesi, E., Harsch, J., Ganzosch, G., Eugster, S.R.: Discrete versus homogenized continuum modeling in finite deformation bias extension test of bi-pantographic fabrics. Continuum Mechanics and Thermodynamics pp. 1–14 (2020)
12. Barchiesi, E., Yang, H., Tran, C., Placidi, L., Müller, W.: Computation of brittle fracture propagation in strain gradient materials by the FEniCS library. Mathematics and Mechanics of Solids pp. 1–16 (2020)
13. Basdevant, J.L.: Feynman's principle in quantum mechanics. In: Variational Principles in Physics, pp. 145–165. Springer (2007)
14. Benvenuto, E.: An introduction to the history of structural mechanics: Part I: Statics and resistance of solids. Springer Science & Business Media (2012)
15. Berrey, M.: Hellenistic science at court, vol. 5. Walter de Gruyter GmbH & Co KG (2017)
16. Burton Russell, J.: Inventing the flat earth: Columbus and modern historians (1991)
17. Capobianco, G., Eugster, S.R.: Time finite element based Moreau-type integrators. International Journal for Numerical Methods in Engineering **114**(3), 215–231 (2018)
18. Capobianco, G., Eugster, S.R., Winandy, T.: Modeling planar pantographic sheets using a nonlinear Euler–Bernoulli beam element based on B-spline functions. Proceedings in Applied Mathematics and Mechanics **18**(1), e201800220:1–2 (2018)
19. Capobianco, G., Winandy, T., Eugster, S.: The principle of virtual work and Hamilton's principle on Galilean manifolds. Journal of Geometric Mechanics pp. 1–27 (2021)
20. Carcaterra, A., dell'Isola, F., Esposito, R., Pulvirenti, M.: Macroscopic description of microscopically strongly inhomogenous systems: A mathematical basis for the synthesis of higher gradients metamaterials. Archive for Rational Mechanics and Analysis **218**(3), 1239–1262 (2015)
21. Cazzani, A., Atluri, S.: Four-noded mixed finite elements, using unsymmetric stresses, for linear analysis of membranes. Computational Mechanics **11**(4), 229–251 (1993)
22. Cazzani, A., Malagù, M., Turco, E.: Isogeometric analysis of plane-curved beams. Mathematics and Mechanics of Solids **21**(5), 562–577 (2016)
23. Cazzani, A., Malagù, M., Turco, E., Stochino, F.: Constitutive models for strongly curved beams in the frame of isogeometric analysis. Mathematics and Mechanics of Solids **21**(2), 182–209 (2016)
24. Cazzani, A., Serra, M., Stochino, F., Turco, E.: A refined assumed strain finite element model for statics and dynamics of laminated plates. Continuum Mechanics and Thermodynamics **32**(3), 665–692 (2020)
25. Cazzani, A., Stochino, F., Turco, E.: An analytical assessment of finite element and isogeometric analyses of the whole spectrum of Timoshenko beams. Zeitschrift für Angewandte Mathematik und Mechanik **96**(10), 1220–1244 (2016)

26. Cohen, L.: Quantization problem and variational principle in the phase-space formulation of quantum mechanics. Journal of Mathematical Physics **17**(10), 1863–1866 (1976)
27. Cosserat, E., Cosserat, F.: Théorie des corps déformables. A. Hermann et fils (1909)
28. Cremaschini, C., Tessarotto, M.: Synchronous Lagrangian variational principles in general relativity. The European Physical Journal Plus **130**(6), 123 (2015)
29. Cuomo, M.: Continuum damage model for strain gradient materials with applications to 1d examples. Continuum Mechanics and Thermodynamics **31**(4), 969–987 (2019)
30. Cuomo, M., Contrafatto, L., Greco, L.: A variational model based on isogeometric interpolation for the analysis of cracked bodies. International Journal of Engineering Science **80**, 173–188 (2014)
31. Cuomo, S.: Pappus of Alexandria and the mathematics of late antiquity. Cambridge University Press (2000)
32. dell'Isola, F.: The academic and scientific activity of a "Maestro" in applied mechanics: Laudatio of Professor Antonio Di Carlo, Università di Roma TRE (Italy). Mathematics and Mechanics of Solids **18**(8), 787 (2013)
33. dell'Isola, F.: Big-(Wo)men, Tyrants, Chiefs, Dictators, Emperors and Presidents: Towards the Mathematical Understanding of Social Groups. Springer (2019)
34. dell'Isola, F., Andreaus, U., Placidi, L.: At the origins and in the vanguard of peridynamics, non-local and higher-gradient continuum mechanics: an underestimated and still topical contribution of Gabrio Piola. Mathematics and Mechanics of Solids **20**(8), 887–928 (2015)
35. dell'Isola, F., Bucci, S., Battista, A.: Against the fragmentation of knowledge: The power of multidisciplinary research for the design of metamaterials. In: Advanced Methods of Continuum Mechanics for Materials and Structures, pp. 523–545. Springer (2016)
36. dell'Isola, F., Corte, A.D., Giorgio, I.: Higher-gradient continua: The legacy of Piola, Mindlin, Sedov and Toupin and some future research perspectives. Mathematics and Mechanics of Solids **22**(4), 852–872 (2017)
37. dell'Isola, F., Della Corte, A., Greco, L., Luongo, A.: Plane bias extension test for a continuum with two inextensible families of fibers: a variational treatment with Lagrange multipliers and a perturbation solution. International Journal of Solids and Structures **81**, 1–12 (2016)
38. dell'Isola, F., Gavrilyuk, S.: Variational models and methods in solid and fluid mechanics, vol. 535. Springer Science & Business Media (2012)
39. dell'Isola, F., Guarascio, M., Hutter, K.: A variational approach for the deformation of a saturated porous solid. A second-gradient theory extending Terzaghi's effective stress principle. Archive of Applied Mechanics **70**(5), 323–337 (2000)
40. dell'Isola, F., Madeo, A., Seppecher, P.: Boundary conditions at fluid-permeable interfaces in porous media: A variational approach. International Journal of Solids and Structures **46**(17), 3150–3164 (2009)
41. dell'Isola, F., Madeo, A., Seppecher, P.: Cauchy tetrahedron argument applied to higher contact interactions. Archive for Rational Mechanics and Analysis **219**(3), 1305–1341 (2016)
42. dell'Isola, F., Maier, G., Perego, U., et al.: The Complete Works of Gabrio Piola: Volume I (Advanced Structured Materials, vol. 38) (2014)
43. dell'Isola, F., Maier, G., Perego, U., et al.: The Complete Works of Gabrio Piola: Volume II. Cham, Switzerland: Springer (2019)
44. dell'Isola, F., Placidi, L.: Variational principles are a powerful tool also for formulating field theories. In: Variational models and methods in solid and fluid mechanics, pp. 1–15. Springer (2011)
45. dell'Isola, F., Seppecher, P.: Edge contact forces and quasi-balanced power. Meccanica **32**(1), 33–52 (1997)
46. dell'Isola, F., Seppecher, P., Madeo, A.: How contact interactions may depend on the shape of Cauchy cuts in nth gradient continua: approach "à la D'Alembert". Zeitschrift für angewandte Mathematik und Physik **63**(6), 1119–1141 (2012)
47. dell'Isola, F., Steigmann, D.J.: Discrete and Continuum Models for Complex Metamaterials. Cambridge University Press (2020)

48. Deutschmann, B., Eugster, S.R., Ott, C.: Reduced models for the static simulation of an elastic continuum mechanism. IFAC-PapersOnLine **51**(2), 403 – 408 (2018). 9th Vienna International Conference on Mathematical Modelling
49. Earman, J., Janssen, M., Norton, J.D.: The attraction of gravitation: new studies in the history of general relativity, vol. 5. Springer Science & Business Media (1993)
50. Eremeyev, V.A.: On non-holonomic boundary conditions within the nonlinear Cosserat continuum. In: New Achievements in Continuum Mechanics and Thermodynamics, pp. 93–104. Springer (2019)
51. Eremeyev, V.A., Lebedev, L.P., Altenbach, H.: Foundations of micropolar mechanics. Springer Science & Business Media (2012)
52. Eremeyev, V.A., Pietraszkiewicz, W.: Material symmetry group and constitutive equations of anisotropic Cosserat continuum. In: H. Altenbach, S. Forest, A. Krivtsov (eds.) Generalized continua as models for materials, *Advanced Structures Materials*, vol. 22. Springer (2012)
53. Eringen, A.C.: Nonlocal polar elastic continua. International Journal of Engineering Science **10**(1), 1–16 (1972)
54. Eringen, A.C., Edelen, D.: On nonlocal elasticity. International Journal of Engineering Science **10**(3), 233–248 (1972)
55. Eugster, S.R.: Geometric Continuum Mechanics and Induced Beam Theories, *Lecture Notes in Applied and Computational Mechanics*, vol. 75. Springer (2015)
56. Eugster, S.R.: Hellinger's encyclopedia article on the fundamentals of the mechanics of continua. In: F. dell'Isola, S.R. Eugster, M. Spagnuolo, E. Barchiesi (eds.) Evaluation of Scientific Sources in Mechanics: Heiberg's Prolegomena to the Works of Archimedes and Hellinger's Encyclopedia Article. Springer (2021)
57. Eugster, S.R., dell'Isola, F.: Exegesis of the introduction and sect. I from "Fundamentals of the mechanics of continua" by E. Hellinger. Zeitschrift für angewandte Mathematik und Mechanik **97**(4), 477–506 (2017)
58. Eugster, S.R., dell'Isola, F.: An ignored source in the foundations of continuum physics "Die Allgemeinen Ansätze der Mechanik der Kontinua" by E. Hellinger. Proceedings in Applied Mathematics and Mechanics **17**(1), 413–414 (2017)
59. Eugster, S.R., dell'Isola, F.: Exegesis of sect. II and III.A from "Fundamentals of the mechanics of continua" by E. Hellinger. Zeitschrift für angewandte Mathematik und Mechanik **98**(1), 31–68 (2018)
60. Eugster, S.R., dell'Isola, F.: Exegesis of sect. III.B from "Fundamentals of the mechanics of continua" by E. Hellinger. Zeitschrift für angewandte Mathematik und Mechanik **98**(1), 69–105 (2018)
61. Eugster, S.R., dell'Isola, F., Steigmann, D.: Continuum theory for mechanical metamaterials with a cubic lattice substructure. Mathematics and Mechanics of Complex Systems **7**(1), 75–98 (2019)
62. Eugster, S.R., Deutschmann, B.: A nonlinear Timoshenko beam formulation for modeling a tendon-driven compliant neck mechanism. Proceedings in Applied Mathematics and Mechanics **18**(1), e201800208:1–2 (2018)
63. Eugster, S.R., Glocker, Ch.: Constraints in structural and rigid body mechanics: a frictional contact problem. Annals of Solid and Structural Mechanics **5**(1-2), 1–13 (2013)
64. Eugster, S.R., Glocker, Ch.: On the notion of stress in classical continuum mechanics. Mathematics and Mechanics of Complex Systems **5**(3-4), 299–338 (2017)
65. Eugster, S.R., Harsch, J.: A variational formulation of classical nonlinear beam theories. In: B.E. Abali, I. Giorgio (eds.) Developments and Novel Approaches in Nonlinear Solid Body Mechanics, pp. 95–121. Springer International Publishing (2020)
66. Eugster, S.R., Hesch, C., Betsch, P., Glocker, Ch.: Director-based beam finite elements relying on the geometrically exact beam theory formulated in skew coordinates. International Journal for Numerical Methods in Engineering **97**(2), 111–129 (2014)
67. Eugster, S.R., Steigmann, D.J.: Variational methods in the theory of beams and lattices. In: H. Altenbach, A. Öchsner (eds.) Encyclopedia of Continuum Mechanics, pp. 1–9. Springer (2018)

68. Feynman, R.P., Brown, L.M.: Feynman's thesis: a new approach to quantum theory. World Scientific (2005)
69. Feynman, R.P., Hibbs, A.R., Styer, D.F.: Quantum mechanics and path integrals. Courier Corporation (2010)
70. Feynman, R.P., Leighton, B.R., Sands, M.: The Feynman Lectures on Physics, Volume I. Basic Books (2018)
71. Forest, S.: Mechanics of Cosserat media – an introduction. Ecole des Mines de Paris, Paris pp. 1–20 (2005)
72. George, D., Spingarn, C., Dissaux, C., Nierenberger, M., Rahman, R.A., Rémond, Y.: Examples of multiscale and multiphysics numerical modeling of biological tissues. Bio-Medical Materials and Engineering **28**(s1), S15–S27 (2017)
73. Germain, P.: Functional concepts in continuum mechanics. Meccanica **33**(5), 433–444 (1998)
74. Germain, P.: My discovery of mechanics. Continuum Thermomechanics, P. Germain's Anniversary Volume pp. 1–24 (2000)
75. Germain, P.: The method of virtual power in the mechanics of continuous media, I: Second-gradient theory. Mathematics and Mechanics of Complex Systems **8**(2), 153–190 (2020)
76. Germain, P., Nayroles, B.: Applications of Methods of Functional Analysis to Problems in Mechanics: Joint Symposium IUTAM/IMU Held in Marseille, Sept. 1-6, 1975, vol. 503. Springer (2006)
77. Gingerich, O.: Did Copernicus owe a debt to Aristarchus? Journal for the History of Astronomy **16**(1), 37–42 (1985)
78. Giorgio, I., Andreaus, U., Dell'Isola, F., Lekszycki, T.: Viscous second gradient porous materials for bones reconstructed with bio-resorbable grafts. Extreme Mechanics Letters **13**, 141–147 (2017)
79. Giorgio, I., Ciallella, A., Scerrato, D.: A study about the impact of the topological arrangement of fibers on fiber-reinforced composites: Some guidelines aiming at the development of new ultra-stiff and ultra-soft metamaterials. International Journal of Solids and Structures **203**, 73–83 (2020)
80. Giorgio, I., Harrison, P., dell'Isola, F., Alsayednoor, J., Turco, E.: Wrinkling in engineering fabrics: a comparison between two different comprehensive modelling approaches. Proceedings of the Royal Society A: Mathematical, Physical and Engineering Sciences **474**(2216), 20180063:1–20 (2018)
81. Giorgio, I., Scerrato, D.: Multi-scale concrete model with rate-dependent internal friction. European Journal of Environmental and Civil Engineering **21**(7-8), 821–839 (2017)
82. Harsch, J., Capobianco, G., Eugster, S.R.: Finite element formulations for constrainted spatial nonlinear beam theories. Mathematics and Mechanics of Solids pp. 1–26 (2021)
83. Harsch, J., Eugster, S.R.: Finite element analysis of planar nonlinear classical beam theories. In: B.E. Abali, I. Giorgio (eds.) Developments and Novel Approaches in Nonlinear Solid Body Mechanics, pp. 123–157. Springer International Publishing (2020)
84. Hartle, J.B., Sharp, D.H.: Variational principle for the equilibrium of a relativistic, rotating star. The Astrophysical Journal **147**, 317 (1967)
85. Heath, T.: Aristarchus of Samos, the Ancient Copernicus: A History of Greek Astronomy to Aristarchus, Together with Aristarchus's Treatise on the Sizes and Distances of the Sun and Moon. Cambridge University Press (2013)
86. Hellinger, E.: Die allgemeinen Ansätze der Mechanik der Kontinua. In: Encyklopädie der mathematischen Wissenschaften, vol. 5. B.G. Teubner Verlag (1913)
87. Hesch, C., Schuß, S., Dittmann, M., Eugster, S.R., Favino, M., Krause, R.: Variational space–time elements for large-scale systems. Computer Methods in Applied Mechanics and Engineering **326**, 541–572 (2017)
88. Huffman, C.: Archytas of Tarentum: Pythagorean, Philosopher and Mathematician King. Cambridge University Press (2005)
89. Irby-Massie, G.L., Keyser, P.T., Rihll, T.E.: Greek science of the Hellenistic era: A sourcebook. Aestimatio: Critical Reviews in the History of Science **1**, 44–50 (2004)

90. Khakalo, S., Niiranen, J.: Anisotropic strain gradient thermoelasticity for cellular structures: Plate models, homogenization and isogeometric analysis. Journal of the Mechanics and Physics of Solids **134**, 103728 (2020)
91. Kuhn, T.S.: The structure of scientific revolutions. University of Chicago press (2012)
92. de Lagrange, J.L.: Mécanique analytique, vol. 1. Mallet-Bachelier (1853)
93. Lagrange, J.L.: Analytical mechanics, vol. 191. Springer Science & Business Media (2013)
94. Landau, L.D., Lifshitz, E.M.: Quantum mechanics: non-relativistic theory, vol. 3. Elsevier (2013)
95. Lloyd, J., Mitchinson, J.: QI: The Book of General Ignorance-The Noticeably Stouter Edition. Faber & Faber (2010)
96. Lloyd, J., Mitchinson, J.: The Second Book of General Ignorance: Everything You Think You Know is (still) Wrong. Crown (2011)
97. Maugin, G., Trimarco, C.: Pseudomomentum and material forces in nonlinear elasticity: variational formulations and application to brittle fracture. Acta mechanica **94**(1-2), 1–28 (1992)
98. Maugin, G.A.: A selection of scientific works and publications by Paul Germain. In: G.A. Maugin, R. Drouot, F. Sidoroff (eds.) Continuum Thermodynamics: The Art and Science of Modelling Material Behaviour, *Solid Mechanics and Its Applications*, vol. 76, pp. 25–27. Kluwer Academic Publishers (2000)
99. Maugin, G.A.: Generalized continuum mechanics: what do we mean by that? In: Mechanics of Generalized Continua, pp. 3–13. Springer (2010)
100. Maugin, G.A.: A historical perspective of generalized continuum mechanics. In: Mechanics of generalized continua, pp. 3–19. Springer (2011)
101. Maugin, G.A.: Continuum mechanics through the twentieth century. Springer (2013)
102. Maugin, G.A., Eringen, A.C.: Variational formulation of the relativistic theory of microelectromagnetism. Journal of Mathematical Physics **15**(9), 1494–1499 (1974)
103. Meissner, H.: Ernst Hellinger. In: Topics in operator theory: Ernst D. Hellinger memorial volume, *Operator Theory: Advances and Applications*, vol. 48. Birkhäuser Verlag, Basel (1990)
104. Misra, A., Poorsolhjouy, P.: Granular micromechanics model for damage and plasticity of cementitious materials based upon thermomechanics. Mathematics and Mechanics of Solids **25**(10), 1778–1803 (2020)
105. Müller, C.H., Timpe, A.: Die Grundgleichungen der mathematische Elastizitätstheorie. In: Mechanik, *Enzyklopädie der Mathematischen Wissenschaften mit Einschluss ihrer Anwendungen*, vol. IV/4, pp. 1–54. B.G. Teubner Verlag (1906)
106. Neugebauer, O.: The history of ancient astronomy problems and methods. Journal of Near Eastern Studies **4**(1), 1–38 (1945)
107. Parks, M.L., Lehoucq, R.B., Plimpton, S.J., Silling, S.A.: Implementing peridynamics within a molecular dynamics code. Computer Physics Communications **179**(11), 777–783 (2008)
108. Pietraszkiewicz, W., Eremeyev, V.: On vectorially parameterized natural strain measures of the non-linear Cosserat continuum. International Journal of Solids and Structures **46**(11-12), 2477–2480 (2009)
109. Piola, G.: Intorno alle equazioni fondamentali del movimento dei corpi qualsivogliono, considerati secondo naturale loro formae costituzione (1845)
110. Piola, G.: Memoria intorno alle equazioni fondamentali del movimento di corpi qualsivogliono considerati secondo la naturale loro forma e costituzione. Modena, Tipi del RD Camera (1846)
111. Pipkin, A.C.: Equilibrium of Tchebychev nets. In: The Breadth and Depth of Continuum Mechanics, pp. 287–303. Springer (1986)
112. Placidi, L.: A variational approach for a nonlinear 1-dimensional second gradient continuum damage model. Continuum Mechanics and Thermodynamics **27**(4-5), 623–638 (2015)
113. Placidi, L., Barchiesi, E.: Energy approach to brittle fracture in strain-gradient modelling. Proceedings of the Royal Society A: Mathematical, Physical and Engineering Sciences **474**(2210), 20170878:1–19 (2018)

114. Placidi, L., Barchiesi, E., Misra, A.: A strain gradient variational approach to damage: a comparison with damage gradient models and numerical results. Mathematics and Mechanics of Complex Systems **6**(2), 77–100 (2018)
115. Placidi, L., Barchiesi, E., Misra, A., Andreaus, U.: Variational methods in continuum damage and fracture mechanics. Encyclopedia of Continuum Mechanics. Springer (2018)
116. Placidi, L., Misra, A., Barchiesi, E.: Two-dimensional strain gradient damage modeling: a variational approach. Zeitschrift für angewandte Mathematik und Physik **69**(3), 56 (2018)
117. Placidi, L., Misra, A., Barchiesi, E.: Simulation results for damage with evolving microstructure and growing strain gradient moduli. Continuum Mechanics and Thermodynamics **31**(4), 1143–1163 (2019)
118. Poisson, M.: Mémoire sur l'équilibre et le mouvement des corps élastique. In: Mémoires de l'académie royale des sciences, pp. 357–570 (1829)
119. Polizzotto, C.: Nonlocal elasticity and related variational principles. International Journal of Solids and Structures **38**(42-43), 7359–7380 (2001)
120. Rodriguez-Ramos, R., Pobedria, B., Padilla, P., Bravo-Castillero, J., Guinovart-Diaz, R., Maugin, G.: Variational principles for nonlinear piezoelectric materials. Archive of Applied Mechanics **74**(3-4), 191–200 (2004)
121. Rosenbrock, H.: A variational principle for quantum mechanics. Physics letters A **110**(7-8), 343–346 (1985)
122. Rowbotham, S.B., et al.: Zetetic astronomy: Earth not a globe. Ravenio Books (2015)
123. Sarton, G.: A history of science: Ancient science through the golden age of Greece. Oxford University Press London (1953)
124. Sarton, G.: Hellenistic science and culture in the last three centuries BC. Courier Corporation (1993)
125. Scholtz, F., Geyer, H., Hahne, F.: Quasi-Hermitian operators in quantum mechanics and the variational principle. Annals of Physics **213**(1), 74–101 (1992)
126. Schulte, J., Dittmann, M., Eugster, S.R., Hesch, S., Reinicke, T., dell'Isola, F., Hesch, C.: Isogeometric analysis of fiber reinforced composites using Kirchhoff–Love shell elements. Computer Methods in Applied Mechanics and Engineering **362**, 112845:1–34 (2020)
127. Segev, R.: Locality and continuity in constitutive theory. Archive for Rational Mechanics and Analysis **101**(1), 29–39 (1988)
128. Silling, S.A.: Reformulation of elasticity theory for discontinuities and long-range forces. Journal of the Mechanics and Physics of Solids **48**(1), 175–209 (2000)
129. Silling, S.A., Epton, M., Weckner, O., Xu, J., Askari, E.: Peridynamic states and constitutive modeling. Journal of Elasticity **88**(2), 151–184 (2007)
130. Silling, S.A., Lehoucq, R.B.: Convergence of peridynamics to classical elasticity theory. Journal of Elasticity **93**(1), 13 (2008)
131. Stahl, W.H.: Aristarchus of Samos. Dictionary of scientific biography **1**, 246–250 (1970)
132. Taub, A.: Variational principles in general relativity. In: Relativistic Fluid Dynamics, pp. 205–300. Springer (2011)
133. Taub, A.H.: General relativistic variational principle for perfect fluids. Physical Review **94**(6), 1468 (1954)
134. Toupin, R.A.: Theories of elasticity with couple-stress. Archive for Rational Mechanics and Analysis **17**(2), 85–112 (1964)
135. Truesdell, C., Toupin, R.: The classical field theories. In: S. Flügge (ed.) Principles of Classical Mechanics and Field Theory, *Encyclopedia of Physics*, vol. III/1. Springer (1960)
136. Turco, E., Barchiesi, E., Giorgio, I., dell'Isola, F.: A Lagrangian Hencky-type non-linear model suitable for metamaterials design of shearable and extensible slender deformable bodies alternative to Timoshenko theory. International Journal of Non-Linear Mechanics p. 103481 (2020)
137. Turco, E., dell'Isola, F., Cazzani, A., Rizzi, N.L.: Hencky-type discrete model for pantographic structures: numerical comparison with second gradient continuum models. Zeitschrift für angewandte Mathematik und Physik **67**(4), 85 (2016)

138. Turco, E., Golaszewski, M., Cazzani, A., Rizzi, N.L.: Large deformations induced in planar pantographic sheets by loads applied on fibers: experimental validation of a discrete Lagrangian model. Mechanics Research Communications **76**, 51–56 (2016)
139. Winter, T.N.: The mechanical problems in the corpus of Aristotle. Faculty Publications, Classics and Religious Studies Department (2007)

# Chapter 5
# A partial report on the controversies about the Principle of Virtual Work: from Archytas of Tarentum to Lagrange, Piola, Mindlin and Toupin.

Emilio Barchiesi, Alessandro Ciallella, and Daria Scerrato

The efforts paid in the philological search of the most ancient sources, in which one can find the formulation of a certain theory, may seem devoid of utility. Especially if such a search is complicated by the infinite formulation variants that a theory can present, in the simplest cases because of change of notations or, in the most difficult ones, because of the presence of equivalent postulations that are chosen as starting points. To further increase the difficulties, the mathematical tools used in the development of the theory, sometimes, may appear to be very different, albeit they produce the same logical consequences.

The most famous case of different, but equivalent, mathematical tools used in a theory is represented by the synthesis operated by Cartesian geometry. The equivalence of algebraic and geometrical concepts obtained by establishing the Cartesian correspondence between geometrical and algebraic objects did allow a major advancement in human capacities to model the physical reality. It proved that one can develop equivalent theories either by using geometrical concepts or by using algebraic ones. As an instance, the concept of physical quantity can be modeled either by means of Dedekind sections (the modern name given to the geometrical definition of real numbers, which was probably due to Eudoxus of Cnidus) or by introducing Cauchy sequences of rational numbers (the modern definition based

Emilio Barchiesi, e-mail: barchiesiemilio@gmail.com
International Research Center on Mathematics and Mechanics of Complex Systems, University of L'Aquila, L'Aquila, Italy.
Ecole Nationale d'Ingénieurs de Brest, ENIB, UMR CNRS 6027, IRDL, F-29200 Brest, France.
Alessandro Ciallella e-mail: alessandro.ciallella@univaq.it
International Research Center on Mathematics and Mechanics of Complex Systems, University of L'Aquila, L'Aquila, Italy.
Dipartimento di Ingegneria Civile, Edile-Architettura e Ambientale (DICEAA), University of L'Aquila, Italy.
Daria Scerrato, e-mail: daria.scerrato@gmail.com
International Research Center on Mathematics and Mechanics of Complex Systems, University of L'Aquila, L'Aquila, Italy.
Dipartimento di Scienze di Base ed Applicate per l'Ingegneria (SBAI), University of Rome La Sapienza, Italy.



F. dell'Isola et al. (eds.), *Evaluation of Scientific Sources in Mechanics*,
Advanced Structured Materials 152, https://doi.org/10.1007/978-3-030-80550-0_5

on algebraic concepts). A theory does not become original simply because it uses Dedekind sections instead of Cauchy sequences!

An often heard criticism against historical research regards its lack of utility in solving any kind of «practical» problem. We believe that this criticism is not well-grounded. In general, one should consider history as a source of phenomenological evidence, to be used in building theories capable to predict the behavior of social groups. In particular, one could daydream about a theory explaining the reasons for which some groups of scientists are polarized to follow some leaders in choosing a postulation scheme instead of another one when they try to found novel theories or to reorganize already existing ones.

Looking for supporters of this point of view in antiquity, one can think about Cicero. He was aware of the important role that history and its study can have in the advancement of human knowledge. One of his most famous sentences is indeed:

> «Historia vera testis temporum, lux veritatis, vita memoriae, magistra vitae, nuntia vetustatis»
> (Cicero, De Oratore, II, 9, 36)

that one can translate as follows:

> «History is true witness of the times, light of truth, life of memory, teacher of life, messenger of antiquity».

We claim that history can be our teacher also in formulating novel theories. In conclusion, albeit the study of the most ancient sources of the theory of generalized continua has surely its own scholarly interest, we are also motivated by a really practical aim: to find the most effective postulation process to be used when one wants to develop physical models.

## 5.1 Some «forgotten» – but not «lost» – sources in mechanical sciences.

Gabrio Piola (see [46, 39, 43, 42, 47]) considered himself a continuator of the scientific endeavors that Lagrange believed had to be performed to firmly found mechanical sciences. Therefore, Piola invested all his intellectual forces to show that the Principle of Virtual Work (or the Principle of Least Action, which can be regarded to be an important particular case of the previous one) has to be used when creating new models. Piola was aware of the request imposed by Occam's razor: a postulation scheme must use the minimum possible set of postulates, and if one has to choose between a long list of ad hoc assumptions it is much more likely to include, in the formulated theory, some logical incongruences. If the aim is to produce models for successfully predicting the observed experimental evidence and some unknown phenomena, a theory based on the fewest possible assumptions has to be preferred.

Unfortunately, Piola's point of view was not shared by the «French Geometers», as he calls the group led by Cauchy, Poisson and Navier. In this group the preferred postulation scheme was based on the balance of force, the balance of moment of forces and on some «ad hoc» (postulated) restrictions on constitutive equations. For this reason, in some groups of scholars and in some places and times, the contributions by Lagrange, Piola and all their followers were simply ignored.

The process of removal of a source, during the long period when original works were still copied manually, implied as a consequence the total loss of that source. If a librarian decided that a book was not useful, he simply did not ask the amanuenses to copy it, and after some time the book was lost. The process was sometimes made more expedite by simply erasing the «useless» book words from a parchment for reusing it to copy a more important text. This was the destiny of at least one of the most deep contributions of Archimedes to science (we refer to the famous Heiberg palimpsest).

Luckily, the commercial value of the paper on which Piola's works were written did not deserve the attention of any librarian: albeit they were ignored by too many scholars for too much long time, they did remain available in many libraries. Another fortunate circumstance is that Piola's works were originally written in Italian: a not so frequently spoken language, that can however be understood by some scholars capable to understand the sophisticated mathematical reasoning exploited in Piola's works.

As we agree with Piola's epistemological point of view and we can read Italian language, we will try in this chapter to briefly examine how the Principle of Virtual Work (called at the beginning *Principle of Virtual Velocities*) has been used to found (continuum) mechanics and how it allowed for the formulation of the theory of higher-gradient continua since its first formulation by Piola.

Piola's theories of second and higher-gradient continua were ignored, or rejected because believed to be logically inconsistent, by some scholars, followers of Poisson, who refused to accept the Lagrangian postulation of mechanics. The existence of a large group of scholars supported by public institutions did allow for the formation of «resistance» subgroups of scholars who did continue, sometimes even without knowing the existence of his works, Piola's research project. One of the leaders of Lagrangian mechanics was Paul Germain, member of the French Academy of Science in Paris (for more details about the personality of Paul Germain see [124]).

His works in continuum mechanics [88, 90, 58] and his leadership inspired a group of scholars to pursue the Lagrangian scientific program (see the works [1, 86, 5, 12, 159, 15, 50, 49, 135, 2, 157, 87, 98, 99, 100, 69, 123, 122, 125, 80]). In fact Paul Germain was not the only scholar who believed that Lagrange program had to be applied in Continuum Mechanics. A very partial list can be attempted: Lev Davidovič Landau [116] and Leonid Ivanovich Sedov [156, 154, 155] in URSS, Heinz Parkus [133] in Austria, Ekkehart Kröner [113, 114] in Germany, Rivlin [143, 144, 145], Green [102], Pipkin [84, 107, 108, 109, 110, 137, 138, 139], Mindlin [126, 127, 128, 129], Steigmann [54, 162, 166, 163] in USA. Paraphrasing Piola: all listed scholars «could manage to accept the use of the powerful abstract concepts given to us by the genius of Lagrange». A special mention needs to be given here to Richard Toupin (for more details see [43, 37]). He could fully master the mathematical techniques needed to handle variational principles, since he had studied the famous textbook by Landau. He managed to include within the book [170] a chapter about variational principles, albeit he had to accept Truesdell's very negative remarks. This move was very wise as, thanks to this chapter, many scholars could learn about and appreciate them, notwithstanding the negative remarks by Truesdell. Probably, this circumstance became clear to Truesdell and Noll: in fact, in the subsequent textbook [169] the authors simply erase any trace of this very important part of continuum mechanics. Another eminent personality in the valorization of variational principles in Continuum Mechanics is Ernst Hellinger: his fundamental entry in the *Encyklopädie der mathematischen Wissenschaften mit Einschluss ihrer Anwendungen* (1913, Bd. IV-4, Hft. 5) has been essential in preserving the scientific heritage of Lagrange and Piola (see [73, 75, 76, 74, 71]). Like Piola was (see the preface of his work written in 1848 as translated in [46]), one must be extremely surprised in discovering that the Principles of Virtual Work and Least Action still need to be supported and that there are ongoing controversies about its validity and role in continuum mechanics. The social phenomenon that led to their rejection in so many scientific groups, notwithstanding that they were adopted as the most fundamental postulate by undisputed scientific authorities like D'Alembert, Lagrange, Hamilton, Landau, Feynman [85], and Sedov, should be investigated. A remark seems necessary in this regard: all the scholars advocating the importance of variational principles in continuum mechanics have a background in Theoretical Physics. Therefore, it seems that in continuum mechanics there are rather distinct cultural subgroups that are competing for affirming their point of view.

As the memory of the works of Piola, Hellinger, Kröner, Sedov, Mindlin, Green, and Pipkin seems to have been somehow removed by some groups of scholars, we believe that it is needed to reaffirm what follows: if one wants to model microscopically strongly inhomogeneous systems, he has (see e.g. [24, 18, 135, 20]) to consider, in the Principle of Virtual Work, internal work functionals (see Germain [89] and Salençon [148]) depending linearly also on second and higher gradients of virtual displacements. Albeit one could remark that this is a statement already found in the work by Piola (published between 1822 and 1856), it is sure that it has been systematically ignored or overlooked. In our opinion, the theory of higher-gradient continua gives a further example of the «erasure, loss or removal and rediscovery» of human (scientific) knowledge. This kind of processes have been observed for many other scientific theories (see e.g. [147]).

Unfortunately, and this is a big loss for science, we do not have enough sources to prove that, already after its first formulation, the Principle of Virtual Work caused a sharp debate. This debate most likely had some bitter moments, as the complete loss of Archytas works may indicate. It is fortunate that an unknown amanuensis did save for posterity the Pseudo-Aristotelian on *The Mechanical Problems*. Attributing it to Aristotle, who could not be in any way his author (in many of his works, Aristotle shows that he is not versed in mathematics), this amanuensis showed to have understood deeply both the importance of the text that he was saving and the nature of human psychology and social behavior. It is sure, however, that the debate about the importance of the Principle of Virtual Work, and its impact on science, has been taking place without interruption until our days.

## 5.2 An Italian secondary source: Vailati. While underestimating Hellenistic mechanics, he recognizes in it two different ways for studying Statics problems.

In his paper [174] on the history of the Principle of Virtual Work, Vailati starts his introduction with the words

> «The works in which we had preserved the memory of the ideas of Greeks on Mechanics and of the degree of elaboration that, because of their contribution, the theories concerning the equilibrium did reach, can be divided neatly in two categories, that correspond to two, radically different, directions in the manner of considering and solving the questions of Statics».[1]

Therefore, it seems that Vailati believes to be able to distinguish two competing methodologies (or postulations) in the Hellenistic texts of mechanics. On this point we agree completely with him: probably, already in Hellenistic Mechanics both the Principle of Virtual Work together with the balances of force and moment of force were systematically used in the study of problems of Mechanics. Vailati's introduction then continues by stating that:

> «Greek writers did not manage to reach even the enunciation of the most elementary among the principles of Dynamics: the law of Inertia».[2]

On this point we do not agree with Vailati. This statement does not seem well-founded. We will not delve here in this subject, and we refer to [147] for a discussion of the big amount of secondary sources of Hellenistic Science, where the concept of inertia is formulated. Refraining from any polemics, we simply comment here that, in available secondary sources of Hellenistic texts (once translated into Latin or a more modern language), one, obviously, cannot find the exact words (force in particular) which Vailati expected to find for describing the concept. This circumstance has been the cause of many misunderstandings in the history of mechanics. Unfortunately, it has to be remarked that Vailati's opinion has been followed by the majority of scholars. For instance, in the foreword of [56] by Louis de Broglie, one finds stated that:

> «The history of mechanics is one of the most important branches of the history of science. From earliest times man has sought to develop tools that would enable him to add to his power of action or to defend himself against the dangers threatening him. Thus he was unconsciously led to consider the problems of mechanics. So we see the first scholars of ancient times thinking about these problems and arriving more or less successfully at a solution. The motion of the stars which, from the Chaldean shepherds to the great Greek and Hellenistic astronomers, was one of the first preoccupations of human thought, led to the discovery of the true laws of dynamics. As is well known, although the principles of statics

[1] «Gli scritti nei quali ci è stata conservata memoria delle idee dei Greci sulla Meccanica e del grado di elaborazione che raggiunsero per opera loro le teorie relative all'equilibrio, si possono, [...] dividere nettamente in due categorie, corrispondenti a due indirizzi radicalmente diversi nel modo di considerare e di risolvere le questioni di Statica».

[2] «..[..] gli scrittori greci [...] non seppero assorgere neppure all'enunciazione del più elementare dei principi della Dinamica: la legge d'inerzia.»

had been correctly presented by the old scholars those of dynamics, obscured by the false conceptions of the Aristotelian school, did not begin to see light until the end of the Middle Ages and the beginning of the modern era».

We believe that the perfection obtained by Hellenistic Astronomy could not be attained without a clear understanding of a version of the law of inertia. (see e.g. [146, 134, 9, 10, 19, 117]). Instead Vailati, still in his introduction, adds that:

«The questions of Statics seem to have been the only ones of which the Greek writers tried to pursue a general treatment, that is scientific in the modern sense of the word;[i)] as, for what that concerns the study of the laws of motions, they seem to having been contented of [ii)] gross descriptions and classifications of phenomena»[3]

[i)] The observations on the composition of movements, which one finds in the works by Aristotles, and the more elaborated theories as elaborated by the astronomers on the same subject (see about this subject the classical monograph by Schiaparelli: *On the homocentric spheres by Eudoxus, Callippus and Aristotle*) belong rather more to Geometry than truly to Mechanics. To the same class belong also the researches of Archimedes «On Spirals», that are also based on kinematical assumptions.[4]

[ii)] And maybe it is only because of the fault of unskillful compilers and commentators that aforementioned theories and researches did take, later, the aspect and the pretension of scientific theories. And it is interesting, in this regard, to consider as close the considerations developed by Aristotle, in his III book Περὶ οὐρανοῦ, on the distinction between heavy and light bodies, with the following sentence with which he concludes his answer to the 33-rd of the Mechanical Problems, in which one demands: Why does anything get carried its own course when the propulsion does not follow along and keep pushing? [as it is very short, we report here the whole answer as translated by Winter] Perhaps it is clear that the first has done such as to push another, and that another, but it stops when what is propelling the carried object is no longer able to push, and when the weight of the object being carried slopes more than the forward force of the pushing.[5]

Aimed at helping the reader to frame the previously reported sentence by Vailati, we recall here that the Aristotelian text: On the Heavens (Περὶ οὐρανοῦ; De Caelo or De Caelo et Mundo), is the Aristotelian effort to describe the universe. It was

[3] «[...] le questioni di Statica [...] sembrano esser state le sole delle quali gli scrittori greci di Meccanica abbiano intrapresa una trattazione generale e scientifica nel senso moderno della parola; [i)] poiché, per ciò che riguarda lo studio delle leggi del moto, essi sembrano essersi accontentati di[ii)] grossolane descrizioni e classificazioni dei fenomeni».

[4] «Le osservazioni sulla composizione dei movimenti, che si trovano nelle opere d'ARISTOTELE, e le teorie più elaborate degli astronomi su questo stesso soggetto (cfr. in proposito la classica monografia dello SCHIAPARELLI: *Sulle sfere omocentriche d'Eudosso, Calippo, ed Aristotele*) appartengono piuttosto alla Geometria che non alla Meccanica propriamente detta. Alla stessa classe appartengono pure le ricerche di ARCHIMEDE «Sulle spirali», basate anch'esse su considerazioni cinematiche.»

[5] «E forse solo per colpa dei compilatori e commentatori imperiti che queste assunsero più tardi l'aspetto e la pretensione di teorie scientifiche. E interessante a questo riguardo riavvicinare le considerazioni svolte da ARISTOTELE, nel III libro Περὶ οὐρανοῦ, sulla distinzione fra i corpi pesanti e i leggeri, colla seguente frase con cui egli chiude la sua risposta alla 33a delle Questioni meccaniche, nella quale si domanda: Perché i corpi scagliati non continuano a muoversi indefinitamente? A Greek sentence follows whose translation by Winters is given in the text».

written in 350 BC and includes an astronomical catalog together with Aristotelian ideas (based on those by Eudoxus) on how the terrestrial world was constituted.

It must be remarked that Winter (see [175]), whose interpretation and analysis we support completely, does interpret the answer given in the pseudo-Aristotelian work in the completely opposite way: Winter recognizes in the answer to the 33-rd mechanical problem a description of the principle of inertia.

Let us now quote the relevant part of Vailati's work on which we agree completely. In fact, Vailati describes in a very involute way the differences of the two approaches to Statics found in Hellenistic textbooks. In his work we read:

> «In the first of previously mentioned directions is characterized by the tendency to proceed to the determination of the equilibrium conditions by directly examining, for each mechanism, the relationships that subsist among the compatible motions of its parts and tracing the analogies that, from this point of view, one can find in the various devices to which the human intelligence recurs to win with small efforts the great resistances. It is represented, first of all, by the short work on Mechanical Problems (Μηχανικά Προβλήματα) and secondly by another work, which is not less important for the history of mechanics, which has been transmitted to us only by means of a Latin compilation, having as title, *De ponderibus*, due to Jordanus Nemorarius, a mathematician working during the XIII century.»[6]

The reader will remark how quickly and roughly the formulation of the Principle of Virtual Work is formulated by Vailati. One has to master it fully to recognize that Vailati is talking really about it. One may suspect that Vailati is attributing to his Hellenistic sources the confusion which most likely is, instead, in his own formulation. Correctly, Vailati attributes to Hellenistic science the *De ponderibus*. In fact, he continues by stating that:

> «The Greek origin of the *De ponderibus*, albeit cannot be considered as completely ascertained, is, nevertheless, admitted by authoritative scholars as really likely [...].)»[7]

The description of the second line along which Statics was developed by Hellenistic Science is then described by Vailati:

> «The peculiarities of the second direction are, instead, from one side, the intention to place as exclusive foundation of Statics the consideration of the centers of gravity, and, on the other side, the preoccupation to build this science following the model of the Euclidean Geometry, by presenting it under the form of a series of theorems one linked to the others that can be obtained by deduction of a certain number of fundamental propositions having the same character of immediate evidence as the axioms of Geometry».[8]

---

[6] «Il primo dei suddetti due indirizzi è caratterizzato dalla tendenza a procedere alla determinazione delle condizioni di equilibrio esaminando direttamente, per ciascun meccanismo, le relazioni che sussistono tra i moti compatibili delle sue parti e rintracciando le analogie che presentano, da questo punto di vista, i vari ordigni a cui l'industria umana ricorre per vincere con piccoli sforzi grandi resistenze. Esso è rappresentato anzitutto dall'operetta sulle Questioni meccaniche (Μηχανικά Προβλήματα). attribuita ad Aristotele [...] e in secondo luogo da un altro scritto, non meno importante per la storia della Meccanica, che ci è giunto solo attraverso a una compilazione latina, portante il titolo *De ponderibus*, dovuta a Giordano Nemorario, matematico del XIII secolo.»

[7] «L'origine greca del De ponderibus, sebbene non possa considerarsi come completamente accertata, è nondimeno ammessa da critici autorevoli come assai probabile [...].»

[8] «Caratteri del secondo indirizzo sono invece da una parte il proposito di porre ad esclusivo fondamento della Statica la considerazione dei centri di gravità, e dall'altra la preoccupazione

Then Vailati attempts attributing to Archimedes the second direction which he is referring to:

> «This second direction that can be attributed to Archimedes, who probably was his first initiator, is represented not only by his works *On the Equilibrium of Planes* and *On Floating Bodies* but also by the fragments concerning the Statics that are found gathered in the eight book of the Συναγωγή by Pappus».[9]

Finally and unfortunately, Vailati concludes with a statement that seems completely devoid of any philological and logical support:

> «The finding, due to the Orientalist Carra de Vaux, of the previously cited oeuvre by Heron, whose Arabic translation was laying forgotten among the manuscripts of the library of Leiden (to which it had arrived by means of the erudite traveler and Dutch mathematician Golius [1596-1667]) [...], albeit supplying to us a document, being as more precious as it is unique, of a treatment of Statics in which the two methods of which I have spoken cooperate one with the other and are applied simultaneously sometimes also to the solution of one and the same question, does not, however, gives any new information for what that concerns the historical relationships of the two aforementioned directions, which seem to have been developed with perfect independence one from the other, albeit it is difficult to believe that they have not had ever any reciprocal influence.»[10]

It is therefore clear that, even though being blurred by the standard prejudices about Hellenistic science, also Vailati finds in the available sources the traces of the very ancient interaction and counterposition between the Postulation based on the Principle of Virtual Work and the Postulation based on The Balance of Forces and Moments of Forces.

---

di costruire questa scienza sul modello della geometria di Euclide esponendola sotto la forma d'una serie di teoremi concatenati ed ottenibili per deduzione da un certo numero di proposizioni fondamentali aventi lo stesso carattere di evidenza immediata che presentano gli assiomi della geometria.»

[9] «Questo secondo indirizzo che fa capo ad Archimede, il quale secondo ogni probabilità ne fu il primo iniziatore, è rappresentato oltreché dalle sue opere *Sull'equilibrio delle figure piane* e *Sui Galleggianti*, anche dai frammenti riguardanti la Statica che si trovano raccolti nel libro ottavo delle Συναγωγή di Pappo.»

[10] «Il ritrovamento, dovuto all'orientalista Carra de Vaux, dell'opera di Erone dianzi citata, la cui traduzione araba giaceva dimenticata tra i manoscritti della biblioteca di Leida (alla quale era pervenuta per mezzo dell'erudito viaggiatore e matematico olandese Golius [1596-1667]) [..] pure fornendoci un documento, tanto piu prezioso in quanto è unico, d'una trattazione della Statica nella quale i due metodi di cui ho parlato cooperano l'uno accanto all'altro e sono promiscuamemte applicati talvolta anche alla soluzione d'una stessa questione, non ha tuttavia recato alcun nuovo dato per ciò che riguarda le relazioni storiche dei due suddetti indirizzi, i quali sembrano essersi svolti con perfetta indipendenza l'uno dall'altro, sebbene sia difficile credere che essi non abbiano mai avuta alcuna influenza reciproca.»

## 5.3 The Principle of Virtual Work as formulated by Archytas of Tarentum in the Mechanical Problems.

We believe, however, that the presumed date of publication attributed to *The Mechanical Problems* in the Corpus of Aristotle by Winter (who argues that the author is Archytas of Tarentum, 428–347 BC) and the fact that this work solves all the studied mechanical problems by using the Principle of Virtual Work indicate that this Principle was formulated before the concepts of force and moment of forces were introduced in any form.

*The Mechanical Problems* (in Greek Μηχανικά) is nearly unanimously considered the most ancient oeuvre in mechanical sciences to be produced by Western Culture of which we have any information. Although it was attributed to Aristotle (and transmitted to us in the Corpus Aristotelicum) – albeit it must be acknowledged that there are some disputes about its attribution to Archytas of Tarentum – it is nowadays nearly universally accepted that its author wrote it at the end of the fourth century before Christ, and that already in antiquity it had been falsely attributed to Aristotle. On this false attribution many speculations can be made, and we leave them to the judgment of the reader, as absolutely no evidence is available.

Its introduction begins with a short and traditional appeal to the "marvel" that must guide any kind of research (this is a *topos* in scientific Hellenistic works) and it is argued that, to understand and describe phenomena, it is necessary linking them to their causes. After that, one finds the range of applicability and investigations of the "mechane", and the principles on which this theory is based. Following again another well-established topos, used since the pre-Hellenistic times and also nowadays in physics textbooks, the presentation of arguments is carried out via the traditional scheme constituted by questions and detailed answers, based always on the same fundamental principle. Here is the translation – by Winter [175] – of the first part of the introduction of *The Mechanical Problems:*

> «One marvels at things that happen according to nature, to the extent the cause is unknown, and at things happening contrary to nature, done through art for the advantage of humanity. Nature, so far as our benefit is concerned, often works just the opposite to it. For nature always has the same bent, simple, while use gets complex. So whenever it is necessary to do something counter to nature, it presents perplexity on account of the difficulty, and art [techne] is required. We call that part of art solving such perplexity a mechane.».

Unfortunately, the translation from an ancient language never manages to be really faithful. However, we believe that Winter's translation manages to transmit the original spirit of the text. *The Mechanical Problems*, as translated into English by Winter (written in Italic), with our comments are the following:

- 1. *So first the circumstances about the yoke are confusing, through what cause are the larger yokes more accurate than the smaller?*
- 2. *Why does a balance beam return when you remove the weight if the string is set from the top, and not return, but stay put when supported from below?*
- 3. *Why is it that small forces can move big weights with a lever?*

These first three questions immediately show the style of Greek investigations. Contrarily to what is read very often in some books of History of science, Greek spirit did not delve only into theoretical questions that are not connected with experimental evidence or applications. The principle of lever is explained and formulated having in mind probably its most ancient application: the balance.

- 4. *Why do the men at the middle of the boat move the boat most? Is it because the oar is a lever?*
- 5. *Why does a steering oar, small as it is, and at the end of the boat have such force that with one little handle and the force of one man, and that gentle, it moves the great bulk of ships?*
- 6. *Why when the yardarm is higher does the boat sail faster, with the same sail and the same wind?*
- 7. *Why when out of the wind they wish to run across, the wind not being at their back, do they tighten [send, furl] the sail toward the steers-man and, having made it a foot wide, let it out toward the prow?*

After having explained the principle of the lever, other applications are proposed to the science of sailing, that was so important for all Greek people. Again, one can ask where the myth of Greeks not being interested in applications comes from.

- 8. *Why are round things easier to move than things of other shapes?*
- 9. *Why, with larger circles, whether wheels, pulleys, or rollers, do we move more easily and quickly the things which are lifted or pulled?*
- 10. *Why is an empty balance beam easier to move than a weighted one?*
- 11. *Why do burdens go easier on rollers than on wagons, despite wagons having large diameter wheels and rollers small?*

Here is a peculiarity of the mechanical problems studied in this textbook: the author wants to underline the power of the use of the mathematical concept of circle in solving mechanical problems.

- 12. *Why are spears or pellets carried farther from the sling than from the hand?*
- 13. *Why, around the same capstan, are longer spikes moved more easily? And likewise, thinner winches, by the same force?*
- 14. *Why is wood the same length broken over the knee more easily if you break it while holding it having set it equidistant from the ends rather than being close alongside the knee? And if you set it on the ground and step into it, you break it farther from the hand rather than near?*
- 15. *Why are pebbles at the seashore rounded?*
- 16. *Why is it that the longer a board is, the weaker it gets? and, lifted, bends more, even if the short one – say, two cubits – is thin, and the long one – say, 100 cubits – is thick?*
- 17. *Why are big heavy bodies split by little wedges?*
- 18. *Why, if someone makes two pulleys working together on two blocks, and puts a rope around them in a circle, one block hanging, the other getting lifted up/let down, and hauls on the end of the rope, does he draw up great weights, even if the lifting force is small?*

- 19. *Why, if you put a large ax on wood, and a large burden on that, doesn't it pull apart the wood, no matter how considerable the burden is? But if you raise the ax and hit with it, you split the wood even if you have less weight than you put on the ax in the first place?*
- 20. *Why is it that phalanxes balance big heavy meats hanging from a stub?*
- 21. *Why do doctors pull out teeth more easily even adding weight – that of the tooth-puller – than with the bare hand?*
- 22. *How do they crack nuts easily, without even hitting, in the tools which they make for cracking them?*
- 23. *Why, when both terminal points of a parallelogram carry two vectors, don't they go an equal straight line, but instead one goes many times the other?*
- 24. *It is confusing why the larger circle describes a line equal to a smaller circle's when they have been put on the same center.*
- 25. *Why do they make beds the way they do, sides two-to-one – one side six feet and little more, the other three? And why don't they web them on the diagonal?*
- 26. *Why, given that the weight is the same in each case, is it more difficult to carry long boards at the end on one's shoulder than by the middle?*
- 27. *Why, given two burdens of equal weight, is the one too long harder to carry on the shoulder – even if one carries it at the middle – than if it were shorter?*
- 28. *Why at water wells do they make shadoofs*[11] *as they do?*
- 29. *Why, when two men carry an equal weight on a board or some such, do they not labor the same unless the weight be at the middle, but it is more work for the one of the carriers who is closer?*
- 30. *Why, standing up, do we all first make an acute angle with calf and thigh, and with thigh and torso, and if we don't we cannot arise?*
- 31. *Why is it easier to move something moving than something at rest? As for instance, they pull wagons faster moving than starting. Because it is most difficult to move a weight which is moving the opposite way?*
- 32. *Why do objects thrown stop?*
- 33. *Why does anything get carried its own course when the propulsion does not follow along and keep pushing?*

Questions 32 and 33 and their answers clearly show that the concept of inertia was not ignored by Greek scientists. A harsh debate on this point has been started since Middle Ages, which is still not settled.

- 34. *Why, when thrown, do neither smaller nor larger objects go further, but always must have some symmetry to the one throwing?*
- 35. *Why in eddying water does everything end up getting carried into the middle?*

The reader will note that not all the questions are dealing with the use of machines employed in engineering applications and are aimed to get a "mechanical gain".

[11] Note of the Authors: The shadoof, or sweep, is an early crane-like tool with a lever mechanism, used in irrigation since around 3000 BCE by the Mesopotamians, 2000 BCE by the ancient Egyptians, and later by the Minoans, Chinese (c. 1600 BCE), and others. The sweep is used to lift water from a water source onto land or into another waterway or basin. The mechanism comprises a long counterbalanced pole on a pivot, with a bucket attached to the end of it.

Many questions are aimed at simply understanding how and why some natural phenomena occur, how some object of common use can be employed and why they are useful as they are. Other questions are aimed at explaining real world situations, the functioning of certain instruments or the evolution of the configurations assumed by certain bodies. There are several questions involving sailing techniques, a very important subject for the trading tradition of Greeks.

In conclusion, we can dare to conjecture that *The Mechanical Problems* seems to be a text including exercises to be solved, like some modern Solved Exercises textbooks that are accompanying textbooks where the theory is fully exposed. If this conjecture is well-grounded, it is a pity that the original theoretical textbook was not transmitted to us. The Authors believe also that it would be interesting to elaborate the answers to all the questions with modern methods.

## 5.4 D'Alembert: the rediscovery of the Principle of Virtual Velocities (or Virtual Work)

While Lagrange, in his *Mécanique analytique* (first edited in 1788) [115] acknowledges Johann Bernoulli as the one who formulated the Principle of Virtual Work already in 1717 in a letter to Varignon, we believe that, most likely, one can track the first modern formulation of the Postulation of Mechanics in which such a principle is the basic postulate in the *Traité de Dynamique* (1768) by Jean-Baptiste Le Rond d'Alembert. As the reader may have already agreed, it is not possible to establish how novel D'Alembert treatise is when compared with Hellenistic sources, which have been lost, and whose existence we can simply conjecture by reading secondary sources. It is in fact rather unlikely, also on the basis of Vailati's considerations, that *The Mechanical Problems* did not produce a large subsequent literature. We can rather safely state that the most relevant Hellenistic sources have been lost (see also [56]): the few fragments and references can be found in very corrupted sources whose interpretation has been very often debated.

Nobody can reasonably doubt, however, that the masterpiece by D'Alembert managed to produce a renewed impetus to mechanics and mechanical sciences. The fame of the *Traité de Dynamique* has been so widespread that the Principle of Virtual Velocity (later and more often called also the Principle of Virtual Work) therein enunciated, is often named after his author: the D'Alembert Principle. It has to be explicitly remarked here that also mechanical systems where a part of energy is dissipated can be described by means of models based on the Principle of Virtual Work: this was probably the reason for which already D'Alembert and Lagrange did prefer to use it, instead of the Principle of Least Action, as their most fundamental postulate.

We believe that it is suitable to quote here an excerpt from the *Traité de Dynamique* (see also [42]). This is an excerpt that reflects the spirit of D'Alembert in considering science. We recognize some stylistic affinities with the words found in Greek scientific texts, notwithstanding how deformed and corrupted they may have reached us. We believe indeed that D'Alembert is providing a "Manifesto" for mechanicians and that, at the same time, he guides mechanicians towards the correct methodologies, techniques and perspective to be used in (Continuum) Mechanics:

> «The certainty of mathematics is an advantage which these sciences owe to the simplicity of their object. [...] the most abstract notions, those which the layman regards as the most inaccessible, are often those which carry with them the greatest light»

Here D'Alembert warns about the need of using abstraction and abstract objects in formulating mathematical models. He then continues stating that:

> «[...] in order to treat following the best possible method [...] any Science whatsoever it is necessary [...] to imagine, in the most abstract and simple way possible, the particular object of this Science, [it is necessary] to suppose and admit in this subject anything else, than the properties which this same Science treats and supposes. »

Subsequently, D'Alembert introduces a "Principle", and this unique Principle is the founding principle of Mechanics, as a deductive theory. This Principle is precisely presented in his treatise together with its important consequences and applications. In his *Mécanique analytique*, Lagrange continues and develops the oeuvre by D'Alembert and continues to call such a principle the Principle of Virtual Velocities, which somehow evokes its ancient Greek origin. Its name was later changed into *Principle of Virtual Work*. Probably, this was not a good choice, it being possibly suggested by some echoes of the debate involving the supporters of the Postulation based on the Laws of Balance.

D'Alembert seems to have suffered because of the criticism formulated by the inductivists. The detractors of falsificationist–deductive physical theories base their arguments always on the same refrain: the deductive theory is too much abstract, "far from experience" and "devoid of physical content". The detractors of "falsificationism"[12] claim that a sound physical theory is based on "solid" experimental grounds and that these solid experimental grounds are encoded within the theory starting from the formulation of its basic principles. Instead, D'Alembert, a follower of the Archimedean vision of science, considers mathematics as a fundamental tool for formulating logically well-posed theories to be, subsequently, verified experimentally.

D'Alembert then continues his *Traité de Dynamique* by stating that:

> «From this standing two advantages result: the principles receive all clarity to which they are susceptible: [and these principles] are finally reduced to the smallest number possible [...] as the object of a Science is necessarily determined, the principles will be more fecund if they will be less numerous [...]. »

[12] Falsificationism is the scientific philosophy claiming that a theory is scientific only if it is based on a set of hypotheses implying a cascade of falsifiable deductions. If a statement cannot be refuted (i.e. it is not falsifiable) then, following the point of view formulated by Karl Raimund Popper – probably the 20th century's most influential epistemologist –, it is not a scientific claim. Popper indeed rejects the classical inductivist description of the scientific method and replaces it with the concept of "empirical falsification". He claims that, while a scientific theory cannot be certainly proven, it should be in principle falsifiable, that is proven to be false by some experiments.

In writing these words D'Alembert refers to an epistemological (meta-)principle dating back to Hellenistic science that is often referred to as the Occam's razor[13]. It has to be remarked here that the supporters of postulations of mechanics based on balance laws are ready to multiply the number of principles on which mechanics should be based.

In the words that we will read next, D'Alembert particularizes his reasoning and refers more specifically to Mechanics, claiming its special need, among all exact sciences, for a clear and solid foundation:

> «Mechanics, above all, seems to be (the Science) which has been more neglected from this point of view: also the great majority of its principles either obscure by them-selves, or enunciated and demonstrated in an obscure way, have given place to several spiny problems [...] I proposed to my-self to move back the limits of Mechanics and to make its approach easier, [I proposed to my-self] not only to deduce the principles of Mechanics from the most clear notions, but also to apply them to new uses, to make it clear at the same time both the inutility of the many and various principles which have been used up to now in Mechanics and the advantage which can be drawn by the combination of others (principles) in order to have the progress of this Science in one word (I want to make clear which is the advantage) of extending the principles by reducing them.»

Choosing some more technical excerpts from the treatise by D'Alembert may need a difficult work of interpretation, as the nomenclature used by D'Alembert is changed nowadays. He applies the Principle of Virtual Velocities to a wide range of cases and shows how powerful it is in inventing new models and providing new explanations for observed phenomena. It will be easier to comment the text written by Lagrange. He indeed spent all his active life to write and rewrite his *Méchanique Analytique*, and his style of presentation is so elegant and precise that one can easily read it even nowadays. Lagrange repeatedly credits D'Alembert for having clarified the importance of the Principle of Virtual Velocities.

We conclude by recalling that, at beginning of D'Alembert's *Traité de Dynamique*, one can find the following statement that is also a scientific program:

> «I have proscribed completely the forces relative to the bodies in motion, entities obscure and metaphysical, which are capable only to throw darkness on a Science which is clear by itself. »

However, D'Alembert is not an extremist. As proven by the following statement that we find in his work:

---

[13] Occam's razor, Ockham's razor, Ocham's razor (Latin: novacula Occami) or law of parsimony (Latin: lex parsimoniae) is a principle on which the meta-theory guiding the formulation of theories is based: "logical entities and fundamental principles should not be multiplied without necessity". This meta-theoretic principle is linked to the name of the English Franciscan William of Ockham (c. 1287–1347). However, this principle can be found in many authors, and probably dates back – at least – to Hellenistic culture. Equivalent statements are often found like, for instance, "the simplest explanation is most likely the right one", which is often used as common sense statement in discussions. This meta-theoretic razor advocates that when choosing among different competing set of hypotheses giving the same consequences, or predictions, the set including the fewest assumptions has to be chosen. Of course, this does not mean that one has to choose a smaller set of hypotheses when these are producing wrong predictions! Unfortunately, this last misunderstanding is still diffused in modern times.

«I must warn [the reader] that in order to avoid circumlocutions, I have used often the obscure term force, and some other terms which are used commonly when treating the motion of bodies; but I never wanted to attach to this term any other idea different from those which are resulting from the Principles which I have established both in the Preface and in the first part of this treatise.»

We approve this use of the term "force": it is a purely mathematical object introduced as a linguistic shortcut for: "linear continuous functional defined on the set of admissible virtual velocities". This last is the precise definition of "force" in terms of kinematical quantities, previously defined or introduced as primitive concepts.

## 5.5 The formulation of Continuum Mechanics by Lagrange

The treatise *Méchanique Analytique* by Lagrange has been his lifelong endeavor. Lagrange continued to write it until the last moments of his life. One reads at the beginning:

«We will use, in general, the word 'force' or 'power' [puissance] for denoting the cause, whatever it will be, which is impressing or tends to impress motion to the bodies to which it is assumed to be applied.»

The nomenclature chosen by Lagrange has to be deciphered: he uses the word "force" as a synonym of the word "power". This choice may cause (and in fact did cause) a lot of misunderstandings. Some scholars, who wanted to discuss the work of Lagrange in their historical accounts of the development of mechanics, while reading Lagrange textbook without having read its first pages, were confused by Lagrange nomenclature and, therefore, concluded that the ideas of Lagrange were not clear about some fundamental concepts. Some of them arrived to conclude that Lagrange was not capable to distinguish between force and power: this can happen when somebody reads a book of mathematics jumping the pages where definitions and notations are introduced.

Instead, Lagrange tried (unfortunately without success!) to introduce a nomenclature paralleling the nomenclature used by Galileo. In fact, Lagrange (following Galileo) chooses the word 'moment' for meaning what, in the modern nomenclature, is called 'power'. In subsequent initial pages of the *Méchanique Analytique* one reads, in fact,

«Galileo uses the word 'moment' of a weight or a power applied to a machine the effort, the action, the energy, the 'impetus' of this power for moving this machine [...] and he proves that the moment is always proportional to the power times the virtual velocity, depending on the way in which the power acts.»

This sentence clarifies the use of words chosen by Lagrange. Lagrange adds some comments to his choice of nomenclature:

«Nowadays one uses more commonly the word 'moment' for the product of a power times the distance along its direction to a point or a line, that is the lever arm by which it acts [...], but it seems to me that the notion of moment given by Galileo and Wallis is much more natural and general, and I do not see why it was abandoned for replacing it by another which expresses only the value of the moment in certain cases».

The readers who have studied generalized continuum theories, and in particular second gradient continua, will recognize how clearly Lagrange understands the concepts that are at the basis of mechanics: with his choice he wants to open the door to future generalizations of his models. Then Lagrange formulates Archytas-D'Alembert general Principle for Mechanics:

«The Principle of virtual velocities can be formulated in a very general way, as follows: If a system whatsoever constituted by bodies or points each of which is pulled by powers whatsoever is in equilibrium and if one impresses to this system a small motion whatsoever, in virtue of which every point will cover an infinitesimally small distance which will express its virtual velocity, then it will be equal to zero the sum of the powers each multiplied times the distance covered by the points where it is applied along the line of application of this same power, when considering as positive the small distances covered in the same direction as the power and as negative the distances covered in the opposite direction.»

Albeit a modern formulation of this principle usually includes the use of concepts from functional analysis, tensor algebra and mathematical analysis, one must agree that: i) Lagrange's formulation seems so general that it actually includes all versions that have been formulated up to now, ii) it uses the minimum possible mathematical concepts, i.e. only concepts from Euclidean geometry, that are sufficient to express in a rigorous way the principle in its full generality.

## 5.6 The controversy between Poisson and Piola about the deduction of the equation of the equilibrium of fluids: Piola's contact interactions in continua

Lagrange himself and then Gabrio Piola applied the Principle of Virtual Velocities to deduce the equations of the motion for compressible first gradient fluids, without viscosity. Unfortunately, as already remarked by Vailati, Archimedes had formulated the concepts relative to the equilibrium of fluids in terms of a postulation approach based on the necessary condition for equilibrium that can be deduced from the principle of minimum energy: i.e. the balance of force.

We do not have the sources describing how the Principle of Virtual Work had been studied in Hellenistic science and how (and if) the necessary condition concerning resultant forces and moments of forces had been deduced. However, we know that Hellenistic scientists were masters in the logical deduction of consequences from "prime principles". Therefore, we can imagine that, as variational principles are usually more easily applicable when the formulation is framed in a Lagrangian description, while fluids are most suitably described in an Eulerian description, Archimedes preferred to develop his fluid mechanics by using some important consequences of the Principle of Virtual Velocities. Of course this is a purely conjectural statement and one cannot rely on it too much.

In any case, Poisson, following the transmitted Archimedean tradition, preferred to base the mechanics of fluids on the postulation making use of the balance of forces. For finding the equations of the motion of a fluid, Poisson uses: i) the principle of equal pressures in all directions and ii) a principle characterizing constitutively fluids, which is formulated in the following vague manner: fluids have the capacity «de se reconstituer toujours semblablement á eux-même autour de chaque point» (i.e. of reconstituting themselves always similarly to themselves in the neighbourhood of every point).

Gabrio Piola, in his work [136], criticizes Poisson's point of view and underlines how it could be difficult to transform unambiguously the second of the listed hypotheses enunciated by Poisson into any kind of formula or, equivalently, that many formulas should be used as its mathematical counterpart.

Moreover, within the volume [47] it is possible to find the translation of Piola's "Riflessioni sulle unità di misura e altre quantità concrete" (Reflections on the Units and on the Measures of the Various Physical Quantities[14]). In this manuscript, while talking about incompressible fluids, Piola states that the deduction of governing equations for incompressible fluids is greatly simplified by making use of the method of Lagrange multipliers:

> «[...] using Lagrange's method, [..] for it is enough to know the constraint equations as a results of internal constraints originating from passive forces; and it is not necessary to imagine how these forces work. Hence the savings of much effort, and the consciousness of greater certainty.»

Then he continue with a criticism of the perspective chosen by "French Geometers":

> «French Geometers of our time, whatever the reason, looked for another way: they wanted to put together the general equations of body motion, doing violence to imagination, so that it could give them a representation of the way of acting of nature in the least, starting with the few data we have around his way of acting at a great extent.»

Piola underlines how French Geometers, by basing their considerations on the concept of pressure, did violence to imagination in "putting together" the general equations of body motion. Piola continues by stating that:

> «So they replaced the study of clear and certain effects, with that of obscure and uncertain causes.»

The reader will recognize that this statement echoes those already quoted by D'Alembert. Then Piola concludes with: i) a clear criticism based on the use of Occam's razor:

> «The effort that accompanies these procedures and, what is more, the little confidence that inspires results obtained in the midst of many weakly reasoned hypotheses, and of a continuous neglecting of supposedly small amounts, in relation to others and others, are arguments that must persuade us to prefer and carefully keep in mind the Italian method.»

---

[14] This was an unpublished manuscript by Gabrio Piola and we are grateful to the courtesy of Politecnico di Milano, Archivi Storici - Area Servizi Bibliotecari di Ateneo, where the original manuscript is conserved into the "Fondo Gabrio Piola", which allowed for its access.

and ii) a well founded indication for future researches, which echoes the one that can be found in the *Method of Mechanical Theorems* by Archimedes:

> «Let it be applied to all the researches attempted by the French scholars: by largely prevailing in effectiveness over all the analyzes they made, let it make open for us a path towards further discoveries: this is a proposition which I do support for a long time with my voice and with my pen.»

While Piola seems to share with Poisson the opinion that considers the "true physical reality" linked to the existence of atoms (or elementary molecules), their opinion becomes divergent for what concerns the concept of force; in fact, Poisson believed that the law of balance of force could give evolution equations for any mechanical system. Piola relates his deduction of evolution equations for a generalized continuum to a homogenization procedure, based on a micro-model where the considered mechanical system is regarded as constituted by interacting material particles. In a sense, Piola accepts to consider the continuum model as a limit of a discrete system. Piola considers this identification process as based on three steps:

1. the expressions of the Principle of Virtual Work for the micro-model and macro-model, once the kinematics of these models are specified;
2. the identification of a specific micro-motion once given a macro-motion, the micro-motion being considered a meaningful representative of all micro-motions which may be correspondent to the given macro-motion;
3. the identification of macro Virtual Work in terms of the micro Virtual Work functionals, and the correspondent identification of macro constitutive equations in terms of the micro geometry and micro material properties.

We prefer leaving Piola ([40], p. 2) to explain the reasons why such a procedure should be accepted:

> «In my opinion it is not safe enough to found the primordial formulas [of a theory] upon hypotheses which, even being very well-thought, do not receive support if not for a far correspondence with some observed phenomena, correspondence obtained by particularizing general statements, [...] indeed the magisterium of nature [i.e. the experimental evidence] at the very small scale, in which we try to conceive the effect of molecular actions, will perhaps actually be very different from what we can mentally realize by means of the images impressed in our senses when experiencing their effects on a larger scale.»

Here Piola envisions the possibility that micro-physics could be ruled by laws that are very different from those that are valid in macro-physics. In this prudent approach he is really safely and wisely prudent. In fact, Quantum Mechanics is based on laws that are very different from those valid at the macro-scale. The reader will however remark that also Quantum Mechanics has been firmly based on Variational Principles. After having warned about the dangers of extending macro-theories to the micro-level, Piola continues as follows:

> «Even let us assume that this difference be very small: a deviation quite insensitive in the fundamental constituents [of matter] – which one needs to consider as multiplied by millions and by billions before one can reach sensible dimensions – can be the ultimate source of notable errors.»

In these words Piola resumes all the difficulties that would have been met by Statistical Mechanics later on. Subsequently, Piola advocates the efficacy of Lagrangian methods:

> «On the contrary, by using Lagrangian methods, one does not consider in the calculations the actions of internal forces but [only] their effects, which are well-known and are not at all influenced by the incertitude about the effects of prime causes, [so that] no doubt can arise regarding the exactitude of the results.»

The idea of Piola seems clear to us: albeit one is neglecting (by ignorance) the details of micro-motions by identifying the macroscopic expression of Virtual Work as an "average" expression of microscopic Virtual Work, the results «are not influenced by the incertitude about the effects of prime causes». Piola then concludes with some epistemological considerations:

> i) «It is true that our imagination may be less satisfied, as [with Lagrangian methods] we do not allow to it to trace the very fundamental origins of the internal motions in bodies: does it really matter? A very large compensation for this deprivation can be found in the certitude of deductions.»
>
> ii) «It has to be remarked that I do not intend for this reason to proscribe the dictation of modern Physics about the internal constitution of bodies and the molecular interactions;»
>
> iii) «When the equations of equilibrium and motion will be established firmly upon indisputable principles, because one has calculated certain effects [i.e. those contained in the Principle of Virtual Velocities] rather than hypothetical expression of forces, I believe to be licit to try to reconstruct anew these equations by means of [suitable] assumptions about such molecular interactions: and if we manage in this way to get results which are identical to those we already know to be true, I believe that these hypotheses will acquire such a high degree of likeliness which one could never hope to get with other methods.»
>
> iv) «Then the molecular Physics will be encouraged to continue with its deductions, under the condition that, being aware of the aberrations of some bald ancient thinkers, it will always mind to look carefully in the experimental observation those hints [coming by the application of Lagrangian macroscopic methods] which are explicit warnings left there to indicate every eventual deviation.»

When dealing with the particular case of the deduction of the equations of fluids, Piola comes back to the essence of his controversy with Poisson:

> «It is now convenient that we hold to think about the difference between our conclusions and those of Poisson.»

Piola clearly states that his analysis gives the same results as those obtained by Euler, and argues about the different results obtained by Poisson:

> «Our analysis, confirming the Eulerian theory, would embrace both the fluid in equilibrium and those in motion, so the liquid as the aeriform fluids. On the contrary, Poisson thought to add new terms to the general equations of fluid motion: and, here is, if I have well understood, the thread of his argument. [Poisson] Begins to say that the equations we already had to express the movement of fluids, were derived using the principle of D'Alembert from those ones of the equilibrium, which presuppose the principle of equal pressure in all directions, a principle experimentally recognized [to be] true only for fluids at rest. [Poisson] continues and asserts that the property to press equally in all directions comes from another property that the fluids fulfill, [that is] always to rebuild themselves similarly to themselves around each of their points. Then [Poisson] rightly reflects that this reconstruction requires a bit

of time to be done: and even if the interval had very short duration, when the fluid is in motion, that reconstruction cannot be at each instant perfect. In the absence of a perfect reconstruction, according to him, the pressure equal in all directions is missing: therefore those equations which originate from such a principle shall be in default.»

One should be surprised of not finding often in the literature any comment about the treatment of the dynamics of perfect fluids proposed by Poisson and the "uncertainties" in his logical argument and final results, that are the consequence of Poisson's preferred postulation scheme. Piola implies that Poisson, while refusing to use the Principle of Virtual Work, is obliged to look for "*ad hoc corrections*" in his deduction process, adjusting and changing nearly at every step the logical flow of his reasoning. The final criticism that Piola formulates about Poisson's deductions deserves to be reported here:

«But is it true that the principle of equal pressure in all directions is intimately linked with the regular distribution of the molecules, so that it can not exist one without the other? (Poisson. Traité de Mécanique. Tome II p. 506). I doubt it very much, and I think that, here as well, one has gone forward a bit too far into the deductions: and this because the ideas around that quantity which we call the internal pressure of the fluid have not yet completely clarified.».

In fact, the whole oeuvre by Piola was dedicated to the clarification of the concept of internal state of tension of a continuum body.

## 5.7 Navier, Cauchy, Poisson, and Saint-Venant *versus* Lagrange, Piola, and George Green *or* postulations based on Balance Laws *versus* postulations based on the Principle of Virtual Velocity.

The discussion developed in this section starts from the analysis developed in [16], which is a really valuable secondary source organizing in an original way the primary sources of Theoretical Mechanics and its applications to Structural Mechanics. Benvenuto transfers in the field of history of science the doctrine developed by William of Moerbeke for the "correct" translation of scientific texts. Benvenuto refers, nearly with the words of the original authors, the principal parts of their results and scientific points of view. The choice of the arguments to be discussed is one of Benvenuto's main contribution to science: and we agree nearly completely with him.

Benvenuto considers the scientific personality of George Green (1793-1841) and his role in the development of continuum mechanics. He clearly states that Green was a follower of Lagrangian mechanics. Therefore, it is not surprising that George Green's contributions to mechanical science support and confirm those already obtained by Piola, which have been up to now. Green's contribution to the problem of determining the most general expression for constitutive equations in linear isotropic elasticity can be found in his works published in between 1834 and 1839 (the interested reader is referred to [104]).

This very particular problem is confronted by Green with a clear and firm epistemological point of view, that can be overlapped to that expressed by Piola (see the excerpts quoted previously). Rephrasing Green we can say that:

> «instead of trying to understand the ultimate root of reality, it is much more prudent and convenient to deduce all possible consequences from some fundamental and general principles whose validity seems well-grounded».

The principle that George Green considers generally valid for all conservative systems is easily formulated. The Principle of Virtual Work involves expressions for expended work that are the first variations of some energy functionals. By quoting the true words by George Green (see [103]):

> «The principle selected as the basis of the reasoning contained in the following paper is this : In whatever way the elements of any material system may act upon each other, if all the internal forces exerted be multiplied by the elements of their respective directions, the total sum for any assigned portion of the mass will always be the exact differential of some function. But, this function being known, we can immediately apply the general method given in the *Mécanique Analytique*, and which appears to be more especially applicable to problems that relate to the motions of systems composed of an immense number of particles mutually acting upon each other. One of the advantages of this method, of great importance, is, that we are necessarily led by the mere process of the calculation, and with little care on our part, to all the equations and conditions which are requisite and sufficient for the complete solution of any problem to which it may be applied.»

The reader will appreciate the Tacitean style (see, e.g., Tacitus' style (as an instrument of thought) in [92]) used by Green. It is possible that one of the reasons for which Green's works had a greater success than Piola's ones, in subsequent literature, is related not only to the fact that Piola used Italian in writing his works while Green English, but also to the difficulties found in reading the complex writing style used by Piola. By rephrasing Benvenuto:

> «The construction by Green is admirable: from a unique principle he deduces all the properties of the constitutive equations for a linear elastic isotropic solid.
>
> Green does not need to introduce further hypotheses on the material particles constituting the considered body, and can avoid all of them. Instead of adding, one after the other, a series of conjectures about the physical reality, the follower of Lagrange manages to account for some general properties that are observed without compromising about unnecessary conjectures about minute details of phenomena.»

As observed again by Benvenuto (who seems not to be aware of the great contribution by Piola to Continuum Mechanics, albeit he cites one time his name in his fundamental historical book) the French champions of the postulation based on balance of forces and molecular mechanics (who were headed by Navier) had to fight a hopeless battle and "had to clash against a physics of solids mainly imaginary". On the other hand, the champions of Lagrangian mechanics (namely Piola, George Green and, later on, Hamilton) could avoid useless and empty discussions about the properties of atoms and use the firm mathematical properties that are common to all conservative phenomena.

It is nowadays clear that French Geometers in contrast with Piola started a controversy against Green and the other British followers of Lagrange. The final result of this scientific controversy could not have a different result than the establishment of the prevalence of variational methods upon the complex postulation scheme started by Navier, Cauchy, Poisson and Saint-Venant. Every effort aimed to found the basic equations of continuum mechanics on molecular microscopic models was abandoned. At that point, a bifurcation occurred: Cauchy, making use of his "tetrahedron argument", managed to formulate a postulation at macro-level which was still based on the balance of force, albeit he had to add the balance of moment of forces as additional postulate. The assumption by Cauchy (see [26]) excluded edge contact forces (see for an in-depth explanation of this fact the papers [52, 45, 51]) and, therefore, his results are limited to continua in which the deformation energy depends on the first gradient of placement only. Instead, and most likely a few years before Cauchy, Piola laid down some more general foundations for Continuum Mechanics.

## 5.8 Nationalistic Science *or* How Piola's legacy has been blurred because of writing in Italian and counter-posing Italian science to French science.

We believe that an interesting subject to be studied using the scientific method concerns the phenomenon of removal and/or erasure of scientific results in the tradition of one discipline. This kind of phenomena do occur rather often, and surely deserve a careful and deep investigation by using an advanced version of sociological theories. Here, we limit ourselves to remark that it's really surprising to notice that very topical and important contributions to continuum mechanics, as those to be credited to Gabrio Piola, could have been ignored or nearly completely neglected (even by Italian authors) for more than 150 years. A superficial analysis attempting to understand why Gabrio Piola's contributions to mechanical sciences were ignored easily leads to conjecture the following concurring reasons:

i) his works were written in Italian. This was a nationalistic choice, as Piola could surely write fluently at least in French;
ii) the Lagrangian school, which he championed, was rather bitterly countered by the French school, which was, in that historical period, the strongest in the world;
iii)the Italian attitude towards compatriots seems to be rather negative: in general, Italians are rather xenophiles.

The reason why Gabrio Piola wanted to write his works in Italian can be related to his leading cultural and scientific role in the Italian Risorgimento (Resurgence): Piola invested many of his intellectual resources in promoting Italian science, in organizing Pan-Italian scientific conferences, in supporting a unitary vision of Italian science and culture. He wanted to prove that Italian language could be the «vector» of advanced scientific theories, and that Italian scientists were capable to keep up scientists of other nationalities. This conjecture can be proven by observing that Piola's eulogy in memoriam of his "Maestro" Vincenzo Brunacci is concluded by this statement:

> «In the life long efforts that produced Brunacci's works one can recognize a strong commitment "for the advancement of SCIENCES, for the glory of the AUTHOR and for the prestige of ITALY"».

Another excerpt of the same eulogy is also meaningful in this context. In fact, Piola writes that

> «It seemed as if the Spirit of Italy, who was in great sufferance because in that time the most brilliant star of all mathematical sciences, the illustrious Lagrangia, had left the Nation, that Spirit wanted to have the rise of another star, which being born on the banks of the river Arno [he refers to Brunacci], was bound to become the successor of the first one.»

It is very important to remark that Piola refers to Lagrange using the original Italian version of his name, Lagrangia, and to Italy as a unique Nation, by evoking its "Spirit". In the eulogy for Brunacci, Piola focuses also some problems in which the Italian school of mechanics managed to give important contributions:

> «I will content myself to indicate here three Memoirs where he [i.e. Brunacci] examines the doctrine of capillary attraction of Monsieur Laplace, comparing it with that of Pessuti and where, with his usual frankness which is originated by his being persuaded of how well-founded was his case, he proves with his firm reasoning, whatever it is said by the French geometers, some propositions which are of great praise for the mentioned Italian geometer.».

The Nationalism of Piola is revealed by the bold statement «whatever is said by the French geometers», in which he pours his courageous pride of being Italian. It has also to be remarked that Brunacci, Pessutti and Piola already engaged themselves in the study of capillary phenomena and that, some centuries later, the Lagrangian French school headed by Pierre Casal [25] and Paul Germain [88, 90] recovered the Italian spirit, as represented by Lagrange's Principle of Virtual Velocities, to firmly found a continuum model for capillary fluids that cannot be framed into the postulation scheme preferred by Cauchy and Navier.

We can affirm that the finality of every work written by Piola and the principal aim of his scientific activity has been to prove that every mechanical theory can be founded by using the Principle of Virtual Work and that, when one is faced with the problem of postulating a novel model, this principle is the best guidance. Piola was surely the first scientist who, using Lagrange's postulation scheme, defined precisely, for a generic continuum, the dual in work of the gradient of virtual displacement in the referential description. This mathematical object conceived by Piola will then be framed later on in the modern theory of distributions (as defined by Schwartz). In fact, once Tensor Calculus will have been developed by Ricci and Levi-Civita, this dual in work will be called the *Piola stress tensor*[15].

Because of the neglect reserved to Piola's work and the dominance of Cauchy postulation scheme, the greatest part of Piola's most original results (in particular his studies about continua whose deformation energy depends on higher gradients of the strain) are, even nowadays, not known to the great majority of scholars.

[15] While, most likely, Kirchhoff studied this tensor later than Piola, Truesdell named it after him because, most likely, he read Müller and Timpe [130, p. 23], according to whom Kirchhoff was the first to formulate continuum mechanics based on the integral balance of forces and moment of forces (on this point see also the introduction of [80, p. 301])

## 5.9 The formulation of N-th Gradient Continuum Mechanics by Piola: an ignored result that is still topical after more than 150 years

**When inventing a theory for describing some phenomena, one must start from specifying its kinematics.**

The chosen space of configurations gives the mathematical model of the space of states of the physical system which is studied. Once the set of admissible configurations is fixed, then the concept of motion can be easily introduced: it is a function defined in a time interval which maps any time instant into the configuration assumed in that time instant. The most important epistemological question that is debated since the times of Archytas concerns the problem of finding some equations and/or algorithm for calculating, under specified external interactions, initial and boundary conditions, the predicted motion for the studied physical system. The most effective meta-theory that has been proposed up to now builds the "dynamics" (that is the part of the model which predicts the system's motion) of any model following what we could call Lagrange-Hamilton-Rayleigh (LHR-)scheme (see e.g. [44, 7, 48, 53, 41, 31, 36, 17, 23, 72, 21]).

In the LHR-scheme, once fixed the space of configurations and the set of admissible motions, the predicted motion is found by formulating the Principle of Virtual Work. In the case of Generalized Continuum Mechanics, the already formulated version of this principle by Lagrange can be made more specific, by using some concepts of functional analysis (for a discussion of this specific point see [88, 90, 89, 91]).

### The Principle of Virtual Work for Generalized Continua

Let us start by postulating the existence of three functionals, each defined on the Cartesian product of the space of admissible motions and the space of variations of admissible motions, respectively called work of internal interactions, work of external interactions and work of inertial interactions. Let us assume that they are linear and continuous with respect to the variations of admissible motions. The work of internal interactions and the work of external interactions are both decomposed into a conservative and non-conservative part. The conservative part is the functional derivative of the mechanical energy of the system with respect to the variations of motions. The non-conservative part is built in terms of the so-called Rayleigh dissipative functional depending on the variations of admissible motions and on the time derivatives of the variations of admissible motions. The non-conservative part of internal and external interactions works are obtained by calculating their functional derivative with respect to the time derivatives of the variations of the admissible motions.

The Principle of Virtual Work states that the predicted motion can be characterized as that motion for which the sum of internal, external and inertial interaction work linear functionals vanish for every admissible variation of motion. The reader is referred for instance to ([6, 53]) for a more technical presentation of the principle that, however, we believe has been presented clearly enough in the previous sentences for a reader who is familiar with the basic ideas of functional analysis. The Principle of Virtual Work reduces any continuum mechanics theory to the formulation of some clear conjectures: the choice of the space of configurations and the choice of conservative, non-conservative and inertia work functionals. There is no *ad hoc* adaptation of the hypotheses while developing the theory, there are not lacking terms that have to be added to the evolution equations *a posteriori* for avoiding logical incongruences and, paraphrasing Lagrange, the mathematical deduction process flows smoothly starting from the initial assumptions to the most detailed predictions. Of course, these predictions must be in agreement with experimental evidence: otherwise, some of the initially postulated expressions for the functionals must be modified accordingly.

The Principle of Virtual Work is very conveniently placed at the basis of continuum mechanics, as it is also the mathematical basis of the analysis of so-called weak solutions for mechanical problems. Weak statements of the boundary-value problems of continuum mechanics and mechanics of structures are unavoidable for the development of finite element techniques, Rayleigh-Ritz and Galerkin-type approximated solutions. Making use of the virtual work principle formulation, based on a clear understanding of the mechanical phenomenology, some problems were profitably studied within the framework of the surface elasticity of energetic boundaries or of the Steigmann-Ogden surface elasticity (see for instance [111, 112, 27, 28, 3, 4, 65, 64, 164, 131]). There are many other fields of modern mechanics that may exploit, or have already exploited, the modeling efficacy of the Principle of Virtual Work. The list could be very long and we limit ourselves here to list some works that have had some influence on our own research efforts

- in the generalization to bio-mechanical and bone growth phenomena of classical mathematical methods used in continuum mechanics [162]
- in the study of large deformations of beams and lattices of beams [165, 167, 70, 83, 81]
- in generalized continuum mechanics [77, 158]
- generalized shell theory [66, 63, 152].

The Principle of Virtual Work, as clearly understood by Piola, has many important consequences, one of which needs to be recalled explicitly here. As the equality to zero of the sum of the internal work and external work functionals must be assured, for instance in mechanical equilibrium configurations, for any admissible variation of placement, it is clear that, given a class of continua characterized by a specific class of internal work functionals, NOT ALL EXTERNAL INTERACTIONS can be applied. Let us make this explicit with an example: if one assumes that the internal work functional is the functional derivative of a deformation energy depending on the Eulerian mass density only (this is the case of Eulerian perfect fluids) then the external interactions which can involve the considered fluid cannot include shear surface contact forces. This fact is well-known and accepted by all mechanicians.

However, the followers of postulations based on the balance of force do not seem to consider this circumstance. Indeed, contact interactions at the external boundaries of continua are NOT determined by "experimental evidence", as sometimes has been claimed, and independently from the postulated form of the internal work functionals. The choice of admissible contact external interactions is implicit in the postulated form of the internal work functional.

Now, following what was done by Piola, we are ready to define N-th gradient continua. These continua have been completely characterized in [52, 45]. We recall here the main results that can be found in the fundamental works by Gabrio Piola and in the aforementioned works, that try to complete his scientific program. First gradient continua are characterized by the validity of the so-called Cauchy postulate, plus Cauchy's implicit assumptions about contact interactions. These implicit assumptions can be stated as simply as follows: contact interactions are expending work only on variations of placement (and therefore they are not expending work, for instance, on the surface normal derivatives of variations of placement) and they are only concentrated on contact surfaces (and therefore there are not, for instance, contact forces per unit line or concentrated on points). Noll's Theorem, which proves the Cauchy postulate, is based on the same assumptions and therefore does not increase really the generality of Cauchy's treatment.

As already proven by Piola, first gradient continua verify Cauchy's so-called postulate, as in this class of continua contact interactions are concentrated on surfaces and are depending on the shape of the contact surface only via its normal. The vice-versa is proven in [50]. More generally, in N-th gradient continua, as envisaged already by Piola (see [42, 39]) the structure of contact forces is (much) more complex and still to be explored. In second (and higher) gradient continua, one can have contact forces concentrated on lines, also. In third (and higher) gradient continua, one can have forces concentrated on points of the contact surface between bodies. However the presence of concentrated forces on points and lines are not the only non-standard (that is: not included in Cauchy continuum mechanics) features of contact interactions in N-th gradient continua.

In fact, as already observed by Germain and fully exploited in [50, 52], internal and external work functionals can be regarded as a particular kind of distributions in the sense of Schwartz. Now, some general theorems by Schwartz (see [153]) prove that, in general, distributions concentrated on embedded manifolds involve not only the values of test functions, but also the values of all derivatives normal to the embedded manifolds of the test functions.

Therefore, as expected, in N-th gradient continua one can have up to N-forces concentrated on contact surfaces, up to (N-1)-forces on contact lines and up to (N-2)-forces on contact points. To be more precise, we will recall that, following Germain, 1-forces are those vectors that expend work on variations of displacements (therefore are the well-known forces), and N-forces are those vectors that expend work on (N-1)-normal derivatives of the variations of displacements. The expression of contact interactions in terms of the many stress tensors needed to describe the state of stress in N-th gradient continua can be found in [52], where the representation theorem for contact forces in terms of the stress tensor and of the normal to the contact surface is generalized.

## 5.10 Research perspectives as suggested by the lesson given by History of Mechanics

As we have already discussed in a previous section, unfortunately, there is not (yet!) a meta-theory telling us how to build new theories. However, relying on the Latin expression *Historia magistra vitae*, we can exploit the experience gathered in the past (partially described above) efforts made to advance the scientific understanding of reality. We underline that new theories are not only demanded in physics, engineering sciences or the other so-called *hard* sciences. In fact, any intellectual activity of the human being should tend to produce predictive knowledge, as also envisaged by [8]. Indeed, the absence of a predictive knowledge tends to produce in the human being a feeling of impotence, sometimes mixed with marvel. As Giambattista Vico already pondered in his masterpiece (for a detailed description of the personality of Vico we refer to [32])

> «The marvel is daughter of ignorance»[16]
> Giambattista Vico, Scienza Nuova (1725, New Science) (libro I, II, 35; p. 45)

It is suggestive, as also imagined in [38], to conjecture that a kind of minimum principle holds also in social sciences, so that the hope to find equilibrium configurations in social groups, and to model the evolution from an equilibrium to another one, will be, one day, realized somehow similarly to what has been done in mechanics. Such generalizations to social sciences have been already demanded by eminent philosophers of science. To quote again Giambattista Vico:

> «Things, outside their natural state, neither tend to remain nor last.»[17]
> Giambattista Vico, Scienza Nuova (1725, New Science) (libro I, II, 8; p. 39)

As remarked by Edmund Wilson [57]:

> «Vico had read Francis Bacon, and had decided that it ought to be possible to the study of human history methods similar to those proposed by Bacon for the study of the natural world.»

To address less ambitious research perspectives, we list here some of the possible fields of research (those in which we feel to be more expert than we are in mathematical sociology) which wait for innovative ideas and models and which can exploit the insight given by the Principle of Virtual Work.

- mathematically singular models arising in the theory of metamaterials. Using a particular class of Sobolev's spaces called anisotropic Sobolev spaces weak solutions were analysed for gradient incomplete strain gradient elasticity

[16] «La maraviglia è figliuola dell'ignoranza».

[17] «Le cose fuori del loro stato naturale né vi si adagiano né vi durano.»

[62, 61, 59, 67];
- formulation of numerical codes based on a formulation of the Principle of Virtual Work based on the mechanical peculiarities of considered mechanical systems [60, 30, 68, 29, 82, 132, 101, 106, 105, 79, 55, 78]. It is worth to remark that some applications of Rayleigh-Ritz techniques to solution of problems within the modeling of various electromagnetic and mechanical couplings were presented in [35, 121, 118, 119, 120];
- rate dependent and rate independent dissipative behaviors in materials with micro-structure [151, 150, 97, 33, 34, 142, 141, 168, 140, 159];
- discrete formulations for the description of elastic materials with micro-structure [93, 171, 14, 172, 173, 11, 22];
- direct or homogenized continuum modeling for the description of mechanical meta-materials [13, 12, 95, 94, 149, 96, 161, 160, 15].

This Chapter must end with some further considerations about the reasons for which the Principle of Virtual Work has found so many opposers in the community of continuum mechanics. We believe that the main reason can be very easily found in the true nature of the Principle, whose formulation needs sophisticated mathematical concepts and tools, as we have seen in the previous sections. In the continuum mechanics formulation, complex concepts from functional analysis, differential geometry of embedded manifolds and theory of distributions must be mastered in order to be capable of capturing the true mathematical essence of the principle and to be able to apply it to "practical" cases.

The Principle of Virtual Work has been systematically used by physicists to guide their researches. It is suitable here to attract the attention of the reader to an interesting quote from a famous astrophysicist (from "I am Neil deGrasse Tyson" – Reddit AMA Session held on November 13, 2011):

> «There are street artists. Street musicians. Street actors. But there are no street physicists. A little known secret is that a physicist is one of the most employable people in the marketplace – a physicist is a trained problem solver. How many times have you heard a person in a workplace say, "I wasn't trained for this!" That's an impossible reaction from a physicist, who would say, instead, "Cool. A problem I've never seen before. Let's see how I can figure out how to solve it!" Oh, and, have fun along the way.»

Unfortunately, there are many mechanicians who refused to adopt the right attitude towards this problem and preferred to look for simplifications and/or shortcuts. The main shortcut that was attempted to overcome the mathematical difficulties implied by the use of the Principle of Virtual Work, that is the use of the law of balance of force, did not even manage to handle the case of solid continua. In passing from Euler's fluids to first gradient solids it was indeed necessary to add the (extra) balance of moment of forces. Then, for every generalization it was necessary to add, one after the other, a series of extra balances of "something". This "something" was chosen to be what was necessary to get the lacking equations. An extra set of constitutive equations had to be introduced to this balance, with some *ad hoc* compatibility or physical consistency demands. Indeed, the entropy inequality was artificially introduced in mechanics exactly to handle this situation, that was simply caused by the absence of an internal energy functional postulated at the beginning, which was instead painfully recovered only *a posteriori*.

Only the school of theoretical physicists, guided by Feynman and Landau, did keep away from this useless effort of avoiding variational principles and remained faithful to the ancient lesson of Archytas. Physicists are continuing the long tradition of looking for the most appropriate model to describe phenomena.

> *«When scientifically investigating the natural world, the only thing worse than a blind believer is a seeing denier.»*
>
> Neil deGrasse Tyson, Death by black hole: And other cosmic quandaries (WW Norton & Company, 2007).

## References


1. Abbas, I.A., Abdalla, A.E.N.N., Alzahrani, F.S., Spagnuolo, M.: Wave propagation in a generalized thermoelastic plate using eigenvalue approach. Journal of Thermal Stresses **39**(11), 1367–1377 (2016)
2. Alibert, J.J., Seppecher, P., dell'Isola, F.: Truss modular beams with deformation energy depending on higher displacement gradients. Mathematics and Mechanics of Solids **8**(1), 51–73 (2003)
3. Altenbach, H., Eremeyev, V.A., Lebedev, L.P.: On the existence of solution in the linear elasticity with surface stresses. Zeitschrift für Angewandte Mathematik und Mechanik **90**(3), 231–240 (2010)
4. Altenbach, H., Eremeyev, V.A., Lebedev, L.P.: On the spectrum and stiffness of an elastic body with surface stresses. Zeitschrift für Angewandte Mathematik und Mechanik **91**(9), 699–710 (2011)
5. Andreaus, U., Spagnuolo, M., Lekszycki, T., Eugster, S.R.: A Ritz approach for the static analysis of planar pantographic structures modeled with nonlinear Euler–Bernoulli beams. Continuum Mechanics and Thermodynamics **30**(5), 1103–1123 (2018)
6. Antman, S.S.: Problems in nonlinear elasticity. Nonlinear Problems of Elasticity pp. 513–584 (2005)
7. Auffray, N., dell'Isola, F., Eremeyev, V.A., Madeo, A., Rosi, G.: Analytical continuum mechanics à la Hamilton–Piola least action principle for second gradient continua and capillary fluids. Mathematics and Mechanics of Solids **20**(4), 375–417 (2015)
8. Avlami, C.: From historia magistra vitae to history as empirical experimentation of progress. trans. Ágnes Bezeczky, in Multiple Antiquities, Multiple Modernities: Ancient Histories in Nineteenth-Century European Cultures, ed. Gábor Klaniczay, Michael Werner and Ottó Gecser (Frankfurt/Main: Campus Verlag, 2011) pp. 135–62 (2011)
9. Barbour, J.B.: Absolute or relative motion?: a study from a Machian point of view of the discovery and the structure of dynamical theories, vol. 1. CUP Archive (1989)
10. Barbour, J.B.: The discovery of dynamics: a study from a Machian point of view of the discovery and the structure of dynamical theories. Oxford University Press (2001)
11. Barchiesi, E., dell'Isola, F., Bersani, A.M., Turco, E.: Equilibria determination of elastic articulated duoskelion beams in 2D via a Riks-type algorithm. International Journal of Non-Linear Mechanics **128** (2021)
12. Barchiesi, E., Eugster, S.R., dell'Isola, F., Hild, F.: Large in-plane elastic deformations of bi-pantographic fabrics: asymptotic homogenization and experimental validation. Mathematics and Mechanics of Solids **25**(3), 739–767 (2020)
13. Barchiesi, E., Eugster, S.R., Placidi, L., dell'Isola, F.: Pantographic beam: A complete second gradient 1D-continuum in plane. Zeitschrift für angewandte Mathematik und Physik **70**(5), 135 (2019)

14. Barchiesi, E., Harsch, J., Ganzosch, G., Eugster, S.R.: Discrete versus homogenized continuum modeling in finite deformation bias extension test of bi-pantographic fabrics. Continuum Mechanics and Thermodynamics pp. 1–14 (2020)
15. Barchiesi, E., Khakalo, S.: Variational asymptotic homogenization of beam-like square lattice structures. Mathematics and Mechanics of Solids **24**(10), 3295–3318 (2019)
16. Benvenuto, E., Corradi, M., Foce, F., Becchi, A.: La scienza delle costruzioni e il suo sviluppo storico. Edizioni di storia e letteratura Roma (2006)
17. Bersani, A.M., Caressa, P.: Lagrangian descriptions of dissipative systems: a review. Mathematics and Mechanics of Solids (2020)
18. Boutin, C., Hans, S., Chesnais, C.: Generalized beams and continua. dynamics of reticulated structures. In: Mechanics of generalized continua, pp. 131–141. Springer (2010)
19. Bowen, A.C., Rochberg, F.: Hellenistic astronomy: the science in its contexts. Brill (2020)
20. Camar-Eddine, M., Seppecher, P.: Determination of the closure of the set of elasticity functionals. Archive for rational mechanics and analysis **170**(3), 211–245 (2003)
21. Capobianco, G., Eugster, S.R.: Time finite element based Moreau-type integrators. International Journal for Numerical Methods in Engineering **114**(3), 215–231 (2018)
22. Capobianco, G., Eugster, S.R., Winandy, T.: Modeling planar pantographic sheets using a nonlinear Euler–Bernoulli beam element based on B-spline functions. Proceedings in Applied Mathematics and Mechanics **18**(1), e201800220:1–2 (2018)
23. Capobianco, G., Winandy, T., Eugster, S.: The principle of virtual work and Hamilton's principle on Galilean manifolds. Journal of Geometric Mechanics pp. 1–27 (2021)
24. Carcaterra, A., dell'Isola, F., Esposito, R., Pulvirenti, M.: Macroscopic description of microscopically strongly inhomogenous systems: A mathematical basis for the synthesis of higher gradients metamaterials. Archive for Rational Mechanics and Analysis **218**(3), 1239–1262 (2015)
25. Casal, P., Gouin, H.: Invariance properties of inviscid fluids of grade n. In: PDEs and Continuum Models of Phase Transitions, pp. 85–98. Springer (1989)
26. Cauchy, A.: Recherches sur l'équilibre et le mouvement intérieur des corps solides ou fluides, élastiques ou non élastiques. In: Œuvres Complètes, *2*, vol. 2, pp. 300–304. Gauthier-Villars et Fils (1823)
27. Chatzigeorgiou, G., Javili, A., Steinmann, P.: Multiscale modelling for composites with energetic interfaces at the micro-or nanoscale. Mathematics and Mechanics of Solids **20**(9), 1130–1145 (2015)
28. Chatzigeorgiou, G., Meraghni, F., Javili, A.: Generalized interfacial energy and size effects in composites. Journal of the Mechanics and Physics of Solids **106**, 257–282 (2017)
29. Chróścielewski, J., dell'Isola, F., Eremeyev, V.A., Sabik, A.: On rotational instability within the nonlinear six-parameter shell theory. International Journal of Solids and Structures (2020)
30. Chróścielewski, J., Schmidt, R., Eremeyev, V.A.: Nonlinear finite element modeling of vibration control of plane rod-type structural members with integrated piezoelectric patches. Continuum Mechanics and Thermodynamics **31**(1), 147–188 (2019)
31. Ciallella, A.: Research perspective on multiphysics and multiscale materials: a paradigmatic case (2020)
32. Costelloe, T.: Giambattista Vico. In: E.N. Zalta (ed.) The Stanford Encyclopedia of Philosophy, fall 2018 edn. Metaphysics Research Lab, Stanford University (2018)
33. Cuomo, M.: Forms of the dissipation function for a class of viscoplastic models. Mathematics and Mechanics of Complex Systems **5**(3), 217–237 (2017)
34. Cuomo, M.: Continuum model of microstructure induced softening for strain gradient materials. Mathematics and Mechanics of Solids **24**(8), 2374–2391 (2019)
35. Dastjerdi, S., Akgöz, B., Civalek, Ö., Malikan, M., Eremeyev, V.A.: On the non-linear dynamics of torus-shaped and cylindrical shell structures. International Journal of Engineering Science **156** (2020)
36. Del Vescovo, D., Giorgio, I.: Dynamic problems for metamaterials: review of existing models and ideas for further research. International Journal of Engineering Science **80**, 153–172 (2014)

37. dell'Isola, F.: In memoriam: Richard A. Toupin, 1926–2017. Mathematics and Mechanics of Solids **23**(2), 133–135 (2018)
38. dell'Isola, F.: Big-(Wo) men, Tyrants, Chiefs, Dictators, Emperors and Presidents: Towards the Mathematical Understanding of Social Groups. Springer (2019)
39. dell'Isola, F., Andreaus, U., Placidi, L.: At the origins and in the vanguard of peridynamics, non-local and higher-gradient continuum mechanics: An underestimated and still topical contribution of Gabrio Piola. Mathematics and Mechanics of Solids **20**(8), 887–928 (2015)
40. dell'Isola, F., Andreaus, U., Placidi, L., Scerrato, D.: Intorno alle equazioni fondamentali del movimento di corpi qualsivogliono, considerati secondo la naturale loro forma e costituzione. In: The complete works of Gabrio Piola: Volume I, pp. 1–370. Springer (2014)
41. dell'Isola, F., Corte, A.D., Giorgio, I.: Higher-gradient continua: The legacy of Piola, Mindlin, Sedov and Toupin and some future research perspectives. Mathematics and Mechanics of Solids **22**(4), 852–872 (2017)
42. dell'Isola, F., Della Corte, A., Esposito, R., Russo, L.: Some cases of unrecognized transmission of scientific knowledge: from antiquity to Gabrio Piola's peridynamics and generalized continuum theories. In: Generalized continua as models for classical and advanced materials, pp. 77–128. Springer (2016)
43. dell'Isola, F., Della Corte, A., Giorgio, I.: Higher-gradient continua: The legacy of Piola, Mindlin, Sedov and Toupin and some future research perspectives. Mathematics and Mechanics of Solids p. 1081286515616034 (2016)
44. dell'Isola, F., Madeo, A., Seppecher, P.: Boundary conditions at fluid-permeable interfaces in porous media: A variational approach. International Journal of Solids and Structures **46**(17), 3150–3164 (2009)
45. dell'Isola, F., Madeo, A., Seppecher, P.: Cauchy tetrahedron argument applied to higher contact interactions. Archive for Rational Mechanics and Analysis **219**(3), 1305–1341 (2016)
46. dell'Isola, F., Maier, G., Perego, U., et al.: The complete works of Gabrio Piola: Volume I (Advanced Structured Materials, vol. 38) (2014)
47. dell'Isola, F., Maier, G., Perego, U., et al.: The Complete Works of Gabrio Piola: Volume II. Cham, Switzerland: Springer (2019)
48. dell'Isola, F., Placidi, L.: Variational principles are a powerful tool also for formulating field theories. In: Variational models and methods in solid and fluid mechanics, pp. 1–15. Springer (2011)
49. dell'Isola, F., Seppecher, P.: The relationship between edge contact forces, double forces and interstitial working allowed by the principle of virtual power. Comptes Rendus- Academie des Sciences Paris Serie 2 Mecanique Physique Astronomie Fascicule B (1995)
50. dell'Isola, F., Seppecher, P.: Edge contact forces and quasi-balanced power. Meccanica **32**(1), 33–52 (1997)
51. dell'Isola, F., Seppecher, P., Corte, A.D.: The postulations á la D'Alembert and á la Cauchy for higher gradient continuum theories are equivalent: a review of existing results. Proceedings of the Royal Society of London A: Mathematical, Physical and Engineering Sciences **471**(2183) (2015)
52. dell'Isola, F., Seppecher, P., Madeo, A.: How contact interactions may depend on the shape of Cauchy cuts in Nth gradient continua: approach "à la D'Alembert". Zeitschrift für angewandte Mathematik und Physik **63**(6), 1119–1141 (2012)
53. dell'Isola, F., Seppecher, P., Placidi, L., Barchiesi, E., Misra, A.: 8. Least Action and Virtual Work Principles for the Formulation of Generalized Continuum Models. Discrete and Continuum Models for Complex Metamaterials p. 327 (2020)
54. dell'Isola, F., Steigmann, D.: A two-dimensional gradient-elasticity theory for woven fabrics. Journal of Elasticity **118**(1), 113–125 (2015)
55. Deutschmann, B., Eugster, S.R., Ott, C.: Reduced models for the static simulation of an elastic continuum mechanism. IFAC-PapersOnLine **51**(2), 403 – 408 (2018). 9th Vienna International Conference on Mathematical Modelling
56. Dugas, R.: Histoire de la mécanique, vol. 16. Editions du griffon (1950)
57. Edmund, W.: To the Finland Station: A Study in the Writing and Acting of History. Collins (1960)

58. Epstein, M., Smelser, R.: An appreciation and discussion of Paul Germain's "The method of virtual power in the mechanics of continuous media, I: Second-gradient theory". Mathematics and Mechanics of Complex Systems **8**(2), 191–199 (2020)
59. Eremeyev, V., dell'Isola, F.: Weak solutions within the gradient-incomplete strain-gradient elasticity. Lobachevskii Journal of Mathematics **41**(10), 1992–1998 (2020)
60. Eremeyev, V., Nasedkin, A., Solov'yev, A., et al.: Partitioned schemes of the finite-element method for dynamic problems of acoustoelectroelasticity. Journal of Applied Mathematics and Mechanics **64**(3), 367–377 (2000)
61. Eremeyev, V.A., Alzahrani, F.S., Cazzani, A., dell'Isola, F., Hayat, T., Turco, E., Konopińska-Zmysłowska, V.: On existence and uniqueness of weak solutions for linear pantographic beam lattices models. Continuum Mechanics and Thermodynamics **31**(6), 1843–1861 (2019)
62. Eremeyev, V.A., dell'Isola, F., Boutin, C., Steigmann, D.: Linear pantographic sheets: existence and uniqueness of weak solutions. Journal of Elasticity **132**(2), 175–196 (2018)
63. Eremeyev, V.A., Lebedev, L.P.: Existence theorems in the linear theory of micropolar shells. Zeitschrift für Angewandte Mathematik und Mechanik **91**(6), 468–476 (2011)
64. Eremeyev, V.A., Lebedev, L.P.: Existence of weak solutions in elasticity. Mathematics and mechanics of solids **18**(2), 204–217 (2013)
65. Eremeyev, V.A., Lebedev, L.P.: Mathematical study of boundary-value problems within the framework of Steigmann–Ogden model of surface elasticity. Continuum Mechanics and Thermodynamics **28**(1), 407–422 (2016)
66. Eremeyev, V.A., Lebedev, L.P., Cloud, M.J.: The Rayleigh and Courant variational principles in the six-parameter shell theory. Mathematics and Mechanics of Solids **20**(7), 806–822 (2015)
67. Eremeyev, V.A., Lurie, S.A., Solyaev, Y.O., dell'Isola, F.: On the well posedness of static boundary value problem within the linear dilatational strain gradient elasticity. Zeitschrift für angewandte Mathematik und Physik **71**(6), 1–16 (2020)
68. Eremeyev, V.A., Turco, E.: Enriched buckling for beam-lattice metamaterials. Mechanics Research Communications **103** (2020)
69. Eringen, A.C., Maugin, G.A.: Electrodynamics of continua I: foundations and solid media. Springer Science & Business Media (2012)
70. Eugster, S.R.: Geometric Continuum Mechanics and Induced Beam Theories, *Lecture Notes in Applied and Computational Mechanics*, vol. 75. Springer (2015)
71. Eugster, S.R.: Hellinger's encyclopedia article on the fundamentals of the mechanics of continua. In: F. dell'Isola, S.R. Eugster, M. Spagnuolo, E. Barchiesi (eds.) Evaluation of Scientific Sources in Mechanics: Heiberg's Prolegomena to the Works of Archimedes and Hellinger's Encyclopedia Article. Springer (2021)
72. Eugster, S.R., Capobianco, G., Winandy, T.: Geometric description of time-dependent finite-dimensional mechanical systems. Mathematics and Mechanics of Solids **25**(11), 2050–2075 (2020)
73. Eugster, S.R., dell'Isola, F.: Exegesis of the introduction and sect. I from "Fundamentals of the mechanics of continua" by E. Hellinger. Zeitschrift für angewandte Mathematik und Mechanik **97**(4), 477–506 (2017)
74. Eugster, S.R., dell'Isola, F.: An ignored source in the foundations of continuum physics "Die Allgemeinen Ansätze der Mechanik der Kontinua" by E. Hellinger. Proceedings in Applied Mathematics and Mechanics **17**(1), 413–414 (2017)
75. Eugster, S.R., dell'Isola, F.: Exegesis of sect. II and III.A from "Fundamentals of the mechanics of continua" by E. Hellinger. Zeitschrift für angewandte Mathematik und Mechanik **98**(1), 31–68 (2018)
76. Eugster, S.R., dell'Isola, F.: Exegesis of sect. III.B from "Fundamentals of the mechanics of continua" by E. Hellinger. Zeitschrift für angewandte Mathematik und Mechanik **98**(1), 69–105 (2018)
77. Eugster, S.R., dell'Isola, F., Steigmann, D.: Continuum theory for mechanical metamaterials with a cubic lattice substructure. Mathematics and Mechanics of Complex Systems **7**(1), 75–98 (2019)

78. Eugster, S.R., Deutschmann, B.: A nonlinear Timoshenko beam formulation for modeling a tendon-driven compliant neck mechanism. Proceedings in Applied Mathematics and Mechanics **18**(1), e201800208:1–2 (2018)
79. Eugster, S.R., Glocker, Ch.: Constraints in structural and rigid body mechanics: a frictional contact problem. Annals of Solid and Structural Mechanics **5**(1-2), 1–13 (2013)
80. Eugster, S.R., Glocker, Ch.: On the notion of stress in classical continuum mechanics. Mathematics and Mechanics of Complex Systems **5**(3-4), 299–338 (2017)
81. Eugster, S.R., Harsch, J.: A variational formulation of classical nonlinear beam theories. In: B.E. Abali, I. Giorgio (eds.) Developments and Novel Approaches in Nonlinear Solid Body Mechanics, pp. 95–121. Springer International Publishing (2020)
82. Eugster, S.R., Hesch, C., Betsch, P., Glocker, Ch.: Director-based beam finite elements relying on the geometrically exact beam theory formulated in skew coordinates. International Journal for Numerical Methods in Engineering **97**(2), 111–129 (2014)
83. Eugster, S.R., Steigmann, D.J.: Variational methods in the theory of beams and lattices. In: H. Altenbach, A. Öchsner (eds.) Encyclopedia of Continuum Mechanics, pp. 1–9. Springer (2018)
84. Everstine, G., Pipkin, A.: Boundary layers in fiber-reinforced materials. Journal of Applied Mechanics **40**, 518–522 (1973)
85. Feynman, R., Hibbs, A.: Quantum Mechanics and Path Integrals. McGraw-Hill, New-York (1965)
86. Franciosi, P., Spagnuolo, M., Salman, O.U.: Mean Green operators of deformable fiber networks embedded in a compliant matrix and property estimates. Continuum Mechanics and Thermodynamics **31**(1), 101–132 (2019)
87. Gavrilyuk, S., Gouin, H., Perepechko, Y.: A variational principle for two-fluid models. Comptes Rendus de l'Académie des Sciences-Series IIB-Mechanics-Physics-Chemistry-Astronomy **324**(8), 483–490 (1997)
88. Germain, P.: The method of virtual power in continuum mechanics. part 2: Microstructure. SIAM Journal on Applied Mathematics **25**(3), 556–575 (1973)
89. Germain, P.: Functional concepts in continuum mechanics. Meccanica **33**(5), 433–444 (1998)
90. Germain, P.: The method of virtual power in the mechanics of continuous media, I: Second-gradient theory. Mathematics and Mechanics of Complex Systems **8**(2), 153–190 (2020)
91. Germain, P., Nayroles, B.: Applications of Methods of Functional Analysis to Problems in Mechanics: Joint Symposium IUTAM/IMU Held in Marseille, Sept. 1-6, 1975, vol. 503. Springer (2006)
92. Gildenhard, I., Owen, M.: Tacitus, Annals, 15.20-23, 33-45: Latin Text, Study Aids with Vocabulary, and Commentary. Open Book Publishers (2013)
93. Giorgio, I.: A discrete formulation of Kirchhoff rods in large-motion dynamics. Mathematics and Mechanics of Solids **25**(5), 1081–1100 (2020)
94. Giorgio, I.: Lattice shells composed of two families of curved Kirchhoff rods: an archetypal example, topology optimization of a cycloidal metamaterial. Continuum Mechanics and Thermodynamics pp. 1–20 (2020)
95. Giorgio, I., Ciallella, A., Scerrato, D.: A study about the impact of the topological arrangement of fibers on fiber-reinforced composites: some guidelines aiming at the development of new ultra-stiff and ultra-soft metamaterials. International Journal of Solids and Structures **203**, 73–83 (2020)
96. Giorgio, I., De Angelo, M., Turco, E., Misra, A.: A Biot–Cosserat two-dimensional elastic nonlinear model for a micromorphic medium. Continuum Mechanics and Thermodynamics pp. 1–13 (2019)
97. Giorgio, I., Scerrato, D.: Multi-scale concrete model with rate-dependent internal friction. European Journal of Environmental and Civil Engineering **21**(7-8), 821–839 (2017)
98. Gouin, H.: Noether theorem in fluid mechanics. Mechanics Research Communications **3**(3), 151–155 (1976)
99. Gouin, H.: Utilization of the second gradient theory in continuum mechanics to study the motion and thermodynamics of liquid-vapor interfaces. In: Physicochemical Hydrodynamics, pp. 667–682. Springer (1988)

100. Gouin, H.: The second gradient theory applied to interfaces: Models of continuum mechanics for fluid interfaces. In: Dynamics of Multiphase Flows Across Interfaces, pp. 8–13. Springer (1996)
101. Greco, L., Cuomo, M.: An isogeometric implicit G1 mixed finite element for Kirchhoff space rods. Comput Methods Appl Mech Eng **298**, 325–349 (2016)
102. Green, A., Naghdi, P.: Thermoelasticity without energy dissipation. Journal of elasticity **31**(3), 189–208 (1993)
103. Green, G.: On the laws of the reflexion and refraction of light at the common surface of two non-crystallized media. Transactions of the Cambridge Philosophical Society **7**, 1 (1848)
104. Green, G.: Mathematical papers of the late George Green. Cambridge University Press (2014)
105. Harsch, J., Capobianco, G., Eugster, S.R.: Finite element formulations for constrainted spatial nonlinear beam theories. Mathematics and Mechanics of Solids pp. 1–26 (2021)
106. Harsch, J., Eugster, S.R.: Finite element analysis of planar nonlinear classical beam theories. In: B.E. Abali, I. Giorgio (eds.) Developments and Novel Approaches in Nonlinear Solid Body Mechanics, pp. 123–157. Springer International Publishing (2020)
107. Hilgers, M., Pipkin, A.: Elastic sheets with bending stiffness. The Quarterly Journal of Mechanics and Applied Mathematics **45**, 57–75 (1992)
108. Hilgers, M., Pipkin, A.: Energy-minimizing deformations of elastic sheets with bending stiffness. Journal of elasticity **31**, 125–139 (1993)
109. Hilgers, M., Pipkin, A.: Bending energy of highly elastic membranes II. Quarterly of applied mathematics **54**, 307–316 (1996)
110. Hu, M., Kolsky, H., Pipkin, A.: Bending theory for fiber-reinforced beams. Journal of composite materials **19**(3), 235–249 (1985)
111. Javili, A., dell'Isola, F., Steinmann, P.: Geometrically nonlinear higher-gradient elasticity with energetic boundaries. Journal of the Mechanics and Physics of Solids **61**(12), 2381–2401 (2013)
112. Javili, A., McBride, A., Mergheim, J., Steinmann, P., Schmidt, U.: Micro-to-macro transitions for continua with surface structure at the microscale. International Journal of Solids and Structures **50**(16-17), 2561–2572 (2013)
113. Kröner, E.: Elasticity theory of materials with long range cohesive forces. International Journal of Solids and Structures **3**(5), 731–742 (1967)
114. Kröner, E.: Mechanics of generalized continua. Springer (1968)
115. de Lagrange, J.L.: Mécanique analytique, vol. 1. Mallet-Bachelier (1853)
116. Landau, L., Lifshitz, E.: Field theory (1973)
117. Luce, J.V.: Greek science in its hellenistic phase. Hermathena (145), 23–38 (1988)
118. Malikan, M., Eremeyev, V.A.: On nonlinear bending study of a piezo-flexomagnetic nanobeam based on an analytical-numerical solution. Nanomaterials **10**(9) (2020)
119. Malikan, M., Eremeyev, V.A.: Post-critical buckling of truncated conical carbon nanotubes considering surface effects embedding in a nonlinear Winkler substrate using the Rayleigh-Ritz method. Materials Research Express **7**(2) (2020)
120. Malikan, M., Krasheninnikov, M., Eremeyev, V.A.: Torsional stability capacity of a nano-composite shell based on a nonlocal strain gradient shell model under a three-dimensional magnetic field. International Journal of Engineering Science **148**, 103210 (2020)
121. Malikan, M., Uglov, N.S., Eremeyev, V.A.: On instabilities and post-buckling of piezomagnetic and flexomagnetic nanostructures. International Journal of Engineering Science **157** (2020)
122. Maugin, G.A.: Wave motion in magnetizable deformable solids. International Journal of Engineering Science **19**(3), 321–388 (1981)
123. Maugin, G.A.: Continuum mechanics of electromagnetic solids. Elsevier (2013)
124. Maugin, G.A.: Paul Germain et la mécanique des fluides (1945–1970). Comptes Rendus Mécanique **345**(9), 605–612 (2017)
125. Maugin, G.A.: Material inhomogeneities in elasticity. CRC Press (2020)
126. Mindlin, R.D.: Force at a point in the interior of a semi-infinite solid. physics **7**(5), 195–202 (1936)

127. Mindlin, R.D.: Microstructure in linear elasticity. Tech. rep., Columbia Univ New York Dept of Civil Engineering and Engineering Mechanics (1963)
128. Mindlin, R.D.: Second gradient of strain and surface-tension in linear elasticity. International Journal of Solids and Structures **1**(4), 417–438 (1965)
129. Mindlin, R.D., Eshel, N.: On first strain-gradient theories in linear elasticity. International Journal of Solids and Structures **4**(1), 109–124 (1968)
130. Müller, C.H., Timpe, A.: Die Grundgleichungen der mathematische Elastizitätstheorie. In: Mechanik, *Enzyklopädie der Mathematischen Wissenschaften mit Einschluss ihrer Anwendungen*, vol. IV/4, pp. 1–54. Teubner (1906)
131. Nasedkin, A.V., Eremeyev, V.A.: Harmonic vibrations of nanosized piezoelectric bodies with surface effects. Zeitschrift für Angewandte Mathematik und Mechanik **94**(10), 878–892 (2014)
132. Niiranen, J., Balobanov, V., Kiendl, J., Hosseini, S.: Variational formulations, model comparisons and numerical methods for Euler–Bernoulli micro-and nano-beam models. Mathematics and Mechanics of Solids **24**(1), 312–335 (2019)
133. Parkus, H., Sedov, L.I.: Irreversible Aspects of Continuum Mechanics and Transfer of Physical Characteristics in Moving Fluids: Symposia Vienna, June 22–28, 1966. Springer Science & Business Media (2012)
134. Pfister, H., King, M.: Inertia and gravitation: The fundamental nature and structure of space-time, vol. 897. Springer (2015)
135. Pideri, C., Seppecher, P.: A second gradient material resulting from the homogenization of an heterogeneous linear elastic medium. Continuum Mechanics and Thermodynamics **9**(5), 241–257 (1997)
136. Piola, G.: Sull'applicazione de'principj della meccanica analitica del Lagrange ai principali problemi. Memoria di Gabrio Piola presentata al concorso del premio e coronata dall'IR Istituto di Scienze, ecc. nella solennità del giorno 4 ottobre 1824. dall'Imp. Regia stamperia (1825)
137. Pipkin, A.: Generalized plane deformations of ideal fiber-reinforced materials. Quarterly of Applied Mathematics **32**, 253–263 (1974)
138. Pipkin, A.: Energy changes in ideal fiber-reinforced composites. Quarterly of Applied Mathematics **35**, 455–463 (1978)
139. Pipkin, A.: Some developments in the theory of inextensible networks. Quarterly of Applied Mathematics **38**, 343–355 (1980)
140. Placidi, L., Barchiesi, E.: Energy approach to brittle fracture in strain-gradient modelling. Proceedings of the Royal Society A: Mathematical, Physical and Engineering Sciences **474**(2210), 20170878 (2018)
141. Placidi, L., Barchiesi, E., Misra, A.: A strain gradient variational approach to damage: a comparison with damage gradient models and numerical results. Mathematics and Mechanics of Complex Systems **6**(2), 77–100 (2018)
142. Placidi, L., Misra, A., Barchiesi, E.: Two-dimensional strain gradient damage modeling: a variational approach. Zeitschrift für angewandte Mathematik und Physik **69**(3), 56 (2018)
143. Rivlin, R.: Large elastic deformations of isotropic materials IV. further developments of the general theory. Philosophical Transactions of the Royal Society of London. Series A, Mathematical and Physical Sciences **241**(835), 379–397 (1948)
144. Rivlin, R.: Plane strain of a net formed by inextensible cords. In: Collected Papers of RS Rivlin, pp. 511–534. Springer (1997)
145. Rivlin, R.S., Ericksen, J.L.: Stress-deformation relations for isotropic materials. In: Collected Papers of RS Rivlin, pp. 911–1013. Springer (1997)
146. Russo, L.: The astronomy of Hipparchus and his time: A study based on pre-ptolemaic sources. Vistas in Astronomy **38**, 207–248 (1994)
147. Russo, L., et al.: The forgotten revolution: how science was born in 300 BC and why it had to be reborn. Springer Science & Business Media (2013)
148. Salençon, J.: Handbook of continuum mechanics: General concepts thermoelasticity. Springer Science & Business Media (2012)

149. Scerrato, D., Giorgio, I.: Equilibrium of two-dimensional cycloidal pantographic metamaterials in three-dimensional deformations. Symmetry **11**(12), 1523 (2019)
150. Scerrato, D., Giorgio, I., Della Corte, A., Madeo, A., Dowling, N., Darve, F.: Towards the design of an enriched concrete with enhanced dissipation performances. Cement and Concrete Research **84**, 48–61 (2016)
151. Scerrato, D., Giorgio, I., Madeo, A., Limam, A., Darve, F.: A simple non-linear model for internal friction in modified concrete. International Journal of Engineering Science **80**, 136–152 (2014)
152. Schulte, J., Dittmann, M., Eugster, S.R., Hesch, S., Reinicke, T., dell'Isola, F., Hesch, C.: Isogeometric analysis of fiber reinforced composites using Kirchhoff–Love shell elements. Computer Methods in Applied Mechanics and Engineering **362**, 112845:1–34 (2020)
153. Schwartz, L.: Théorie des distributions, vol. 2. Hermann Paris (1966)
154. Sedov, L.I.: Similarity and dimensional methods in mechanics. CRC press (1993)
155. Sedov, L.I.: Mechanics of continuous media (IN 2 VOLS), vol. 4. World Scientific (1997)
156. Sedov, L.I., Chu, C., Cohen, H., Seckler, B., Gillis, J.: Two-dimensional problems in hydrodynamics and aerodynamics. Physics Today **18**(12), 62 (1965)
157. Seppecher, P.: Moving contact lines in the Cahn-Hilliard theory. International journal of engineering science **34**(9), 977–992 (1996)
158. Shirani, M., Luo, C., Steigmann, D.J.: Cosserat elasticity of lattice shells with kinematically independent flexure and twist. Continuum Mechanics and Thermodynamics **31**(4), 1087–1097 (2019)
159. Solyaev, Y., Lurie, S., Barchiesi, E., Placidi, L.: On the dependence of standard and gradient elastic material constants on a field of defects. Mathematics and Mechanics of Solids **25**(1), 35–45 (2020)
160. Spagnuolo, M., Andreaus, U.: A targeted review on large deformations of planar elastic beams: extensibility, distributed loads, buckling and post-buckling. Mathematics and Mechanics of Solids **24**(1), 258–280 (2019)
161. Spagnuolo, M., Yildizdag, M.E., Andreaus, U., Cazzani, A.M.: Are higher-gradient models also capable of predicting mechanical behavior in the case of wide-knit pantographic structures? Mathematics and Mechanics of Solids (2020)
162. Steigmann, D., Baesu, E., Rudd, R.E., Belak, J., McElfresh, M.: On the variational theory of cell-membrane equilibria. Interfaces and Free Boundaries **5**(4), 357–366 (2003)
163. Steigmann, D., dell'Isola, F.: Mechanical response of fabric sheets to three-dimensional bending, twisting, and stretching. Acta Mechanica Sinica **31**(3), 373–382 (2015)
164. Steigmann, D., Ogden, R.: Elastic surface-substrate interactions. Proceedings of the Royal Society of London. Series A: Mathematical, Physical and Engineering Sciences **455**(1982), 437–474 (1999)
165. Steigmann, D.J.: The variational structure of a nonlinear theory for spatial lattices. Meccanica **31**(4), 441–455 (1996)
166. Steigmann, D.J.: Koiter's shell theory from the perspective of three-dimensional nonlinear elasticity. Journal of Elasticity **111**(1), 91–107 (2013)
167. Steigmann, D.J., Faulkner, M.: Variational theory for spatial rods. Journal of Elasticity **33**(1), 1–26 (1993)
168. Timofeev, D., Barchiesi, E., Misra, A., Placidi, L.: Hemivariational continuum approach for granular solids with damage-induced anisotropy evolution. Mathematics and Mechanics of Solids (2020)
169. Truesdell, C., Noll, W.: The non-linear field theories of mechanics. In: The non-linear field theories of mechanics, pp. 1–579. Springer (2004)
170. Truesdell, C., Toupin, R.: The classical field theories. In: Principles of classical mechanics and field theory/Prinzipien der Klassischen Mechanik und Feldtheorie, pp. 226–858. Springer (1960)
171. Turco, E.: Discrete is it enough? the revival of Piola–Hencky keynotes to analyze three-dimensional Elastica. Continuum Mechanics and Thermodynamics **30**(5), 1039–1057 (2018)
172. Turco, E.: Stepwise analysis of pantographic beams subjected to impulsive loads. Mathematics and Mechanics of Solids (2020)

173. Turco, E., Barchiesi, E.: Equilibrium paths of Hencky pantographic beams in a three-point bending problem. Mathematics and Mechanics of Complex Systems **7**(4), 287–310 (2019)
174. Vailati, G.: Il principio dei lavori virtuali da Aristotele a Erone d'Alessandria, Scritti (Bologna, Forni, 1987), vol. II, pp. 113&128. Atti della R. Accademia delle Scienze di Torino **32** (1897)
175. Winter, T.N.: The mechanical problems in the corpus of Aristotle. Faculty Publications, Classics and Religious Studies Department (2007)